MICROCALORIMETRY OF MACROMOLECULES

MICROCALORIMETRY OF MACROMOLECULES
The Physical Basis of Biological Structures

PETER L. PRIVALOV

Department of Biology
Johns Hopkins University
Baltimore, Maryland

WILEY

A JOHN WILEY & SONS, INC., PUBLICATION

Published by John Wiley & Sons, Inc., Hoboken, New Jersey.
Published simultaneously in Canada.

For general information on our other products and services or for technical support, please contact our Customer Care Department within the United States at (800) 762-2974, outside the United States at (317) 572-3993 or fax (317) 572-4002.

Wiley also publishes its books in a variety of electronic formats. Some content that appears in print may not be available in electronic formats. For more information about Wiley products, visit our web site at www.wiley.com.

Library of Congress Cataloging-in-Publication Data

9781118104514

10 9 8 7 6 5 4 3 2 1

CONTENTS

1

INTRODUCTION

Talking and contention of Arguments must soon be turned into labours; all the fine dreams of Opinions and universal metaphysical natures, which the luxury of subtle brains has devised would quickly vanish and give place to solid Histories, Experiments and Works.

Hooke (1665)

The microcalorimetry of biological molecules is attracting increasing attention for several reasons. First, it was finally realized that proteins and nucleic acids, consisting of thousands of atoms participating in thermal motion, represent individual quasi-macroscopic systems. Correspondingly, they are usually called **macromolecules**. As with other macroscopic systems, understanding individual macromolecules requires knowledge of their thermodynamics, since that determines their most general properties.

Second, the thermodynamics of biological macromolecules is expected to be very abnormal because of the unusual spatial organization of these objects: every atom in their structure occupies a definite place, as in a crystal—but in contrast to a crystal these macromolecules have no symmetry and no periodicity in the disposition of

Microcalorimetry of Macromolecules: The Physical Basis of Biological Structures, First Edition.
Peter L. Privalov.

their atoms. Such ordered aperiodic macroscopic systems have never before been dealt with in physics. Therefore, one cannot *a priori* predict the thermodynamic properties of biological macromolecules. In consequence, without knowing their thermodynamics one cannot engineer new macromolecules with defined properties. Without knowledge of their energetic basis, all discussion of the principles of organization of these macromolecules, of the mechanism of their formation and the stabilization of their three-dimensional structure, and therefore of their function (which assumes certain rearrangements of their structure), is mere speculation. This has become apparent only after many years of unsuccessful attempts to solve these problems by just analyzing the known structures of macromolecules. This failure has made it clear that structural information represents only one facet of a macromolecule; the other facet is its energetic basis, that is, its thermodynamics. These two fundamental information sets cannot be deduced from one another: each has to be obtained experimentally using very different methods.

Third, new and efficient experimental methods have been developed to obtain the necessary thermodynamic information on individual macromolecules in solution. Of special importance has been the development of supersensitive calorimetric instruments, isothermal reaction and heat capacity microcalorimeters, for studying the thermodynamic properties of biological macromolecules to measure the energetic bases of these molecular constructs. These properties of individual macromolecules need to be studied in highly dilute solutions—using, moreover, minimal quantities of these expensive objects: this has required especially sensitive and precise instruments.

In this book we start by reminding readers of the basics of thermodynamics useful for calorimetry and by giving relevant physicochemical information on the aqueous solutions of organic compounds. Then we describe the calorimetric techniques used for thermodynamic studies of biological macromolecules: the instruments for measuring the heat effects of various processes, namely, the heats of isothermal reactions between various reagents, the heats of temperature-induced changes in the samples being studied, that is, the heat capacities at constant pressure, and the heats associated with the pressure-induced changes at constant temperature. Calorimetry is a classical method that has been used extensively in science for a long time. However, studies of the thermodynamics of biological macromolecules, which are available in very limited amounts and can be studied only in highly dilute solutions, required development of supersensitive calorimetric instruments—microcalorimeters and even nanocalorimeters—to measure heats of isothermal reaction (isothermal titration nanocalorimeter), heat capacities over a broad temperature range (scanning nanocalorimeter), and pressure effects (pressure perturbation nanocalorimeter). Chapter 3 gives advice on how to use these techniques effectively in experiments with biological macromolecules, that is, proteins, nucleic acids, and their complexes.

Chapter 4 condenses general information on the structure of biological macromolecules—proteins and nucleic acids—to focus attention on the key thermodynamic problems relating to their structure. The results of calorimetric studies of various types of biological macromolecules and their complexes are then considered in the following chapters. We start from the two simplest, but highly important and far from fully understood, structural elements: the α and polyproline helices and their complexes, the α-helical coiled-coil and the polyproline coiled-coils. We then continue with more complicated macromolecular formations: small globular proteins; multidomain proteins and their complexes, particularly with DNA; and finally nucleic acids themselves. As will be seen, these calorimetric studies have led to serious reconsideration of many widely accepted dogmas concerning the roles of hydrogen bonding, hydrophobic interactions, and water in the formation of macromolecular structures.

Finally, I thank all my collaborators who worked with me during almost half a century on creation of a new experimental technique, microcalorimetry, and developing with such instruments a new field in experimental biophysics—the energetics of biological macromolecules. Among my numerous collaborators I have to mention particularly Vincent Cavina, Colyn Crane-Robinson, Anatoly Dragan, Vladimir Filimonov, Ernesto Freire, Hans Hinz, Nick Khechinashvili, George Makhatadze, Leonid Medved, Jamlet Monaselidze, Valery Novokhatni, Valerian Plotnikov, Wolfgang Pfeil, Sergei Potekhin, George Privalov, Oleg Ptitsyn, Rusty Russel, Tamara Tsalkova, and Paul Vaitiekunas. I have to mention specially my late friends who stimulated my involvement in studying the thermodynamics of biological macromolecules: Chris Anfinsen, John Edsall, Stanley Gill, Julian Sturtevant, and Jeffries Wyman.

I thank Thermo Analytical Instruments for their excellent manufacture of the calorimeters designed by my group and for providing photos of their parts for this book. The availability of these supersensitive instruments, nanocalorimeters, has opened a wide prospect for the experimental investigation of the thermodynamics of biological macromolecules and their complexes.

2

METHODOLOGY

2.1. THERMODYNAMIC BASICS OF CALORIMETRY

2.1.1. Energy

Energy is one of the most abstract notions. The energy conservation law is one of the greatest generalizations in science:

> *Energy does not disappear or appear; it only changes its form and appears in the form of mechanical energy, thermal energy, electrostatic energy, and so on.*

We cannot sense energy, cannot measure it directly, but can only judge it by the manifestation of its changes, which appear in the forms of work (W) done and heat (Q) evolved:

$$\Delta E = Q + W \tag{2.1}$$

In the case of chemical reactions, particularly those involving proteins and those connected with changes of protein structure, the mechanical work done is associated

Microcalorimetry of Macromolecules: The Physical Basis of Biological Structures, First Edition.
Peter L. Privalov.

with the change of volume ΔV of the system being considered; at constant external pressure P this is

$$W = -P\,\Delta V \tag{2.2}$$

If the volume of a system decreases under the process being considered (i.e., the system is compressed under external pressure), then the work done on the system is positive, and the energy of the system increases:

$$\Delta E = Q - P\,\Delta V \tag{2.3}$$

2.1.2. Enthalpy

Rewriting Equation (2.3) as

$$Q = \Delta E + P\,\Delta V \tag{2.4}$$

it appears that the heat provided to the system changes its energy and does the work. This heat could be regarded as some invisible liquid substance that is poured into the system—for a long time heat was regarded as a liquid substance, **phlogiston**. It was assumed that the phlogiston poured into a system raised the heat content of a system. The heat content of a system is called **enthalpy** and is designated by the symbol H. Thus, in providing heat to the system we are increasing its enthalpy:

$$Q = \Delta H = \Delta E + P\,\Delta V \tag{2.5}$$

Enthalpy is the energy of the extended system: it includes not only the internal energy of a system itself (ΔE) but also the energy of its surroundings; that is, it also includes the external work (PΔV) that is done by the system or is performed on the system by its surroundings.

Since the systems with which we usually deal are in some environment, it is clear that all changes of such systems are associated with a change of this extended energy, the enthalpy. The change of enthalpy of a system is measured by the heat that is released or absorbed by the system under the process being considered. This is just what calorimetry does: measuring the heat of reaction determines the enthalpy change of the system in which the reaction takes place.

2.1.3. Temperature

Temperature is a measure of the warmth or coldness of an object with reference to some standard value. The temperature of two macroscopic systems is the same when the systems are in thermal equilibrium. There is no heat flow between the systems when they are in equilibrium, and the heat flow between them increases with increasing difference in temperature between these systems, flowing from the warmer to the cooler system.

Temperature does not depend on the size of the system; it is an intensive characteristic of the thermal state of any macroscopic system.

There are several scales for temperature. The most popular in the United States, the **Fahrenheit scale**, is the vaguest because it does not have clear reference temperatures: 0°F is the freezing point of mercury and 100°F is the physiological temperature of a human, which certainly is not a good reference because different parts of the body have different temperatures. Much better and more practical is the **Celsius scale** because it has clear reference points: 0°C is the temperature of water freezing at normal pressure, that is, at 1 atmosphere (1 atm = 1 kg/cm^2 = 9.8 N/cm^2); 100°C is the temperature of water boiling at normal pressure. This scale has particular importance for biology since liquid water is a natural internal component of all living systems. The most appropriate for science is the **absolute scale**, or the **Kelvin scale**, because this scale has a clear physical meaning: at 0 K all thermal motions are frozen—the system is at the lowest energy level. In this scale the temperature appears as an absolute characteristic of the intensity of thermal motion of the atoms constituting the macroscopic object. Therefore, the Kelvin scale is the only scale that can be used in thermodynamic analysis.

The units of temperature (degrees) in the Celsius and Kelvin scales are identical and are designated as °C or as K, respectively. Denoting temperature in the Kelvin scale by °K is incorrect. The temperature 0°C corresponds to 273.16 K, and 0 K corresponds to −273.16°C. Temperature difference in the Celsius scale is also expressed in kelvins (the unit K); thus the difference between T_2 = 37°C and T_1 = 25°C is 12 K, but not 12°K or 12°C.

2.1.4. Energy Units

The heat effect, that is, the amount of thermal energy, is measured in **calories**. One **calorie** is the amount of heat necessary to raise the temperature of one gram of water from 14.5°C to 15.5°C. Calories are useful for measuring heat in practical life and in biology because of the dominant role of water in our external and internal media. However, the calorie is not a practical unit for measuring energy in science, particularly in the physical sciences. The International Union of Pure and Applied Chemistry (IUPAC) recommends using the joule as the energy unit in science.

***Joule (J)** is the unit of work done by a force of one newton when its point of application moves through a distance of one meter in the direction of the applied force.*

***Newton (N)** is the unit of force that produces an acceleration of one meter per square second when applied to a mass of one kilogram.*

The joule is about 4 times smaller a unit of energy than the calorie:

1 cal = 4.16 J

In atomic physics a much smaller unit of energy, the **erg**, is used:

1 joule = 10^7 ergs

> ***Erg*** *is the work of 1 dyne of force applied over the distance of one centimeter.*
>
> ***Dyne*** *is the unit of force that produces an acceleration of one centimeter per square second on a mass of one gram.*

In biological physical chemistry the erg is not used because it is too small a unit of energy.

2.1.5. Heat Capacity

The amount of heat needed to heat a body from temperature T_1 to temperature T_2 depends on a thermal property of the body called the **heat capacity**. The heat capacity of a body is determined by the amount of heat that is required to increase its temperature by 1 K. If the body is heated at constant volume, the heat provided to the body will be used to increase its internal energy. Correspondingly, the heat capacity at constant volume is

$$C_V = \left(\frac{dE}{dT}\right)_V \qquad (2.6)$$

If the body is heated at constant pressure, its heating would be associated with thermal expansion, that is, with the external work. Therefore in that case the heat capacity is determined by the change of the enthalpy of the system when its temperature is increased by 1 K:

$$C_p = \frac{\partial H}{\partial T} \qquad (2.7)$$

The heat capacity at constant pressure is a more important characteristic than the heat capacity at constant volume because direct measurement of the heat capacity of solid or liquid bodies at constant volume is impossible due to their thermal expansion.

> *Heat capacity is an extensive characteristic of a body because it is proportional to the body's size (i.e., its mass):*

$$C_p = c_p \times m \qquad (2.8)$$

The coefficient of proportionality, c_p, is called **specific heat capacity**.

> *Specific heat capacity is an intensive characteristic of the material constituting a body since it does not depend on the body's size.*

It is determined by the amount of heat (i.e., the amount of enthalpy) required to raise the temperature of 1 g of material by 1 K.

The unit of specific heat capacity is joules per kelvin per gram (written J/K·g or $J \cdot K^{-1} \cdot g^{-1}$). The heat capacity also can be specified per mole of substance; in that case it is called a *molar heat capacity* and is expressed as joules per kelvin per mole (written J/K·mol or $J \cdot K^{-1} \cdot mol^{-1}$). The sequence of these symbols is important. To write J/mol·K is incorrect because J/mol does not have the meaning of heat capacity.

Usually the heat capacity is a temperature-dependent function:

$$C_p(T) = \frac{\partial H(T)}{\partial T} \tag{2.9}$$

Integrating this expression from temperature T_1 to T_2, one gets

$$\int_{T_1}^{T_2} C_p(T)dT = \int_{T_1}^{T_2} \frac{\partial H(T)}{\partial T} dT$$

Rearranging this, we get

$$H(T_2) - H(T_1) = \int_{T_1}^{T_2} C_p(T)dT$$

or

$$H(T_2) = H(T_1) + \int_{T_1}^{T_2} C_p(T)dT \tag{2.10}$$

Thus, if we know the enthalpy of a system at temperature T_1 and the heat capacity of this system in the whole temperature range from T_1 to T_2, we can estimate what the enthalpy of this system would be at temperature T_2. If the heat capacity does not depend on temperature, we simply have

$$H(T_2) = H(T_1) + C_p(T_2 - T_1) \tag{2.11}$$

2.1.6. Kirchhoff's Relation

Consider the following process: heat the system in state A from temperature T_1 to T_t. At temperature T_t some transformation of this system (e.g., melting) takes place, with its state changing from A to B. This transformation is accompanied by the change in heat capacity

$$\Delta_A^B C_p(T_t) = C_p^B(T_t) - C_p^A(T_t)$$

and enthalpy

$$\Delta_A^B H(T_t) = H^B(T_t) - H^A(T_t)$$

What will the enthalpy of the system be at temperature T_2?

$$H^B(T_2) = H^A(T_1) + \int_{T_1}^{T_2} C_p^A(T)\mathrm{d}T + \Delta_A^B H(T_t) + \int_{T_1}^{T_2} C_p^B(T)\mathrm{d}T \tag{2.12}$$

If $T_1 = T_2 = T$, then

$$H^B(T) - H^A(T) = \Delta_A^B H(T) \quad \text{and} \quad H^B(T_t) - H^A(T_t) = \Delta_A^B H(T_t)$$

and

$$\Delta_A^B H(T) = \Delta_A^B H(T_t) - \int_{T_1}^{T_2} \Delta C_p(T)\mathrm{d}T \tag{2.13}$$

If $\Delta_A^B C_p$ = constant, then

$$\Delta_A^B H(T) = \Delta_A^B H(T_t) - \Delta C_p \times (T_t - T) \tag{2.14}$$

Thus, if we know the enthalpy of the reaction at some temperature T_2*, we can calculate it for any other temperature* T *(Fig. 2.1). All we need to know are the heat capacities of the initial and final states in the temperature range being considered.*

If $T_2 = T_1 + \delta T$ and $T_2 - T_1 = \delta T$, then $\Delta H(T_2) - \Delta H(T_1) = \delta\Delta H(T)$ and $\delta\Delta H(T) = \Delta C_p \times \delta T$, or

$$\frac{\partial \Delta H}{\partial T} = \Delta C_p \tag{2.15}$$

This is the **Kirchhoff** relation, which means the following:

If the heat capacity increases upon some reaction, the enthalpy of the reaction is an increasing function of temperature and vice versa—if the enthalpy of reaction

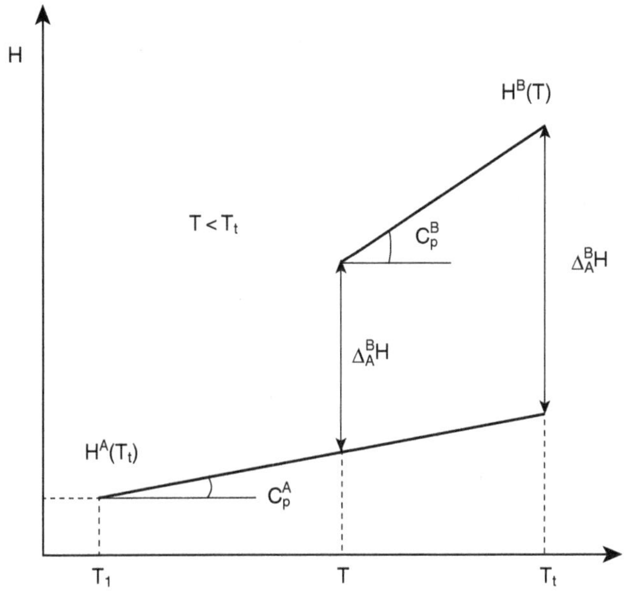

Figure 2.1. If the enthalpy of a reaction at some temperature T_t is known, one can calculate it for any other temperature T using the difference between the heat capacities of the initial and final states.

is an increasing function of temperature, this means that the heat capacity of the final state is larger than the heat capacity of the initial state.

2.1.7. Entropy

The entropy of a system represents the part of the energy of a system that has been dissipated in thermal motion. Therefore, the entropy can be considered as a **measure of a system's disorder**. According to the second principle of thermodynamics:

All processes in an isolated system develop in the direction of an increase of its entropy, that is, of raising its disorder.

In contrast to enthalpy, the change in the entropy of a system in some reaction is determined by the heat Q that is received by the system, divided by the absolute temperature at which the reaction takes place:

$$\delta S(T)\delta S = \frac{\delta Q}{T} = \frac{C_p \delta T}{T} \tag{2.16}$$

If we want to determine the change in entropy of a system upon heating from T_1 to T_2, we have to integrate the following equation:

$$S(T_2) = S(T_1) + \int_{T_1}^{T_2} \frac{C_p(T)}{T} dT = S(T_1) + S(T_2) = S(T_1) + \int_{T_1}^{T_2} \frac{C_p}{T} dT \qquad (2.17)$$

According to the third principle of thermodynamics:

> *At absolute zero temperature (0 K) the entropy of all macroscopic systems is zero; that is, the systems are in complete order.*

Therefore, for the absolute entropy of a system at any other temperature we have

$$S(T) = \int_{0}^{T} \frac{C_p(T)}{T} dT = \int_{0}^{T} C_p(T) d\ln T \qquad (2.18)$$

The absolute entropy of a system is an absolute measure of the system's disorder. However, measurement of the heat capacity function of an object from absolute zero temperature to room temperature is not easy. It is especially difficult for aqueous solutions, which freeze below 273 K with large decrease in entropy. Also the entropy of an aqueous solution at 0 K does not become zero because water at that temperature still has residual disorder from the undetermined location of its two protons between two possible positions for each. Therefore, the absolute entropy of aqueous solutions cannot be determined. Because of that, in considering aqueous solutions **the standard entropy** is usually used, choosing some state as a reference.

Suppose we know that upon heating a system changes state from A to B at the transition temperature T_t, and this results in a change of entropy $\Delta_A^B S(T_t)$; the heat capacities of the initial and final states are $C_p(T)^A$ and $C_p(T)^B$. What will the entropy of this system be at temperature T compared with its entropy at temperature T_t?

We have

$$S(T)^B = S(T_1)^A + \int_{T_1}^{T_t} C_p^A(T) d\ln T + \Delta_A^B S(T_t) + \int_{T_t}^{T} C_p^B(T) d\ln T \qquad (2.19)$$

If $T = T_1$,

$$\Delta_A^B S(T) = S(T)^B - S(T)^A = \Delta_A^B S(T_t) - \int_{T}^{T_t} \Delta_A^B C_p(T) d\ln T \qquad (2.20)$$

If $\Delta_A^B C_p = \text{constant}$,

$$\Delta_A^B S(T) = \Delta_A^B S(T_t) - \Delta_A^B C_p \times \ln(T_t/T) \qquad (2.21)$$

This shows that the entropy of a reaction is a function of temperature and depends on the sign of the heat capacity difference between the final and initial states, ΔC_p.

If $\Delta C_p > 0$*, then the entropy of reaction is an increasing function of temperature.*
If $\Delta C_p < 0$*, then the entropy is a decreasing function of temperature.*

The unit of entropy is joules per kelvin (J/K) or calories per kelvin (cal/K); the latter is also called the **entropy unit** and is designated simply as "e.u." Correspondingly, the molar entropy is expressed in joules per kelvin per mole (J/K·mol) or calories per kelvin per mole (cal/K·mol) and the specific entropy in joules per kelvin per gram (J/K·g) or calories per kelvin per gram (cal/K·g), which can be also written as $J \cdot K^{-1} \cdot g^{-1}$ and $cal \cdot K^{-1} \cdot g^{-1}$. In the case of proteins it could be also expressed per mole of amino acid residues, J/K·(mol-res) or $cal \cdot K^{-1} \cdot (mol\text{-}res)^{-1}$; in the case of nucleic acids the specific entropy could be expressed per mole of base pairs, that is, J/K·(mol-bp) or cal/K·(mol-bp).

2.1.8. Gibbs Free Energy

Change of the Gibbs free energy, ΔG, shows the part of the energy of an extended system that can be converted into work at a constant temperature.

Since

$$\Delta H = Q + W, \quad W = \Delta H - Q, \quad \text{and} \quad \Delta S = Q/T$$

we have the following for the Gibbs energy:

$$\Delta G = W = \Delta H - T\,\Delta S \tag{2.22}$$

The Gibbs free energy is a very important function because it determines the equilibrium constant for any reaction:

$$K = \exp(-\Delta G/RT) \tag{2.23}$$

Correspondingly, the Gibbs energy of reaction can be determined from the equilibrium constant:

$$\Delta G(T) = -RT\ln(K), \tag{2.24}$$

Here R is the universal gas constant: $R = 2.0$ cal/K·mol = 8.31 J/K·mol (or 8.31 $JK^{-1} \cdot mol^{-1}$).

2.2. EQUILIBRIUM ANALYSIS

2.2.1. Two-State Transition

Thermodynamics of some process in a system occurring upon a variation of conditions is more or less easily described if the observed changes are due to transitions between two definite states of the system. In this case, and only in this case, all the observed effects can be specified through the equilibrium constant:

$$K = \frac{\theta_x - \theta_a}{\theta_b - \theta_a} \tag{2.25}$$

where θ_a and θ_b are values of any observed indices characterizing the initial and final states of the system being considered under the given conditions. Studying the dependence of equilibrium constant on external variables (such as temperature, pressure, and ion activity) one can derive the effective parameters characterizing the process.

Consider equilibrium of two phases at the temperature of the phase transition, T_t. In the case of a monomolecular reaction the equilibrium constant is determined by the ratio of the fractions of the molecules in the final and initial states:

$$K = \frac{f_2}{f_1} = \frac{f_2}{1 - f_1} \tag{2.26}$$

where $K = 1$ at the temperature of the transition midpoint, T_t. Bearing in mind Equation (2.24), we have that at the transition midpoint $\Delta G(T_t) = 0$. This means that at the transition temperature the transfer from one phase to another does not lead to a gain or loss of energy—no work is done. Then, since

$$\Delta G(T_t) = \Delta H(T_t) - T_t \Delta S(T_t) = 0$$

we get the following for temperature T_t:

$$\frac{\Delta_t H(T_t)}{T_t} = \Delta_t S(T_t) \tag{2.27}$$

It is important that T be given in the absolute scale.

Thus, we can determine the entropy at the transition temperature just by measuring the heat effect of a transition, that is, the enthalpy of a transition. If we know the heat capacity of the initial state and of the final state, that is, the difference of these heat capacities ΔC_p, then we can determine the entropy difference of two phases at any other temperature:

$$\Delta S(T) = \frac{\Delta H(T_t)}{T_t} + \int_{T_t}^{T} \frac{\Delta C_p}{T} dT \tag{2.28}$$

If ΔC_p does not depend on temperature, then

$$\Delta S(T) = \frac{\Delta H(T_t)}{T_t} + \Delta C_p \ln\left(\frac{T}{T_t}\right) \tag{2.29}$$

2.2.2. Derivatives of the Equilibrium Constant

The equilibrium constant of any transition depends on the intensive variables determining the environmental conditions, particularly temperature (T), pressure (P), and ligand activity (a).

If in the monomolecular two-state transition A $\Leftrightarrow$ B only the temperature is variable and the pressure and ligand activity are constant, then, bearing in mind that

$$K = \exp\left(-\frac{\Delta G}{RT}\right) = \exp\left(-\frac{\Delta H}{RT} + \frac{\Delta S}{R}\right) \tag{2.30}$$

and

$$\ln K = -\frac{\Delta H}{RT} + \frac{\Delta S}{R} \tag{2.31}$$

we have the following for the temperature derivative of the equilibrium constant:

$$\frac{\partial \ln K}{\partial T} = -\frac{1}{RT}\frac{\partial \Delta H}{\partial T} + \frac{\Delta H}{RT^2} + \frac{1}{R}\frac{\partial S}{\partial T} \tag{2.32}$$

However, since

$$\frac{\partial \Delta H}{\partial T} = \Delta C_p, \quad \frac{\partial \Delta S}{\partial T} = \frac{\Delta C_p}{T}$$

and

$$\frac{\partial \ln K}{\partial T} = -\frac{\Delta C_p}{RT} + \frac{\Delta H}{RT^2} + \frac{\Delta C_p}{RT} = \frac{\Delta H}{RT^2}$$

we get

$$\Delta H^{\mathrm{vH}} = RT^2\left(\frac{\partial \ln K}{\partial T}\right)_{P,a} \tag{2.33}$$

This is called the **van't Hoff equation**. It can be rewritten as

$$\Delta H^{\mathrm{vH}} = -R\frac{\partial \ln K}{\partial 1/T}, \quad \text{or} \quad -\frac{\Delta H^{\mathrm{vH}}}{R} = \frac{\partial \ln K}{\partial 1/T} \tag{2.34}$$

Thus, analyzing the temperature dependence of the equilibrium constant of a reaction, one can determine the enthalpy of the reaction. This enthalpy is usually called the **van't Hoff enthalpy** or the **effective enthalpy** because it is valid only if the reaction represents a two-state transition.

To get the value of the van't Hoff enthalpy, $\ln K$ is plotted against $1/T$. The slope of this function is equal to

$$-\frac{\Delta H^{\mathrm{vH}}}{R}$$

If the function thus obtained is linear, the enthalpy does not depend on temperature; if it has noticeable curvature, the enthalpy depends on temperature; that is, the reaction proceeds with heat capacity change.

When pressure is a variable parameter in a reaction, bearing in mind that $\Delta H = \Delta E + P\Delta V$ and $\Delta G = -RT\ln K = \Delta H - T\Delta S = \Delta E + P\Delta V - T\Delta S$, one finds

$$-RT(\mathrm{d}\ln K/\mathrm{d}P)_{T,a} = \Delta V \tag{2.35}$$

Thus, analyzing the dependence of the logarithm of the equilibrium constant on pressure, at constant temperature and ligand concentration, one finds the volume effect of the reaction.

If the variable parameter is ligand concentration, bearing in mind that

$$\Delta G(a)_{T,P} = -RT\ln(K) = \Delta G(a_0)_{T,P} - \Delta n\ln(a/a_0)_{T,P} \tag{2.36}$$

one finds

$$RT\left(\frac{\partial \ln K}{\partial \ln a}\right)_{P,T} = \Delta n \tag{2.37}$$

At low concentration of a ligand its activity is close to its molar concentration. Therefore, analyzing the dependence of the logarithm of the equilibrium constant on the logarithm of ligand concentration, one can determine the quantity of bound or released ligand in a process.

2.3. AQUEOUS SOLUTIONS

2.3.1. Specificity of Water as a Solvent

Water represents a universal averment for all biological species and a unique solvent for its components, particularly proteins and nucleic acids. The specificity of water proceeds from its very particular structure: The two hydrogens of water and two lone electrons form a highly polar tetrahedron (Fig. 2.2). Therefore water has two hydrogen donors and two hydrogen acceptors: the hydrogen acceptors are the two lone electrons; the role of donor is played by the oxygen, which has two covalently bound hydrogens. As a result, in ice each water molecule forms four hydrogen bonds with four of its neighbors (Fig. 2.3).

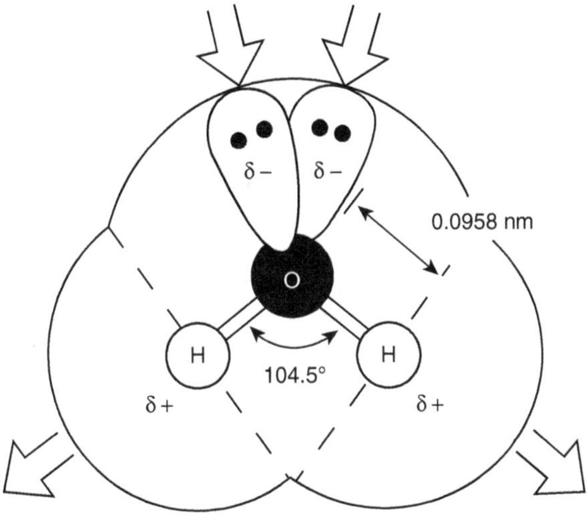

Figure 2.2. The electronic structure of a water molecule.

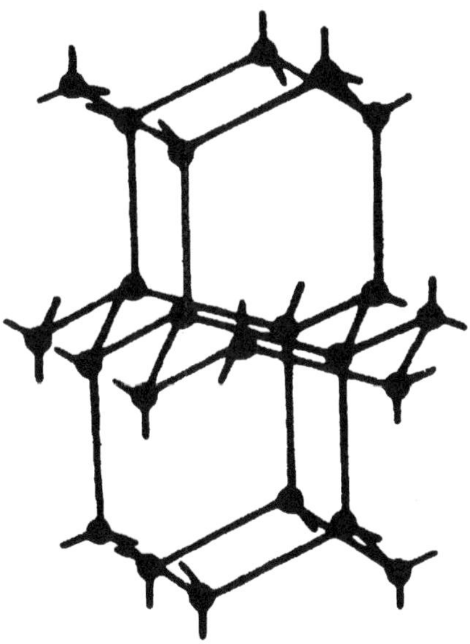

Figure 2.3. The structure of normal ice.

An important property of the water molecule is **cooperativity in formation of hydrogen bonds**:

> *Formation of one hydrogen bond with a neighboring water molecule increases the probability of formation of a bond with another neighbor because it increases the electronegativity of the acceptor groups.*

The ability to form four hydrogen bonds, and cooperativity in their formation, results in unique properties of water: it is a liquid with high tendency to have an ordered structure. The structure of water below 0°C (i.e., of ice) is rather open, transparent, with large cavities. Density of ice is significantly lower than that of liquid water. But even above 0°C, water molecules have a tendency to form an icelike structure. Therefore, water represents a highly structured liquid. The combination of "liquid" and "structured" sounds like a paradox. This structure, however, is not fixed; it largely fluctuates. Thus, its orderliness appears as flickering clusters of the icelike structure. With temperature increase the average amount of the ordered clusters decreases. It appears as if they melt gradually and this results in a **very high heat capacity of water**. The heat capacity of water is almost 3 times higher than that of organic liquids.

Due to its polarity water forms hydrogen bonds with all polar solutes and strongly interacts with charged molecules. Correspondingly, polar and charged molecules are highly soluble in water. Because of the high polarity of the water molecule, liquid water has a very high dielectric constant, 78. Therefore, water efficiently screens electrostatic interaction between charged groups and promotes dissociation of ionic pairs, particularly dissociation of the removable hydrogen in acids.

2.3.2. Acid–Base Equilibrium

The acid–base properties of any substance in aqueous solution are connected with the presence of a removable proton. We call a substance an **acid** if it tends to release its proton and a **base** if it tends to accept a proton. The structure B, which remains after a hydrogen ion is removed from the acid BH, represents the base conjugate to the acid:

$$\mathrm{BH} \leftrightarrow \mathrm{B}^- + \mathrm{H}^+ \tag{2.38}$$

Since the hydrogen ion is positively charged, it is clear that either the BH or B, or both, must be electrically charged, and that the charge on B must be more negative (or less positive) than that on BH by one proton unit. If the net charge of BH is Z, then the charge of B is $Z - 1$.

$$\mathrm{BH}^Z \leftrightarrow \mathrm{B}^{(Z-1)} + \mathrm{H}^+ \tag{2.39}$$

The equilibrium constant of Reaction (2.38) is

$$K = \frac{a_\mathrm{H} a_\mathrm{B}}{a_\mathrm{BH}} \tag{2.40}$$

where a_H, a_B, and a_BH are the corresponding activities.

Replacing activities by concentrations, since at low concentrations they do not differ much, one gets

$$K \approx \frac{[\mathrm{H}][\mathrm{B}]}{[\mathrm{BH}]} \tag{2.41}$$

In the case of water, we have

$$H_2O \leftrightarrow H^+ + OH^- \tag{2.42}$$

The hydrogen ion (proton) is extremely reactive. It immediately joins another water molecule, forming the **hydronium** ion, H_3O^+.

Concentrations of H^+ and OH^- in pure water are equal and very low, 10^{-7} M. This is usually presented in a logarithmic scale: $\log([H^+]) = -7.0$. Here $[H^+]$ is the dimensionless relative concentration of protons $[H^+]/[H^+]^0$, where $[H^+]^0 = 1$ M is a standard concentration of protons. The "–log" is usually denoted as "p"; therefore $-\log([H^+]) = \text{pH}$.

The pH of the neutral aqueous solution is 7.0. In this notation, Reaction (2.42) can be rewritten as

$$\text{pH} = \text{p}K + \log([\text{B}]/[\text{BH}]) \tag{2.43}$$

Let α be the fraction of the deprotonated form B:

$$\alpha = \frac{[\text{B}]}{[\text{B}]+[\text{BH}]} \tag{2.44}$$

Correspondingly, the fraction of the protonated form is

$$1-\alpha = \frac{[\text{BH}]}{[\text{B}]+[\text{BH}]} \tag{2.45}$$

and Equation (2.43) appears as

$$\text{pH} = \text{p}K + \log\left(\frac{\alpha}{1-\alpha}\right) \tag{2.46}$$

This can be rewritten as

$$\alpha = \frac{10^{\text{pH}-\text{p}K}}{1+10^{\text{pH}-\text{p}K}} \tag{2.47}$$

It shows that when pH = pK, then α = 1/2; when pH << pK, then α decreases to 0; when pH >> pK, then α approaches 1.0. Figure 2.4 illustrates this for titration of the amino acid glycine. This amino acid has two titratable groups: the α-carboxy, with pK = 2.4; and the α-amino, with pK = 9.8. At low pH both groups are protonated and the molecule has one positive charge. With increasing pH, that is, decreasing hydrogen concentration in the solution, first the α-carboxy group starts to deprotonate, becoming negatively charged. Therefore, at neutral pH glycine has one positive and one negative charge. With further increase of pH (decrease of hydrogen

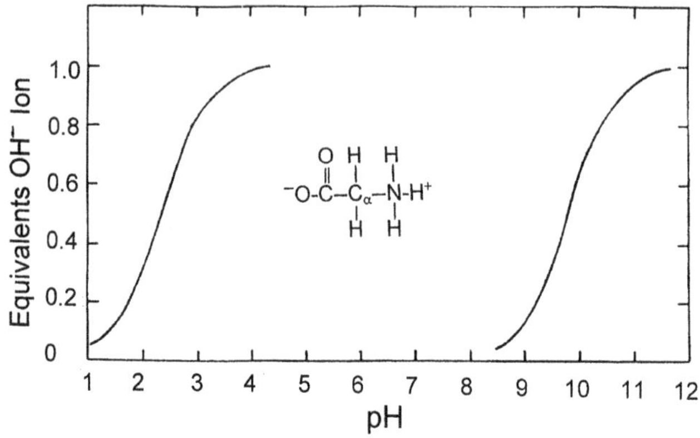

Figure 2.4. Deprotonation of glycine at increasing pH.

concentration), the α-amino group deprotonates, that is, loses its positive charge, and the molecule becomes negatively charged.

The work required to remove hydrogen is expressed by the Gibbs energy of this reaction:

$$\Delta G = -RT \ln(K) \tag{2.48}$$

For the reaction BH $\leftrightarrow$ $B^- + H^+$, $-\ln K = -2.3 \log K = 2.3\,\text{p}K$, and we have

$$\Delta G = 2.3RT\ \text{p}K \tag{2.49}$$

Using this equation one finds that at room temperature (25°C = 298.15 K) removal of the first hydrogen from glycine costs $\Delta G^{\text{I}} = 8.36 \times 298.15 \times 2.4$ kJ/mol = 6.0 kJ/mol; the second hydrogen costs $\Delta G^{\text{II}} = 24.4$ kJ/mol. Thus, the α-amino group holds onto the extra hydrogen much more strongly than the α-carboxyl group. The thermodynamic characteristics of protonation of the most common groups in proteins are listed in Table 2.1.

2.3.3. Partial Quantities

In biochemistry, molecular biology, and physical chemistry of biopolymers, we usually deal with aqueous solutions, and mostly with rather dilute solutions. However, we are particularly interested in the properties of individual components in the solution. Separation of the properties of a certain component from the overall properties of the solution is not a simple problem because interactions between the solute components of the solution are not simple. However, at low concentrations, where interactions between the solute molecules are negligible, their contribution to the overall properties of the solution is defined by the **partial characteristic** of a component. In particular, in the case of heat capacity, the partial heat capacity $C_p^{\phi_i}$ of the ith component is defined as a change in the total heat capacity of a unit mass of

TABLE 2.1. Thermodynamic Characteristics of Deprotonation of Protein Groups at 25°C

Group	pK	ΔG (kJ/mol)	ΔH (kJ/mol)	ΔS (J/K·mol)
α-Carboxy, $-CO-O^{-}H^{+}$	2.8	15.6	1.7	−47
Asp, side chain, $-CH_2-CO-O^{-}H^{+}$	3.9	21.8	4.6	−58
Glu, side chain, $-CH_2-CH_2-CO-O^{-}H^{+}$	4.3	24.0	1.6	−75
His, side chain, $-CH_2-C_3NH_4-NH^{+}$	6.0	33.5	28.8	−16
α-Amino, $-NH_2H^{+}$	9.8	54.7	43.8	−40
Tyr, side chain, $-CH_2-C_6H_4-OH^{+}$	10.1	56.4	25.1	−105
Lys, side chain, $-CH_2-CH_2-CH_2-CH_2-NH_3^{+}$	10.8	60.3	53.8	−22
Arg, side chain, $-CH_2-CH_2-CH_2-NH-CNH_2-NH_3^{+}$	12.5	69.8	51.8	−60

solution upon addition of component i at a constant concentration of all other components j:

$$C_p^{\phi_i} = \lim_{\omega_i \to 0} \left(\frac{\partial C_p}{\partial \omega_i} \right)_{T,P,j} \tag{2.50}$$

The heat capacities of a series of solutions, in which only the concentration of the ith component ω_i is varied, are measured and plotted against ω_i; by extrapolating the slope of this function to low concentrations, the partial heat capacity of the ith component is determined. However, if the heat capacity of a macromolecular solution is measured at concentrations less than 0.1%, the partial heat capacity of the solute can be determined just by comparing the heat capacity of the solution with the heat capacity of the solvent (see Section 3.2.3). Figure 2.5 presents examples of the calorimetric determination of partial molar heat capacity of linear alcohols.

Combinatorial analysis of the partial molar heat capacities of various linear molecules led to understanding that the contribution of various groups to the overall heat capacity of the compound is additive and quite definite (Fig. 2.6). It showed that the contribution of the nonpolar group to the overall heat capacity of the molecule is positive and decreases with increasing temperature, in contrast to the contribution of the polar and charged groups, which decreases with decreasing temperature and even changes sign at low temperatures, becoming negative (Makhatadze and Privalov, 1988, 1989, 1990; Murphy and Gill, 1991; Livingstone et al., 1991; Spolar et al., 1989, 1992).

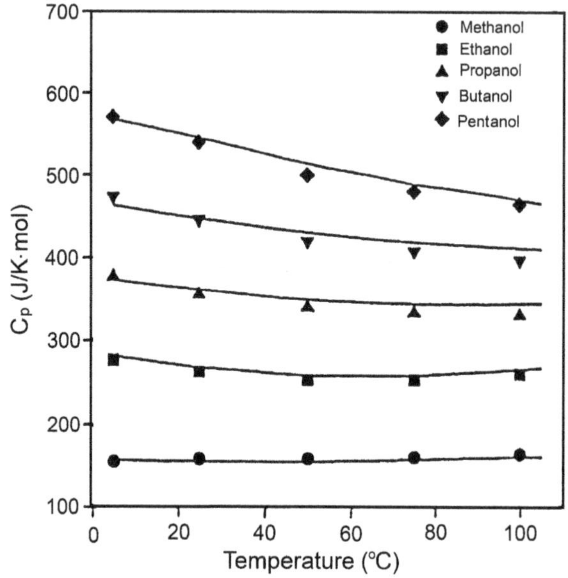

Figure 2.5. The calorimetrically measured partial molar heat capacity of linear alcohols (Makhatadze et al., 1997).

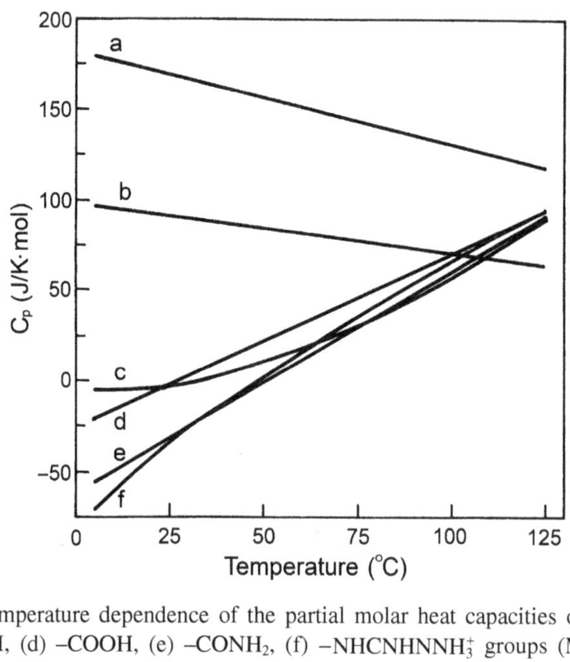

Figure 2.6. Temperature dependence of the partial molar heat capacities of (a) $-CH_3$, (b) $-CH_2^-$, (c) $-OH$, (d) $-COOH$, (e) $-CONH_2$, (f) $-NHCNHNNH_3^+$ groups (Makhatadze and Privalov, 1990).

2.4. TRANSFER OF SOLUTES INTO THE AQUEOUS PHASE

2.4.1. Hydration Effects

Transfer of molecules from the nonaqueous phase into the aqueous phase results in hydration of these molecules. The molecules could be transferred into the liquid aqueous phase from their condensed phase, liquid phase, or gaseous phase. However:

> *The net hydration effects include the enthalpy, entropy, and Gibbs energy of transfer of a molecule from a fixed position in an ideal gas phase into a fixed position in water.*

These can be determined from the heats of desolation of the liquid or condensed phase in water, taking into account the enthalpy and entropy of melting and vaporization of these phases (Fig. 2.7). The thermodynamic characteristics of this process do not include the effects associated with the difference in volume of the two phases, that is, the difference in the translational motion of the molecule in the phases. They do not include the interactions between the transferred molecules, but only the effects associated with the insertion of the solute molecule into solvent, usually water. The insertion of a solute molecule into water is in itself a complex process. It includes cavity formation in water, interaction of the water molecules with the inserted molecule, and reorganization of the water molecules caused by the inserted solute (Lee, 1985).

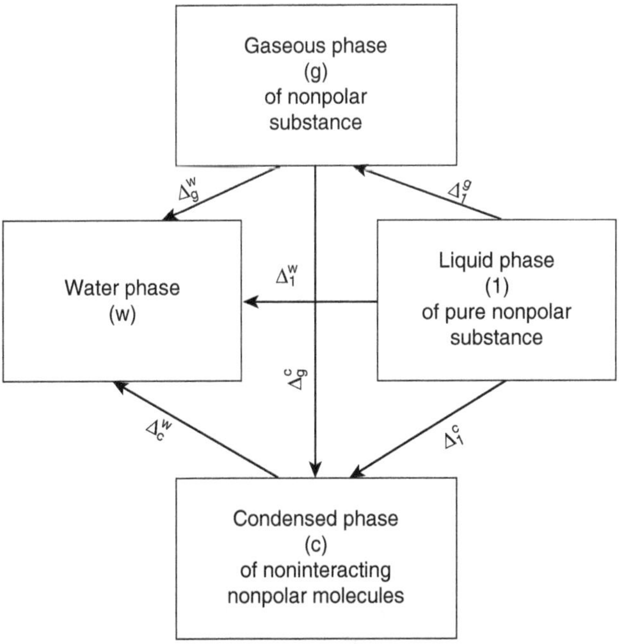

Figure 2.7. Scheme of various ways to transfer solutes into water.

The chemical potential μ of a solute in any phase can be separated into two parts: $\mu = \mu_1 + \mu_2$, where μ_1 is the work involved in inserting the solute molecule at a fixed point in the given phase, and $\mu_2 = RT\ln(\Lambda^3\rho)$ is the work involved in "liberating" the molecule from the fixed point so that it can roam about entire volume of the phase. Here Λ is the momentum partition function, and ρ is the molar concentration (Ben-Naim, 1980, 1987). Because the hydration effect is associated with μ_1, it follows that all experimentally obtained transfer characteristics should be corrected for the effect of liberation.

When the concentration is expressed in the molar scale, the Gibbs energy of hydration is just the energy of transfer of the solute molecule from the gaseous phase into water:

$$\Delta G^{\text{hyd}} = \Delta G^{\text{trs}} = RT\ln(\rho_{\text{g}}/\rho_{\text{w}}) \tag{2.51}$$

where ρ_{g} and ρ_{w} are the molar concentrations of the molecules in the gaseous phase and in the water solution. For the enthalpy of hydration the correction for thermal liberation gives (Ben-Naim, 1987):

$$\Delta H^{\text{hyd}} = \Delta_{\text{g}}^{\text{w}} H + \delta H^{\text{lib}} = \Delta_{\text{g}}^{\text{w}} H + RT(1-\alpha^{\text{w}}T) \tag{2.52}$$

where α^{w} is the thermal expansion coefficient of water at constant pressure.

For the entropy of hydration at 25°C we have

$$\Delta S^{\text{hyd}}(25^\circ) = [\Delta H^{\text{hyd}}(25^\circ) - \Delta G^{\text{hyd}}(25^\circ)]/298.15 \tag{2.53}$$

The heat capacity effect of hydration is determined as

$$\Delta C_p^{\text{hyd}}(T) = \text{d}(\Delta H^{\text{hyd}})/\text{d}T = \Delta_{\text{g}}^{\text{w}} C_p(T) + R[1-2\alpha^{\text{w}}T - (\text{d}\alpha^{\text{w}}/\text{d}T)T^2] \tag{2.54}$$

Here $\Delta_{\text{g}}^{\text{w}} C_p(T)$ is the heat capacity change on transfer of the molecule from the gaseous phase into water.

Studies of dissolution of various compounds in water showed the following:

The solubility of polar and charged substances is very high (i.e., they are hydrophilic), and the Gibbs energies of their transfer are negative.

In contrast, solubility of nonpolar substances is very low (i.e., they are hydrophobic); correspondingly the Gibbs energies of their transfer into water are positive. Surprisingly, it was found that the enthalpy of transfer of nonpolar substances into water at room temperature is close to zero, which means that the entropy of their transfer should be large and negative.

The other remarkable feature of the transfer of nonpolar compounds into water is that it proceeds with positive heat capacity increment—in contrast to the transfer of polar groups, which proceeds with heat capacity decrement (Table 2.2).

TABLE 2.2. Solubility, Gibbs Energy, Enthalpy, Entropy, and Heat Capacity Increment of Transfer of a Nonpolar Substance from Pure Liquid Phase to Water at 25°C[a]

Substance and Its Surface Area (Å^2/mol)	Solubility in Mole Fractions ($\times 10^4$)	ΔG (kJ/mol)	ΔH (kJ/mol)	ΔS (kJ/K·mol)	ΔC_p (kJ/K·mol)
Benzene, 240	4.010	19.4	2.08	−58.06	225
Toluene, 275	1.010	22.8	1.73	−70.7	263
Ethyl benzene, 291	0.258	26.2	2.02	−81.0	318
Cyclohexane, 273	0.117	28.2	−0.10	−94.8	360
Pentane, 272	0.095	28.7	−2.00	−102.8	400
Hexane, 282	0.020	32.5	0	−109.1	440

[a]Privalov and Gill (1989).

It is remarkable that the heat capacity increment of transfer increases with increasing surface area of the compound, suggesting that it results from hydration. Since decrease of entropy is thermodynamically unfavorable, it was believed that water expells the nonpolar substance. This expelling action of water on the nonpolar substance was called the **hydrophobic force**.

Decrease of entropy induced by the presence of a nonpolar substance in water was explained by increasing microscopic "icebergs" of water around the nonpolar substance (Frank and Evans, 1945). Later, these icebergs were considered as "flickering clusters" (Frank and Wen, 1957). However, it was unclear why ordering of water in the presence of nonpolar molecules, which means increase of hydrogen bonding between water molecules, does not proceed with decreasing enthalpy. One would expect an exact compensation for the entropy and enthalpy changes that arise from water ordering, and thus this effect would contribute nothing to the hydration Gibbs energy (Lumry et al., 1982; Privalov and Gill, 1988).

2.4.2. Hydrophobic Force

The heat capacity increment with the transfer of nonpolar molecules into water seemed to confirm an assumption that the presence of a nonpolar compound increases order in the surrounding water, which with increasing temperature starts to melt gradually and the excess heat of melting appears as a heat capacity increment. It was found indeed that transfer of nonpolar substance into water results in an increase of heat capacity. This, however, means that if the enthalpy of transfer of nonpolar molecules into water is zero at room temperature, it should increase with increasing temperature. Correspondingly, if the entropy of transfer of nonpolar molecules into water is negative at room temperature, it should decrease in magnitude with increasing temperature and at some high temperature should drop to zero (Fig. 2.8). For all nonpolar molecules the entropy of transfer from liquid phase to water becomes equal to zero in a rather limited temperature range above 100°C (Privalov and Gill, 1988). However, zero entropy is a condition for the ΔG extremum:

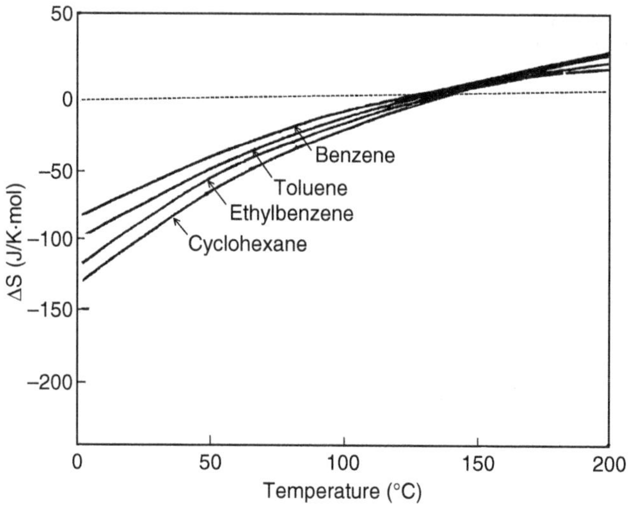

Figure 2.8. Temperature dependence of the entropy of transfer of various nonpolar substances from the pure liquid phase to water (Privalov and Gill, 1988).

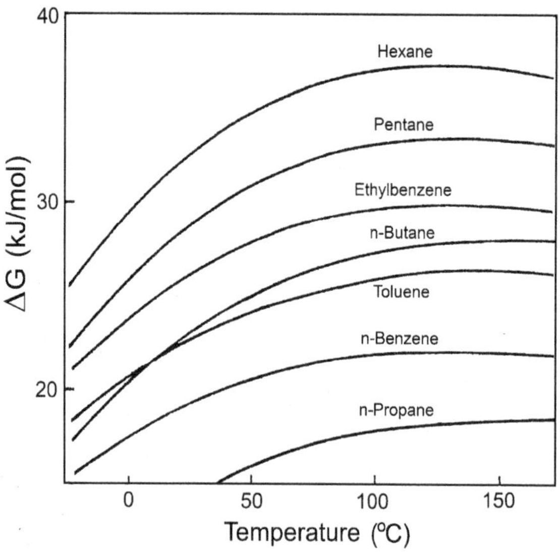

Figure 2.9. Temperature dependence of the Gibbs energy of transfer for various liquid hydrocarbons to water (Privalov and Gill, 1988).

$$\Delta S(T_S)^{\max} = -\partial \Delta G/\partial T = 0 \tag{2.55}$$

Thus, at this temperature T_S, the ΔG of transfer of a nonpolar substance from the liquid phase into water reaches its maximum value (Fig. 2.9). This maximal value of ΔG is provided only by the enthalpy of transfer, which is positive and large at temperature T_S. This enthalpy of transfer at T_S appeared to be close in value to the enthalpy of vaporization of the pure nonpolar substance; that is, it is provided by the van der Waals interaction between the nonpolar molecules in the liquid phase (Baldwin, 1986; Privalov and Gill, 1989).

It appears that there are two temperatures of a universal nature that describe the thermodynamic properties for the dissolution of nonpolar substances (liquid hydrocarbons) into water. The first of these, T_H, is the temperature at which the heat of solution is zero and has a value of approximately 20°C for a variety of nonpolar substances. The second universal temperature is T_S, where the standard entropy change is zero; T_S is about 140°C. The standard-state Gibbs free energy change can be expressed in terms of these two temperatures, requiring knowledge only of the heat capacity change for an individual substance:

$$\Delta_l^w G = \int_{T_H}^{T} \Delta_l^w C_p dT - T\int_{T_S}^{T} (\Delta_l^w C_p / T) dT \tag{2.56}$$

In the approximation where $\Delta_l^W C_p$ is constant, this equation can be integrated to yield

$$\Delta_l^w G = \Delta_l^w C_p (T - T_H) - T\Delta_l^w C_p \ln(T / T_S) \tag{2.57}$$

This function is plotted in Figure 2.10a. Since at T_S the entropy of transfer of nonpolar solutes from their liquid phase into water is zero and the Gibbs energy of

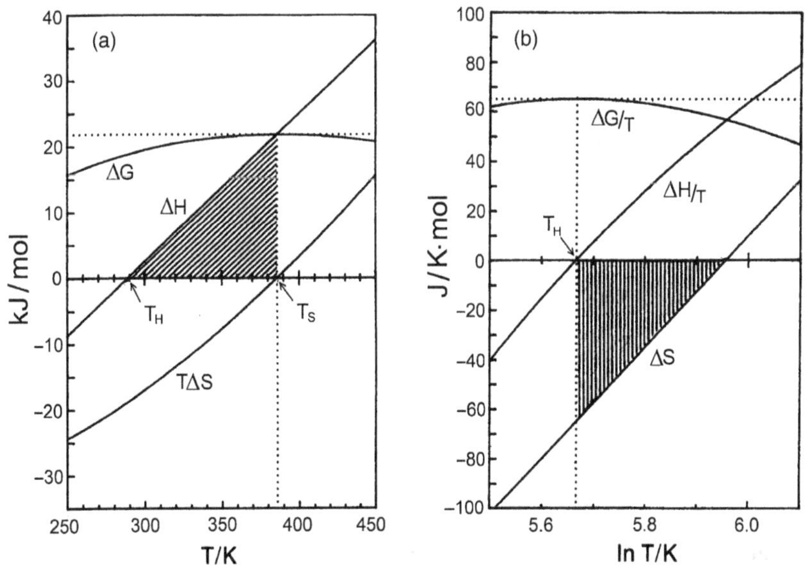

Figure 2.10. Two thermodynamic representations of dissolution of a liquid hydrocarbon (benzene) into water, assuming constant heat capacity effect of hydration: (a) free Gibbs energy of transfer and underlying contributions as a function of temperature; (b) solubility function $(\Delta_l^w G / T)$ and underlying contributions as a function of $\ln(T)$. Dotted lines are shown at the respective maxima, T_H and T_S (Privalov and Gill, 1988).

transfer reaches its maximal value, one can conclude that, although at this temperature the nonpolar solute molecules do not induce water ordering, transfer of these substances into water is most unfavorable. It appears thus, that:

The positive Gibbs energy of transfer of nonpolar substances into water is provided by van der Waals interactions between these molecules, whereas the contribution of water ordering is negative (for details see Privalov and Gill, 1988, 1989).

On the other hand, Equation (2.57) can be rearranged into the following:

$$\Delta_l^w G / T = \Delta_l^w C_p (1 - T_H / T) - \Delta_l^w C_p \ln(T / T_S) \qquad (2.58)$$

which actually expresses the solubility of the nonpolar substance in water, since $\Delta_l^w G / T = -R$(solubility). As Figure 2.10b shows, the solubility reaches its minimal value at T_H. However, one can hardly say that its minimal value results from unfavorable water ordering because at this temperature the enthalpy of van der Waals interactions between nonpolar molecules is compensated by the enthalpy of water ordering, that is, of increasing hydrogen bonding between the water molecules.

An important consequence of the fact that the van der Waals and hydration effects contribute to the hydrophobic effect with opposite signs is the biphasic character of their combined effect.

Indeed, the van der Waals interactions are short range, in contrast to the water ordering effect. It is evident, therefore, that the "hydrophobic interactions" should be attractive at a short distance and repulsive at a long distance (exceeding the size of water molecule).

It appears that this biphasic character of the hydrophobic effect might be one of the reasons for the extreme cooperativity of protein folding–unfolding process (see Section 9.2).

2.4.3. Hydration of Polar and Nonpolar Groups

Analysis of the transfer characteristics of various linear molecules led to understanding that the hydration effects are proportional to the number of groups in the molecule; that is, they are additive (Fig. 2.11). Comparing the effects of transfer of various molecules into water, where the molecules have different combinations of some groups, one can determine the contribution of an individual group to the hydration effect. As Table 2.3 shows, hydration enthalpies and entropies of polar and nonpolar groups are negative at room temperature, but they are much more negative for polar groups than for aliphatic nonpolar groups. As a result, the Gibbs energies of polar groups are negative and large in magnitude, whereas for the nonpolar groups they are small, or even positive. Correspondingly, solubility of the polar molecules in water is high (i.e., they are hydrophilic), in contrast to the solubility of nonpolar molecules in water, which is low (i.e., they are hydrophobic).

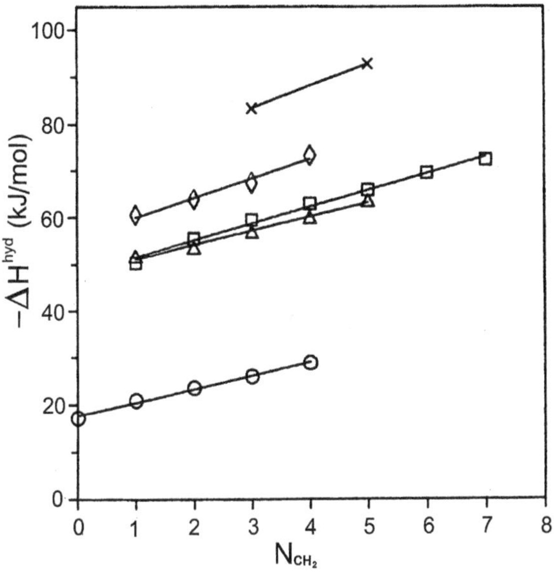

Figure 2.11. Dependence of the enthalpy of transfer from gaseous phase to water for alkanes (○), alcohols (□), amines (△), secondary alcohols (◇), and N-substituted amines (×) on the number of $-CH_2^-$ groups in molecule, N_{-CH2-} (Makhatadze and Privalov, 1993).

TABLE 2.3. Hydration Enthalpy and Gibbs Energy of Various Groups at 25°C[a]

Polarity	Group	ΔH^{hyd} (kJ/mol)	ΔG^{hyd} (kJ/mol)
Nonpolar	$-CH_2-$	−3.39	0.72
	$-CH_3$	−8.28	3.72
	−CH<	2.83	−1.62
	>C<	8.45	−4.86
	−CH=	−4.91	−0.60
	>C=	2.63	−4.58
Polar	−OH	−39.60	−25.62
	NH_2	−38.74	−25.74
	>C=O	−25.05	−23.13
	−COOH	−43.94	−32.71
	−NH−	−38.74	−25.74
	−CONH	−59.57	−48.87
	−S−	−19.56	−14.41

[a]Makhatadze and Privalov (1995).

The remarkable difference between the hydration effects of polar and nonpolar groups is that, with increasing temperature, the hydration enthalpy and entropy of nonpolar groups decrease in absolute value (Fig. 2.12), whereas for polar groups they increase in absolute value (Fig. 2.13). The situation with the aromatic groups is somewhat in between (Fig. 2.14). The surface normalized hydration effects of the aliphatic

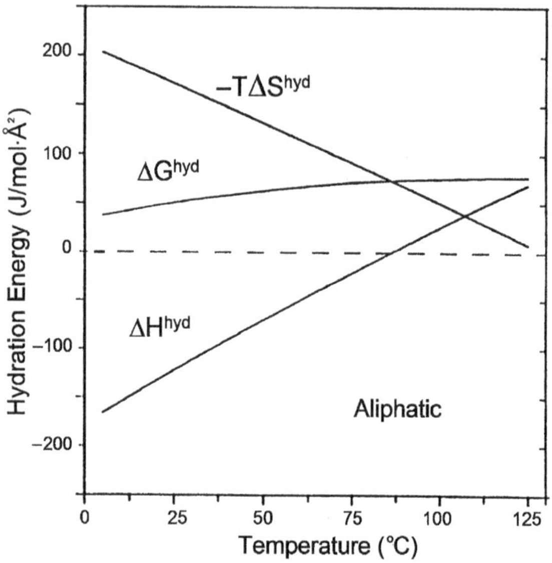

Figure 2.12. Temperature dependence of the hydration enthalpy, Gibbs energy, and entropy factor for the aliphatic groups (Makhatadze and Privalov, 1995).

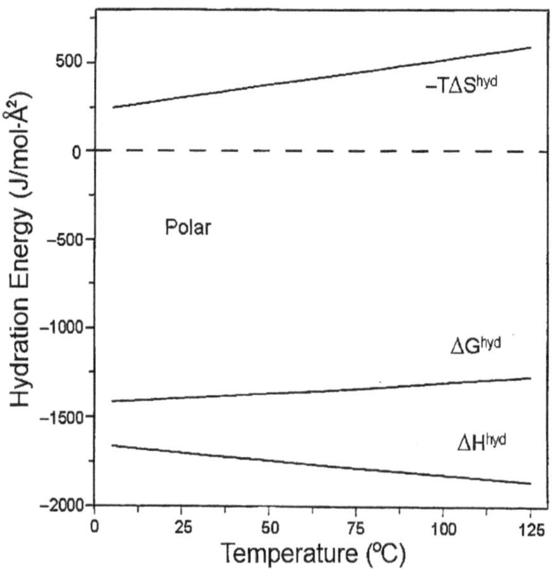

Figure 2.13. Temperature dependence of the hydration enthalpy, Gibbs energy, and entropy factor for the polar groups (peptide unit, –CONH$^-$) (Makhatadze and Privalov, 1995).

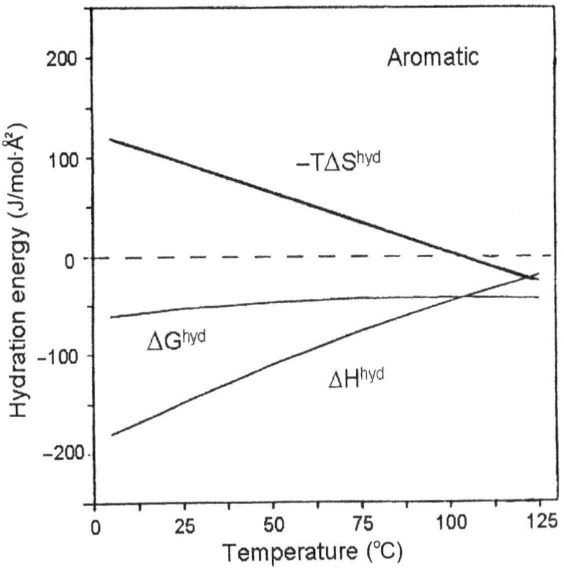

Figure 2.14. Temperature dependence of the hydration enthalpy, Gibbs energy, and entropy factor for the aromatic groups (Makhatadze and Privalov, 1995).

and aromatic groups differ considerably both in magnitude and in their dependence on temperature. Correspondingly, the hydration effects of these compounds reach zero values at different temperatures. These temperatures are different for the enthalpy and entropy of hydration of the aliphatic and aromatic groups: for the aliphatic groups, the temperature at which hydration enthalpy becomes zero, T_H, is 81°C; for the aromatic groups it is 125°C. The temperature T_S at which the hydration entropy for the aliphatic groups becomes zero is 122°C; for aromatic groups it is 104°C.

In contrast to the enthalpy and entropy of hydration, the Gibbs energy does not depend much on temperature; that is, the temperature dependencies of the enthalpy and entropy compensate each other effectively. Most important, however, is that the Gibbs energy of hydration of polar groups is large and negative, and its absolute value increases in magnitude with decreasing temperature. Therefore, these groups are clearly hydrophilic, and their hydrophilicity increases with decreasing temperature. The Gibbs energy of hydration of aliphatic groups is positive and its absolute value decreases with decreasing temperature. These groups are therefore hydrophobic, and their hydrophobicity decreases with decreasing temperature. The Gibbs energy of hydration of aromatic groups, which have been traditionally regarded as typical hydrophobic groups, is negative, although it is much smaller in magnitude than the Gibbs energy of hydration for polar groups (Privalov and Makhatadze, 1993). Therefore, these groups cannot be classified as hydrophobic. Moreover, they can hardly be classified as nonpolar, as is usually assumed.

According to Kirchhoff's relation, $d(\Delta H)/dT = \Delta C_p$, the enthalpy dependence on temperature is determined by the heat capacity change upon reaction: it is positive for apolar groups and negative for polar groups. It appears thus that the inclusion of an apolar solute in water results in an increase of the heat capacity; the inclusion of a polar solute in water results in a decrease of the heat capacity. The question is then why the heat capacity increases upon insertion of the apolar group in water and decreases upon insertion of the polar group?

It is supposed that insertion of the polar groups into water increases order in water by enhancing the hydrogen bonding among surrounding water molecules. Therefore, the flickering clusters of ordered water are more stable in the presence of polar groups and melt less upon increasing temperature. This results in overall decrease of the solution heat capacity. The apolar groups also increase the order in water but the heat capacity effect of their presence is opposite in sign. It appears then that the water ordering in these two cases is different: the apolar groups increase the order in water by occupying the cavities in its icelike structure. One would expect then that the order in water, which is induced by the presence of nonpolar molecules, is not stable and melts gradually upon heating. This diffused heat effect appears as a heat capacity increment.

REFERENCES

Baldwin R. (1986). *Proc. Natl. Acad. Sci. USA*, **83**, 8069–8072.

Ben-Naim A. (1980). *Hydrophobic Interactions*. Plenum Press, New York and London.

Ben-Naim A. (1987). *Solvation Thermodynamics*. Plenum Press, New York and London.

Frank H.S. and Evans M.W. (1945). *J. Chem. Phys.*, **13**, 507–532.

Frank H.S. and Wen W.Y. (1957). *Discuss. Faraday Soc.*, **24**, 133–140.

Lee B. (1985). *Biopolymers*, **24**, 813–823.

Livingstone J.R., Spolar R.S., and Record M.T., Jr. (1991). *Biochemistry*, **30**, 4237–4244.

Lumry R., Battisel E., and Jolicoeur C. (1982). *Faraday Discuss. Chem. Soc.*, **17**, 93–108.

Makhatadze G.I. and Privalov P.L. (1988). *J. Chem. Thermodynam.*, **20**, 405–412.

Makhatadze G.I. and Privalov P.L. (1989). *J. Solution Chem.*, **18**, 927–936.

Makhatadze G.I. and Privalov P.L. (1990). *J. Mol. Biol.*, **213**, 375–384.

Makhatadze G.I. and Privalov P.L. (1993). *J. Mol. Biol.*, **232**, 639–659.

Makhatadze G.I. and Privalov P.L. (1995). *Adv. Protein Chem.*, **47**, 307–425.

Makhatadze G.I., Lopez M.M., and Privalov P.L. (1997). *Biophys. Chem.*, **64**, 93–101.

Murphy K.P. and Gill S.J. (1991). *J. Mol. Biol.*, **222**, 699–709.

Privalov P.L. and Gill S.J. (1988). *Adv. Protein Chem.*, **139**, 191–234.

Privalov P.L. and Gill S.J. (1989). *Pure Appl. Chem.*, **61**, 1097–1104.

Privalov P.L. and Makhatadze G.I. (1993). *J. Mol. Biol.*, **232**, 660–679.

Spolar R.S., Ha J.-H, and Record M.T., Jr. (1989). *Proc. Natl. Acad. Sci. USA*, **86**, 8382–8385.

Spolar R.S., Livingstone J.R., and Record M.T., Jr. (1992). *Biochemistry*, **31**, 3947–3955.

3

CALORIMETRY

3.1. ISOTHERMAL REACTION MICROCALORIMETRY

3.1.1. The Heat of Mixing Reaction

In the simplest binding reaction, macromolecule M has one binding site for the ligand L:

$$\mathrm{M} + \mathrm{L} \Leftrightarrow \mathrm{ML}$$

The association constant of this reaction is

$$K_a = \frac{[\mathrm{ML}]}{[\mathrm{M}] \times [\mathrm{L}]} \tag{3.1}$$

where [ML], [M], and [L] are the corresponding concentrations, which in dilute solutions are close to the activities.

Microcalorimetry of Macromolecules: The Physical Basis of Biological Structures, First Edition.
Peter L. Privalov.

In considering the dissociation process

$$\mathrm{ML} \Leftrightarrow \mathrm{M} + \mathrm{L}$$

the dissociation constant is

$$K_d =\approx \frac{[\mathrm{M}] \times [\mathrm{L}]}{[\mathrm{ML}]} \tag{3.2}$$

Thus, $K_d = 1/K_a$. Correspondingly, the dimension of K_a is M^{-1}, and of K_d is M.

The fraction F of macromolecules with the bound ligand is

$$F = \frac{[\mathrm{ML}]}{[\mathrm{M}] + [\mathrm{ML}]} = \frac{K[\mathrm{M}][\mathrm{L}]}{[\mathrm{M}] + K[\mathrm{M}][\mathrm{L}]} = \frac{K[\mathrm{L}]}{1 + K[\mathrm{L}]} \tag{3.3}$$

The fraction of macromolecules without ligands is

$$1 - F = 1 - \frac{K_a[\mathrm{L}]}{1 + K_a[\mathrm{L}]} = \frac{1}{1 + K_a[\mathrm{L}]} \tag{3.4}$$

From Equations (3.3) and (3.4) one gets

$$\frac{F}{1 - F} = K_a[\mathrm{L}] \tag{3.5}$$

At the concentration where $F = 1/2$,

$$K_a = 1/[\mathrm{L}], \quad \text{or} \quad 1/K_a = K_d = [\mathrm{L}] \tag{3.6}$$

Thus, the dissociation constant, which is the reciprocal of the association constant, is just the concentration of free ligand at which half of the macromolecules have a bound ligand and half do not. The larger the binding constant, the lower the dissociation constant—and the lower the concentration of ligand at which the half-saturation is reached.

The heat of the association reaction depends on the molar enthalpy of association ΔH_a, the association constant K_a, and the ligand concentration [L]. The heat effect of binding can be positive or negative, but usually for protein–protein or protein–DNA interactions it is not large in magnitude, less than 100 kJ/mol. If K_a is large, then all ligands would be bound even at small concentrations. Therefore, to get the complete binding curve, the **isotherm of reaction**, we need to use a very low concentration of macromolecules. A high association constant is specific for many reactions of biological macromolecules, which requires precise recognition of the partner; for example, the binding constant of gene regulating proteins is above 10^8 M^{-1} (i.e., the dissociation constant is below 10^{-8} M). Correspondingly, to get a complete picture of their binding, one needs to work with very dilute solutions of macromolecules, of the order of 10^{-8} M. On the other hand, there are many nonspecific binding reactions, which are characterized by low binding constants. For example, the interaction of denaturants (urea or guanidine hydrochloride) with proteins is characterized by a very low binding constant. Therefore, a significant effect of these reagents on proteins is observed only in concentrated solutions. However, the solubility of denaturants is limited—less than 8 M. Therefore, to achieve a high concentration of denaturant in a solution with protein, one has to mix a comparable amount of the protein solution with the concentrated solution of the denaturant. In contrast, when the binding constant is high, we have to add a very small volume of the reagent to a large volume of protein solution and repeat that many times to get a binding isotherm. Correspondingly, for studying biopolymers two different types of reaction calorimetric instruments are needed: one that permits mixing two solutions in comparable volumes, and another that permits mixing a small portion of one solution in a large portion of another one and repeating that many times.

3.1.2. Mixing of Reagents in Comparable Volumes

There are two methods of mixing reagents in comparable volumes: batch and flow-mix.

In **batch calorimeters** comparable volumes of two reagents are placed in the two equal compartments of the rotating calorimetric cell and are mixed by turning this cell upside down. The heat of mixing is measured by the electric signal that is produced by the heat flowing through the Peltier element (a battery of semiconductors) to the thermostat. The main disadvantage of batch calorimeters is that the noise resulting from the rotating cell and from mixing the reagents is too high and does not reproduce well. The main reason is the air phase in the calorimetric cell—which, upon turning the cell upside-down, produces bubbles in the mixture that is formed. These bubbles induce some heat effects, which cannot be taken into account. Therefore these calorimeters are no longer much used for studying reactions with biological macromolecules.

In the **flow-mix calorimeter** continuous laminar flows of two reagents are mixed in the reactor (Fig. 3.1). The mixture flows through a heat exchanger with the Peltier element, which converts the heat flowing to the thermostat into an electric current that is amplified and measured. The advantage of this instrument is that the liquids being studied do not contact the gas phase and the laminar constant flow of the reagent does not produce much noise. Efficient mixing in this case requires a certain turbulence; however, the Joule heat resulting from this turbulence is rather constant and can be taken into account. The main disadvantage of the existing commercial flow-mix calorimeters is that they require quite a considerable amount of the reagents to realize their constant flow through the instrument. Nevertheless, the most thorough calorimetric study of the thermodynamics of protein interaction with urea and guanidinium chloride was done using an LKB FLOW-MIX calorimeter (Pfeil and Privalov, 1976; Makhatadze and Privalov, 1992).

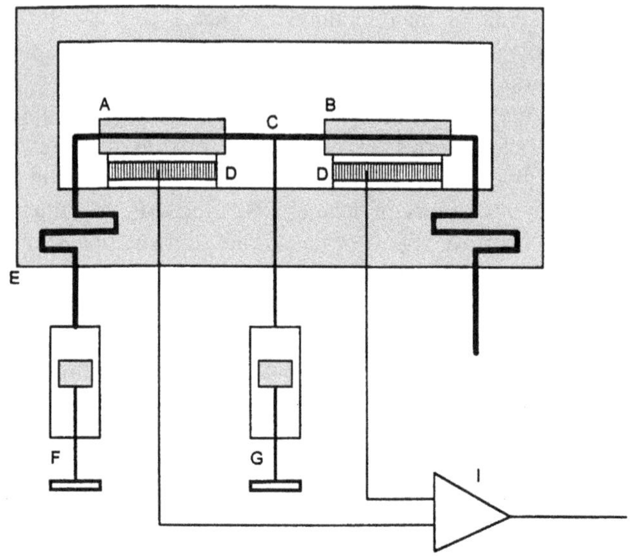

Figure 3.1. Schematic of the flow-mix calorimeter. The solution of one reactant is injected continuously into the calorimetric channel by the syringe F or a peristaltic pump; syringe G injects the other reagent into the channel at a point after the first heat-measuring part A and before the second heat-measuring part B. The difference between the heat effects recorded by B and A represents the heat effect of mixing of these two solutions, which is recorded.

3.1.3. Isothermal Titration Microcalorimeter

Isothermal titration calorimeters (ITCs), which permit calorimetric titration of one reagent by another, were widely used in physical chemistry. However, the reactors of conventional titration calorimeters were of rather large volume (of the order of deciliters) needed to place a mechanical stirrer to homogenize the mixture and also to minimize the influence of the surroundings on the measured heat effect. Therefore these instruments cannot be used for studying the heats of reactions between biological molecular objects, which are available in very limited amounts. The first microscale modification of the ITC that could be practically used for studying biological reactions was designed by S.J. Gill (McKinnon et al., 1984) and is shown in Figure 3.2. In this differential instrument two identical cells of 0.5 mL volume are placed into the thermal equilibrator block and a few microliters of titrant are injected periodically into the reactor cell by the rotating microsyringe, which also serves as a stirrer. The heat effect of mixing thus produced is detected by the sensor (the copper–constantan battery) placed between the reactor and the thermal equilibrator block and is compared with the heat effect between the reference cell and this block. After the appearance of this first micro-ITC, several companies (Microcal, Calorimetry Science Corporation CSC, TA Instruments) started production of various modifications differing from the original mainly in the sensors, electronics, and thermostatization system used.

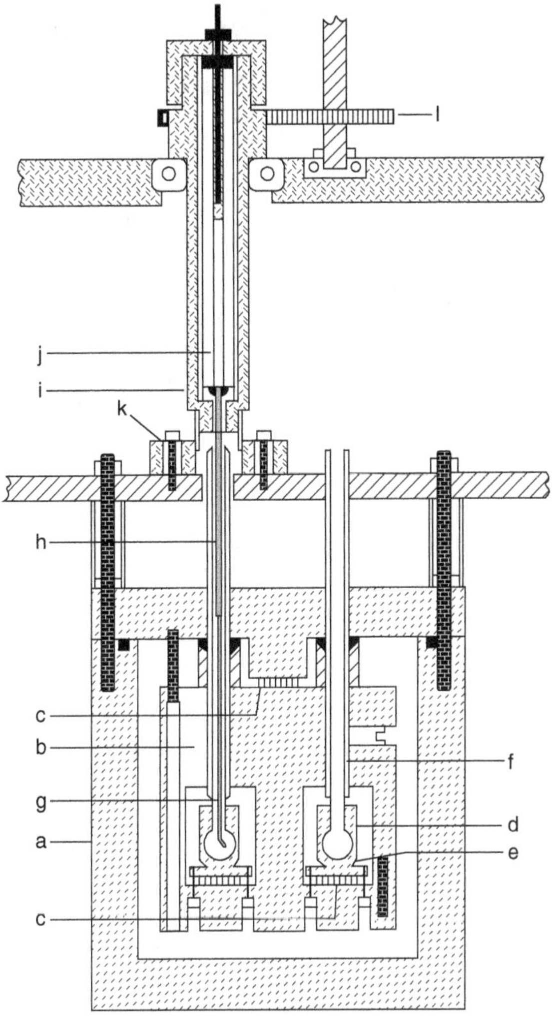

Figure 3.2. The first twin isothermal titration microcalorimeter, designed by Stanley Gill: (a) submarine container; (b) thermal equilibration block; (c) thermoelectric module; (d) calorimetric cell (glass bulb embedded in copper cylinder with low-melting metal); (e) electric heaters; (f) close-fitting stainless-steel tube with inside glass tube; (g) glass capillary stirrer and injection needle; (h) stainless-steel centering sleeve; (i) syringe holder; (j) microsyringe; (k) sleeve bearing assembly for syringe centering; (l) stirrer drive gear (McKinnon et al., 1984).

The operational volume of commercial ITC instruments is usually on the order of 1 mL. A smaller operational volume is not appropriate since the most critical part for an ITC experiment is not the sample volume but the concentration of the reagents, whereas decreasing the operational volume increases the optimal concentrations for the ITC experiment.

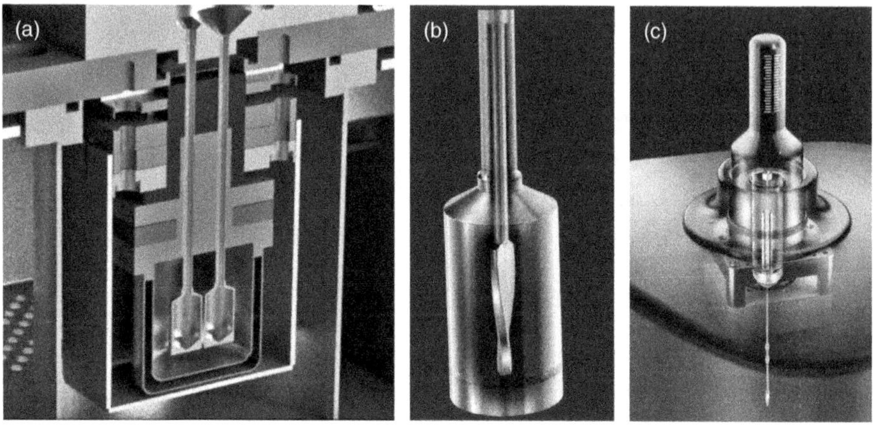

Figure 3.3. Construction of the Nano-ITC of TA Instruments: (a) cross section of the calorimetric block showing the twin reaction cells within the thermostatization system; (b) the 1-mL golden reactor cell with syringe needle ending in a propeller for steering the liquid; (c) the block that rotates the syringe/stirrer and moves its piston stepwise to inject highly precise portions of titrant into the cell. Courtesy of TA Instruments.

Figure 3.3 shows the construction of some important parts of the commercial Nano-ITC of TA Instruments: (a) the calorimetric block, showing the locations of the twin reaction cells within the thermostatization system; (b) the reactor cell with the syringe/stirrer inside; and (c) the block rotating syringe/stirrer and its piston, which moves stepwise to inject precise portions of the titrant into the reactor cell. Perfect thermostatization and electronics of this instrument permitted reduction of the noise level to 2.5 nW and the baseline stability to ±20 nW per hour. Other important characteristics of the instrument are its short response time and the high accuracy of the portions of reagent injected into the reactor cell. The cells are made from pure gold, which is the best material for calorimetric cells not only because of its high chemical inertness but also because of its high thermal conductivity, which is important for the temperature uniformity in the cell. Figure 3.4 shows the Nano-ITC of TA Instruments.

Figure 3.5 illustrates a typical titration experiment performed by the Nano-ITC: titration of the 12-base-pair (bp) oligonucleotide by its complementary strand, which results in the formation of the DNA duplex. The heat effect upon injecting only 5 μL of the reagent (i.e., 1.6 μg of the oligonucleotide complementary to that in the cell) is quite considerable and permits determination of not only the enthalpy but the whole isotherm of the binding process.

3.1.4. ITC Experiments

In carrying out an ITC experiment the solution that is titrated and the titrant solution must differ only in the reacting components and it is therefore essential that both solutions be dialyzed carefully in the same solvent. After dialysis both solutions

Figure 3.4. View of Nano-ITC of TA Instruments. The main characteristics of this instrument are the reactor cell volume (1 mL), operational temperature range (2°C–80°C), temperature stability (±0.0002 K), noise level (2.5 nW), baseline stability (±20 nW/hr), and minimum detectable heat (100 nJ). Courtesy of TA Instruments.

should be placed for a short time under vacuum to remove dissolved gases, since bubbles, which might appear upon the titration experiment, will produce undesirable heat effects, which appear as noise in the recording. The degassing procedure should proceed for a very short time, otherwise evaporation of the solvent might change the concentration of the solution.

Another important point in the ITC experiment is that the measured heat of mixing the two solutions with the studied reagents includes the intrinsic heat of reaction between these reagents and the heat of changing the concentration of the reagents upon mixing the two solutions. This concerns, first of all, the change in concentration of titrant, since a small volume of titrant is injected into the much larger volume of the other reagent in the calorimetric cell, resulting in a large change in concentration of the titrant. The concentration of the reagent that is in the calorimetric cell does not change much upon injecting the small volume of titrant. Therefore, to determine the net heat of reaction between the components, one has to determine experimentally the heat of dilution of the titrant. This is measured by injecting the titrant into a solvent that does not contain the reagent being studied. The measured heat effect of dilution is then excluded from the heat effect measured upon injecting the titrant into the solution containing the reagent being studied.

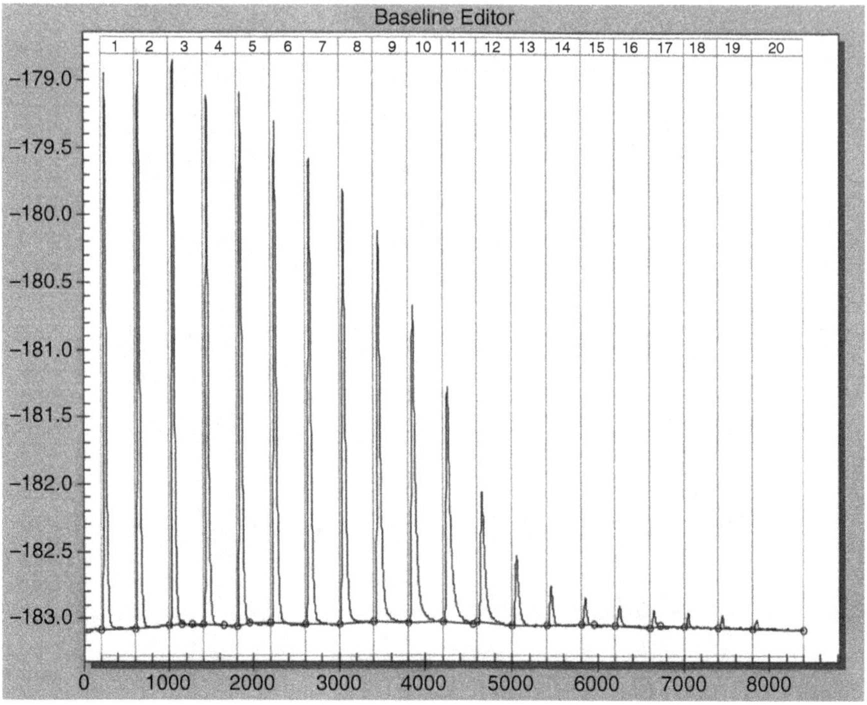

Figure 3.5. The Nano-ITC titration of the 12-bp oligonucleotide 5′-GCGAAATTTGCG-3′ with its complementary strand 3′-CGCTTTAAACGC-5′ at 45°C: concentration of the oligonucleotides in the cell, 0.022 mg/mL = 0.006 mM; concentration in the syringe, 0.36 mg/mL = 0.1 mM (i.e., 20 times higher); solvent, 5 mM NaCl, 5 mM sodium phosphate, pH 7.4. The titrant is injected in 5-μL portions at 300-second intervals.

It should be noted that the optimal concentration of reagent to obtain a well-resolved binding isotherm is on the order of the dissociation constant, $K^d = 1/K^a$. If the association constant is too high, that is, the dissociation constant is too low, the concentration of the solution being titrated must also be low. Correspondingly, the heat effect of mixing with the titrant might be too small for accurate measurement. Contemporary ITC instruments work best with micromolar concentrations of reagents, that is, studying reactions with dissociation constants greater than micromolar. When the dissociation constant is less than micromolar, one should use some other method (e.g., optical) to determine the association constant and use the ITC only for determination of the association enthalpy. This is because measurement of the enthalpy does not require determination of the full binding isotherm: the enthalpy is estimated from the heat effect of the first few injections when the process is far from saturation and all injected titrant binds to the reagent. The concentration of titrant should be one or two orders of magnitude higher than that of the reagent, and it depends on how many injections are planned in the experiment: is the plan to get the full isotherm of binding (which needs about 20 injections) or just to determine the enthalpy of binding, which requires a few injections. Modern ITC instruments (e.g., the Nano-ITC, TA Instruments) are equipped with a program that helps in *a priori* estimation of the optimal concentrations of both reagents for the planned experiment.

It is notable also that among the series of injections the first one is usually deficient and should not be taken into account. This is because during the equilibration of the instrument, after loading the reagents, there is diffusion of the reagent from the open end of the syringe. Therefore, concentration of the reagent in the first portion of injected titrant is usually lower.

3.1.5. Analysis of the ITC Data

Let Q be the observed heat effect of the reaction in the calorimetric cell. If the total molar concentration of macromolecules in the cell is $[\mathrm{M}]_{\mathrm{tot}} = [\mathrm{M}] + [\mathrm{ML}]$ and the volume of the cell is V, then bearing in mind Equation (3.3) we have

$$Q = \Delta H \cdot V \cdot [\mathrm{ML}] = \Delta H \cdot V \cdot [\mathrm{M}]_{\mathrm{tot}} \frac{K[\mathrm{L}]}{1+K[\mathrm{L}]} \tag{3.7}$$

When the protein has n binding sites:

$$Q = \Delta H \cdot V \cdot [\mathrm{M}]_{\mathrm{tot}} \frac{nK[\mathrm{L}]}{1+K[\mathrm{L}]} \tag{3.8}$$

If the binding sites are different and have different binding constants, then

$$Q = \Delta H \cdot V \cdot [\mathrm{M}]_{\mathrm{tot}} \sum_i n_i \frac{K_i[\mathrm{L}]}{1+K_i[\mathrm{L}]} \tag{3.9}$$

The problem is, however, that the concentration of free ligand [L] cannot be determined directly. Therefore, determination of the equilibrium constant from the number of heat effects obtained in the course of calorimetric titration requires rather complex approximation of all experimental data. The method of approximation depends on the number of binding sites, n.

The simple case when n = 1. Bearing in mind that

$$[\mathrm{L}]_{\mathrm{tot}} = [\mathrm{L}]+[\mathrm{ML}] \quad \text{and} \quad [\mathrm{L}] = [\mathrm{L}]_{\mathrm{tot}} - [\mathrm{ML}]$$
$$[\mathrm{M}]_{\mathrm{tot}} = [\mathrm{M}]+[\mathrm{ML}] \quad \text{and} \quad [\mathrm{M}] = [\mathrm{M}]_{\mathrm{tot}} - [\mathrm{ML}]$$

we have

$$K([\mathrm{L}]_{\mathrm{tot}} - [\mathrm{ML}])([\mathrm{M}]_{\mathrm{tot}} - [\mathrm{ML}]) = [\mathrm{ML}]$$

and

$$K[\mathrm{M}]_{\mathrm{tot}}[\mathrm{L}]_{\mathrm{tot}} - K[\mathrm{M}]_{\mathrm{tot}}[\mathrm{ML}] - [\mathrm{ML}] + K[\mathrm{L}]_{\mathrm{tot}}[\mathrm{ML}] + K[\mathrm{ML}]^2 - [\mathrm{ML}] = 0$$

Thus

$$[\mathrm{ML}]^2 + (-[\mathrm{M}]_{\mathrm{tot}} - [\mathrm{L}]_{\mathrm{tot}} - 1/K)[\mathrm{ML}] + [\mathrm{M}]_{\mathrm{tot}}[\mathrm{L}]_{\mathrm{tot}} = 0$$

Or

$$[\mathrm{ML}]^2 + b[\mathrm{ML}] + c = 0 \tag{3.10}$$

where

$$b = -[\mathrm{M}]_{\mathrm{tot}} - [\mathrm{L}]_{\mathrm{tot}} - 1/K$$
$$c = [\mathrm{M}]_{\mathrm{tot}}[\mathrm{L}]_{\mathrm{tot}}$$

Solving Equation (3.10) one gets

$$[\mathrm{ML}] = \frac{-b - \sqrt{b^2 - 4c}}{2} \tag{3.11}$$

The cases when $n \neq 1$. The equilibrium constant is given by

$$K = \frac{[\mathrm{L}]_{\mathrm{bound}}}{([\mathrm{M}]_{\mathrm{tot}} n - [\mathrm{L}]_{\mathrm{bound}})[\mathrm{L}]} \tag{3.12}$$

where $[\mathrm{L}]_{\mathrm{bound}} = [\mathrm{L}]_{\mathrm{tot}} + [\mathrm{L}]$ is the concentration of bound ligands. Here again we come to the quadratic equation:

$$[\mathrm{L}]_{\mathrm{bound}}^2 + b[\mathrm{L}]_{\mathrm{bound}} + c = 0 \tag{3.13}$$

and again we get

$$[\mathrm{L}]_{\mathrm{bound}} = \frac{-b - \sqrt{b^2 - 4c}}{2} \tag{3.14}$$

However, in that case:

$$b = -n[\mathrm{M}]_{\mathrm{tot}} - [\mathrm{L}]_{\mathrm{tot}} - 1/K$$
$$c = n[\mathrm{M}]_{\mathrm{tot}}[\mathrm{L}]_{\mathrm{tot}}$$

If $[\mathrm{L}]_{\mathrm{bound}}$ is known for the given $[\mathrm{M}]_{\mathrm{tot}}$ and $[\mathrm{L}]_{\mathrm{tot}}$, one can calculate the heat effect of binding as follows:

$$Q_k = \sum_{i=1}^{k} q_i = \Delta H \cdot V \cdot [\mathrm{L}]_{\mathrm{bound}} \tag{3.15}$$

Varying n, K, and ΔH one can simulate the results of a titration experiment and get the values of these parameters.

Modern ITC instruments are equipped with a program for computer analysis of the experimental data.

3.2. HEAT CAPACITY CALORIMETRY

3.2.1. Technical Problems

Unlike isothermal reaction calorimeters, which operate at constant temperature, heat capacity calorimeters operate at changing temperature and this induces certain technical problems. According to the definition of heat capacity, it is measured by the heat required to raise a sample's temperature by 1 K. The easiest way to produce this heat is by using an electric heater:

$$Q = RI^2\Delta\tau$$

where $\Delta\tau$ is the time of heating in seconds, I is the electric current in amperes, and R is the resistance of the heater. If ΔT is the observed change of temperature, then $C_p(T) = Q/\Delta T$, where T is the middle temperature between the initial and final temperatures upon heating. Repeating this procedure many times, the temperature dependence of the heat capacity could be determined step by step in a broad temperature range. However, realization of this procedure when the sample is a liquid and is available in small amounts faces serious technical problems:

(a) **Equilibrium.** The temperature increase induced by heating should be measured at equilibrium; that is, the temperature should be homogeneous over the entire sample. However, the heat conductivity of liquids is very low. Therefore, if the sample volume is considerable, the passive redistribution of the added heat over the whole sample takes a long time. One can accelerate this process using a mechanical stirrer, but this produces Joule heat, which is not easy to take into account because it depends on the viscosity of the liquid—which changes upon heating.

(b) **Influence of the surroundings.** The calorimetric cell containing the sample exchanges heat with its surroundings, and this exchange will increase with increasing temperature difference between them. To minimize this heat exchange, or to make it constant, the temperature of the surroundings should follow the temperature of the cell; that is, the cell's surroundings should have the same temperature as the cell during the whole experiment, or the difference in their temperatures should not change upon heating.

(c) **Sensitivity.** The heat capacities of biological macromolecules can be studied only in very dilute solutions in which they do not interact. But in dilute solutions the heat capacity effect associated with the presence of the biological substance is very low. For example: in 0.1% aqueous solution of protein the heat capacity effect of the protein will be less than 0.03% of the total heat capacity of solution, because the heat capacity of organic compounds is 3 times smaller than that of water. Therefore, measurement of the heat capacity contribution of protein to its dilute solution requires extremely high accuracy in measuring the solution heat capacity. The precision of conventional heat capacity calorimeters for liquids is about 0.5%. But the heat capacity effect of protein present in the 0.1% solution is 20 times lower; that is, a conventional heat capacity calorimeter is unable even to sense the presence of protein in such a solution.

(d) **Volume.** Most biological molecules are not easily available and are very expensive. The operational volume of the cells of conventional heat capacity calorimeters for liquids used in thermochemistry is on the order of deciliters. Thus, in the case of a 0.1% protein solution one will need for one experiment about 100 mg of sample, which is incredible for most proteins and nucleic acids.

Because of these complications all attempts to use conventional heat capacity calorimeters for liquids in studying the heat capacities of biological molecules have failed. This required development of a special technique, which is now called differential scanning calorimetry (DSC).

3.2.2. Differential Scanning Microcalorimeter

The superprecise technique needed for studying the heat capacity of small volumes of dilute solutions of biopolymers has been developed by combining several methods (Fig. 3.6):

(a) **Differential measurements.** The differential method measures not the absolute value of the heat effect but the difference between the effects in the object and some standard used for this object.For dilute biopolymer solutions, their solvents are usually used as standards. In the case of calorimetry, the sample solution and the standard solution (solvent) are placed in two identical, twin calorimetric cells. Both cells are heated with the same power and the difference in their heating effects is measured.

(b) **Compensation.** Instead of measuring the temperature difference between the cells upon heating, the additional energy needed to compensate for the difference in temperature of the heated cells is measured.

(c) **Adiabatization.** To exclude influence of the surroundings, both cells are placed in a single surround (the thermal shield), which is heated with the cells to exclude heat exchange between them.

(d) **Continuous heating.** To have efficient temperature control of the cells and the thermal shield, the cells are not heated stepwise but continuously; that is, they are scanned over the temperature scale at a fixed rate.

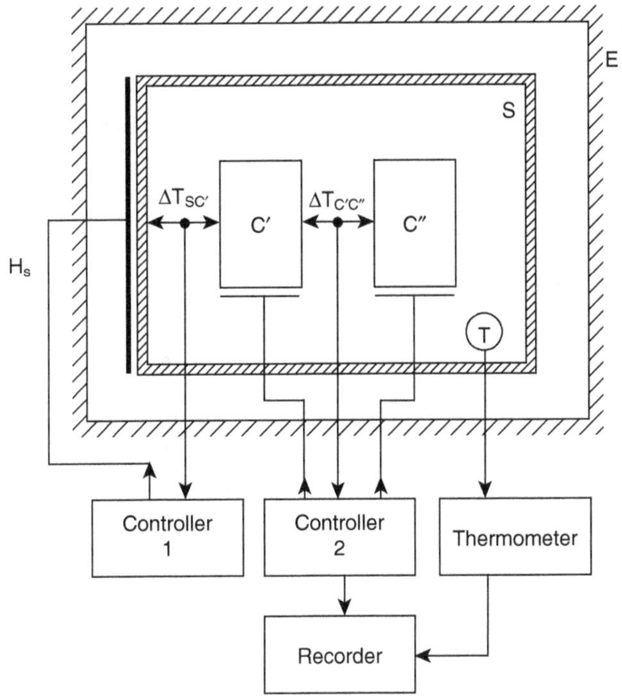

Figure 3.6. The principle of the differential adiabatic scanning calorimeter. The twin calorimetric cells (C′, C″) placed into the adiabatization system of electrically controlled thermal shields are continuously heated. The temperature difference between the two cells is measured by a Peltier thermal battery and is compensated by additional power, which is recorded as a function of temperature.

(e) **Thermal gradients.** Continuous heating results in a temperature gradient in the cells; that is, the temperature of the continuously heated sample would not be uniform. The nonuniformity of temperature could be decreased by decreasing the cell volume and choosing a proper shape for it that provides minimal distances of the liquid from the heater.

Each of these methods was well known but they were never used together in calorimetry. However, just combining all five in a single instrument leads to an essential, thousand-times-higher sensitivity of measuring the heat capacity of liquids. The first differential scanning microcalorimeter constructed by P. Privalov et al. (1965) is shown in Figure 3.7. This was a vacuum adiabatic instrument with several thermal shields and a water jacket surrounding the 1-mL twin differential cells with internal electric heater. It permitted, for the first time, measurement of the heat of DNA melting (Fig. 3.8). However, although the sensitivity of this instrument was high, it was not easy to use because reloading the sample required dismounting the whole adiabatization system: the water jacket, the vacuum container, the thermal shields.

Loading the sample also presented a problem in superprecise heat capacity calorimetry. If one removes the cell from the adiabatization system, the initial

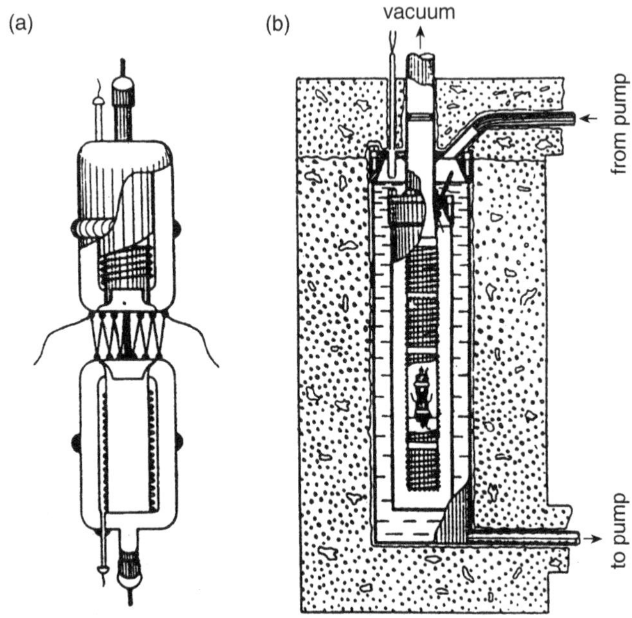

Figure 3.7. The first differential scanning microcalorimeter with twin calorimetric cells of 1-mL volume surrounded by thermal shields that follow the cells in temperature, all enclosed in a vacuum tube and water jacket: (a) one of the twin golden cells, with electric heater inside, connected by the chromel–constantan thermal battery; (b) vacuum tube and water jacket (P. Privalov et al., 1965).

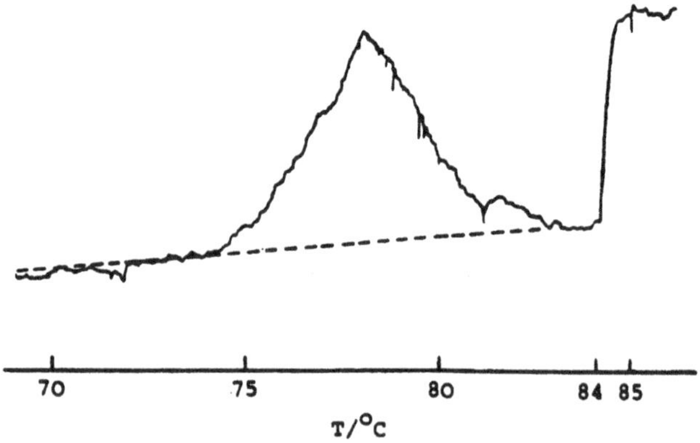

Figure 3.8. The very first recording of the heat effect of the phage T_2 DNA melting upon heating at 1 K/min. The amount of DNA in the cell is 0.685 mg; a calibration mark of 21 μW is given at 84°C (P. Privalov et al., 1965).

environmental conditions will never be reproduced with sufficient accuracy. Then, if the sensitivity of the instrument is of the order of 10^{-7} W (which means that its sensitivity to heat capacity at heating rate 1 K/min is on the order of 10^{-6} cal/K), the accuracy of loading the cell should be within one microgram of the sample, which is not easy to achieve.

The other problem is with the free volume in the loaded cell. One cannot fill the cell completely and seal it because the thermal expansion of liquids is larger than that of gold, which is usually used for making the microcalorimetric cells. However, upon heating, the liquid will evaporate into the free volume and, since the operational volumes of the two cells are not absolutely identical and the solutions in both cells are not identical (i.e., the vapor pressures are slightly different), the heats of vaporization will be different. All these will be sufficient to produce a large error in the measured heat capacity. These problems with loading the sample were solved by inventing the closed, undismountable calorimetric block with inlet tubes for filling the cells and by shifting from measuring the heat capacity of a sample of a certain mass to measuring the heat capacity of a sample of a certain volume (P. Privalov, 1974, 1980; P. Privalov and Potekhin, 1986; P. Privalov and Plotnikov, 1989; G. Privalov et al., 1995).

In a scanning calorimeter with closed calorimetric block the operational volume is fixed by a thermal shunt, which cuts the heat effects in both inlet tubes (Fig. 3.9).

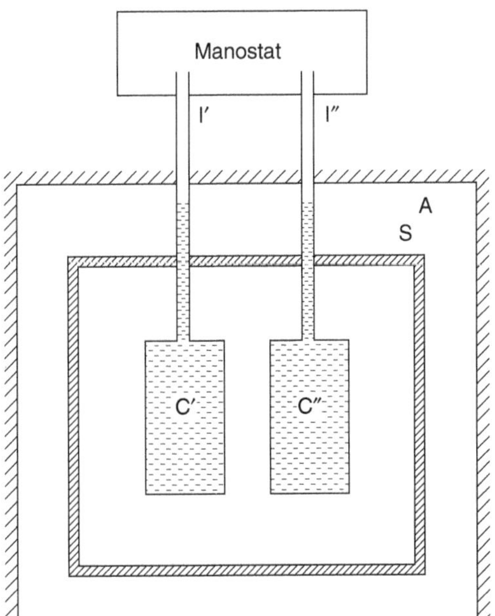

Figure 3.9. The basic concept of the closed calorimetric block with its cells completely filled with the liquid samples through the capillary inlets I′ and I″. The operational volume of the cells is fixed by the point at which the thermal shield (S) contacts the capillary inlets. A manostat is used to apply excess pressure on the liquid in the cells in order to dissolve any bubbles that might be left in the cells and to prevent their reappearance upon heating.

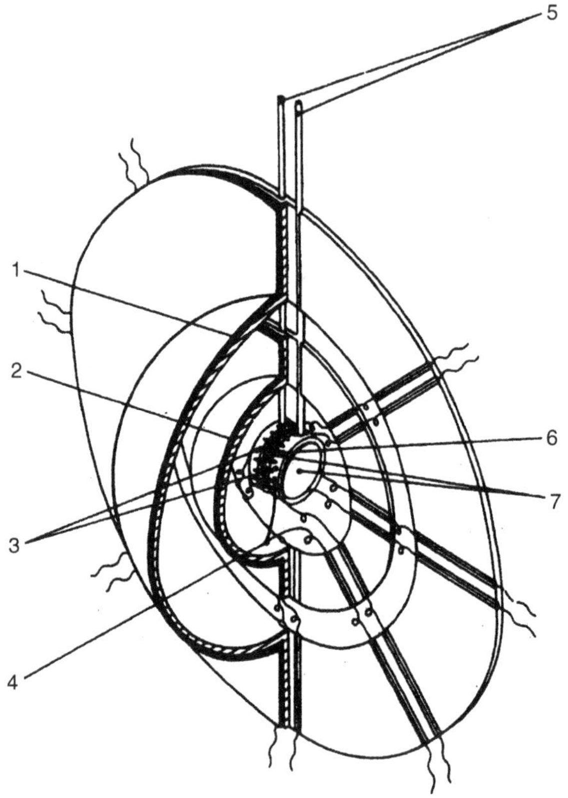

Figure 3.10. Undismountable calorimetric block: (1, 2) "sombrero"-type internal and external adiabatization thermal shields of silver with electric heaters distributed over them; (3) golden flat cylindrical cells; (4) thermosensor between the cell and shield; (5) platinum capillary inlets; (6) chromel–constantan thermopile of 50 junctions; (7) electric heaters of the cells. This instrument was developed at the Protein Research Institute of the Russian Academy of Sciences in 1970 and was a prototype of the first serial scanning microcalorimeter DASM1 manufactured by the Russian Academy of Science (P. Privalov et al., 1975).

The cells and inlet tubes are completely filled by the solution and solvent. Upon heating, the mass of liquid in a fixed volume of the cells decreases because of thermal expansion. However, the thermal expansions of the dilute protein solution and the solvent are very similar and the volume of the cell also increases with heating and that decreases the volume of the liquid leaving the cell upon heating. Thus the error caused by thermal expansion is practically negligible in the determination of the difference heat capacity.

To avoid the appearance of bubbles upon heating, to depress boiling and therefore expand the temperature range of heating of the aqueous solution, excess pressure is applied to both cells by the manostat (Fig. 3.9). This is essential because even microscopic bubbles in the cell prevent measurement of a real heat capacity difference of the solution being studied and the reference liquid, thus preventing an accurate determination of the contribution of the solute to the heat capacity of the solution. The excess pressure of 2 atm allows the aqueous solution to be heated to 120°C without boiling. This is essential because many proteins, particularly of the thermophiles, unfold above 100°C.

The construction of the first closed calorimetric block is shown in Figure 3.10. Here the golden calorimetric cells are made as flat discs with the heaters distributed over the surface and the thermosensor of 50 chromel–constantan junction batteries in between. Most remarkable in this construction are the two layers of "sombrero"-type silver thermal shields with electric heaters distributed over them. Electrically heating the flat cells from both sides permits the temperature gradient to be decreased in the continuously heated cell and the scanning rate to be raised in calorimetric experiments, thus increasing the sensitivity in measuring heat capacity.

The next essential step in developing scanning microcalorimetry was the replacement of cylindrical cells by the through capillary cells wound into coils (Fig. 3.11). The advantage in using the capillary cells is not only that they are easier to wash

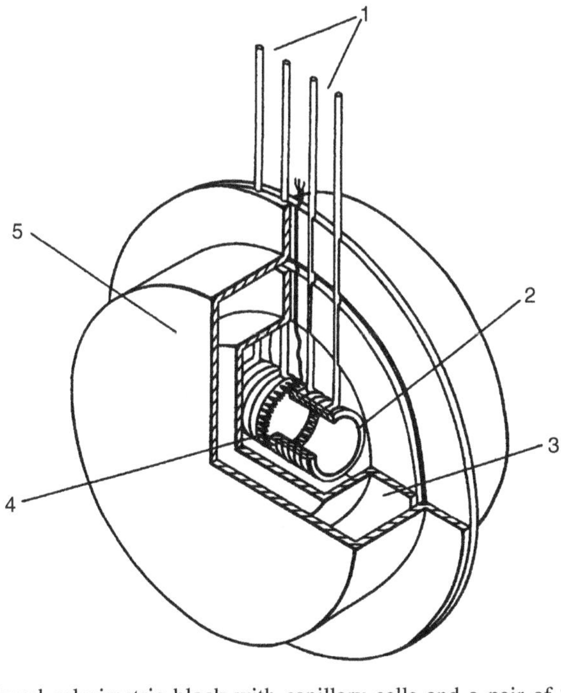

Figure 3.11. Closed calorimetric block with capillary cells and a pair of "sombrero"-type silver thermal shields: (1, 2) capillary cells with (4) chromel–constantan thermosensor; (3) internal shield; (5) external shield. This instrument was developed at the Protein Research Institute of the Russian Academy of Sciences in 1980 and was a prototype of the scanning microcalorimeter DASM4 manufactured by the Russian Academy of Science (P. Privalov and Plotnikov, 1989).

and reload with new samples; the small diameter of the capillary tubes essentially decreases the nonuniformity of temperature in the cell upon continuous scanning and again increases the heating rate, that is, the sensitivity in measuring the heat capacity. It is important that capillary cells do not have convection of the fluid upon continuous heating, which is specific to cylindrical cells. Since convection of fluid depends on its viscosity, and viscosity of protein and DNA solutions changes dramatically upon heating, this results in changes in the temperature gradients within the cell, which mimics the heat effect. Thus, the capillary cells are free from this artifact and therefore are especially useful for calorimetric studies of solutions of fibrillar macromolecules such as collagens and DNA. The other advantages of capillary cells are their much smaller volume (0.3 mL) and that their volume does not change under the excess pressure.

The other essential change in scanning calorimetry was in shifting from adiabatic conditions, when there is no heat exchange between the heated cells and their surroundings, to nonadiabatic conditions with certain heat exchange (Fig. 3.12). In contrast to an adiabatic device, in a nonadiabatic device it is not the temperature of the cell that is controlled but the temperature of the surroundings of the cells, namely, the thermal shunt (B) with the shield (E), using the Peltier batteries (A). This required development of an absolutely new computer-controlled operating system.

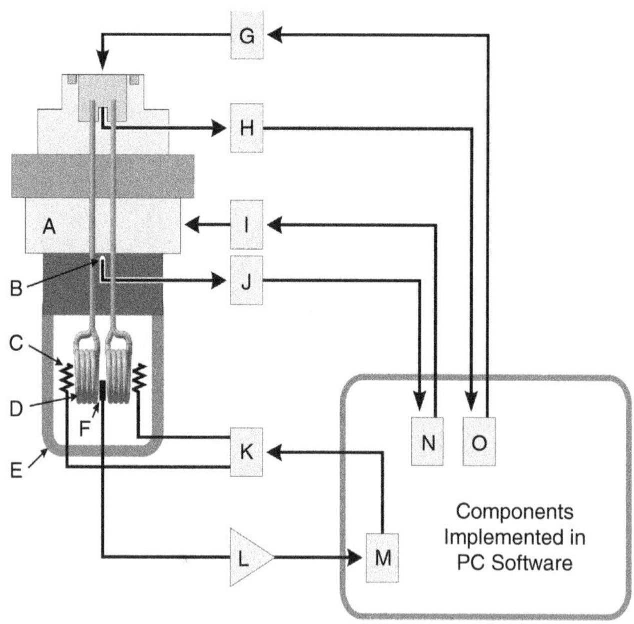

Figure 3.12. Nonadiabatic Nano-DSC with capillary cells, developed by the Calorimetric Center of the Johns Hopkins University and the Calorimetric Sciences Corporation (CSC) and now manufactured by TA Instruments: (A) semiconductor Peltier battery; (B) thermal shunt; (C) heaters of the cells; (D) platinum coiled capillary cells; (E) two-layer silver thermal shield.

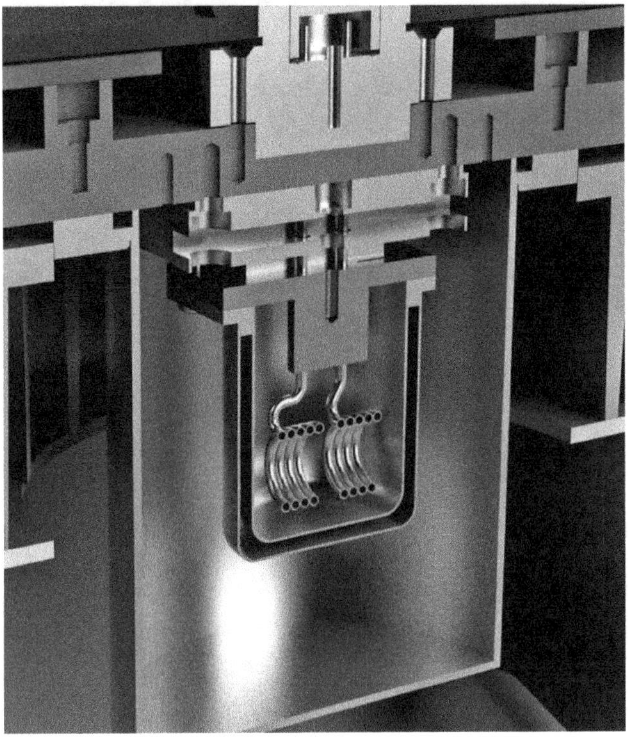

Figure 3.13. Cross section of the Nano-DSC calorimetric block. Courtesy of TA Instruments.

Figure 3.13 shows the design of such a nonadiabatic calorimetric block with capillary twin cells that is used now in the Nano-DSC manufactured by TA Instruments (Fig. 3.14). In this system the cell follows the temperature of its surroundings as a result of heat flow between them. If the temperature of the surroundings constantly exceeds that of the cells, the cells are heated at a fixed rate determined by the required temperature shift between the cells and their surroundings. If the temperature of the surroundings is lower that that of the cells, the cells are cooled at a fixed rate. Thus, this method permits measurements of heat capacity not only upon heating but also upon cooling, that is, scanning up and down in the temperature scale at any desired fixed rate.

Scanning up and down in the temperature scale is important for several reasons. First, it is needed to check the reversibility of temperature-induced processes in molecules being studied (Fig. 3.15). Then, it permits determination of heat capacity at a lower temperature than is possible with the heating experiment. This is so because of instrument relaxation—reliable recording starts with some delay after starting a scan; for example, if we start scanning at 5°C, the recording becomes reliable only from 10°C. However, in cooling experiments, if scanning starts at 25°C,

Figure 3.14. View of the Nano DSC of TA Instruments. The main characteristics of this instrument are its cell volume (0.3 mL), cell composition (platinum), short-term noise (15 nW), baseline stability (±30 nW), operating temperature range (from −10°C to 130°C), and scanning rates (from 0.1 to 2.0 K/min). Courtesy of TA Instruments.

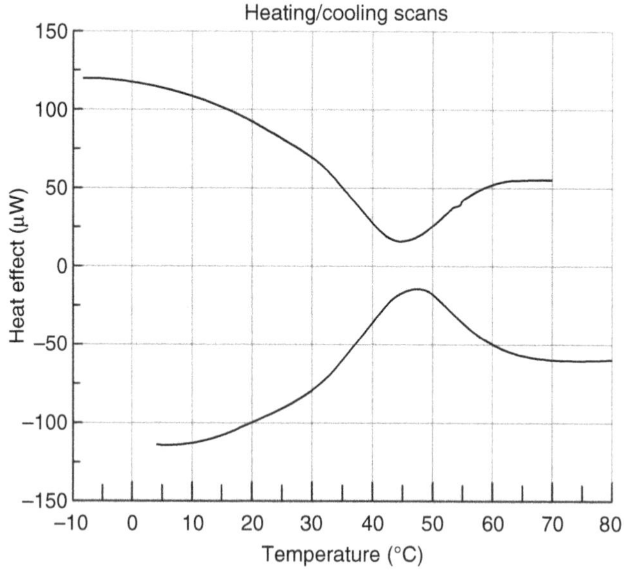

Figure 3.15. DSC recording upon heating (to 80°C) and subsequent cooling (to −10°C) at the same rate (1 K/min) of solution of SOX-5HMG protein. Protein concentration 4.04 mg/mL; solvent 10 mM potassium phosphate (pH 6.0), 100 mM KCl.

the recording becomes reliable from 20°C down to 0°C or even below 0°C. The last is particularly important for studying the cold denaturation of proteins in aqueous solutions. A dust-free aqueous solution can be supercooled down to −15°C, but freezing could start spontaneously any moment and damage the instrument. Therefore the instrument for such experiments must be equipped with an automatic system for rapid switching upon initiation of freezing (which is sensed by a sudden large heat evolution) from the cooling to the heating mode.

3.2.3. Determination of the Partial Heat Capacity of Solute Molecules

In studying protein solutions, the solvent is usually used as a reference sample. Filling both calorimetric cells of a differential calorimeter with the solvent, one first gets the baseline (Fig. 3.16). If the cells are refilled with the same solvent, there should be no change in this baseline. Such stability and reproducibility of the baseline over the whole operational temperature range is one of the most important characteristics of a scanning calorimeter and is particularly high for instruments with capillary cells. The baseline of an instrument might be not close to zero over the whole temperature range because the twin cells are not absolutely identical. Therefore in contemporary Nano-DSC instruments the recorded original nonzero baseline is stored in the memory of the instrument and is used for automatic correction of all further recordings. The corrected recording for the solvent is very close to zero over the whole operational temperature range. This corrected solvent–solvent recording is used as a baseline in all further experiments with the protein or DNA solutions in the same solvent.

On replacing the solvent in one of the cells by the protein solution, the calorimetric recording shifts down by $\Delta C_p^{\mathrm{app}}(T)$. This shift is a consequence of the heat

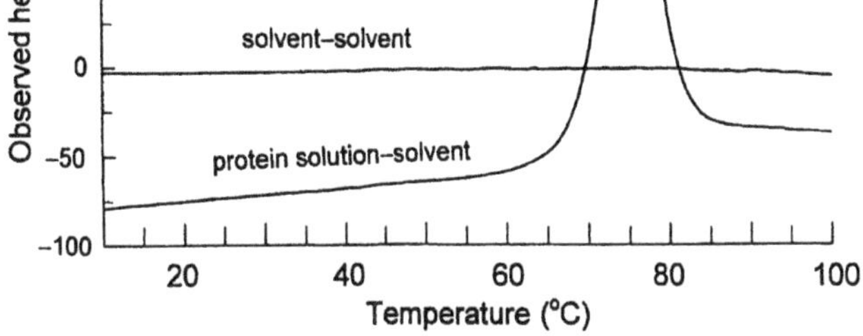

Figure 3.16. DSC recordings of the baseline obtained with the solvent (100 mM NaCl, 20 mM acetate buffer, pH 4.0) in both cells and of the effect of replacing solvent in one of the cells by the protein solution (e.g., lysozyme, 2 mg/mL in the same solvent).

capacity of dissolved protein, C_p^{pr}, being lower than the heat capacity of the solvent that is replaced by protein. One can evaluate this difference in the heat capacity units bearing in mind that

$$-\Delta C_p^{app}(T)_{pr.sol/solv} = C_p(T)_{pr} \times m(T)_{pr} - C_p(T)_{solv}\,\Delta m(T)_{solv} \qquad (3.16)$$

However,

$$\Delta m(T)_{solv} = m(T)_{pr}\frac{V(T)_{pr}}{V(T)_{solv}}$$

Therefore:

$$C_p(T)_{pr} = C_p(T)_{solv}\frac{V(T)_{pr}}{V(T)_{solv}} - \frac{\Delta C_p^{app}(T)_{pr.sol/solv}}{m(T)_{pr}} \qquad (3.17)$$

The partial molar heat capacity determined for lysozyme is shown in Figure 3.17.

In determining the partial specific heat capacity of macromolecules in solution with high accuracy, it is necessary to use precise data on the heat capacity of the solvent and to take into account the change of the mass of macromolecules in the operational volume of the calorimetric cell. The specific heat capacity of the solvent, $[C_p(T)_{solv}]$, can be measured over the desired temperature range with the required accuracy using the same scanning microcalorimeter. To do that, one has to measure the difference in heat capacity of the solvent in relation to distilled water $[C_p(T)_{water}]$:

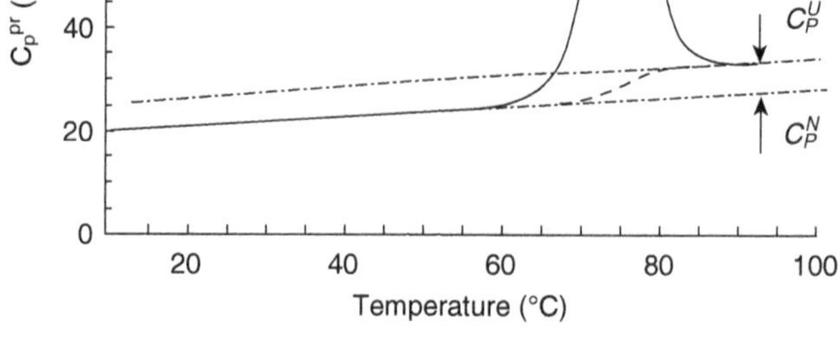

Figure 3.17. The partial molar heat capacity function of lysozyme in 20 mM acetate buffer pH 4.0, 100 mM NaCl solution.

$$\Delta C_p^{\mathrm{app}}(T)_{\mathrm{solv/water}} = C_p(T)_{\mathrm{water}}\, m(T)_{\mathrm{water}} - C_p(T)_{\mathrm{solv}}\, m(T)_{\mathrm{solv}}$$
$$= C_p(T)_{\mathrm{water}}\, \upsilon(T)/V(T)_{\mathrm{water}} - C_p(T)_{\mathrm{solv}}\, \upsilon(T)/V(T)_{\mathrm{solv}} \quad (3.18)$$

where $\upsilon(T)$ is the operational volume of a calorimetric cell at temperature T, and $V(T)_{\mathrm{water}}$ and $V(T)_{\mathrm{solv}}$ are the specific volumes of water and solvent at this temperature. From Equation (3.18) we have

$$C_p(T)_{\mathrm{solv}}/V(T)_{\mathrm{solv}} = C_p(T)_{\mathrm{water}}/V(T)_{\mathrm{water}} - \Delta C_p^{\mathrm{app}}(T)_{\mathrm{solv/water}}/\upsilon(T) \quad (3.19)$$

Substituting Equation (3.18) in Equation (3.19), we get

$$C_p(T)_{\mathrm{pr}} = [C_p(T)_{\mathrm{water}}/V(T)_{\mathrm{water}} - \Delta C_p^{\mathrm{app}}(T)_{\mathrm{solv/water}}/\upsilon(T)]V(T)_{\mathrm{pr}}$$
$$- \Delta C_p^{\mathrm{app}}(T)_{\mathrm{pr.sol/solv}}/m(T)_{\mathrm{pr}} \quad (3.20)$$

The specific volume and the heat capacity of pure water (i.e., $V(T)_{\mathrm{water}}$ and $C_p(T)_{\mathrm{water}}$) are known with high accuracy over entire temperature range of its existence in the liquid phase and can be found in handbooks. As for the operational volume of a calorimetric cell $\upsilon(T)$ and the mass of protein in the cell $m(T)_{\mathrm{pr}}$ at temperature T, they can be calculated from their values at room temperature and known coefficients of thermal expansion of the solution and cell material.

Since this apparent specific heat capacity of a protein is determined in dilute solutions, it does not usually show dependence on concentration and can be considered as a partial specific heat capacity. To distinguish it from the heat capacity of pure protein, $C_{p,\mathrm{pr}}$, it has to be denoted as $C^{\phi}_{p,\mathrm{pr}}$. However, since the proteins studied in biology are mostly in solutions and not the dry, anhydrous proteins, the symbol ϕ is usually omitted. The difference between the specific heat capacity of dry protein, $C_{p,\mathrm{pr}}$, and the partial specific heat capacity of protein, $C^{\phi}_{p,\mathrm{pr}}$, specifies the hydration heat capacity effect:

$$\Delta C_p^{\mathrm{hyd}} = C^{\phi}_{p,\mathrm{pr}} - C_{p,\mathrm{pr}} \quad (3.21)$$

3.2.4. DSC Experiments

The calorimetrically determined partial specific heat capacity function of a macromolecule is one of the most important characteristics of its thermal properties. To obtain this function one has to:

(a) Choose the appropriate solvent (pH, ionic strength, and buffer) in which the macromolecules (proteins, nucleic acids, etc.) to be studied do not aggregate over the whole temperature range studied and the pH of the solution does not change much upon heating.

(b) Prepare a solution of the substances to be studied, in the required concentration in the chosen solvent, and equilibrate it with careful dialysis against the solvent. For the accurate determination of partial specific heat capacities of the substance, the error in determination of its concentration should not exceed 3%.

(c) If the aqueous solution has to be supercooled below 0°C, both the solution and solvent must be filtered using a 0.2-μm filter to remove dust particles that can initiate freezing.
(d) Fill both calorimetric cells with the solvent that will be used in the experiment, and scan the DSC up and down several times until the baseline stabilizes.
(e) Replace the solvent in one of the calorimetric cells with the solution to be studied.—Note that this should be done **without stopping scanning**, just at the moment when the instrument is scanning down in temperature and the temperature of the cells is close to room temperature.—On replacing solvent with solution, the cell should not be dried but just washed with the solution being studied. Drying the cells, moreover washing them with some aggressive liquid, will change the adhesive properties of their surface and re-establishing a stable baseline will require numerous further nonstop scans of the instrument with the cells filled with solvent.
(f) All DSC experiments with solvent and then with studied solutions should be carried out under an excess pressure not lower than 2 atm to prevent appearance of bubbles upon heating.

(For details, see P. Privalov, 2009.)

3.2.5. Determination of the Enthalpy of a Temperature-Induced Process

Determination of the enthalpy of a temperature-induced process, that is, evaluation of the area of the observed heat capacity peak, is not simple if the initial and final states of the protein differ in heat capacities. This is usually done by:

(a) Connecting the visible beginning and the end of the excess heat absorption with a line (Fig. 3.18a)
(b) Extrapolating the initial and the final heat capacities to the middle temperature of the observed transition, T_t (Fig. 3.18b)
(c) Connecting the beginning and the end of the heat absorption curve with a curve that rises proportionally to the excess heat which is absorbed (Fig. 3.18c)
(d) Using as a baseline the partial specific heat capacity of some solute that does not undergo changes upon heating, for example, of a very thermostable protein (Fig. 3.18d)

The last method is of particular importance when the protein being studied undergoes several not well-resolved transitions upon heating. It appears that the most practical standard to use for a fully folded protein is the partial heat capacity function of bovine pancreatic trypsin inhibitor (BPTI), a small highly stable protein that unfolds

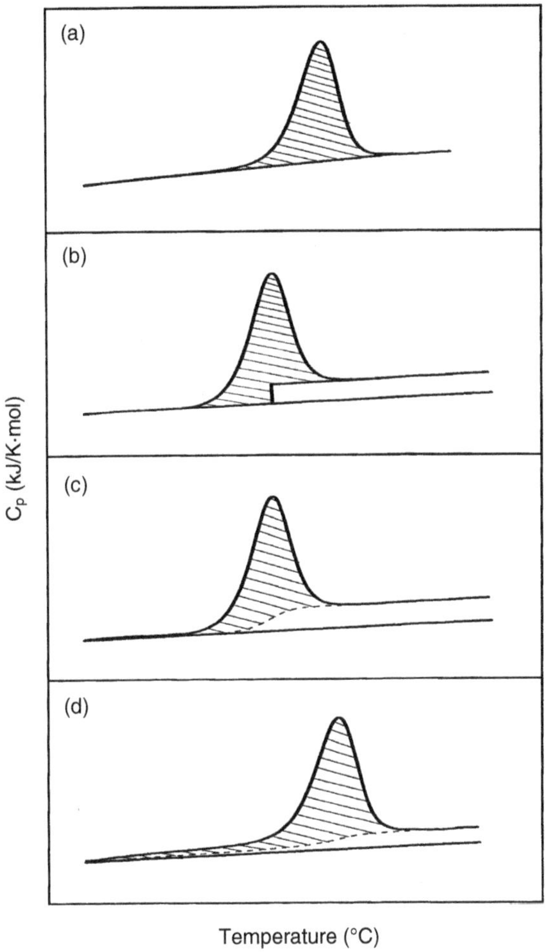

Figure 3.18. The four methods of determining the heat effect of a temperature-induced cooperative process: (a) connecting the pretransition and post-transition heat capacities with a line limiting the heat absorption peak area; (b) extrapolating the initial and final heat capacities to the midpoint of the transition; (c) connecting the beginning and end of the heat absorption curve with a curve that rises proportionally to the excess heat which is absorbed; (d) using as a baseline the partial specific heat capacity of some solute that does not undergo changes upon heating, for example, a very thermostable protein.

upon heating above 100°C (Makhatadze et al., 1993). The heat capacity of this protein in the Celsius scale is expressed by the following formula:

$$C_p(T)^{\text{BPTI}} = [1.295 + 5.971^{-3} \times T]\ \text{J/K·g} \tag{3.22}$$

It is notable that when temperature-induced unfolding of a protein proceeds with a heat capacity increment, which means that the unfolding enthalpy is an increasing function of temperature, the heat absorption peak is asymmetric. In that case the midtransition temperature, T_t, does not coincide with the temperature of the peak at maximum heat absorption, T_{max}, but is lower by δT. This shift in temperature depends on the heat absorption peak height ΔC_p^{max} and the distance in degrees between the centers of masses of both halves of the heat absorption peak ΔT. For practical determination, the following approximate equation might be used:

$$\delta T = T_{max} - T_t = \Delta T \cdot \Delta C_p / \Delta C_p^{max}$$

For globular proteins, which unfold with significant heat capacity increment and significant enthalpy (i.e., the denaturation heat absorption peak is rather sharp and high), T_t differs from T_{max} by approximately 0.7 K, that is, by 0.2% on the absolute scale, which in itself is insignificant. However, this difference becomes quite significant in estimating both the calorimetric and the van't Hoff enthalpies.

3.2.6. Determination of the van't Hoff Enthalpy

Calorimetric measurements of the excess heat absorption give us a real enthalpy of the temperature-induced process. Assuming, however, that the observed heat effect results from a simple and reversible transition, one could use it also for the equilibrium analysis, particularly for estimation of the van't Hoff enthalpy of the process being considered.

The portion of heat absorbed up to a given temperature, $\theta = Q(T)/Q^{tot}$, can be regarded as a parameter indicating the progress of the reaction and giving the equilibrium constant as $K = \theta/(1 - \theta)$. Then, using the van't Hoff equation (Eq. 2.33), and bearing in mind that

$$\partial \ln[\theta/(1-\theta)]\partial T = [1/\theta(1-\theta)](\partial\theta/\partial T)$$

and that at the midpoint of transition $\theta = 1/2$ and $\partial\theta/\partial T = (\partial Q/\partial T)/Q_{tot} = \Delta C_p^{max}/\Delta H_{tot}$, one gets

$$\Delta H^{vH} = 4RT_t^2[\Delta C_p^{max}/\Delta H_{tot}] \quad (3.23)$$

where $\Delta T_{1/2}$ is a width of the heat absorption peak at its half height. Here $\Delta C_p^{max}/\Delta H_{tot} = \Delta C_p^{max}/Q_{tot}$ depends only on the shape of the excess heat absorption peak and does not depend on the concentration of the solution being studied. Approximating the heat absorption peak by a triangle, $Q = \Delta C_p^{max}\Delta T_{1/2}$, one gets the following simple expression:

$$\Delta H^{vH} = 4RT^2/\Delta T_{1/2} \quad (3.24)$$

It should be noted that the van't Hoff enthalpy is determined under the assumption that the observed temperature-induced process is a two-state transition. Thus, it is valid if the process indeed represents a two-state transition and only in that case will it be equal to the real calorimetrically measured enthalpy of the process. Therefore, in that case the ratio $\Delta H^{vH}/\Delta H^{cal} = 1.0$. If this ratio is less than 1.0, this might mean that the observed temperature-induced process does not represent a single two-state transition; if it is higher than 1.0, this is an indication that the process is irreversible.

For the n-meric transition, which proceeds with the heat capacity increment, the van't Hoff enthalpy is determined by the following equation (P. Privalov and Potekhin, 1986):

$$\Delta H^{vH} = \left(\sqrt{n+1}\right)T_m\sqrt{R\left(\Delta C_p^{max} - \frac{\Delta C_p\sqrt{n}}{\sqrt{n+1}}\right)} \tag{3.25}$$

3.2.7. Multimolecular Two-State Transition

Consider the temperature-induced unfolding of a complex consisting of several subunits (P. Privalov and Potekhin, 1986):

$$\mathrm{N} \leftrightarrow m_1\mathrm{D}_1 + m_2\mathrm{D}_2 + m_3\mathrm{D}_3 + \cdots + m_k\mathrm{D}_k$$

The order of this reaction is determined as

$$n = \sum_{i=1}^{k} m_i$$

Let $[\mathrm{N}]_0 = \mathrm{N}^0/\mathrm{N}^{st}$ be the dimensionless concentration of the complex, where N^0 is its initial concentration and N^{st} the concentration of the standard state, which is usually taken as 1 M. If F is the fraction of molecules that has undergone a transition at temperature T, then the concentration of the native state at T can be presented as $[\mathrm{N}]_T = [\mathrm{N}]_0(1 - F)$.

The concentration of each subunit in the denatured state will be $[\mathrm{D}_i] = m_i[\mathrm{N}]_0F$. Therefore, for the equilibrium constant we will have

$$K = \frac{1}{[\mathrm{N}]} \times \prod_{i=1}^{k}[\mathrm{D}_i]^{m_i} = \frac{[\mathrm{N}]_0^{n-1}}{1-F}\prod_{i=1}^{k}[m_i]^{m_i} \tag{3.26}$$

And at the transition midpoint ($F = 1/2$) we will have

$$K = ([\mathrm{N}]_0/2)^{n-1}\prod_{i=1}^{k}[m_i]^{m_i} \tag{3.27}$$

Correspondingly,

$$\Delta G = -RT(n-1)\ln\left\{([\mathrm{N}]_0/2)^{n-1}\prod_{i=1}^{k}[m_i]^{m_i}\right\} \tag{3.28}$$

For the first-order monomolecular reaction, e.g., N $\Leftrightarrow$ D, we will have the following at transition temperature T_t:

$$\mathrm{M}=1, \quad n=1, \quad K=1, \quad \text{and} \quad \Delta G_t=0$$

For the second-order bimolecular reaction we will have the following:

(a) In the case of a homodimer (e.g., N $\Leftrightarrow$ 2D), $m = 2$, $n = 2$, $K = 2[\mathrm{N}_0]$, and $\Delta G_t = -RT\ln(2[\mathrm{N}_0])]$.
(b) In the case of a heterodimer, (e.g., N $\Leftrightarrow$ $\mathrm{D}_1 + \mathrm{D}_2$), $m_1 = 1$, $m^2 = 1$, $n = 2$, $K = [\mathrm{N}_0]/2$, and $\Delta G_t = -RT\ln([\mathrm{N}_0]/2)$.

Correspondingly, since $\Delta G = \Delta H - T\Delta S$ and thus $T\Delta S = \Delta H - \Delta G$, the entropy of homodimer dissociation at the transition temperature is

$$\Delta S(T_t)=\frac{\Delta H_t-\Delta G_t}{T_t}=\frac{\Delta H_t}{T_t}+R\ln\{2[\mathrm{N}_0]\} \tag{3.29}$$

and the entropy of heterodimer dissociation is

$$\Delta S(T_t)=\frac{\Delta H_t-\Delta G_t}{T_t}=\frac{\Delta H_t}{T_t}+R\ln\left(\frac{[\mathrm{N}_0]}{2}\right) \tag{3.30}$$

3.2.8. Analysis of the Complex Heat Capacity Profile

If a macromolecule undergoes several transitions upon heating, the overlap of the heat effects of these transitions results in a rather complex profile of its heat capacity function. Analysis of the heat capacity profiles of macromolecules is based on the unique relation between enthalpy and temperature, that is, the functional dependence of two fundamental conjugate extensive and intensive parameters, which includes all information on the population of the states in the temperature range being considered. Thus, this analysis does not require any assumptions or additional information and, if the partial molar heat capacity function is accurate enough, the results of analysis are quite certain. This is one of the greatest advantages of DSC experiments, which provide this unique opportunity to analyze complex melting profiles.

Several programs have been suggested for deconvolution of the complex heat capacity function of biological macromolecules (Freire and Biltonen, 1978; Filimonov et al., 1982; P. Privalov and Potekhin, 1986; G. Privalov and P. Privalov, 2000). The most practical one to appear models the calorimetrically determined partial heat capacity function with a number of simple cooperative transitions:

$$C_p(T)=C_p^0(T)+\frac{\partial}{\partial T}\left(\sum_{i=1}^{N}P_i\Delta H_i\right) \tag{3.31}$$

where $C_p^0(T)$ is the partial molar heat capacity of the reference state. The second term accounts for any temperature-induced transitions and the possibility that the transition involves an arbitrary number of states, N. In this equation P_i represents the population of molecules in the ith state and ΔH_i is its enthalpy. Expanding Equation (3.31) and bearing in mind that the population of each state is a function of the Gibbs free energy of the state, ΔG_i, we have

$$C_p(T) = C_p^0(T) + \left\{ \left[\sum_{i=1}^{N} \Delta H_i^2 \frac{\exp(-\Delta G_i / RT)}{Q} \right] - \left[\sum_{i=1}^{N} \Delta H_i \frac{\exp(-\Delta G_i / RT)}{Q} \right]^2 \right\} RT^2 \tag{3.32}$$

where Q is the partition function, $\Delta G_i(T) = \Delta H_i(T) - T\Delta S_i(T)$;

$$\Delta H_i(T) = \Delta H_i(T_t) + \int_{T_{t_t}^{i^i}}^{T} C_p^i(T)dT \tag{3.33}$$

and

$$\Delta S_i(T) = \Delta S_i(T_t) + \int_{T_{t_t}^{i^i}}^{T} C_p^i(T)d\ln T \tag{3.34}$$

The computer program CpCalc for modeling the heat capacity function upon denaturation of a protein is provided with the Nano-DSC. Figure 3.19 shows an example of deconvolution of the apparent complex function on the component transition. It is remarkable that the observed temperature-induced process, denaturation of the SOX-5 protein, is highly reversible and the similarity of the components found in these two cases is a strong argument for the correctness of the deconvolution analysis. For more examples see Chapter 10.

3.2.9. Correction for Components Refolding

In studying associations of macromolecules there is a situation in which one of the components is unstable and is partly unfolded at the temperature of ITC experiments but refolds upon association with its partner (see Chapter 12 for examples). If we are interested in the enthalpy of association of fully folded partners, one should take into account this enthalpy of refolding and exclude it from the ITC-measured enthalpy of association. The correction for this refolding energy is described formally as follows (P. Privalov et al., 1999).

Consider the association reaction for the example of protein and DNA at two different temperatures, T_0 and some other temperature T. The enthalpies of protein, DNA, and their complex at these two temperatures can be presented as follows:

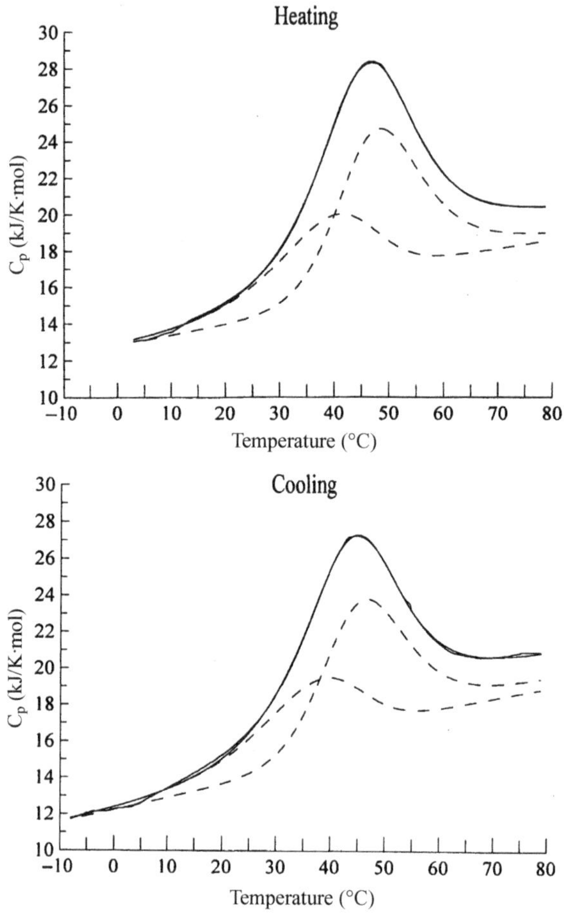

Figure 3.19. Deconvolution of the apparent partial specific heat capacity of SOX-5 protein obtained in heating and cooling experiments (see Fig. 3.15) (Crane-Robinson et al., 1998).

$$H(T)^{\text{pr}} = H(T_0)^{\text{pr}} + \int_{T_0}^{T} C_p(T)^{\text{pr}}\, dT = H(T_0)^{\text{pr}} + C_p(T_0)^{\text{pr}}(T - T_0) + \int_{T_0}^{T} [C_p(T) - C_p(T_0)]^{\text{pr}}\, dT \tag{3.35}$$

$$H(T)^{\text{DNA}} = H(T_0)^{\text{DNA}} + \int_{T_0}^{T} C_p(T)^{\text{DNA}}\, dt = H(T_0)^{\text{DNA}} + C_p(T_0)^{\text{DNA}}(T - T_0) + \int_{T_0}^{T} [C_p(T) - C_p(T_0)]^{\text{DNA}}\, dT \tag{3.36}$$

$$H(T)^{\text{com}} = H(T_0)^{\text{com}} + \int C_p(T)^{\text{com}}\,dt = H(T_0)^{\text{com}} + C_p(T_0)^{\text{com}}(T - T_0) + \int_{T_0}^{T} [C_p(T) - C_p(T_0)]^{\text{com}}\,dT \quad (3.37)$$

Therefore for the enthalpy of association we have

$$\Delta H(T)^{\text{assoc}} = \Delta H(T_0)^{\text{assoc}} - \Delta C_p(T_0)^{\text{assoc}}(T - T_0) + \int_{T_0}^{T} \Big([C_p(T) - C_p(T_0)]^{\text{com}} - [C_p(T) - C_p(T_0)]^{\text{pr}} - [C_p(T) - C_p(T_0)]^{\text{DNA}}\Big)\,dT \quad (3.38)$$

where

$$\Delta C(T_0)^{\text{assoc}} = C_p(T_0)^{\text{com}} - C_p(T_0)^{\text{pr}} - C_p(T_0)^{\text{DNA}}$$

3.3. PRESSURE PERTURBATION CALORIMETRY

3.3.1. Heat Effect of Changing Pressure

The reaction and heat capacity calorimeters considered above work at constant pressure. Variation of pressure, however, also induces a heat effect, which bears important information on the thermal properties of the solutions and macromolecular solutes being studied, particularly their thermal expansion (Fig. 3.20). Pressure perturbation calorimetry (PPC) presents a new application of scanning calorimetry for studying the thermal expansion of liquid solutions (Lin et al., 2002; Ravindra and Winter, 2004; Boehm et al., 2006; Rosgen and Hinz, 2006). The PPC studies carried out in a number of laboratories on aqueous solutions of low-molecular-weight compounds, proteins, and lipids led to understanding that the thermal expansion coefficient and its temperature dependence bear some information on hydration effects.

Compressing a liquid by the increased pressure induces the heat effect δQ and the entropy effect $\delta S = \delta Q/T$. Differentiation with respect to pressure at constant temperature gives

$$\left(\frac{\partial Q}{\partial P}\right)_T = T\left(\frac{\partial S}{\partial P}\right)_T \quad (3.39)$$

Using the Maxwell relation

$$\left(\frac{\partial S}{\partial P}\right)_T = -\left(\frac{\partial V}{\partial T}\right)_P \quad (3.40)$$

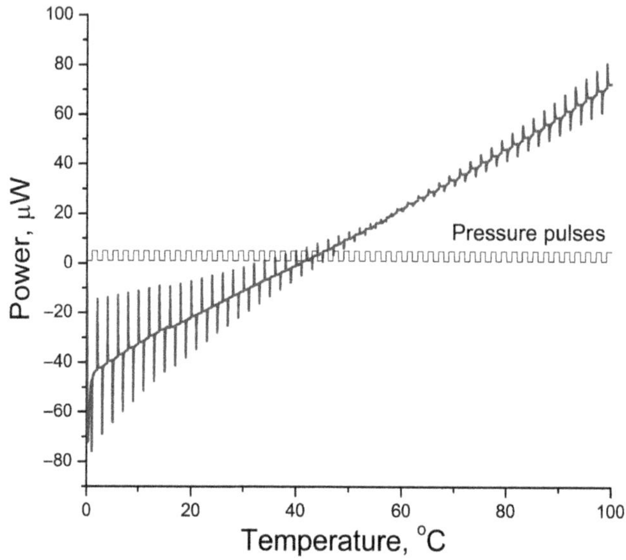

Figure 3.20. The heat effect of pressure pulses measured by PP DSC with buffer (100 mM NaCl, 20 mM sodium phosphate, pH 7.4) in the measuring cell and water in the reference cell: heating rate 0.1 K/min; interval between pressure pulses, 600 sec; the pressure changes from 1 to 5 atm.

one finds that

$$\left(\frac{\partial Q}{\partial P}\right)_T = -T\left(\frac{\partial V}{\partial T}\right)_P = -TV \times \frac{1}{V}\left(\frac{\partial V}{\partial T}\right)_P = -TV\alpha \tag{3.41}$$

Here V is the volume and α is the coefficient of thermal expansion.

Integration of Equation (3.41) at constant temperature over a short pressure range ΔP leads to

$$Q = -TV\alpha\ \Delta P \tag{3.42}$$

If a solution is composed of m_s grams of a solute dissolved in m_0 grams of a solvent, the total solution volume V_{tot} is

$$V_{tot} = m_s V_s + m_0 V_0 \tag{3.43}$$

Differentiating Equation (3.43) with the respect to temperature at constant pressure gives

$$\left(\frac{\partial V_{tot}}{\partial T}\right)_P = m_0\left(\frac{\partial V_0}{\partial T}\right)_P + m_s\left(\frac{\partial V_s}{\partial T}\right)_P \tag{3.44}$$

Substitution of the right-hand side of Equation (3.44) into Equation (3.41) gives

$$\left(\frac{\partial Q}{\partial P}\right)_T = -T\left[m_0\left(\frac{\partial V_0}{\partial T}\right)_P + m_s\left(\frac{\partial V_s}{\partial T}\right)_P\right] \tag{3.45}$$

Multiplying and dividing the first term in brackets by V_0 and the second one by V_s gives

$$\left(\frac{\partial Q}{\partial P}\right)_T = -T\left[m_0 V_0 \alpha_0 + m_s V_s \alpha_s\right] \tag{3.46}$$

Integration of Equation (3.46) over a pressure range ΔP leads to

$$Q = -T[m_0 V_0 \alpha_0 + m_s V_s \alpha_s]\Delta P \tag{3.47}$$

In a differential experiment when the sample and reference solutions are placed in cells with identical volumes and subjected to the same pressure ΔP, the difference heat effect would be

$$\Delta Q = -T\ \Delta P[m_s V_s \alpha_s - m_s V_s \alpha_0] \tag{3.48}$$

which then rearranges to

$$\alpha_s = \alpha_0 - \Delta Q/[T\ \Delta P m_s V_s] \tag{3.49}$$

The pressure perturbation experiment could be performed by applying a series of pressure pulses to the sample and reference solution as they are being continuously heated in a scanning calorimeter. The most suitable for such experiments is the capillary DSC since the operational volume of the capillary cells is less vulnerable to pressure changes. Recently TA Instruments has modified their Nano-DSC instrument, equipping it with a pressure device that periodically monitors pressure pulses by the chosen program.

3.3.2. Pressure Perturbation Experiment

Since the heat effect appears at the increase or decrease of pressure and is proportional to the time derivative of pressure, it is rational to investigate the pressure effect at periodic application of constant pressure pulses. As shown in Figure 3.20, the application of pressure induces a sharp heat effect while removal of pressure results in an effect similar in magnitude but having the opposite sign. The relaxation time of the pressure pulses is quite considerable and depends on the properties of both the instrument and the studied object. Also, since the heat effect depends on the value of the pressure pulse, it is appropriate to use a pressure pulse as large as possible for the given calorimetric instrument. The modification of Nano-DSC with capillary cells designed for pressure perturbation experiments permits the use of pressure pulses up to 5 atm. The pressure perturbation experiment could be carried at constant temperature but since such experiments are performed using a DSC instrument it is appropriate to do it scanning up or down the temperature scale to get information on the thermal expansion coefficient and temperature-induced volume effect over a broad temperature range. However, it is important for the scanning rate to be low because the relaxation time at a pressure jump is rather long and because the equations given previously for estimation of the thermal expansion coefficient were obtained under the assumption that pressure variation takes place at isothermal conditions. Decreasing the heating–cooling rate decreases the sensitivity of scanning calorimetry. Therefore, PPC experiments require rather high concentrations of the reagent: if an ordinary scanning calorimetric experiment uses solutions with concentration about 0.1%, the PPC experiment would require concentrations at least one order of magnitude higher.

Figure 3.21 shows the results of a pressure perturbation experiment with a 12-bp DNA duplex (5.4 mg/mL concentration) in 100 mM NaCl, 20 mM sodium phosphate, pH 7.4, at heating rate 0.1 K/min using periodic pressure pulses from 1 to 5 atm and back to 1 atm with time intervals of 300 sec. In this experiment the DNA solution is in the measuring cell and in the reference cell is solvent. But to get the absolute values for the thermal expansion coefficient of DNA it is necessary to perform similar measurements with the solvent in the measuring cell and pure water in the reference cell. Since the thermal expansion coefficient of pure water is known

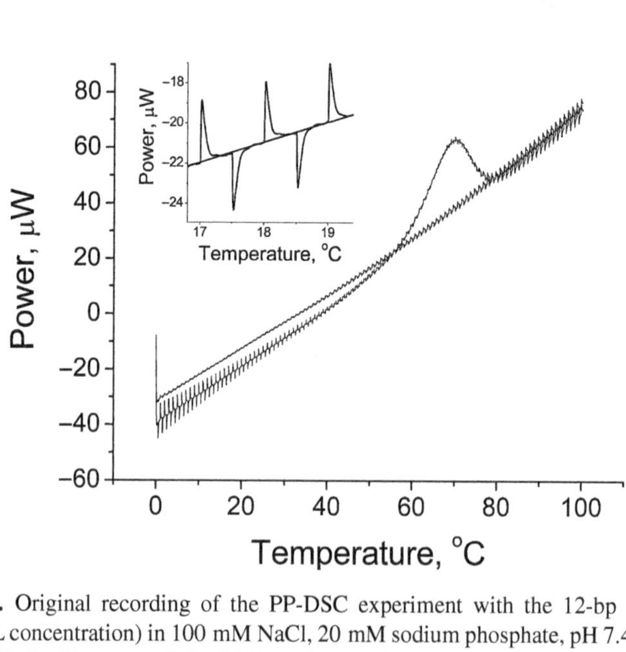

Figure 3.21. Original recording of the PP-DSC experiment with the 12-bp DNA duplex (13.4 mg/mL concentration) in 100 mM NaCl, 20 mM sodium phosphate, pH 7.4, at a heating rate of 0.1 K/min. The scan of the DNA against buffer is shown in red; the scan of buffer against buffer is in blue. The inset gives an expansion of the DNA/buffer scan between 17°C and 19°C (Dragan et al., 2009).

for the whole temperature range being considered, from this experiment one can estimate the thermal expansion coefficient of the solvent used for the DNA solution and then calculate the thermal expansion coefficient of DNA in all of this temperature range (see Section 13.3.3).

REFERENCES

Boehm K., Rösgen J., and Hinz H.J. (2006). *Anal. Chem.*, **78**, 984–990.

Crane-Robinson C., Read C.M., Cary P.D., Driscoll P.C., Dragan A.I., and Privalov P.L. (1998). *J. Mol. Biol.*, **281**, 705–717.

Dragan A.I., Russell D.J., and Privalov P.L. (2009). *Biopolymers*, **91**, 95–101.

Filimonov V.V., Potekhin S.A., Matveyev S.V., and Privalov P.L. (1982). *Molek. Biol. (USSR)*, **16**, 551–562.

Freire E. and Biltonen R.L. (1978). *Biopolymers*, **17**, 463–479.

Lin L.N., Brandts J.F., Brandts J.M., and Plotnikov V.V. (2002). *Anal. Biochem.*, **302**, 144–160.

Makhatadze G.I. and Privalov P.L. (1992). *J. Mol. Biol.*, **226**, 491–505.

Makhatadze G.I., Kim K.S., Woodward C., and Privalov P.L. (1993). *Protein Sci.*, **2**, 2028–2036.

McKinnon I.R., Fall L., Parody-Morreale A., and Gill S.J. (1984). *Anal. Biochem.*, **139**, 134–139.

Pfeil W. and Privalov P.L. (1976). *Biophys. Chem.*, **4**, 33–40.

Privalov G.P. and Privalov P.L. (2000). *Methods Enzymol.*, **323**, 31–62.

Privalov G.P., Kavina V., Freire, E., and Privalov P.L. (1995). *Anal Biochem.*, **232**, 79–85.

Privalov P.L. (1974). *FEBS Lett.*, **40S**, S140-S153.

Privalov P.L. (1980). *Pure Appl. Chem.*, **52**, 479–497.

Privalov P.L. (2009). In: *Protein Structure, Stability, and Interactions* (J.W. Shriver, Ed.), Humana Press, Totowa, NJ, pp. 1–39.

Privalov P.L. and Plotnikov V.V. (1989). *Thermochim. Acta*, **139**, 257–277.

Privalov P.L. and Potekhin S.A. (1986). In: *Methods in Enzymology*, **131**, pp. 4–51. Academic Press, New York.

Privalov P.L., Jelesarov I., Read C.M., Dragan A.I., and Crane-Robinson C. (1999). *J. Mol. Biol.*, **294**, 997–1013.

Privalov P.L., Monaselidze D.R., Mrevlishvili G.M., and Magaldadze V.A. (1965). *Soviet Phys.—JETP*, **20**, 1393–1396.

Privalov P.L., Plotnikov V.V., and Filimonov V.V. (1975). *J. Chem. Thermodynam.*, **7**, 41–47.

Ravindra R. and Winter R. (2004). *Chemphyschem.*, **5**, 566–571.

Rosgen J. and Hinz H.-J. (2006). *Anal. Chem.*, **78**, 991–996.

4

MACROMOLECULES

4.1. EVOLUTION OF THE CONCEPT

The term macromolecules usually includes two very different classes of high-molecular-weight organic compounds: the proteins and the nucleic acids, which are found in all living organisms and only in these organisms. Therefore, these giant molecules evidently represent the most essential components of living systems.

The term **protein** was coined by **Garret Jan Mulder** in 1838: "Protein appears to be a principal substance of animal nutrition which the plants prepare for the herbivores, and which these then supply to the carnivores." According to *Webster's New Dictionary*, "Protein—any of the class of complex nitrogenous substances occurring in all animals and vegetables and essential to diet of animals." Understanding that proteins are not just food for carnivores came rather late.

For a long time proteins were considered as colloidal particles, that is, aggregates of low-molecular-weight organic compounds. By the middle of the nineteenth century it became clear that proteins consisted of smaller components that had certain chemical characteristics in common: they always contained an amino (NH) group and an acidic carboxylic (COOH) group. The term **amino acid** was used for

Microcalorimetry of Macromolecules: The Physical Basis of Biological Structures, First Edition.
Peter L. Privalov.

compounds containing these two groups. Several hundreds of various amino acids were synthesized by chemists; however, living organisms use only 20 quite definite amino acids. It appeared that the difference in properties between proteins results from difference in relative contents of their amino acid constituents. It remained unclear, however, how these small components are integrated.

Substantial progress was made in solving these problems with the invention of the ultracentrifuge by **Theodor Svedberg** in 1924. Ultracentrifugation experiments showed that the mass of each given type of protein is quite definite and unique. Proteins appeared to be monodisperse heteropolymers each consisting of quite a definite number of certain amino acids. On the other hand, it was found that some of the proteins were forming **crystals**. It was also known that proteins change drastically upon heating: at high temperatures they loose their enzymatic activity and precipitate; that is, they **denature**. It was also found that the heat-denatured proteins do not form crystals. It appeared that the natural, enzymatically active protein (i.e., the native protein) should have some structure, which is disrupted upon heating, that is, upon its denaturation. With the rapid success of crystallography in Cambridge under **Lawrence Bragg** in the 1930s this led to an idea of determining protein structure crystallographically. Realization of this idea took two decades. In 1958 the first protein structure appeared, done by **John Kendrew**.

The other great player, the nucleic acid, appeared in science much later than the protein. In 1869 the medical doctor **Friedrich Miescher** (1844–1895) decided to study the chemistry of the cell nucleus from lymphocytes. Unusually large content of phosphorus in the isolated compound led him to the conclusion that: "It is more likely that we have a substance not comparable with any other groups at present known." Miescher called this material the **nucleic acid**.

After Miescher the work on the determination of the chemical structure of nucleic acids was continued by **Phoebus Levene** (1869–1940) in the United States. This led to the understanding by 1935 that nucleic acids are polymers consisting of four different types of compound: (1) purine bases (guanine and adenine); (2) pyrimidine bases (thymine and cytosine); (3) sugars; and (4) phosphates. Levene was particularly interested in the base composition of DNA from various sources. The main conclusion of Levene's more than 10 years' study was that:

> *Adenine, guanine, cytosine, and thymine were present in nucleic acids in equal amounts in the so-called deoxyribonucleic acid (DNA).*

This conclusion was confirmed by the other leading authority in biochemistry, **Alfred Mirsky** from the Rockefeller Institute. This gave rise to the **tetranucleotide hypothesis** of nucleic acid structure:

–[T–A–C–G]–[T–A–C–G]–[T–A–C–G]–

A polymer of identical tetranucleotide units could not match the chemical diversity of the proteins, the variations in amino acid content of which had been recognized. This tetranucleotide hypothesis reinforced the opinion of many biochemists that nucleic acids were uninteresting molecules: Such a molecule could act only as a structural element—"the wooden stretcher behind the Rembrandt" (Horace Judson).

This would delay for half a century the recognition of DNA as the material of heredity.

The fact that DNA is the **material of heredity** was demonstrated by **Avery, MacLeod,** and **McCarty** only in 1943. However, they never received a Nobel Prize for one of the century's greatest discovery. Also not much noticed was the discovery of the surprising equivalence of the adenine–thymine and guanine–cytosine pairs in DNA, made by **Erwin Chargaff** in 1951 (see Chargaff, 1951). Nevertheless, these discoveries made by Avery, MacLeod, McCarty, and Chargaff, and the crystallographic DNA data of **Rosalynd Franklin** obtained in 1951, all condensed by **James Watson** and **Francis Crick** in 1953 into the double-helical model of DNA, made a revolution in biology as it suggested the clear mechanism of multiplication of the genetic information. As Watson and Crick noted in their paper in *Nature*: "The phosphate–sugar backbone of our model is completely regular, but any sequence of the pair of bases can fit the structure. It follows that in a long molecule many different permutations are possible, and it therefore seems likely that the precise sequence of the bases is the code, which carries the general information. If the actual order of the bases on one of the pair of chains were given, one could write down the exact order of the bases on the other one, because of the specific pairing. Thus one chain is, as it were, the complement of the other, and it is this feature which suggests how the deoxyribonucleic acid might duplicate itself" (Watson and Crick, 1953).

Publication of this paper resulted in an avalanche of studies on the nucleic acids and particularly DNA. One of the topics, which attracted most of the attention, was the energetic basis of the double helix, that is, forces between the complementary strands, because the dissociation–association of these strands was central in the mechanism of functioning of this bearer of genetic information. Since this required direct measurements of the heat of dissociation of the DNA strands, calorimetry gained a certain importance in studying DNA. Moreover, the need for this information was the main driving force in developing the microcalorimetric technique.

Later other modifications of the polymeric nucleic acids, the ribonucleic acids (RNAs), attracted attention. Unlike DNA, these macromolecules are single stranded and are able to form a variety of spatial structures fulfilling large spectra of functions.

4.2. PROTEINS

4.2.1. Chemical Structure

Chemically, proteins are linear polymers assembled from the covalently linked standard blocks. The average molecular weight of the block is about 100 Da. Since the molecular weight of proteins varies between 10,000 and 10,000,000 Da, it appears that they are built from 100 to 100,000 blocks.

The blocks constituting proteins have a standard part, which represents the repetitive element in the polymer chain, and nonstandard side chains. The blocks of the polypeptide chain are the parts of the low-molecular-weight compounds, called **α-amino acids** because they have an amino group, and acidic carboxylic group and the central atom in these compounds is an alpha-carbon, C_α:

$$\mathrm{NH_2{-}HC_\alpha R_i{-}CO{-}OH}$$

Here R_i is the nonstandard side chain of type i. Therefore, the amino acids differ in the side chains.

The hydrogens in the terminal groups of amino acids are removable. In acidic solution with high concentration of protons both the carboxyl and amino groups are protonated. With increasing pH the carboxylic group first loses its hydrogen ($pK = 2.4$) and becomes negatively charged. At higher pH the amino group loses its hydrogen ($pK = 9.8$) and becomes neutral. Thus, in neutral aqueous solutions, amino acids are in the form of **zwitterions**—its two sides are oppositely charged:

$$\mathrm{{}^+NH_3{-}HC_\alpha R_j{-}CO{-}O^-}$$

Two amino acids can join together, releasing a water molecule:

$$\mathrm{{}^+NH_3{-}HC_\alpha R_i{-}CO{-}O^- + {}^+NH_3{-}HC_a R_j{-}CO{-}O^-}$$
$$\Leftrightarrow \mathrm{{}^+NH_3{-}HC_\alpha R_i{-}CO{=}NH{-}HC_a R_j{-}CO{-}O^- + H_2O}$$

The covalent bond connecting two amino acids (=) is called a **peptide bond**. The group –CO=NH– formed by this bond is called a **peptide unit**, and the product of joining two amino acids is called a **dipeptide**. By joining one more amino acid a tripeptide is formed:

$$\mathrm{{}^+NH_3{-}HC_\alpha R_i{-}CO{=}NH{-}HC_\alpha R_j{-}CO{-}O^- + {}^+NH_3{-}HC_\alpha R_j{-}CO{-}O^-}$$
$$\Leftrightarrow \mathrm{{}^+NH_3{-}HC_\alpha R_i{-}CO{=}NH{-}HC_\alpha R_j{-}CO{=}NH{-}HC_\alpha R_j{-}CO{-}O^- + H_2O}$$

Correspondingly, the assembled polymer is called a **polypeptide**. The end of a polypeptide chain carrying the primary amino group is called the amino-terminal or just the **N-terminal**; the end with the carboxyl group is called the carboxy-terminal, or the **C-terminal**. The part of the amino acid that is included in the polypeptide $-\{\mathrm{NH{-}HC_\alpha R_j{-}CO}\}-$ is called an **amino acid residue**.

Formation of the peptide bond proceeds with release of one water molecule. Correspondingly, the reverse reaction of disruption of the peptide bond proceeds with the uptake of a water molecule and is called **hydrolysis**. Formation of water molecules from –OH and H– groups is energetically favorable. However, the appearance of extra water molecules with the excess of water in the surroundings is thermodynamically unfavorable. Also, formation of the peptide link requires energy for dehydration of the contacting groups. In total, formation of the peptide link is **thermodynamically unfavorable**. As a result, a **polypeptide chain is unstable** in aqueous solution. It does not degrade only because of a large **activation barrier**.

Although organic chemistry knows several hundreds of amino acids differing in their side chains, living Nature uses only 20 of them for proteins (Fig. 4.1). The arrangement of amino acids in the polypeptide chain or, in fact, the arrangement of the side chains in the polypeptide chain, is called **the primary structure of protein**.

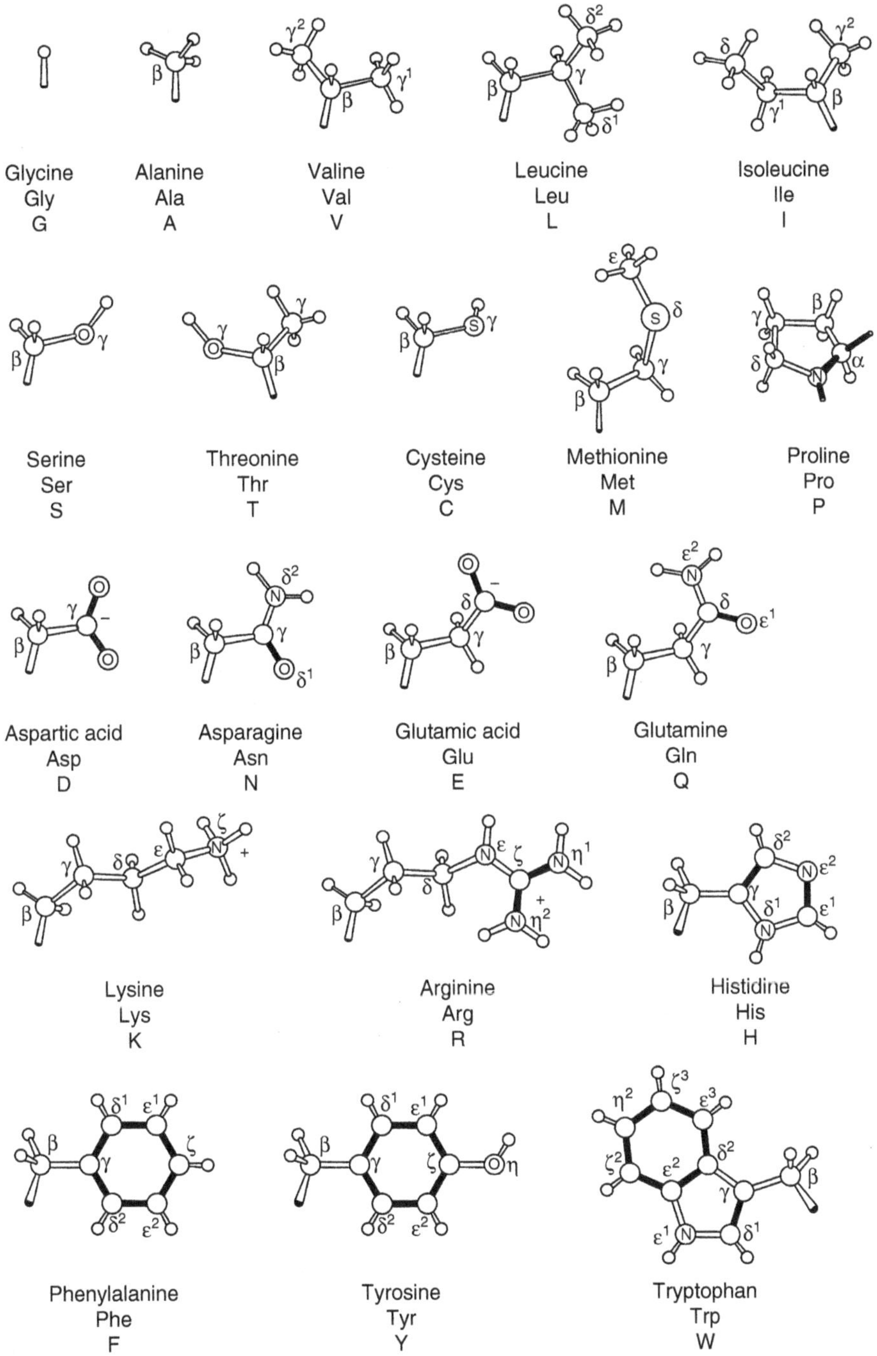

Figure 4.1. Side chains of the 20 amino acids that occur naturally in proteins. Small unlabeled spheres are hydrogen atoms, and large unlabeled spheres are carbon atoms; other atoms are labeled. Double bonds are black and partial double bonds are shaded. Below the name of the amino acid are the three-letter and one-letter abbreviations commonly used. (Reproduced from Creighton, 1993.)

The primary structure of protein is genetically determined and in itself it determines all properties of the given protein.

The side chains of amino acids differ in their size, structure, solubility in water, and ability to keep the hydrogen atom, that is, to dissociate or associate hydrogen at certain solvent conditions, thus changing their charges (Table 4.1).

The easily removable hydrogens have side chains of aspartic and glutamic acids ($pK^{Asp} = 3.91$ and $pK^{Glu} = 4.28$). Therefore, at neutral pH they are positively charged and become neutral in the acidic solution. In contrast, the pK of tyrosine, lysine, and arginine side chains are very high ($pK^{Tyr} = 10.1$, $pK^{Lys} = 10.8$, $pK^{Arg} = 12.5$). In neutral solutions (i.e., solutions with pH close to 7.0) tyrosine is neutral but loses a proton, and it becomes positively charged in an alkaline solution. In contrast, the lysine and arginine are positively charged in a neutral solution and lose their charge with increasing pH. The His is positively charged below pH 6, but above pH 6 it loses this proton and becomes neutral. Thus, the polypeptide in aqueous solution represents a polyion, which is positively charged at low pH (due to protonation of the N-terminal α-amino group and the side chains of Lys, Arg, and His) and negatively charged at high pH (due to deprotonation of the C-terminal α-carboxyl and the side chains of Asp, Glu, His, and, Tyr).

From the pK of the amino acid groups one can determine the Gibbs energy of their protonation–deprotonation, that is, the work needed to remove the bound proton from the group (see Eq. 2.49). As Table 4.1 shows, this work increases with

TABLE 4.1. Specification of the α-Terminal Groups and the Side Chains of Amino Acids for 25°C

		Amino Acid	Side Chains	pK	ΔG (kJ/mol)	ΔH (kJ/mol)
Charged	Negatively	**Carboxy terminal**	$-CO-OH^+$	2.8	15.6	1.7
		HAspartic acid Asp A	$-CH_2-CO-O^-H^+$	3.9	21.8	!
		Glutaminic acid Glu E	$-CH_2-CH_2-CO-O^-H^+$	4.3	24.0	1.6
	Positively	**Amino terminal**	$-NH_2H^+$	9.8	54.7	43.8
		Histidine His H	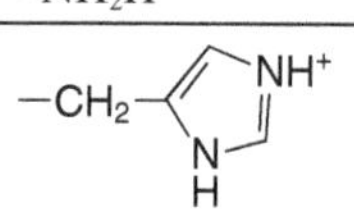	6.0	33.5	28.8
		Lysine Lys K	$-CH_2-CH_2-CH_2-NH_3^+$	10.8	60.3	53.8
		Arginine Arg R	$-CH_2-CH_2-CH_2-NH-C(NH_2)=NH_2^+$	12.5	69.8	51.8

TABLE 4.1. (*Continued*)

	Amino Acid	Side Chains	p*K*	Δ*G* (kJ/mol)	Δ*H* (kJ/mol)
Polar	Serine Ser S	$–CH_2–OH$			
	Threonine Thr T	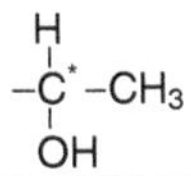			
	Cysteine Cys C	$–CH_2–SH$			
	Tyrosine Tyr Y	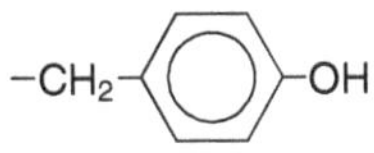			
	Asparagine Asn N	$–CH_2–CO–NH_2$			
	Glutamine Gln Q	$–CH_2–CH_2–CO–NH_2$			
Nonpolar	Glycine Gly G	–H			
	Alanine Ala A	$–CH_3$			
	Valine Val V	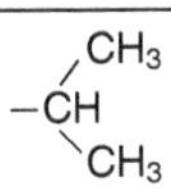			
	Leucine Leu L	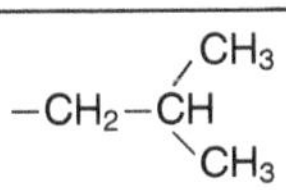			
	Isoleucine Ile I	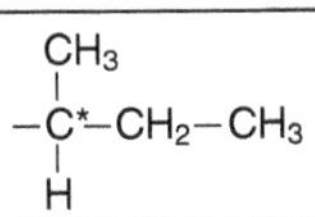			
	Proline Pro P	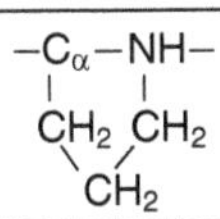			
	Methionine Met M	$–CH_2–CH_2–S–CH_3$			
	Phenylalanine Phe F	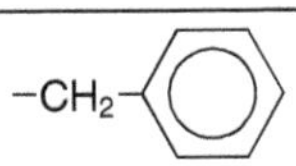			
	Tryptophan Trp W	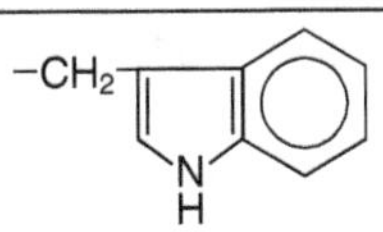			

increasing p*K* of the group. In parallel the enthalpy of deprotonation increases, suggesting that enthalpic interactions are playing an essential role in keeping the proton attached to its donor group.

Another essential characteristic of amino acids is their interaction with water, which is a universal environment for all biological species and their molecular components, particularly proteins. Water is a polar molecule (see Section 2.3). Therefore, the polar and charged molecules are highly soluble in water. In contrast, solubility of nonpolar molecules in water is very low (see Section 2.4). Therefore, interactions of the groups of amino acid residues with water have paramount importance for the physical structure of polypeptide chains.

4.2.2. Physical Structure

An α-carbon is a tetrahedron and the side chain can be attached equally either on its left or right sides, if the amino acid is considered from the CO– to the –NH ends, forming the left-handed or the right-handed amino acids. Correspondingly, the first is denoted by L (*levulo*, Latin) and the second is denoted by D (*dextro*, Latin). In chemical synthesis L and D amino acids are synthesized with the same efficiency. However, Nature uses only the L-isomers for protein synthesis, completely neglecting the D-isomers. The D-isomers are used by various fungi and bacteria in synthesis of some short peptides, for example, antibiotics.

The backbone of a polypeptide chain appears as a rigid, precisely fixed construction (Fig. 4.2). It is such, however, only at 0 K, that is, at the temperature where all motions

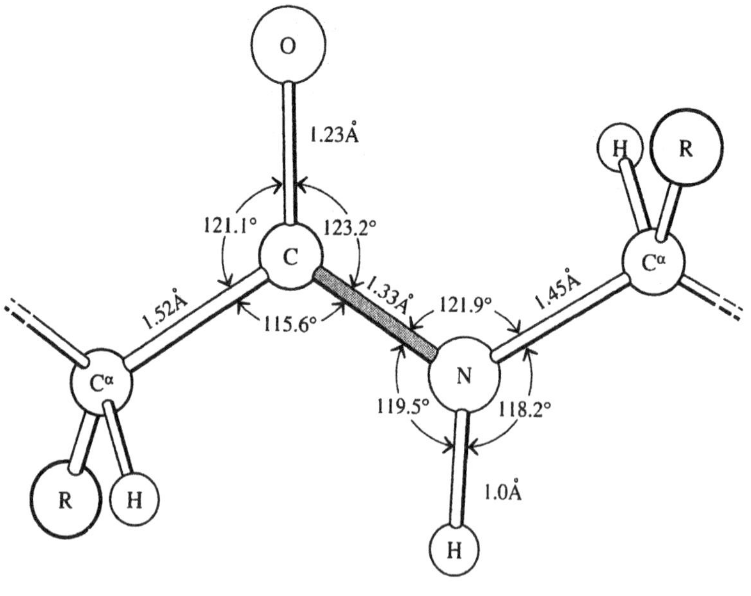

Figure 4.2. Structure of the peptide backbone, with the *trans* peptide bond (shaded), according to Ramachandran (1974).

are frozen and the polypeptide chain is in its lowest energy level. With increasing temperature, all bonds start to fluctuate in length and angles. Covalent bonds in the peptides are strong and at room temperature their length does not fluctuate much, but the angles oscillate within 5°. More pronounced are fluctuations in the torsion angles of the single bonds, which can rotate. Therefore these bonds are regarded as rotors. Thus, at room temperature all this construction is in rapid dynamic motion. Various three-dimensional arrangements of the backbone, which are interconvertable without breaking covalent bonds, are generally described as **conformations**.

The peptide bond connecting neighboring amino acid residues in the polypeptide chain

$$^{+}NH_3\text{–}HC_\alpha R_i\text{–}CO{=}NH\text{–}HC_\alpha R_j\text{–}CO \Leftrightarrow NH\text{–}HC_\alpha R_k\text{–}CO\text{–}$$

has a partly double-bonded character due to resonances in the electronic structures, delocalization of the electrons between the CO and NH groups:

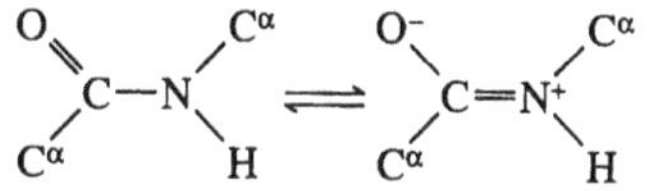

As a result the oxygen atom acquires a net partial negative charge, whereas the hydrogen in the NH group is somewhat positively charged. Thus the peptide unit forms three dipoles oriented in the same direction (Fig. 4.3). The overall dipole moment of the peptide unit is 3.5 Debye. Because of the double-bonded character

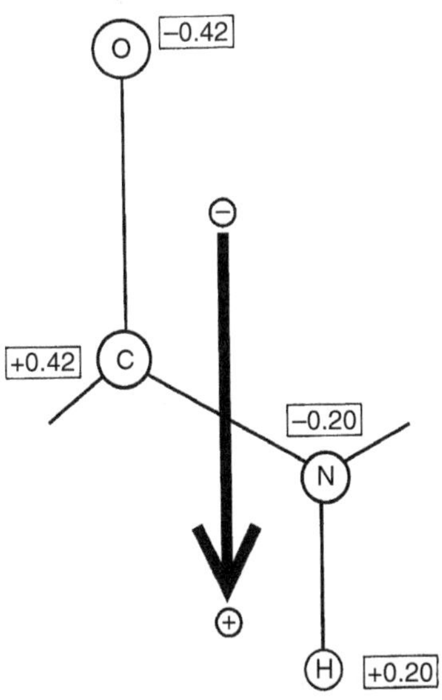

Figure 4.3. Distribution of charges in the peptide unit produces a dipole moment of 3.5 Debye units.

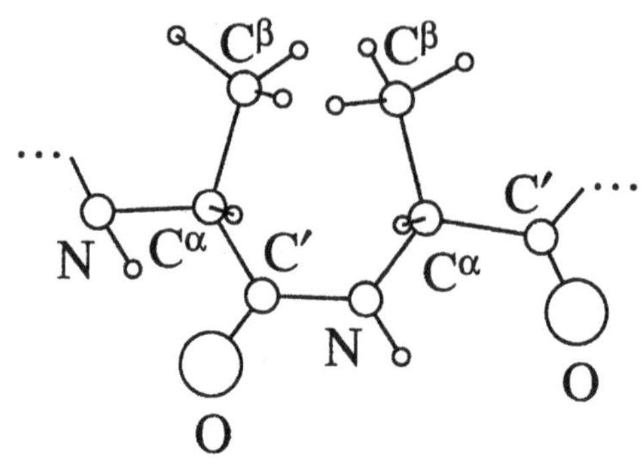

cis

Figure 4.4. The *cis* form is strongly disfavored because of repulsion between the side chains attached to the C_α carbon.

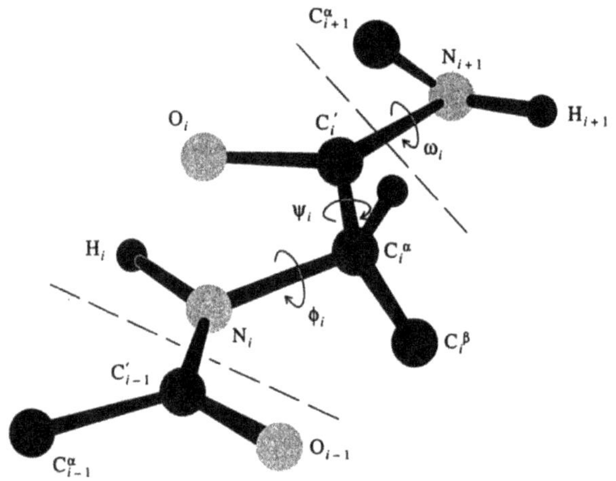

Figure 4.5. Perspective drawing of a segment of polypeptide chain showing rotation of the covalent bonds linked to the C_α atom specified by the ϕ and ψ angles and of the peptide bond specified by the ω angle, (Reproduced from Creighton, 1993.)

of the peptide bond, rotation of this bond is restricted, and the six atoms forming the peptide unit have a strong tendency to be coplanar. Thus, the peptide unit has only two conformations, ***cis*** and ***trans***, with the torsion angle of the peptide bond, ω, of 0° and 180°, respectively. The *trans* form is intrinsically favored energetically by a ratio of approximately 1:1000, because of less repulsion between the side chains attached to the C_α carbon (Fig. 4.4). The *cis* form is strongly disfavored except for residues preceding proline.

The rotations around two other bonds linked to the C_α atom are specified by angles ϕ and ψ (Fig. 4.5). These are single bonds and they can rotate freely.

Therefore, each of the amino acid residues in the polypeptide chain can have many conformations. However, although the single bonds linked to C_α atom can rotate over 360°, three of these conformations are the most populated because in these conformations the rotating groups are the most separated from each other. Assuming that the ϕ and ψ angles have only three values and the ω angles can have only one value, that is, only the *trans* conformation is populated, the number of conformations contributed by one residue to the polypeptide backbone will be $3 \times 3 = 9$. A more realistic number of conformations is about 10, since ω has some probability of having a second value.

The 20 side chains used in proteins differ in size and, correspondingly, in the number of rotatable bonds (Fig. 4.1). For the usual assortment of amino acids in globular proteins the average amino acid side chain has about two rotatable bonds, each with about three rotamers. It appears therefore that $3 \times 3 \approx 10$ conformations are available for the average amino acid side chain. Thus, in total the number of conformations, N, of the polypeptide chain consisting of n amino acid residues amounts to about $N \cong (10 \times 10)^n = 100^n$.

The conformational freedom of a protein is specified usually not by the number of accessible conformations, N, but by the natural logarithm of this number:

$$\ln N = 2.3 \log N = 2.3 \times n \times \log(100) = 2.3 \times 2 \times n = 4.6 \times n$$

At 0 K, where all motions vanish and the polypeptide freezes in the conformation with the minimal energy, $N = 1$ and $\ln N = 0$. With increasing temperature the number of conformations increases and correspondingly the disorder of the polypeptide chain increases. This disorder is specified by the logarithm of the number of conformations, which is called the **conformational entropy**:

$$S^{\text{conf}} = R \ln N = 2.3 R \log N = 2.3 \times 8.36 \times \log N \text{ J/K·mol}$$

where R is the gas constant, $R = 8.36$ J/K·mol.

The conformational entropy multiplied by the absolute temperature—the factor TS^{conf}—specifies the energy that is dissipated in the thermal vibrations of the molecule. For room temperature, 25°C = 298.15 K, this gives the following values for the conformational entropy and the entropy factor of the average side chain:

$$\Delta S^{\text{sch}} = R \ln(3 \times 3) == 18.4 \text{ J/K·mol}$$

$$T\,\Delta S^{\text{sch}} = 18.4 \times 298.15 \text{ K} \times \text{J/K·mol} = 5.5 \text{ kJ/mol}$$

The contribution of each amino acid to the backbone entropy (two rotors each with three rotamers) and the entropy factor at 25°C = 298.15 K are

$$\Delta S^{\text{bb}} = R \ln(3 \times 3) \text{ J/K·mol} = 18.4 \text{ J/K·mol}$$

$$T\,\Delta S^{\text{bb}} = 18.4 \times 298.15 \text{ K} \times \text{J/K·mol} = 5.5 \text{ kJ/mol}$$

Thus, it appears that the total entropy factor of the amino acid residue should amount to 11 kJ/mol at 25°C.

This estimation of the number of conformations and the conformational entropy is highly simplified since in reality the population of rotamers is not identical but depends on the interactions with neighboring groups. Denoting the fractional population of each rotamer by p_i, the conformational entropy can be expressed as

$$S^{\text{conf}} = -R\sum p_i \ln(p_i) \tag{4.1}$$

where p_i is determined by the energy E_i of each rotamer as a function of angle χ_i, weighting the contribution of all neighboring states:

$$p_i = \frac{\exp(-E_i / RT)}{\sum_k \exp(-E_k / RT)} \tag{4.2}$$

The results of the entropy factor calculations for all amino acid side chains, carried out by various authors, are listed in Table 4.2. It appears that the averaged

TABLE 4.2. Theoretically Estimated Changes in Side-Chain Conformational Entropy Factor ($T\Delta S^{\text{conf}}$) on Protein Unfolding[a]

Residue	Pickett and Sternberg (1993)	Abegian and Tortov (1994)	Koehl and Delarue (1994)	Mean
Ala	0	0	0	0
Arg	35	37	21	32
Asn	27	14	13	18
Asp	22	10	11	14
Cys	10	20	11	15
Gln	37	35	22	30
Glu	32	29	23	25
Gly	0	0	0	0
His	17	17	16	17
Ile	15	13	16	13
Leu	14	13	16	13
Lys	34	39	28	33
Met	28	27	22	26
Phe	10	10	11	10
Pro	0	0	5	1
Ser	5	20	8	13
Thr	28	20	10	19
Trp	17	17	20	17
Tyr	17	17	19	17
Val	9	9	11	7
Averaged	19	18	14	17

[a]At 300 K in kJ/mol.

contribution of the amino acids' side chain to the entropy factor upon protein unfolding, as calculated by Equation (4.2), is $T\Delta S^{\text{sch}}$ = (4 ± 1) kJ/mol. This value is remarkably close to that estimated above by simply counting the number of possible rotamers, 5.5 kJ/mol.

The bond rotations are supposed to be chaotic. The chaotic conformation of a polypeptide chain is called the **random coiled conformation**. The random coiled conformation is dynamic; it changes all the time because of thermal motion. The physical dimension of a random coil has only statistical meaning and is specified by some averaged parameters: the root-mean-square value of the distance r between two atoms of the hypothetical random-coiled chain separated by n bonds of length l:

$$\langle r^2 \rangle^{1/2} = l \times n^{1/2} \tag{4.3}$$

and gyration radius R_G, representing the root-mean-square distance of the collection of atoms from their common center of gravity:

$$\langle R_G^2 \rangle^{1/2} = n \times l^{2/6} \tag{4.4}$$

However, movements in polymers are complex and not entirely understood. Each conformational parameter of an ideal random coil is supposed to be independent of all others at equilibrium, but this cannot be the case for fluctuations on a short time scale: if only one bond near the middle of a chain has to rotate by 180°, the end of the chain would have to undergo extremely large movements, which is impossible. It seems that the rotations of all bonds must be coordinated in such a way as to produce more plausible types of movements, but it is unclear how this might occur. Therefore, the random coiled conformation of a protein polypeptide chain has not been modeled yet, even using the most powerful computers.

4.2.3. Restrictions on the Conformation of Polypeptide Chains

There are two circumstances that restrict considerably the number of possible conformations of a polypeptide chain: local and long range. The **local restrictions** proceed from the limitations on the attainable torsion angles ϕ and ψ due to steric clashes between non-neighboring atoms. The permitted values of ϕ and ψ were first determined by Ramachandran and colleagues, using hard–sphere models of the atoms and fixed geometries of the bonds (Ramachandran, 1963; Ramachandran and Sasisekharan, 1968). An example for an Ala residue is illustrated in Figure 4.6. The normally allowed values, for which there is no steric overlap, are contoured; symbols indicate the ϕ and ψ values specific for certain regular conformations. More precise Ramachandran plots were determined using molecular dynamic simulations. An example for *N*-acetyl-Ala-Ala-amide is given in Figure 4.7.

The **long-range restrictions** are provided by covalent cross-links of the polypeptide chain. The cross-linking might occur as a result of oxidation of two contacting Cys residues (–CH_2–SH) containing a sulfhydryl group:

$$2[(-SH)+(-SH)]+O_2 \rightarrow 2[-S-S-]+2H_2O$$

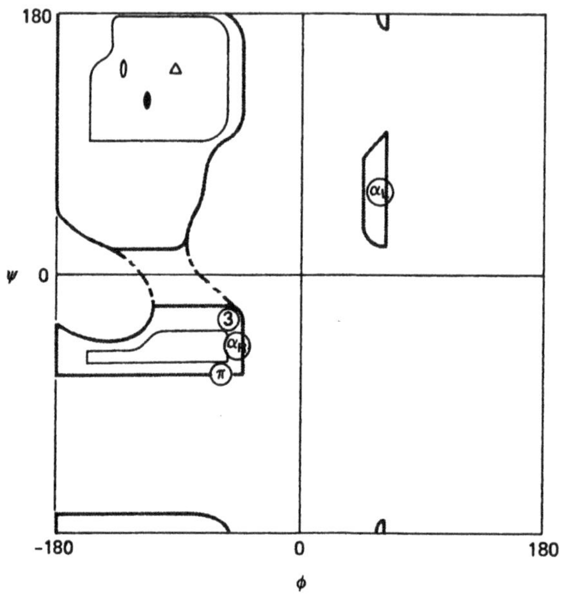

Figure 4.6. Ramachandran plot using hard-sphere models of the atoms for Ala and Gly residues. The fully allowed regions are contoured. The symbols indicate the regular conformation: (α_L) the left-handed α-helix; (α_R) the right-handed α-helix; (0) the antiparallel β-sheet; (•) the parallel β-sheet; (3) the 3_{10}-helix; (π) the π-helix (Ramachandran and Sasisekharan, 1968).

The S–S bond cross-linking two parts of the polypeptide chain decreases the conformational entropy of the disordered polypeptide chain by (Pace et al., 1988)

$$\Delta S^{SS} = -A - (3/2)R\ln(\mathbf{n}) \tag{4.5}$$

Here A is an empirical parameter (about 37 cal/K·mol); **n** is the number of amino acid residues in the loop formed by the –S–S– cross-link. Thus:

$$\Delta S^{SS} = -3[3 + \ln(n)]\ \text{cal/K·mol} = -12.5[3 + \ln(n)]\ \text{J/K·mol} \tag{4.6}$$

4.2.4. Regular Conformations of Polypeptide Chain Proteins

If the ω, ϕ, and ψ torsion angles take certain specific values for all amino acid residues in the polypeptide chain, the conformation of chain becomes regular (Table 4.3). Antiparallel and parallel β-sheets are rather frequent structural elements in proteins but in the isolated form polypeptides cannot be obtained in this conformation because of their strong tendency to associate—forming extended sheets, which aggregate. The situation is the same for polyglycine conformations: the polypeptides with regularly arranged glycine and alanine form strong aggregates—ropes, particularly silk fibroin.

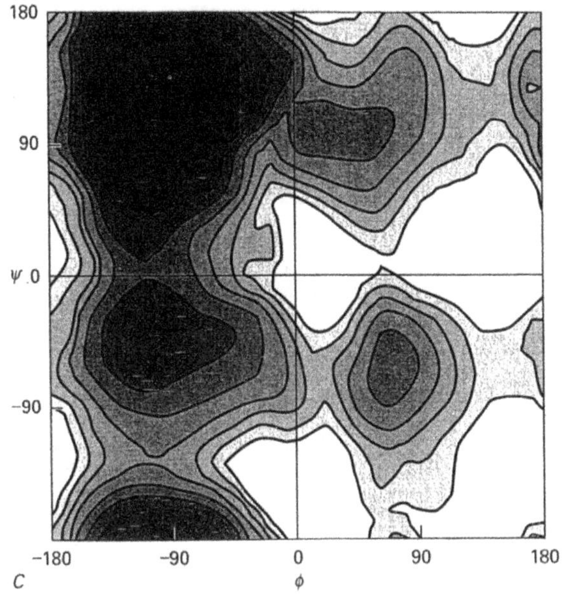

Figure 4.7. Ramachandran plot computed for the dipeptide *N*-acetyl-Ala-Ala-amide using molecular dynamic simulation. The apparent free energy for the various values of ϕ and ψ angles are given in contours of 2 kJ/mol relative to the lowest free energy in the upper left-hand corner, which is shaded black, showing that the difference between allowed and disallowed regions is much less distinct that it seemed when only repulsions between hard-sphere atoms were considered. (Reproduced from Creighton, 1993.)

TABLE 4.3. Parameters for Regular Polypeptide Conformations

	Bond Angle (deg)			Residues	Translation per
Conformation	ϕ	ψ	ω	per Turn	Residue (Å)
Antiparallel β-sheet	−139	+135	−178	2.0	3.40
Parallel β-sheet	−119	+113	180	2.0	3.20
Right-handed α-helix	−57	−47	180	3.6	1.50
3_{10}-Helix	−49	−26	180	3.0	2.00
π-Helix	−57	−70	180	4.4	1.15
Polyproline I	−83	+158	0	3.3	1.90
Polyproline II	−78	+149	180	3.0	3.12
Polyglycine II	−80	+150	180	3.0	3.1

The term **3_{10}-helix** refers to the occurrence of 3 residues per turn and the 10 atoms between the hydrogen-bond donor and acceptor. The packing of the backbone atoms in this helix is somewhat too tight and hydrogen bonds are nonlinear. In contrast, the **π-helix**, which has 4 residues per turn, is less tight and has a hole down the middle, so the backbone atoms are not in contact. Therefore, these two helices are rather unstable and in the isolated state they have not yet been studied experimentally.

A **right-handed α-helix** is the best known of the polypeptide regular structures. It has 3.6 residues per turn and a translation of 1.50 Å per residue, which gives a translation of 5.41 Å per turn (Pauling and Corey, 1953). The α-helix is packed very efficiently, without stresses and without an empty hole inside. In this helix the backbone carbonyl oxygen of each residue is near NH– of the fourth residue along the chain (2.86 Å). Therefore these two groups of opposite polarity can be hydrogen bonded (Fig. 4.8). It is remarkable that formation of the very first hydrogen bond with the fourth residue immobilizes all four peptide units in between, facilitating propagation of the helical conformation along the chain. Also, folding of the α-helix

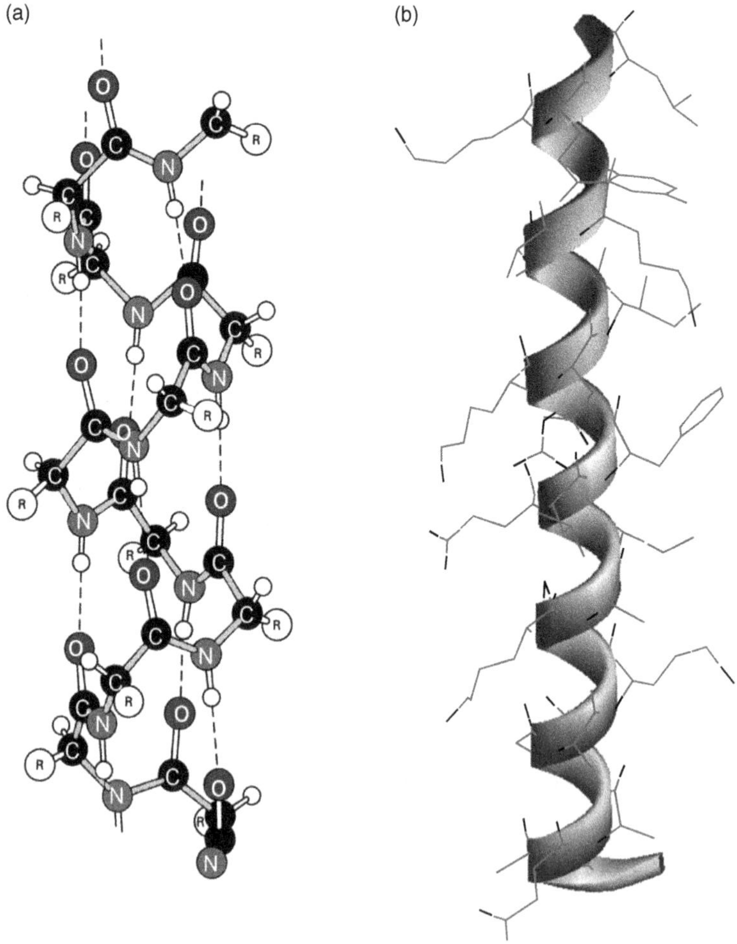

Figure 4.8. (a) The classical right-handed α-helix with stabilizing hydrogen bonds shown by the dashed lines. (b) Ribbon drawing of the G α-helix in myoglobin.

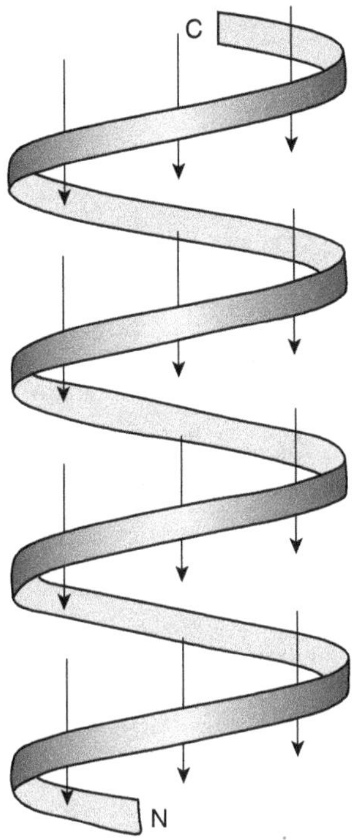

Figure 4.9. The parallel alignment of the individual dipoles in the α-helix.

results in alignment of all dipoles of the peptide units in the same direction, toward the N-terminus (Fig. 4.9). Moreover, formation of these hydrogen bonds in the α-helix results in the assembled shifting of thc hydrogens, also toward the N-terminus. The parallel alignment of the individual dipoles results in a large overall dipole of the α-helix. One can expect that the cooperation of these dipoles stabilizes the α-helical conformation.

A **left-handed α-helix** is also structurally possible: its ϕ and ψ torsion angles have the same absolute values as the right-handed helix, but opposite signs. In this conformation, however, the side chains are too close to the backbone; therefore this conformation is less stable than the right-handed α-helix. As a result, the α-L conformations do not form long helices characteristic of the secondary structure of proteins, but they do play singular roles in effecting chain reversal (Eswar and Ramakrishnan, 1999).

Polyproline forms regular conformations known as poly(Pro) I and poly(Pro) II. Poly(Pro) I contains all *cis* peptide bonds whereas poly(Pro) II has all *trans* (Fig. 4.10). However, in both these conformations the residues in the chain do not interact with each other. Therefore these conformations are highly unstable in aqueous solution. They become stable if the polypeptide chain associates with similar partners, forming the collagen triple coiled-coil (see Chapter 6).

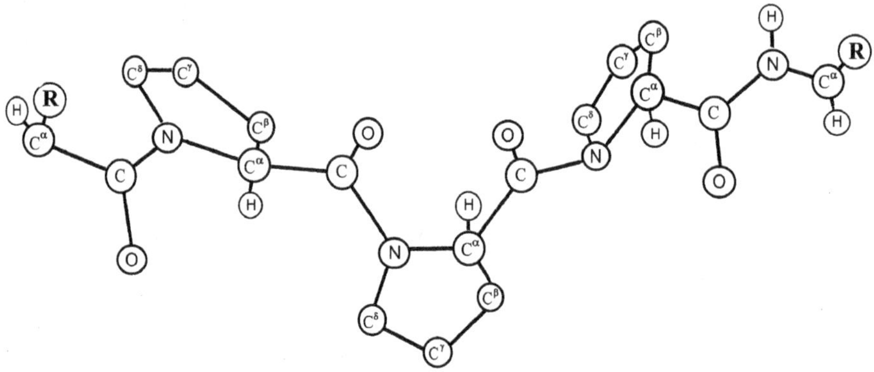

Figure 4.10. The poly(Pro) II helix.

All of the regular molecular constructions considered here are widely used by living organisms. However, in the isolated form only a right-handed α-helix is studied thermodynamically.

4.3. HIERARCHY IN PROTEIN STRUCTURE

4.3.1. Tertiary Structure of Proteins

Protein is synthesized as a linear polymer, than it folds in some specific, "native," conformation. After pioneering work of Anson (1945) followed by experiments of Eisenberg and Schwert (1951) it became clear that folding of a polypeptide chain into its native conformation is a thermodynamically driven process that is determined by the sequence of amino acids in the chain (Anfinsen, 1956). Determination of the very first protein structure, the structure of sperm-whale myoglobin, had a shocking effect even on its author, John Kendrew (Kendrew, 1958) (Fig. 4.11). As he noted: "Perhaps the most remarkable features of the molecule are the complexity and its lack of symmetry. The arrangement seems to be almost totally lacking in the kind of regularities which one instinctively anticipates, and it is more complicated than has been predicted by any theory of protein structure."

Since then the three-dimensional structures of hundreds of proteins have been determined with atomic resolution (Fig. 4.12), but the initial surprise remained: how could such complicated, unique conformations be formed? What are the forces driving a polypeptide chain into a compact unique conformation that withstands the dissipative action of thermal motion?

The mechanism of transformation of the linear information included in the primary structure of a polypeptide chain into a three-dimensional structure was and still is a great problem in protein science. In Nature this transformation occurs very fast and with minimal errors. However, all attempts to find the algorithm of its

Figure 4.11. The first crystallographic structure of protein, myoglobin, determined at 5-Å resolution by John Kendrew in 1958.

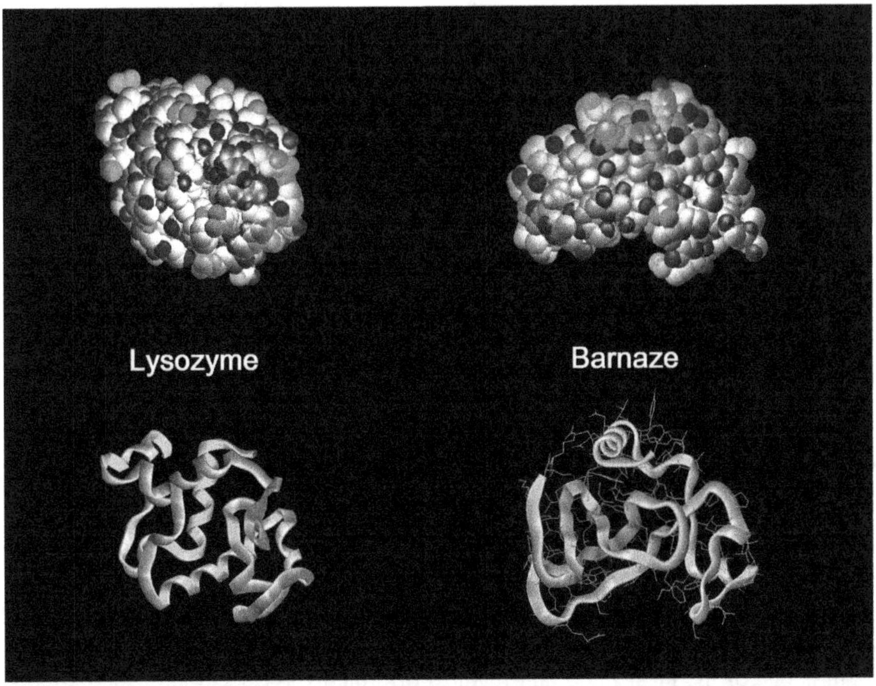

Figure 4.12. The atomic model and the backbone conformation of the typical globular proteins: hen egg white lysozyme and barnase. (See insert for color representation of the figure.)

realization, that is, the mechanism of folding, have been unsuccessful so far. Correspondingly, all attempts to predict the three-dimensional structure of a protein from its primary structure have been unsuccessful, as well as the attempts to synthesize artificial proteins with the needed properties. The main reason for this failure is the lack of experimental information on the interactions between the various groups of protein in aqueous media and their cooperation, which lead to swinging between various extremes.

First it was assumed that an essential role in maintaining the native protein structure is played by hydrogen bonds between the polar groups of the protein. But the protein polar groups could form hydrogen bonds not only between themselves but also with polar water molecules surrounding the protein in aqueous solution. Therefore, it was decided that hydrogen bonds could hardly be essential in the energetics of protein folding. After Kauzmann (1959) it became widely accepted that folding of a polypeptide into the compact native conformation is caused by hydrophobic force repelling the hydrophobic nonpolar groups of protein from water (see Section 2.4.2). This force was assumed to result from water ordering in the presence of nonpolar groups, that is, from the thermodynamically unfavorable entropy decrease of water in the presence of these groups. It seemed most interesting that the entropy increase upon removing the nonpolar groups of protein from water might compensate the loss of conformational entropy of the polypeptide chain upon its folding, so that protein folding should not cost much in entropy. However, if the main cause of protein folding is the entropic hydrophobic force, then why do proteins unfold upon heating? Unfolding of proteins upon heating might mean either that hydrophobic force is not a dominating force in protein folding or that this force is not entropic, as it was believed to be. Solution of this paradox required experimental estimation of the unfolding–refolding thermodynamics of proteins, that is, full-scale investigation of the physical grounds of protein structure (see Chapter 9).

4.3.2. Quaternary Structure of Proteins

According to the accepted nomenclature in chemistry, a molecule is understood to be a definite grouping of atoms held together by covalent bonds. Thus the polypeptide chain represents a molecule. However, not all proteins are formed by a single polypeptide chain. Many proteins are formed by several polypeptide chains (Fig. 4.13). Therefore, from the purist's point of view they represent molecular complexes. Nevertheless, proteins consisting of several polypeptide chains held together are usually regarded as a molecule, assuming that it is the complex as a whole that represents a single functional unit. The spatial organization of such complexes is classified as a **quaternary structure of proteins**. Recognition of their partners with the highest precision and formation of specific macromolecular complexes with them is one of the most important features of proteins. Association of the partners usually proceeds with certain changes and in many cases it is just these changes have functional significance.

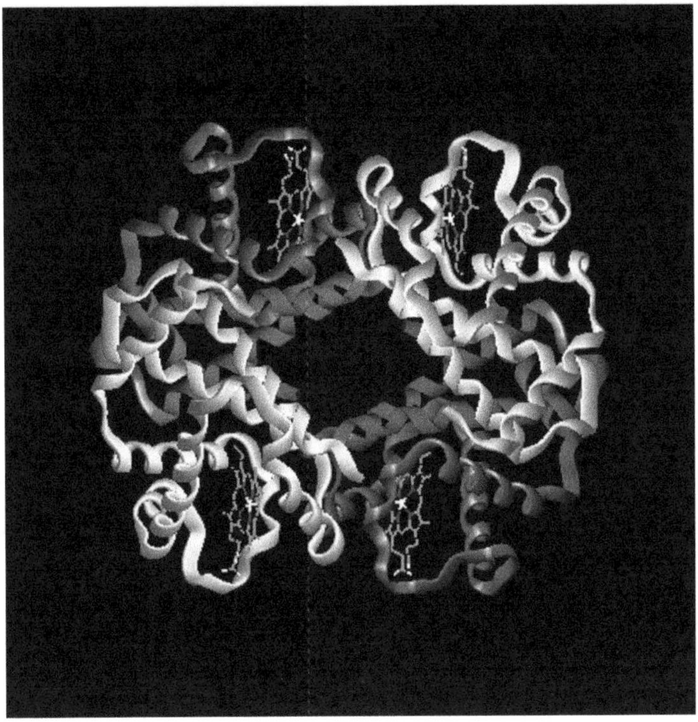

Figure 4.13. Hemoglobin represents a complex of four noncovalently held subunits. Note that these subunits also represent the complexes of protein with the low-molecular-weight heme group. (See insert for color representation of the figure.)

4.4. NUCLEIC ACIDS

4.4.1. Chemical Structure

Nucleic acids are linear polymers made up of monomers called **nucleotides**. Different nucleic acids can have from about 10^2 to over 10^8 nucleotides. As an average molecular weight of nucleotides amounts to 300, the molecular weight of a single polynucleotide chain might be from 3×10^4 to 3×10^{10}.

All nucleotides are constructed from three components: a nitrogen heterocyclic base, a pentose sugar, and a phosphate residue. The major bases are the monocyclic pyrimidines (cytosine, thymine, and uracil) and bicyclic purines (adenine and guanine) (Fig. 4.14).

In nucleosides, the purine or pyrimidine base is joined from a ring nitrogen to carbon-1 of sugar ribose or deoxyribose (Fig. 4.15). The first are used in constructing ribonucleic acids (RNA) the second in constructing deoxyribonucleic acids (DNA). Those nucleotides joined by the phosphates form the linear polymer, with the sequence of the attached bases forming its primary structure. The phosphate esters are strong acids and exist as anions at neutral pH. The bases are very weakly basic

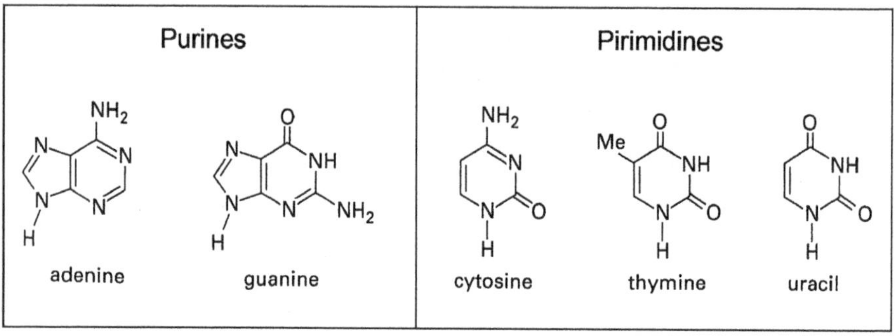

Figure 4.14. Structure of the major purine and pyrimidine bases of DNA.

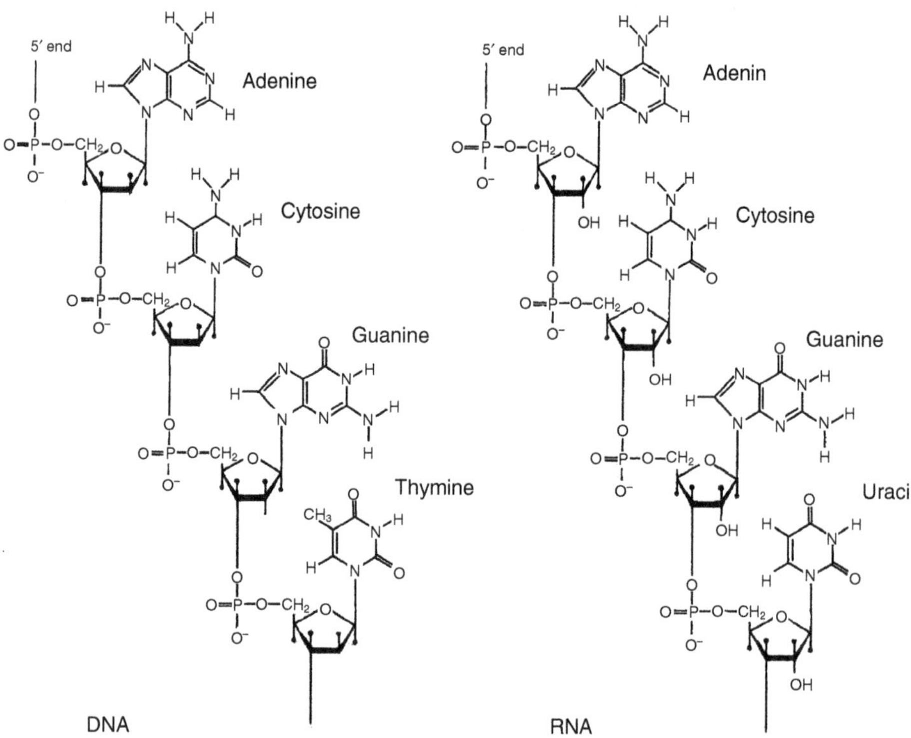

Figure 4.15. The chemical structure of deoxyribo and ribo polynucleotides. Two differences in their chemical groups distinguish RNA from DNA: (a) the sugar in RNA is ribose, while in DNA is deoxyribose, which lacks hydroxyl; (b) in RNA instead of thymine is used closely related pyrimidine uracil.

and adenosine, cytosine, and guanine become protonated only below pH 4. The amide NH groups in guanine, thymine, and uracil are deprotonated at pH values above 9.

4.4.2. Physical Structure

The most remarkable feature of the purine and pyrimidine nucleotides is that they are complementary and can form hydrogen-bonded base pairs with rather similar steric properties (Fig. 4.16). In the case of DNA these are the pairs adenine–thymine and guanine–cytosine. The first of these pairs is bound by two hydrogen bonds and the second pair is bound by three hydrogen bonds. This permits two polynucleotide strands that are complementary sequences to wind together into a double helix.

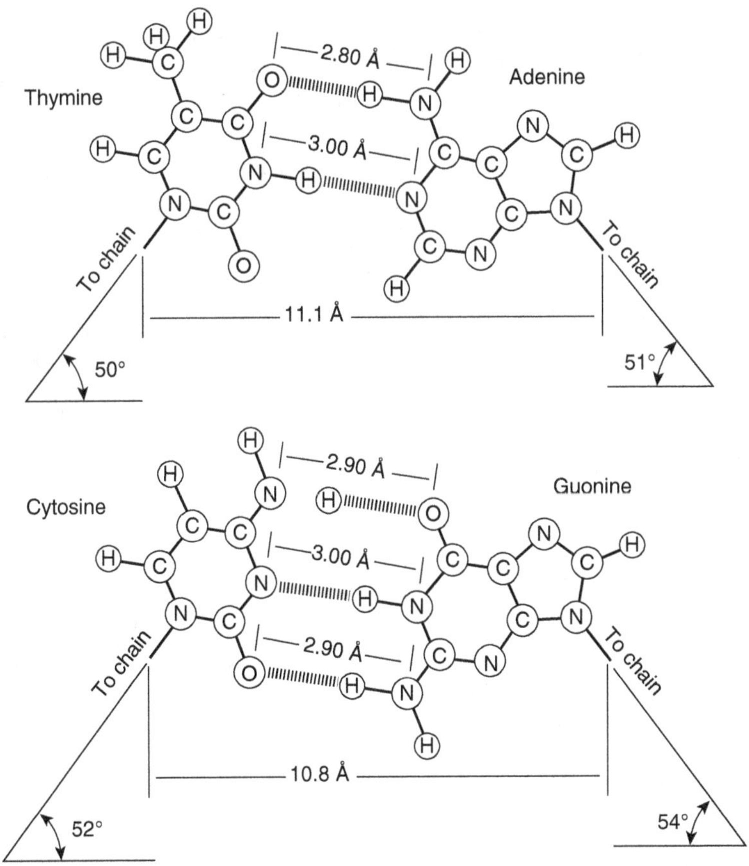

Figure 4.16. Hydrogen-bonded base pairs in DNA. Adenine always attaches to thymine by two hydrogen bonds, whereas guanine always bonds to cysteine by three hydrogen bonds. The obligatory pairing of the smaller pyrimidine with the larger purine allows the two sugar–phosphate backbones to have identical helical configurations. (Reproduced from Watson (1976).)

TABLE 4.4. Average Helix Parameters for the A and B DNA Conformations

Structure	Residues per Turn	Twister per Turn	Rise per bp	Base Tilt $\tau°$	Groove Width (Å)		Groove Depth (Å)	
					Minor	Major	Minor	Major
A-DNA	11	32.7	2.9	20	11.0	2.7	2.8	13.5
B-DNA	10	32.6	3.4	−6	5.7	11.7	7.5	8.8

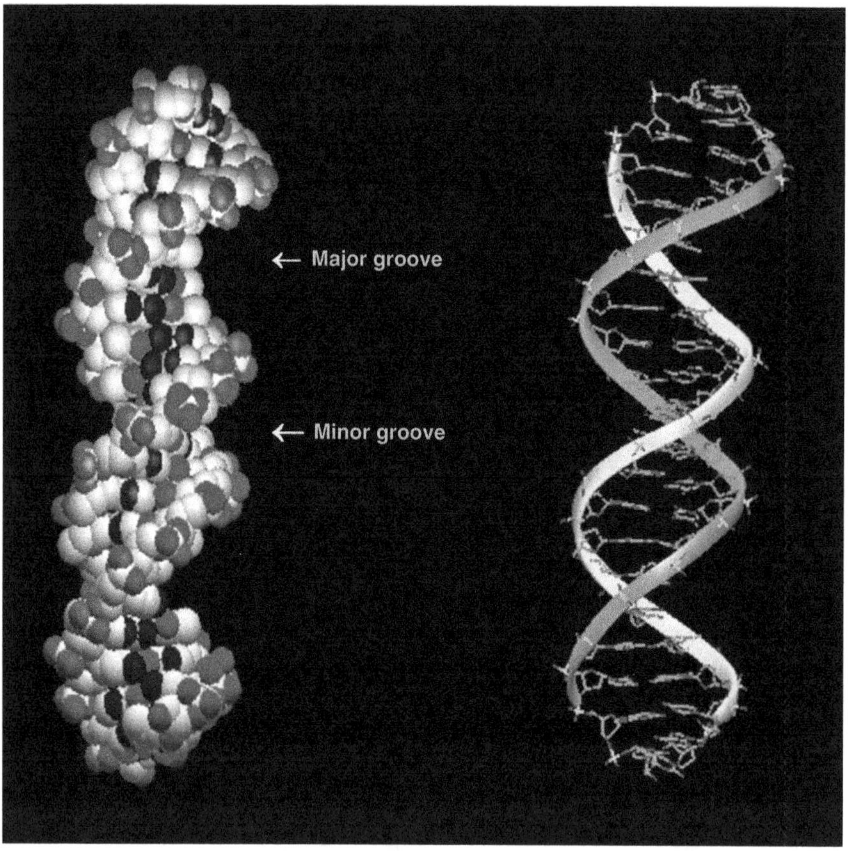

Figure 4.17. The B-DNA double helix. (See insert for color representation of the figure.)

The structure of double helix depends on the presence of water. At low humidity the DNA helix is in the A conformation, which is specified by 11 bases per turn, with the bases tilted 20° (which enhances stacking) and lying 4.4 Å away from the helix axis (Table 4.4). This gives a fairly stiff helix. The major groove is deep and narrow; the minor groove is broad and shallow.

At relatively high humidity (95% relative humidity, RH) the helix is in B conformation (B-DNA), which has 10 bases per turn with little tilting of the bases. In this conformation the base pairs sit directly on the helix axis so that the major and minor grooves are of similar depth (Fig. 4.17). The wide major groove and narrow

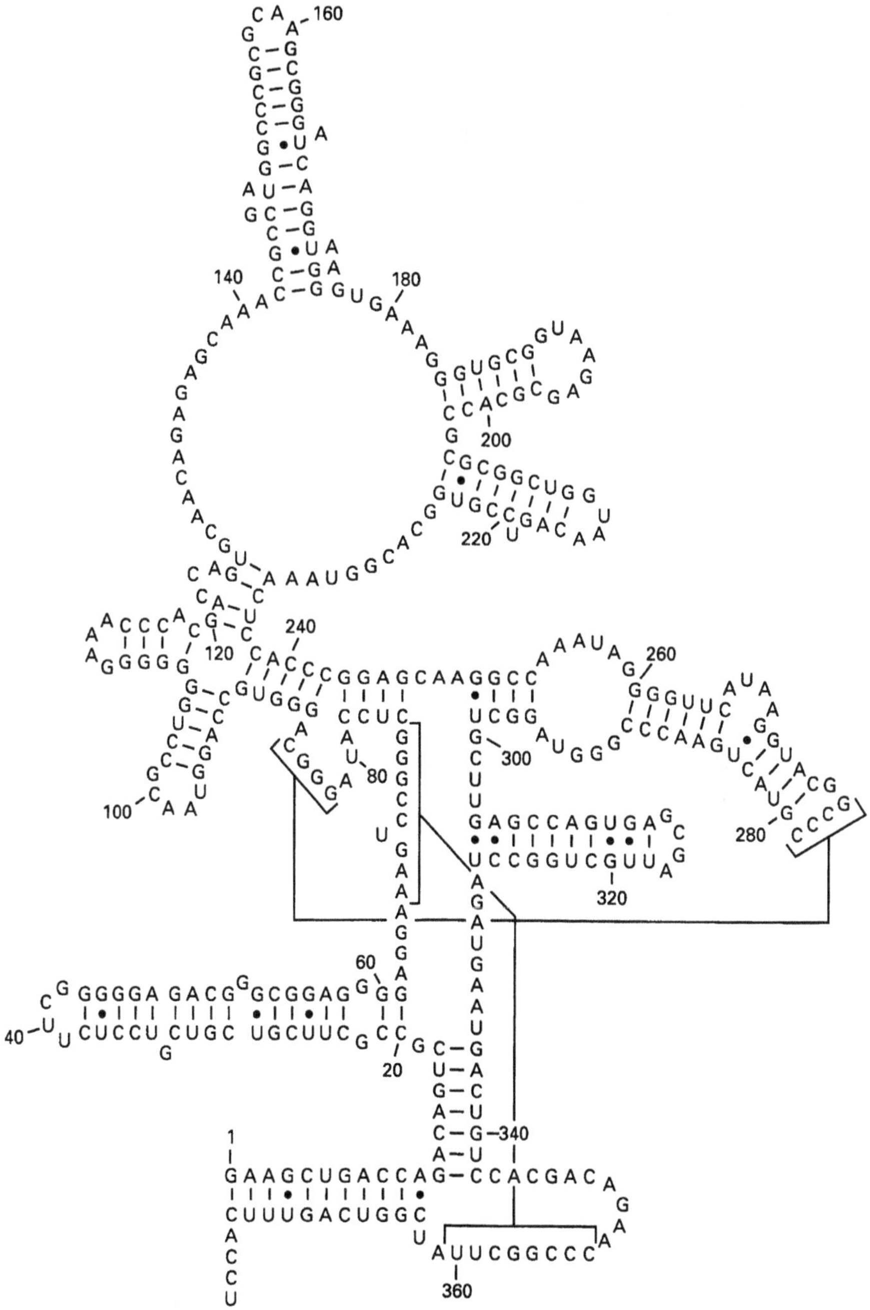

Figure 4.18. Secondary structure of the P RNA from *Escherichia coli*. Long-range base pairs are shown by extended lines. (Reproduced from Haas et al., 1994.)

minor groove are both of moderate depth and both grooves are well solvated by water molecules. The B-form is sufficiently flexible.

The double helix formed by RNA strands is rather similar to that formed by DNA strands. Therefore, the small difference in the chemical structure of these two polynucleotides (i.e., that DNA has deoxyribose instead of ribose and thymine instead of uracil) does not much affect the structural characteristics of these two double helices.

The main specificity of natural RNA that distinguishes it from DNA is that it works in living cells as a single strand. Therefore, the single-strand RNA molecules do not have the complementary strands with which they could form a double helix; however, in each strand there are short complementary sequences (not necessarily nearby along the chain) that could associate, thus winding the strand into a rather complicated, unique construction. A long RNA molecule usually has many complementary regions. Figure 4.18 presents an example of a secondary structure in RNA. One would expect that the unfolding–refolding of such a construction would proceed in several discrete steps.

REFERENCES

Abegian R. and Tortov M. (1994). *J. Mol. Biol.*, **235**, 983–1002.

Anfinsen C.B. (1956). *J. Biol. Chem.*, **221**, 405–412.

Anson M.L. (1945). *Adv. Protein Chem.*, **2**, 361–386.

Chargaff E. (1951). *Exp. Cell Res.*, **2**, 201–240.

Creighton T.E. (1993). *Proteins. Structure and Molecular Properties*. Freeman, New York.

Eisenberg M.A. and Schwert G.W. (1951). *J. Gen. Physiol.*, **34**, 583–606.

Eswar N. and Ramakrishnan C. (1999). In: *Perspectives in Structural Biology* (M. Vijayan, N. Yathindra, and A.S. Kolaskar, Eds.). Universities Press, Hyderabad, India.

Haas E.S., Brown J.W., Pitulle C., and Pace N.R. (1994). *Proc. Natl. Acad. Sci. USA*, **91**, 2527–2531.

Kauzmann W. (1959). *Adv. Protein Chem.*, **14**, 1–63.

Kendrew J.C. (1958). *Nature*, **182**, 764–767.

Koehl P. and Delarue M. (1994). *Proteins*, **20**, 264–278.

Pace C.N., Grimsley G.R., Thomson J.A., and Barnett B.J. (1988). *J. Biol. Chem.*, **263**, 1820–1825.

Pauling L. and Corey R.B. (1953). *Nature*, **171**, 59–61.

Pickett S.D. and Sternberg M.J.E. (1993). *J. Mol. Biol.*, **231**, 825–839.

Ramachandran G.N. (1963). *Science*, **141**, 288–291.

Ramachandran G.N. (1974). *Biochim. Biophys. Acta*, **359**, 298–302.

Ramachandran G.N. and Sasisekharan V. (1968). *Adv. Protein Chem.*, **23**, 283–438.

Watson J.D. (1976). *Molecular Biology of the Gene*. W.A. Benjamin, London, Amsterdam.

Watson J.D. and Crick F.H.C. (1953). *Nature*, **171**, 964–967.

5

THE α-HELIX AND α-HELICAL COILED-COIL

5.1. THE α-HELIX

5.1.1. Calorimetric Studies of α-Helix Unfolding–Refolding

The α-helix is the simplest regular structure formed by polypeptides without stresses and stabilized, as believed after Pauling et al. (1951), just by the hydrogen bonds between the –CO of the *i*th peptide unit and the –NH group of the $(i + 4)$th peptide units (Fig. 4.8). Thus, it seemed that by analyzing the folding and unfolding of the α-helix one could determine the thermodynamic characteristics of hydrogen bonding in aqueous solution. However, the very first experiments showed that investigation of the folding–unfolding of the isolated α-helix is far from being so simple: the single α-helix is marginally stable and its folding–unfolding proceeds over a very broad temperature range: invariably it starts well below 0°C and ends well above 100°C. Therefore, unfolding of the α-helix does not appear to be a highly cooperative process, but is likely to propagate stepwise from both of its ends. Since it cannot be considered as a two-state transition, the temperature-induced unfolding of an α-helix cannot be analyzed using the van't Hoff equation to determine the enthalpy

Microcalorimetry of Macromolecules: The Physical Basis of Biological Structures, First Edition.
Peter L. Privalov.

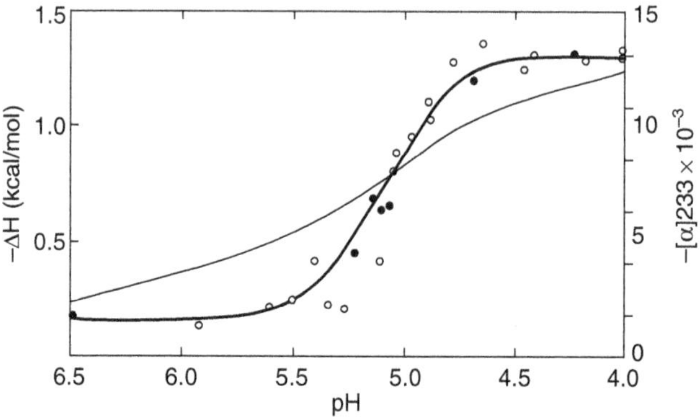

Figure 5.1. Calorimetric titration of poly-L-glutamic acid: open circle, molar enthalpy; closed circle, specific rotation; dashed line, degree of ionization (Rialdi and Hermans, 1966).

of this process. Various theoretical models have been proposed to describe unfolding of an α-helix (Zimm and Bragg, 1959; Lifson and Roig, 1961), but these models are all based on assumptions that, in themselves, require justification by comparing the calculated thermodynamic characteristics, particularly the unfolding enthalpy, with the directly measured ones. Therefore, direct calorimetric measurement of the enthalpy of unfolding of the isolated α-helix has acquired a special importance.

The first calorimetric estimates of the enthalpy of α-helix folding were done on ionizable polypeptides that form random coils when fully ionized, fold into an α-helix under pH conditions of partial ionization, and aggregate when fully uncharged. Using an isothermal reaction calorimeter, Rialdi and Hermans (1966) measured the heat effect upon acidification of the poly-L-glutamate solution (Fig. 5.1) and concluded that the enthalpy of helix formation is −(4.5 ± 0.8) kJ per mole of amino acid residue. A similar value, −5.0 kJ/mol, was obtained by Chou and Scheraga (1971) for poly-L-lysine in calorimetric study of helix formation by this polypeptide under alkaline conditions.

The temperature-induced unfolding of the α-helix was studied calorimetrically first by Scholtz et al. (1991) using an alanine-rich 50-residue helical polypeptide. According to circular dichroism (CD) studies at 220 nm, the helicity of this peptide decreases from the very beginning of heating (0°C) and ends above 80°C (Fig. 5.2). Calorimetric study of the heat capacity of this polypeptide using a MicroCal MC-2 instrument, gave only the fragment of this process in the temperature range from 10°C to 75°C (Fig. 5.3). Therefore, without the pre- and post-transition parts of the heat capacity function, it was impossible to determine the excess heat effect of unfolding. To analyze this calorimetric data it was assumed that the heat capacity of the polypeptide does not change at transition, $\Delta C_p = 0$, and the baseline was estimated by means of a least-squares minimization procedure systematically lowering the assumed baseline, allowing ΔH^{cal} and ΔH^{vH} to float. This led to the transition function that starts from about −40°C and reaches completion at about 150°C. According to such estimates, melting of the considered α-helix is specified

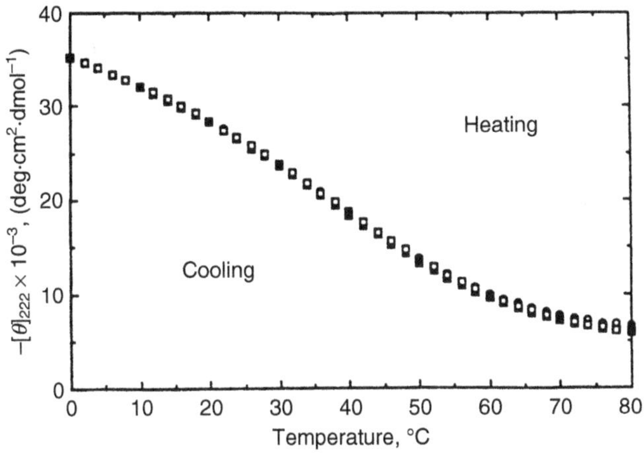

Figure 5.2. Change of ellipticity upon heating and subsequent cooling of the α-helix formed by the polypeptide Ac-Y(AEAAKA)$_8$F-NH$_2$ (Scholtz et al., 1991).

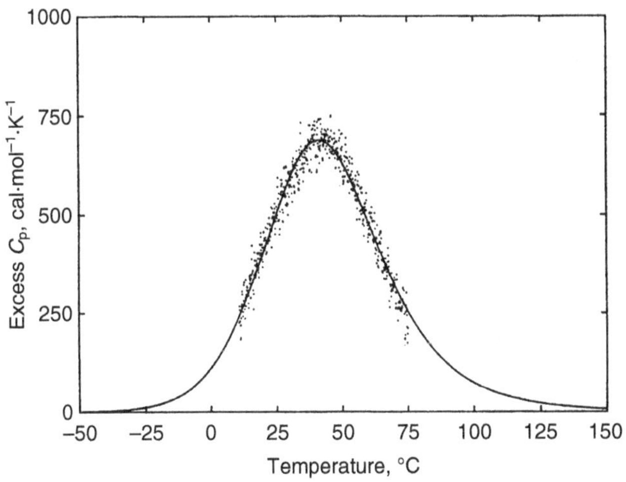

Figure 5.3. Calorimetrically measured excess heat capacity of the heated α-helix formed by the polypeptide Ac-Y(AEAAKA)$_8$F-NH$_2$ (Scholtz et al., 1991).

by $\Delta H^{cal} = (4.5 \pm 1.0)$ kJ/(mol-res) and $\Delta H^{vH} = (1.0 \pm 0.1)$ kJ/(mol-res). Therefore, the temperature-induced unfolding of this helix does not appear to be a two-state transition.

The result obtained seemed to be in agreement with the theory of the α-helix ↔ random coil transition based on a statistical mechanical model for helix formation. For α-helix formation by an infinite-chain polypeptide, the $\Delta H^{cal}/\Delta H^{vH}$ ratio was expected to be $\sigma^{-1/2}$ (Applequist, 1963), where σ is the nucleation constant in the range 10^{-3}–10^{-4} (Ptitsyn, 1972). The $\sigma^{-1/2}$ is thought of as the length of the cooperative unit. For short peptides, $\Delta H^{cal}/\Delta H^{vH}$ is a function of a chain length. However, the accuracy of estimation of ΔH^{cal} and ΔH^{vH} in this first calorimetric experiment was insufficient for thermodynamic specification of the α-helix unfolding–refolding.

Another attempt to measure calorimetrically the enthalpy of melting of the α-helical conformation of a variety of oligopeptides by Richardson and Makhatadze (2004) has faced the same complication: temperature-induced unfolding of the studied helices were not cooperative and therefore too broad, starting much below 0°C and ending above 80°C. Therefore, only the upper part of the melting profile was recorded calorimetrically. Moreover, the helix-forming abilities of the very different peptides being studied were not checked but can hardly be identical. Therefore, the conclusion drawn from these experiments, that the enthalpy of α-helix unfolding decreases with increasing transition temperature, that is, that the heat capacity increment of α-helix unfolding is negative, was not convincing.

Increasing the accuracy of the enthalpy determination required expanding the operational temperature range of the DSC and shortening the transition temperature range of the α-helix, that is, increasing the cooperativity of α-helix unfolding upon heating. The first was achieved by using DSC, which not only scans up the temperature scale, but also down, even below 0°C, measuring the heat capacity in the supercooled aqueous solution (see Section 3.2.2). The second was achieved by covalently closing the terminal loops of the short, 29-residue α-helix formed by the basic segment of GCN4 bZIP (Fig. 5.4). In this polypeptide, designated as GCN-4brNC, the side chains of the first amino acid residue (Lys) and the fifth residue (Asp) at the N-terminal end, as well as those of the 25th residue (Lys) and the 29th residue (Asp) at the C-terminal end, were covalently connected to form closed loops (Taylor et al., 1999). The four amino acid residues enclosed in the cross-linked loops at each end of the polypeptide should make a minimal entropic contribution to the helix-unfolding in this peptide. This should significantly stabilize the ends of the helix. One could expect, therefore, that the temperature-induced unfolding of this short helix will not start from the ends, but will involve the whole helix, that is, will proceed cooperatively.

The CD spectrum of this peptide in 0.1 M NaCl, phosphate buffered solution, pH 7.4, was found to be concentration independent, indicating that the peptide is a monomer, and exhibited the ellipticity minima at 208 and 222 nm that are characteristic of the α-helical conformation (Fig. 5.5): at 0°C, the mean residue ellipticity at 222 nm, $[\theta]_{222}$, was 37,000 deg·cm^2/dmol, which is close to the value of 36,000 deg·cm^2/dmol estimated for 100% helical structure in a 29-residue peptide using the length-dependent formalism $[\theta]_{222} = 39{,}500(1 - [2.57/n])$ deg·cm^2/dmol, where n is the peptide chain length (Chen et al., 1974). At 95°C the $[\theta]_{222}$ decreases in magnetite to 5,700 deg·cm^2/dmol, which is slightly more negative than the value expected for a heat-denatured peptide at this temperature, 4000 deg·cm^2/dmol (Privalov et al., 1989).

In DSC experiments the solutions were heated from 0°C to 100°C and cooled from 100°C to −10°C at rates 1 K/min and 2 K/min (Fig. 5.6). In this temperature range unfolding–refolding of this peptide is fully reversible. The dashed line in this

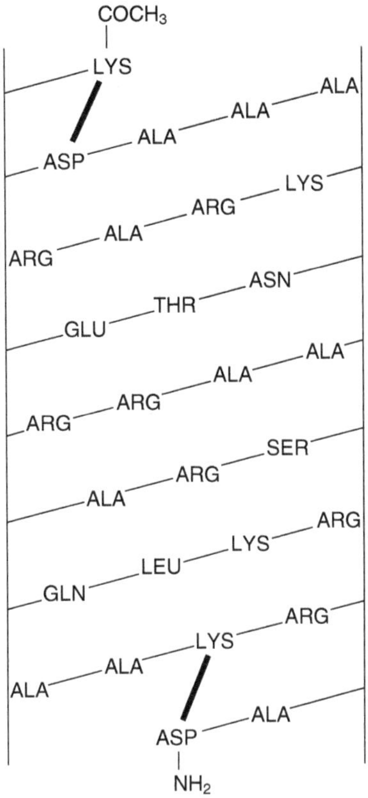

Figure 5.4. Calorimetrically studied peptide GCN4brNC, with covalently closed terminal loops. The 29-residue peptide is acetylated at its N-terminus and amidated at its C-terminus. The lactam bridges covalently linking the side chains of Lys^1 to Asp^5 and Lys^{25} to Asp^{29} are indicated by thick diagonal bars (Taylor et al., 1999).

figure shows the heat capacity function expected for the fully unfolded GCN4brNC polypeptide. This was calculated by summing up the partial heat capacities of the individual amino acid residues in accordance with the sequence of the peptide being studied (see Section 7.2.4). It is remarkable that, at temperatures above 90°C, the calculated heat capacity for the fully unfolded GCN4brNC is in excellent correspondence with the DSC-measured heat capacity of this peptide. Therefore, one can conclude that the considered polypeptide is fully unfolded at temperatures above 90°C, as the CD experiment also showed.

5.1.2. Analysis of the Heat Capacity Function

If the heat capacity increment upon unfolding of the helix is small and can be neglected, then the calculated heat capacity function for the unfolded helix (dashed line in Fig. 5.6) can be taken as a baseline and the heat capacity above this line can

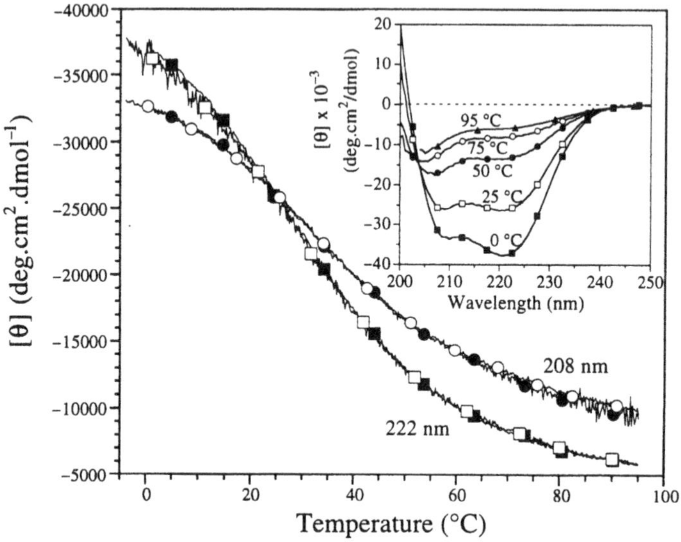

Figure 5.5. Temperature-dependence of the ellipticities at 208 nm (circles) and 222 nm (squares) at peptide concentrations of 200 μM (open symbols) and 10 μM (filled symbols). The insert shows CD spectra of a GCN4brNC solution (70 μM) measured between 200 and 250 nm at different temperatures (Taylor et al., 1999).

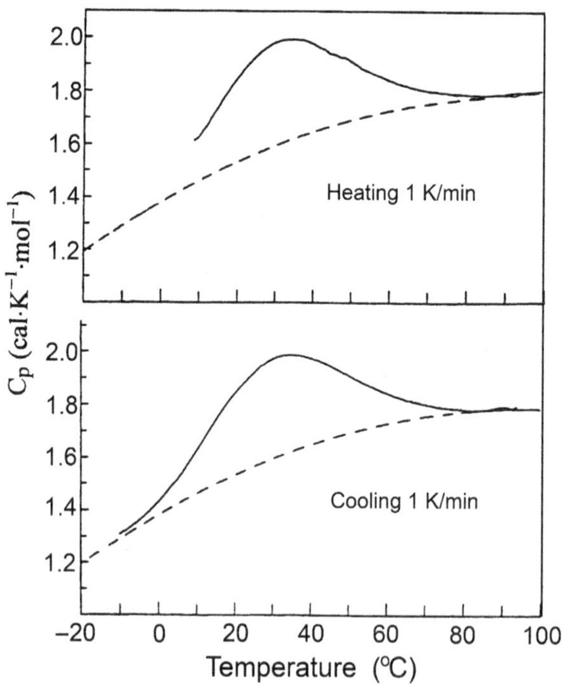

Figure 5.6. Partial heat capacity functions of the GCN4brNC determined by DSC in heating (top) and cooling (bottom) experiments at peptide concentration 3.8 mg/mL = 1.17 mM in 0.1 M NaCl solution, pH 7.4. The dashed line is the calculated heat capacity for the unfolded peptide (Taylor et al., 1999).

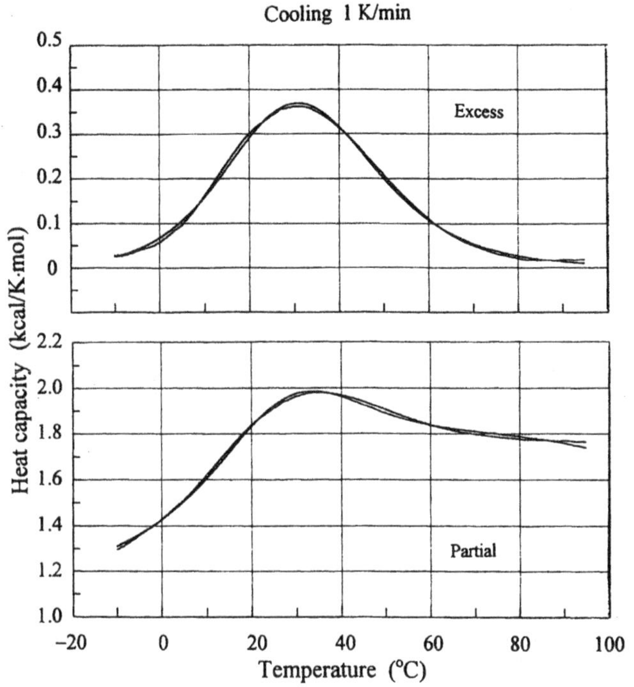

Figure 5.7. Simulation of the excess and partial molar heat capacity functions of GCN4brNC by a two-state transition. The excess heat capacity function is analyzed assuming that the cooperative transition proceeds with no heat capacity increment. The partial molar heat capacity function is analyzed assuming that the cooperative transition proceeds with a heat capacity increment that does not depend on temperature. In both cases, the experimental heat capacity function was determined in cooling experiments, which permitted its determination over a broader temperature range down to −10°C by suppercooling the aqueous solution (Taylor et al., 1999).

be considered as the excess heat capacity function. The example of the excess functions determined from the cooling experiment is given in Figure 5.7 (top). It appeared that these excess heat absorption profiles are rather well approximated by the function for a two-state transition:

$$C_p(T)^{\text{exc}} = \Delta H \frac{\partial P}{\partial T} \tag{5.1}$$

where P is the population of the unfolded molecules:

$$P(T) = \exp(-\Delta G/RT)/[1+\exp(-\Delta G/RT)] \tag{5.2}$$

The fitting parameters found for the heating and cooling experiments at two different rates and different concentrations of the peptide are given in Table 5.1. As can be

TABLE 5.1. Thermodynamic Parameters of GCN44brNC α-Helix Unfolding[a]

Condition, Heating Rate, Concentration	$\Delta C_p = 0$			$\Delta C_p \neq 0$			
	T_t	ΔH_t	χ^2	T_t	ΔH_t	ΔC_p	χ^2
Heating 1 K/min 1.17 mM	31.6	70.6	1.09	27.6	63.1	481	0.23
Cooling 1 K/min 1.17 mM	32.5	70.6	0.39	25.5	69.3	418	0.51
Heating 2 K/min 1.17 mM	33.5	70.6	0.81	27.7	69.0	510	0.63
Cooling 2 K/min 1.17 mM	32.5	67.0	0.51	26.1	64.8	543	0.47
Mean	32.3 ± 0.6	69.7 ± 1.3.0	0.70	28.0 ± 1.6	66.5 ± 1.9	460 ± 80	0.46

[a]T_t in °C; ΔH_t in kJ/mol; ΔC_p in J/K·mol.

seen, they are all in good correspondence, showing that the observed process is close to equilibrium and does not depend on concentration; that is, it represents a fast monomolecular process.

If unfolding of the helix proceeds as a two-state transition and the heat capacity increment of unfolding is not zero, the partial molar heat capacity function should be approximated by the following equation:

$$C_p(T) = C_p(T)^{\text{folded}} + \frac{\partial \Delta HP}{\partial T} = C_p(T)^{\text{folded}} + \Delta H \frac{\partial P}{\partial T} + P\,\Delta C_p \tag{5.3}$$

The example of simulation of the molar partial heat capacity functions according to Equation (5.3), assuming that the heat capacity increment of unfolding does not depend on temperature, is presented in Figure 5.7 (bottom). As Table 5.1 shows, in this case the fit is significantly better than in the case when the heat capacity increment was not considered, which means that the second model is more reliable. According to this fitting procedure, unfolding of the helix proceeds with a heat capacity increment ΔC_p = (110 ± 20) cal/K·mol = (460 ± 80) J/K·mol = (0.14 ± 3) J/K g. This is a rather small effect: it is 5 times smaller than the increment specific for globular proteins (Section 7.2.4). Because of that, one cannot see it at the low temperatures in Figure 5.6, where the heat capacity function expected for the fully unfolded peptide and the measured heat capacity of studied helix are compared. This fitting procedure suggests the following transition characteristics:

$$T_t = (28.0 \pm 1.6)\text{°C}, \quad \Delta H_t = (66.5 \pm 1.9)\ \text{kJ/mol}$$
$$\text{and} \quad \Delta C_p = (460 \pm 80)\ \text{J/K·mol}$$

Perhaps the most surprising conclusion of this experiment is that **unfolding of the α-helix is associated with positive heat capacity increment**. Although this heat capacity increment is rather small, it is highly meaningful since **it shows that unfolding of α-helix results in exposure of nonpolar groups** and thus the enthalpy of unfolding depends on temperature.

The observed transition is highly cooperative; it actually represents a two-state transition. Therefore, using the mean values listed in Table 5.1, one finds by Equations (2.14) and (2.21) the temperature functions of enthalpy, $\Delta H(T)$, the entropy factor, $T\Delta S(T)$, and thus the Gibbs energy, $\Delta G(T) = \Delta H(T) - T\Delta S(T)$, of unfolding for the helix being considered. It appears that the stability of the considered helix at −10°C is about $\Delta G(-10°\text{C}) = 2.0$ kcal/mol = 8 kJ/mol. With this level of stability, the population of unfolded helices at that temperature should be less than 5%. At 30°C half of the molecules are unfolded, while at 90°C the population of the unfolded molecules should be about 99%. These values are in good correspondence with the results of CD determination of the state of the considered polypeptide.

To determine the enthalpy and entropy change on a per-residue basis, one needs to consider that a peptide backbone chain length equivalent to eight residues is still partly constrained by the cyclic covalent bonding of the loops in the fully unfolded peptide. Considering the closed loop as equivalent entropically to one amino acid residue, it appears that only 23 residues are contributing to the helix unfolding. On the other hand, the two side-chain bridges linking Lys and Asp pairs have a well-defined conformation on the helix surface and will gain significant conformational freedom upon unfolding, although they are still constrained by the cyclic structure. Therefore, conversion of the entropy and enthalpy per-residue basis requires division by a number that is probably greater than 23 and possibly lower than 27, that is, a value of (25 ± 2) seems reasonable. Then, using the calorimetric data (Table 5.1) one can calculate that at the transition temperature, $T_t = 28°\text{C} = 301$ K, the transition enthalpy is $\Delta H_t = 2.66$ kJ/(mol-res) and the transition entropy is $\Delta S_t = \Delta H_t/T_t = 8.84$ J/K·(mol-res). Their temperature dependence is shown in Figure 5.8.

The entropy found for helix unfolding includes the conformational entropy of helix unfolding and the hydration entropy of the groups exposed to water upon unfolding. Assuming that the conformational entropy of the backbone unfolding amounts to $\Delta S_{\text{bb}}^{\text{conf}} = R\ln(3\times3) = 18.4$ J/K·(mol-res) and the entropy of unfreezing the side chains, $\Delta S_{\text{SCh}}^{\text{conf}}$, is of the same value (see Section 4.2.2), the overall conformational entropy of the 25-residue helix unfolding would be $\Delta S^{\text{conf}}(28°\text{C}) = (18.4 + 18.4) \times 25$ J/K·mol = 920 J/K·mol, or somewhat less if the side chains in the helix are not completely frozen. Then one finds that the hydration entropy upon helix unfolding amounts on $\Delta S^{\text{hyd}}(28°\text{C}) = \Delta S^{\text{cal}} - \Delta S^{\text{conf}} = (221-920)$ J/K·mol = −699 J/K·mol. It is notable that this entropy is negative; that is, it opposes the helix unfolding. The helix-stabilizing effect of this hydration entropy factor is then $T_t\Delta S^{\text{hyd}}(28°\text{C}) = -301 \times 699$kJ/K·mol = −210 kJ/K·mol. This hydration effect, with conformational entropy factor $T_t\Delta S^{\text{conf}} = 301 \times 920$ J/K·mol = 277 kJ/K·mol, balances the enthalpy of unfolding at the transition temperature, $\Delta H^{\text{cal}}(28°\text{C}) = 66.5$ J/K·mol.

Analysis of the GCN4 α-helix structure shows that the change of water-accessible surface area of nonpolar groups ($\Delta\text{WASA}^{\text{np}}$) upon helix unfolding amounts to 600 Å²/mol. The surface normalized entropy of hydration of nonpolar groups at 20°C is

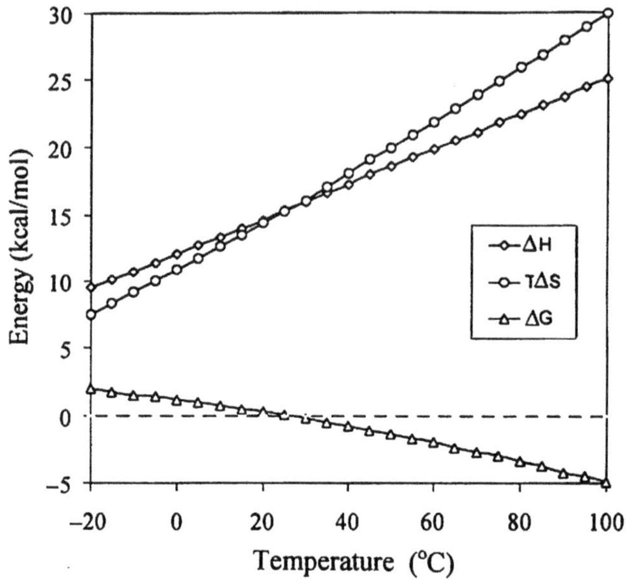

Figure 5.8. Temperature dependence of the enthalpy (ΔH), entropic factor ($-T\Delta S$), and Gibbs energy (ΔG) of unfolding of the considered α-helix, calculated with the assumption that the heat capacity increment of helix unfolding equals ΔC_p = 110 cal/K·mol = 464 J/K·mol, T_t = 28.0°C, ΔH_t = 16.4 kcal/mol = 68 kJ/mol, and ΔS_t = 54.0 cal/K·mol = 226 J/K·mol (Taylor et al., 1999).

about $\delta S^{\text{hyd}}(\text{np}) = -0.59$ J/K·Å^2 (Makhatadze and Privalov, 1995). It appears, then, that the contribution of the nonpolar groups to the dehydration entropy upon helix formation is about $\Delta S^{\text{dhyd}}(\text{np}) = 0.59 \times 600$ J/K·mol = 354 J/K·mol. Correspondingly, since the total entropy of dehydration per residue is about $S^{\text{dhyd}}(28°\text{C}) \approx$ 700 J/K·mol, it appears that the entropies of dehydration of polar and apolar groups upon formation of the α-helix are comparable in value, that is, $\Delta S^{\text{dhyd}}(\text{pol}) \approx \Delta S^{\text{dhyd}}(\text{apol}) \approx 350$ J/K·mol, or recalculating per residue: $\Delta S^{\text{dhyd}}(\text{pol, apol}) =$ 14 J/K·(mol-res).

Since the enthalpy of transfer of a nonpolar substance from the liquid phase into water is close to zero at 25°C (see Fig. 2.10), the enthalpy of formation of the α-helix, $\Delta H^{\text{cal}} = -2.7$ kJ/(mol-res), should be assigned to hydrogen bonding, together with the entropic factor of dehydration of polar groups. Then for the contribution of hydrogen bonding to the stabilization of the α-helix one gets

$$\begin{aligned}\Delta G^{\text{HB}}(T_t) &= \Delta H^{\text{cal}} - T_t\, \Delta S^{\text{dhyd}}(\text{pol}) = (-2700 - 301 \times 14)\ \text{kJ/mol} \\ &= -6.9\ \text{kJ/K·mol} = -1.65\ \text{kcal/mol}\end{aligned}$$

a value that is expected for hydrogen bonding between the polar groups in aqueous media.

It is remarkable that the main contributor in this helix-stabilizing effect is the entropy factor associated with the hydration effects, which amounts to $T\Delta S^{hyd} = 4.2$ *kJ/mol.*

As follows from this, the hydrogen bond is not the only interaction responsible for the formation and stabilization of the α-helix, as is usually believed; a significant role is played by the hydrophobic effect, in fact by the van der Waals interactions between the groups forming the apolar core of the α-helix (see Section 2.4.2).

The presence of this apolar core in the α-helix was surprisingly ignored, although it was known that the π-helix is unstable just because this wider helix does not have a tightly packed apolar core. Bearing in mind that at room temperature the enthalpy of formation of apolar contact in aqueous solution is close to zero, and using the previously estimated entropy of dehydration of apolar groups upon folding of an α-helix, we find that the contribution of this apolar core to helix stabilization is about $T\Delta S^{\text{hyd}}(\text{apol}) = 301 \times 14$ kJ/mol = 4.2 kJ/mol, that is, not much less than the hydrogen bonding.

As follows from the preceding discussion, the enthalpic contribution of hydrogen bonds to the stabilization of an α-helix in aqueous solution is positive at room temperature. With increasing temperature it should decrease and even change its sign due to the negative increment in heat capacity of hydration of polar groups (see Section 2.4.1). Therefore, at higher temperatures the enthalpy of hydrogen bonding converts from a factor stabilizing the folded conformation into a factor destabilizing this conformation. This, however, does not mean that the overall contribution of hydrogen bonding at high temperatures destabilizes the α-helix. It does not because disruption of the hydrogen bond results in negative entropy of hydration of polar groups, which increases with increasing temperature and contributes significantly to the stabilization of the helical conformation.

5.2. α-HELICAL COILED-COILS

5.2.1. Two-Stranded Coiled-Coils

The α-helices of certain compositions could associate, forming a two-stranded or a three-stranded superhelix, the coiled-coils (Fig. 5.9). The general principles of formation of the two-stranded α-helical coiled-coil were formulated by Crick (1952, 1953a,b) and independently by Pauling and Corey (1953). According to Crick, two strands in the α-helical conformation are interwined in a right-handed twist, forming a superhelix with 137-Å pitch, but this winding of the helices results in their distortion. This distortion was envisaged by Pauling and Corey as resulting from a repeating sequence in which the individual residues formed hydrogen bonds of slightly different lengths. The origin of the distortions was thus considered to be intrahelical, while Crick attributed this distortion to interhelical interactions.

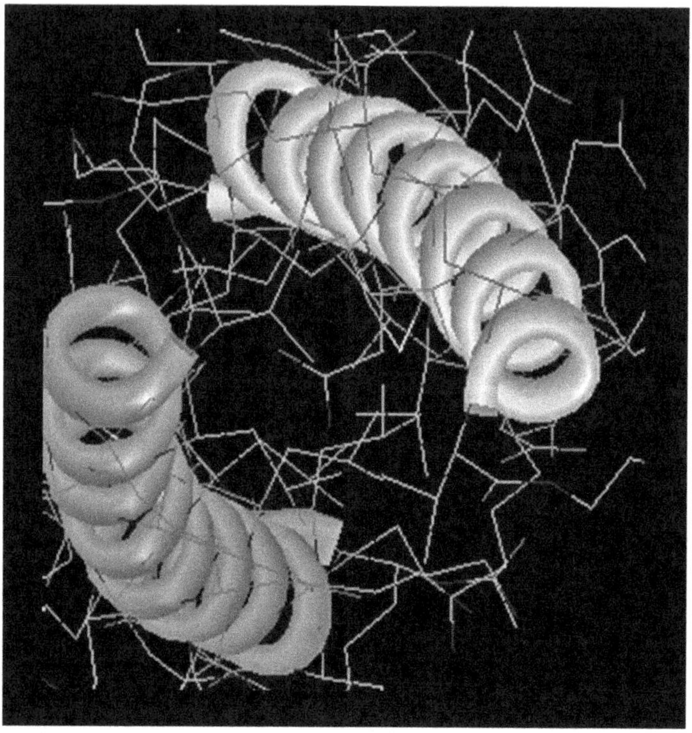

Figure 5.9. Two α-helices interwined into the α-helical coiled-coil. Fragment of the GCN-4 protein.

It was supposed that the two-stranded coiled-coil could be stabilized by the presence of nonpolar residues between the two strands in helical conformation with polar and ionic residues on the outside. This assumes a certain periodical arrangement of the amino acids in heptads along the polypeptide chain (Fig. 5.10).

Calorimetric studies of the temperature-induced unfolding of the synthetic regular two-stranded α-helical coiled-coil showed that it proceeds in a rather broad temperature range (Yu et al., 1996a,b), as is expected for a bimolecular process proceeding with dissociation of two strands (Fig. 5.11, bottom). This prevents precise determination of the enthalpy and the heat capacity effect of formation of the coiled-coil. To avoid this complication, most of the calorimetric studies of two-stranded coiled-coils were done with the S–S cross-linked strands. For this purpose cysteine residue was included in the synthesized peptide. The synthetic peptides usually included tyrosine or tryptophan also, to facilitate determination of their concentration in the solution being studied. An example of such peptides is

ac–YKCKSLESKVKSLESKAKSLESKVKSLESKVKSLESKVKSLES–am
   |
ac–YKCKSLESKVKSLESKAKSLESKVKSLESKVKSLESKVKSLES–am

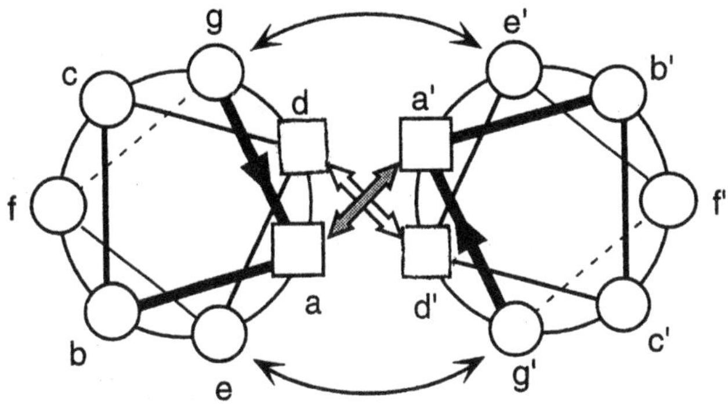

Figure 5.10. Helical-wheel representation of the heptide repeats in the two-stranded coiled-coil. Squares **a** and **d** represent the apolar residues providing hydrophobic contact between two strands in α-helical conformation; circles **e** and **g** are the oppositely charged residues providing the electrostatic attractive forces additionally stabilizing the duplex.

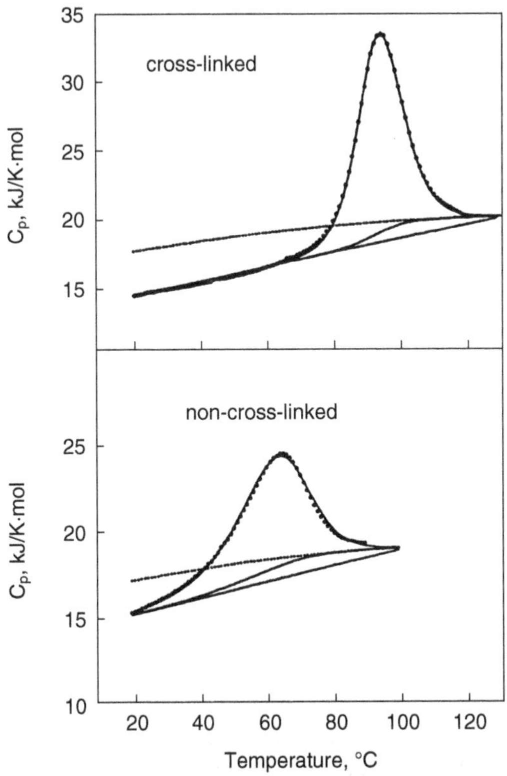

Figure 5.11. DSC-observed melting profiles of α-helical coiled-coils with S–S cross-linked strands and non-cross-linked strands of 40 residues each (Yu et al., 1999).

Figure 5.11 (top) shows the DSC observed melting profiles of two α-helical coiled-coils with and without S–S cross-linked strands. The main difference between the non-cross-linked and cross-linked coiled-coils is that melting of the cross-linked coiled-coil is a first-order monomolecular reaction, whereas unfolding of the non-cross-linked coiled-coil is a second-order bimolecular reaction. Correspondingly the first one does not show any dependence on the concentration of peptides, while the second one shows this dependence. Since unfolding of the cross-linked coiled-coil proceeds without dissociation of strands, that is, without translational entropy, it is much more stable than the non-cross-linked coiled-coil and thus unfolds upon heating at considerably higher temperatures than the non-cross-linked coiled-coil.

Comparing the results of unfolding of the cross-linked coiled-coils with that of an individual α-helix, which is discussed above, showed that the specific enthalpies of their unfolding (calculated per mole of residue) extrapolated to the room temperature (in assumption that the heat capacity increment of unfolding does not depend on temperature) are very similar but the slopes of these functions differ considerably. For the individual α-helix and the two-stranded coiled-coil the enthalpy and entropy functions are shown in Figure 5.12. It is remarkable that, while the specific enthalpy values for the individual α-helix and the two-stranded coiled-coil are identical at 25°C and deviate at higher temperature, their specific entropies differ significantly at 25°C but converge above 100°C. It is notable also that the enthalpy and entropy of the coiled-coil are steeper functions of temperature than those of the individual α-helix. This is because the heat capacity increment at unfolding of the coiled-coil is higher than that of the individual α-helix. Since the positive heat capacity effect is specific for transfer of nonpolar groups into water (see Section 2.4), one can conclude that unfolding of the coiled-coil results in more extensive exposure of nonpolar groups. This is just what is expected since the coiled-coil is held mostly by the contacts between nonpolar groups at the **a** and **d** positions (Fig. 5.10). On the other hand, unfolding enthalpies of an α-helix and a coiled-coil are similar at room temperature because at that temperature the enthalpy of transfer of nonpolar groups from the nonpolar phase into water is zero. This happens because at 25°C the enthalpy of disruption of van der Waals contact between apolar groups is equal in magnitude but opposite in sign to the enthalpy of hydration of these groups and these two effects cancel each other (see Section 2.4). Therefore, one can consider the enthalpy of unfolding at that temperature as the enthalpy of breaking the hydrogen bonds maintaining the helix, which in aqueous media is associated with the hydration of the released polar groups. Since the number of hydrogen bonds per residue is the same in the individual α-helix and the α-helical coiled-coil, the enthalpy of unfolding of the individual α-helix and the α-helical coiled-coil, calculated per residue base, are identical at room temperature.

In contrast to the enthalpy, the entropies of unfolding of the α-helix and the disulfide-cross-linked coiled-coil, which unfolds without dissociation of the strands, differ significantly at room temperature, but converge as the temperature increases. This is also the result of the hydration effect of the nonpolar groups, which is more extensive in the case of the coiled-coil than of the α-helix. As discussed in Section

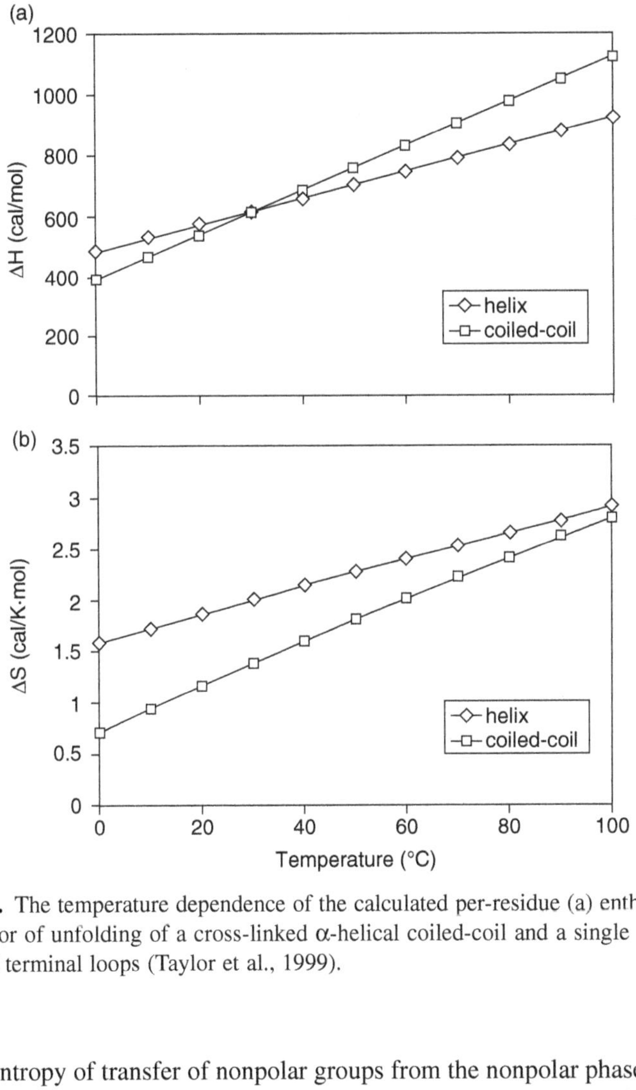

Figure 5.12. The temperature dependence of the calculated per-residue (a) enthalpy and (b) entropy factor of unfolding of a cross-linked α-helical coiled-coil and a single α-helix with cross-linked terminal loops (Taylor et al., 1999).

2.4.2, the entropy of transfer of nonpolar groups from the nonpolar phase into water is large and negative at room temperature, but it decreases with increasing temperature and is likely to vanish at about 120°C (Baldwin, 1986; Privalov and Gill, 1988; Makhatadze and Privalov, 1995). This is just what is observed: the entropies of unfolding of the isolated α-helix and the coiled-coil, which differ in the extent of their nonpolar contacts, differ at room temperature but this difference vanishes with the increase of temperature.

It is notable that, although the entropy of unfolding of α-helix increases with increasing temperature (Fig. 5.12b), even at 100°C it is significantly lower than 4 cal/K·mol = 16 J/K·mol, which is the value expected for the conformational entropy of the backbone chain (Schellman, 1955; Nemethy et al., 1966). The reason is evident: at temperatures above 100°C the entropic effect of hydration of nonpolar groups vanishes (see Section 2.4.2), but the negative entropic effect of hydration of polar groups remains. Therefore, the entropy of unfolding of the α-helix at this temperature is not the conformational entropy of the polypeptide chain only, but it also includes a large negative component of hydration of the polar groups that are exposed upon unfolding of the α-helix.

It should be noted that association of two α-helices into the coiled-coil is mainly driven by the hydrophobic forces between the strands. The role of electrostatic interactions between the strands is of second order: the coiled-coil is formed even without these interactions but they affect somehow the stability of the coiled-coil (Yu et al., 1996a,b). This effect of charges on the stability of the coiled-coil is mostly entropic and is associated with water ordering by the exposed charged groups.

5.2.2. Three-Stranded Coiled-Coils

Three-stranded coiled-coils are more abundant in nature and much less studied than the two-stranded coiled-coils. Perhaps the most studied example presents the three-stranded α-helical coiled-coil, Lpp-56, which is a fragment of a lipoprotein from the outer membrane of *Escherichia coli* (Braun, 1975). Crystallographic study of this protein (Liu et al., 2002; Shu et al., 2000) showed that its three helical strands form an apolar core inside the coiled-coil and the polar external surface (Fig. 5.13).

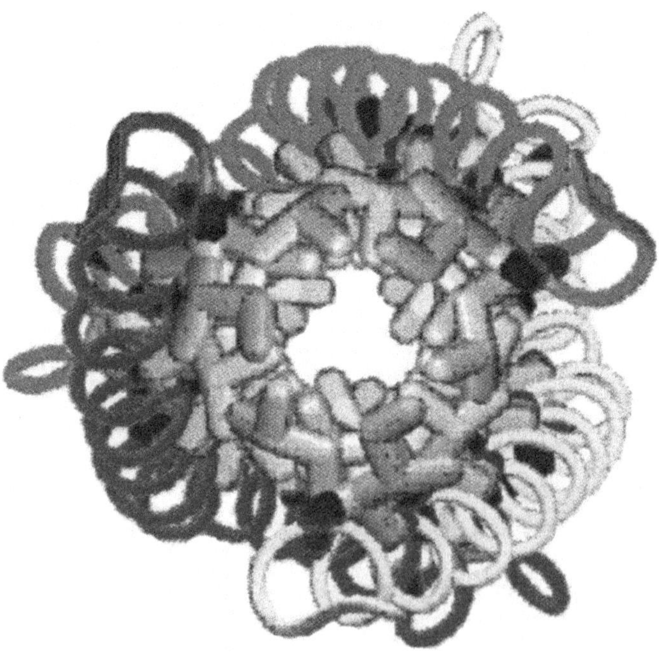

Figure 5.13. Axial slice of Lpp-56 showing formation of the apolar core. (See insert for color representation of the figure.)

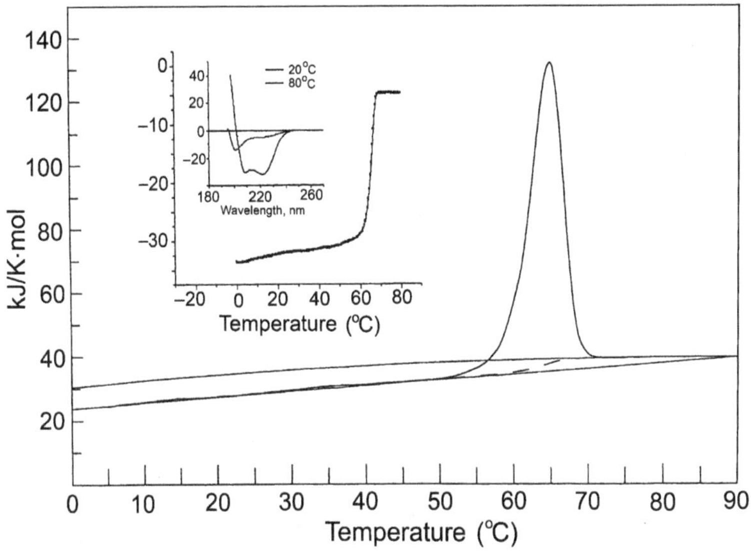

Figure 5.14. Temperature-induced unfolding of the three-stranded α-helical coiled-coil Lpp-56 observed by CD and DSC in 100 mM KCl solution, pH 7.4, and protein concentration 210 μM; DSC scans at 1 K/min heating rate (Dragan et al., 2004).

However, at neutral pH, the three helical strands are held together not only by the hydrophobic forces maintaining the apolar core but also by the salt links between the negatively and positively charged Asp and Lys residues, which become spatially close in the trimeric structure.

The calorimetrically measured temperature-induced unfolding of the three-stranded α-helical coiled-coil of Lpp-56 is presented in Figure 5.14. At neutral pH it is significantly more stable than the two-stranded α-helical coiled-coils and unfolds with a sharp heat absorption peak and noticeable heat capacity increment. Extrapolating the heat capacities of folded and unfolded states into the transition zone, one can determine the heat capacity increment upon protein unfolding, ΔC_p; from the heat absorption peak, one can estimate the calorimetric enthalpy of unfolding, ΔH^{cal} (Table 5.2).

The simplest assumption would be that the process of dissociation and unfolding of the three helical strands of Lpp-56 is a cooperative third-order reaction: $F_3 \leftrightarrow 3U$. One can determine the van't Hoff enthalpy of this process from its sharpness at the melting temperature T_m, from the height of the heat absorption peak, $C_p^{\max}$, and its area, ΔH^{cal}, assuming that $n = 3$ (see Eq. 3.26). The values obtained for ΔH^{vH} are listed in Table 5.2. Surprisingly, they are in striking conflict with the calorimetrically measured ΔH^{cal}: the van't Hoff enthalpy is almost twice the calorimetric enthalpy. Moreover, Table 5.2 shows that the unfolding temperature does not depend on concentration, as expected for a multimeric reaction. On the other hand, it was found that the melting profiles showed a substantial dependence on the heating rate and,

TABLE 5.2. Calorimetrically Measured Apparent Thermodynamic Characteristics of the Lpp56 Trimer and Its Unfolding at pH 7.4 and 3.0

pH	Heating rate (K/min)	Concentration (μM)	$\partial C_p/\partial T$ (kJ/K^2·mol)	ΔC_p (kJ/K·mol)	T_t (°C)	ΔH^{cal} (kJ/mol)	ΔH^{vH} (kJ/mol)
7.0	1.0	69.3	0.17	2.5	65.0	660	1150
	0.5	67.7	0.17	2.5	64.3	676	1080
	0.25	63.7	0.17	2.5	63.1	657	1073
	0.125	138	–	–	61.5	699	1090
3.0	1.0	360	0.25	2.2	43.0	350	430
σ			±0.01	±0.2	0.1±	±20	50±

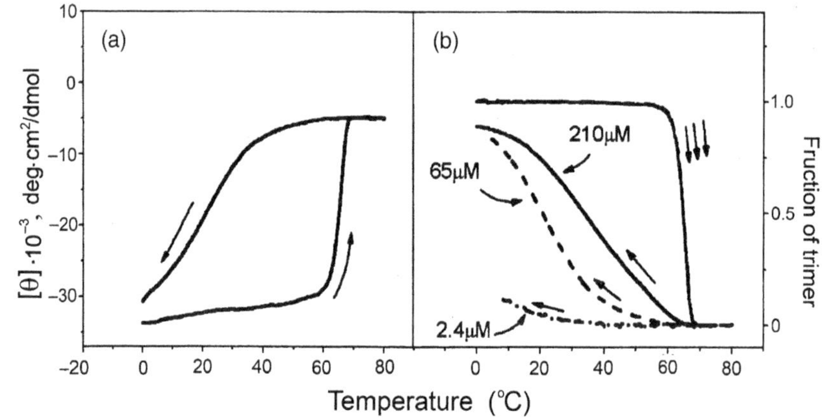

Figure 5.15. (a) Ellipticity profiles of Lpp-56 upon heating and cooling at 65 μM concentration in solution at pH 7.4. The inset shows the ellipticity spectrum of Lpp-56 at 20°C and 83°C. (b) Fraction of trimers in heating and cooling experiments with a rate 1 K/min at different protein concentration (Dragan et al., 2004).

especially, on the cooling rate. Also, the heat effects in the cooling experiments depend significantly on the concentration, in contrast to the heating experiment. This became especially clear using ellipticity as an index of the reaction progress, since the CD spectrometer permits variation of protein concentration over a significantly wider range than the DSC instrument (Fig. 5.15).

The significant difference between the profiles obtained on heating and cooling, that is, the large hysteresis, demonstrates that we do not have thermodynamic equilibrium. This conclusion was confirmed by studying kinetics of unfolding and refolding of Lpp-56 at different temperatures. The slow unfolding and refolding rates of Lpp-56 indicate that the activation barriers in these reactions are unusually high.

The calorimetric and optical experiments led to the model according to which unfolding and refolding of this trimeric complex proceeds without dimeric intermediate. During the transition from the activated state to the unfolded state, water penetrates between the separated helices hydrating the exposed apolar groups and, by decreasing the strength of intrahelical hydrogen bonds at the apolar face, initiates unfolding of the helices. In the folding reaction, the transition into the activated state consists of folding the separated strands into a helical conformation and the proper alignment of the three helices at a distance, which excludes water, but is insufficient to form van der Waals contacts. The probability of proper alignment of three strands in an α-helical conformation is very low, but as soon as it is achieved, all three strands cooperatively collapse into a three-stranded coiled coil (Dragan et al., 2004).

5.3. α-HELICAL COILED-COIL PROTEINS

5.3.1. Muscle Proteins

The tropomyosin, paramyosin, and the myosin rod present the example of natural two-stranded α-helical coiled-coils (Fig. 5.16). The molecular mass of tropomyosin is 65,500 (Stone et al., 1975), of paramyosin is 210,000, and of the myosin rod is

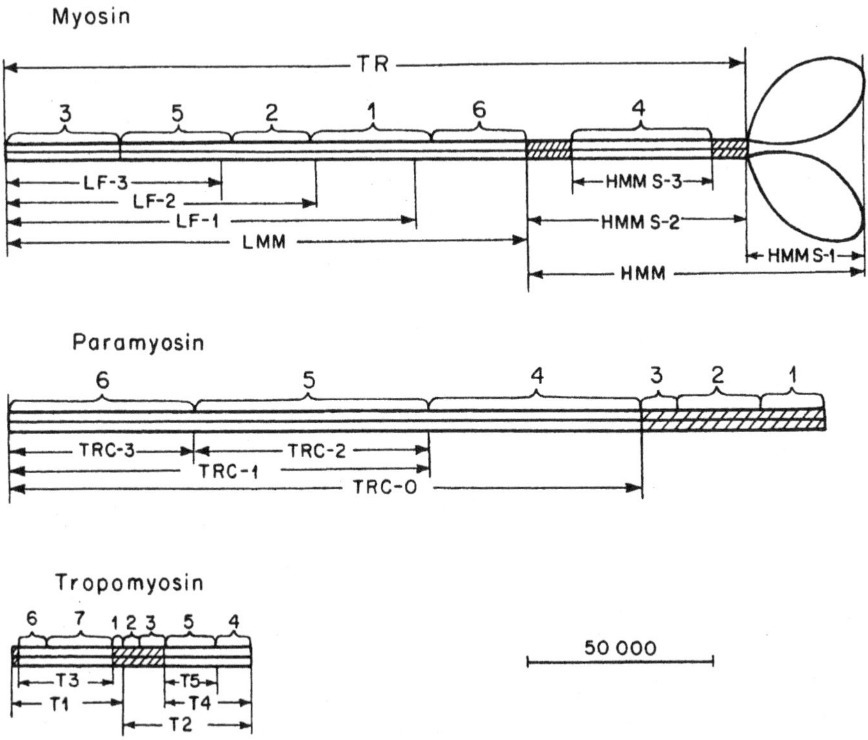

Figure 5.16. Schematic representation of myosin, paramyosin, and tropomyosin and their proteolytic fragmentation. The parts of the molecule susceptible to proteolysis are hatched braces indicate the thermodynamically revealed cooperative blocks. The numeration corresponds to the increasing order of stability (Privalov, 1982).

200,000 (Lowey et al., 1969). In solution they are usually considered as rigid rods with an average diameter of 20 Å. Correspondingly, their length varies from 410 Å for tropomyosin to about 1300 Å for the myosin rod. According to optical studies, their helicity in solution is close to 100%.

These molecules consist of two identical, or almost identical, polypeptide chains adjusted parallel and in an axial register. The regularity for double-stranded α-coiled-coils was found in both stretches—the high probability of nonpolar residues in the a and d positions and the acidic and basic residues in the e and g positions of heptapeptide repeat (Fig. 5.10). However, some irregularity exceptions to this general regularity have been also found perhaps necessary for the specific functions of these molecules (Parry, 1981). It appears, as in the sequence of all coiled-coil proteins, there are two features in the distribution of various amino acid residues—the periodic and the aperiodic; the first determines their structural class, and the second determines their functional specificity. The problem is then how the superposition of these two features reveals itself in the stability of these molecules.

In experimental studies of the stability of the double-stranded coiled-coils two approaches have been used: digestion by proteolytic enzymes and denaturation by various agents or temperature. The first of these approaches yields structural information on the location of the most resistant parts along the molecule, but it does not permit defining their stability quantitatively. The second could provide this information on stability, but in particular the temperature-induced changes of these long fibrilar molecules appear to be too complicated for quantitative thermodynamic analysis of this process. Therefore, a combination of two methods is used for these thermodynamic studies: fragmentation by proteolytic enzymes and calorimetric and optical studies of the temperature-induced changes of the fragments thus obtained.

Digestion studies showed that proteolytic enzymes attack coiled-coil molecules in a very specific manner: eliminating some regions, while leaving the other parts intact (Fig. 5.16). The myosin molecule is split by triptic treatment into two fragments: light meromyosin (LMM) and heavy meromyosin (HMM) (Lowey et al., 1969). The space between these two fragments is very susceptible to proteolysis and is assumed to be flexible (Burke et al., 1973). The HMM fragment can be split by papain into two subfragments: S1, which corresponds to the globular head of myosin; and S2, which corresponds to the rodlike part of HMM from which the smallest fragment, S3, can be obtained. The LMM fragment can be split by trypsin into smaller fragments: LF-1, LF-2, and LF-3.

Studies on limited proteolysis of these coiled-coil proteins led to the conclusion that the accessibility of their backbone for proteolytic enzymes is not uniform along the molecule; that is, various parts of the molecule fluctuate at different rates so the peptide bonds are, to various extents, exposed to enzyme attack. Studies on denaturation of the double-stranded coiled-coil proteins by various optical methods also showed that this process is much more complicated than expected for molecules with linearly homogeneous structure (Burke et al., 1973; Woods, 1976; Privalov, 1982). The complexity of the temperature-induced unfolding or dissociation of coiled-coil proteins was investigated in detail by DSC.

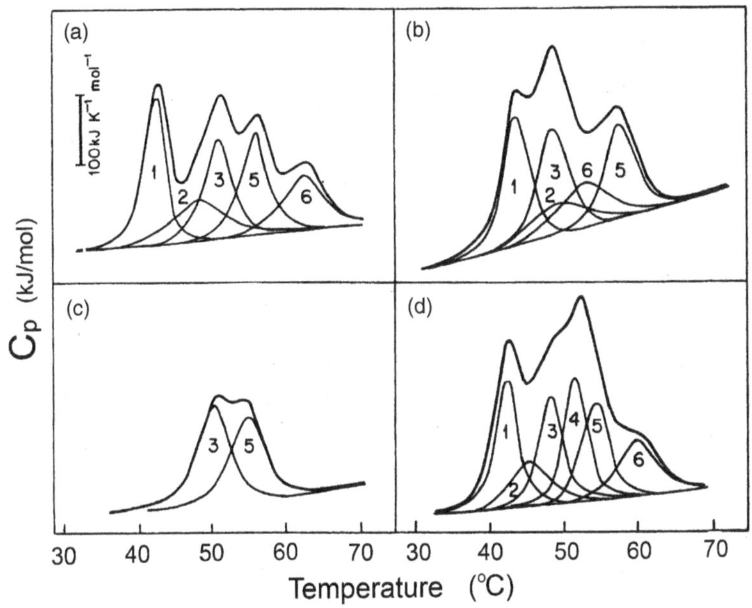

Figure 5.17. Partial molar heat capacity functions of myosin fragments in 0.5 M KCl, 25 mM phosphate buffer, pH 6.5: (a) LMMt (trypsin fragment); (b) LMMp (pepsin fragment); (c) LF-3; (d) TR (total road) (Privalov, 1982).

5.3.2. Myosin Rod

The partial heat capacity functions of four coiled-coil fragments of myosin are presented in Figure 5.17. One can see that complexity of the melting profile increases with increasing fragment size. From the shape of the profile it becomes evident that melting of all the coiled-coil fragments considered is a complex process consisting of several cooperative stages. The most distinct stage is the first one, especially in the case of the LMMt (trypsin) fragment.

Correlation analysis of the results of deconvolution of all fragments of myosin lead to the conclusion that there are six simple transitions and that in various combinations they form the observed melting profiles found for fragments of the myosin molecule (Potekhin and Privalov, 1979). The enthalpies of these six transitions are 820, 440, 710, 760, 680, and 490 kJ/mol. Assuming that the specific melting enthalpy of the coiled-coil structure is the same along the myosin rod, it is possible to evaluate from the estimated enthalpy values of the cooperative transitions the approximate size of the cooperative blocks in the myosin rod.

Taking into account the diagram of fragmentation of the myosin rod, it is not difficult to locate the cooperative blocks revealed along the myosin rod. These blocks are indicated by the braces in the diagram in Figure 5.16. The correspondence of the boundaries of the cooperative blocks with those of fragments obtained by limited proteolysis is most remarkable. Another notable and quite unexpected fact is that the most unstable cooperative block (N1) is positioned in the middle part of the myosin rod; that is, the unzipping model is absolutely inapplicable to the melting of this coiled-coil macromolecule. It is also noteworthy that one of the cooperative blocks is somewhat apart from the others and is located between the regions of the molecule that do not take part in the melting process and hence might be supposed to be unstructured. One of these regions is the "hinge" of the myosin molecule, and the other is the terminal region of the myosin rod, which connects it with the globular heads of this molecule—the "swivel." It is known now that in this region the myosin bends upon functioning.

5.3.3. Paramyosin

Figure 5.18 presents the partial heat capacity function of the intact paramyosin in solutions with various pH at which this molecule is readily soluble and melts reversibly (Potekhin and Privalov, 1979). From the complicated shape of these functions one can conclude that the melting of this molecule passes through several stages.

Analysis of the heat capacity function of the whole molecule and of its two fragments TRC-1 and TRC-2 has shown that melting of the paramyosin molecule proceeds through six distinct stages. Thus, it follows that the structure of the paramyosin molecule is not a continuous double-stranded coiled-coil, but is subdivided into discrete cooperative blocks. The melting enthalpies of the blocks arranged by increasing stability were found to be 380, 530, 180, 1000, 1810, and 1260 kJ/mol.

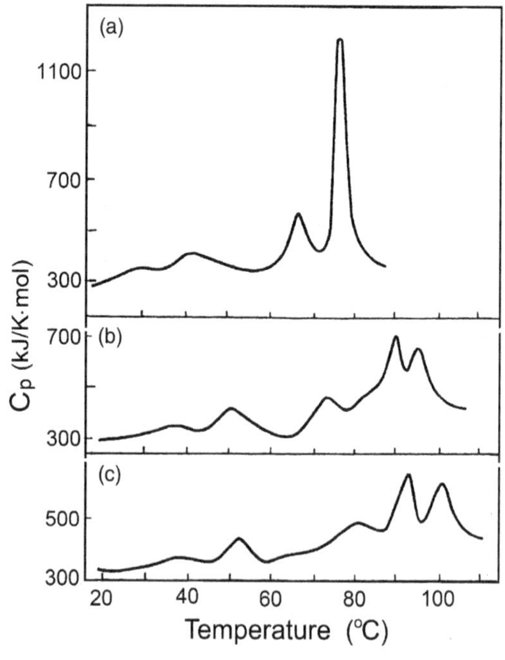

Figure 5.18. Partial molar heat capacity functions of paramyosin in 10 mM glycine buffer at (a) pH 2, (b) pH 3, and (c) pH 3.5 (Potekhin and Privalov, 1979).

The approximate size of these blocks estimated from these enthalpy values are indicated by braces in Figure 5.16. As seen, the situation is similar to that found for the myosin rod: the boundaries between the revealed cooperative blocks are in remarkable correspondence with the boundaries of the proteolytic fragments.

5.3.4. Tropomyosin

The melting of the tropomyosin molecule was studied calorimetrically by several authors, who were attracted mainly by its small size and known sequence (Potekhin and Privalov, 1978, 1982; Krishnan et al., 1978). However, the simplicity of this molecule was only apparent, since the difficulties in the analysis of its melting profile were more serious than in the case of the much larger myosin rod and paramyosin. One of the reasons for this difficulty was the smoothness of the melting profiles of tropomyosin molecule that made it hard to resolve the discrete stages (Fig. 5.19). The cooperative blocks revealed by deconvolution analysis of the partial heat capacities of tropomyosin at various conditions are indicated by braces in Figure 5.16. A remarkable distinction between the blocks in myosin rod and paramyosin and the blocks in tropomyosin is in their size: the cooperative blocks in tropomyosin are much smaller. Correspondingly, the melting enthalpies are smaller,

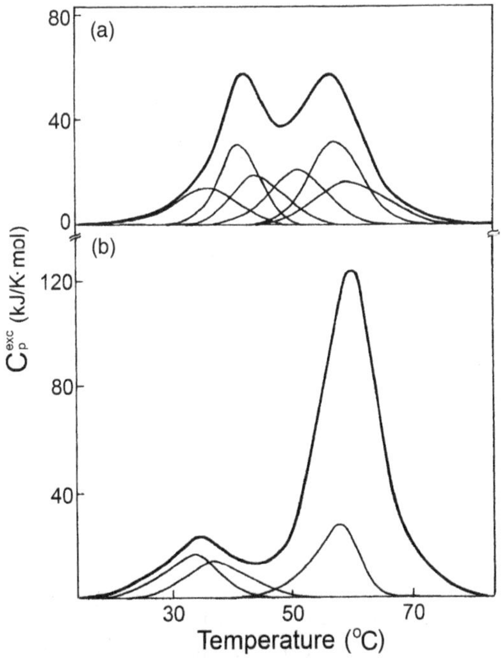

Figure 5.19. Deconvolution of the excess heat capacity function of a-tropomyosin in 1 M KCl, 25 mM potassium phosphate, pH 7.1: (a) carboxymethylated a-tropomyosin; (b) a-tropomyosin containing S–S cross-link (Potekhin and Privalov, 1982).

and, consequently, the transitions in tropomyosin are much more extended over a range of temperature.

The strong overlap of many broad transitions upon melting of tropomyosin is one of the reasons why analysis of tropomyosin melting is not easy. The other complication in the analysis of tropomyosin melting is that in the case of small size of cooperative blocks the energies of interactions between the blocks is not insignificant compared with the energies of melting of the block. As a result, the stability of a block depends on whether its neighboring block is melted or not.

> *In the case of tropomyosin, however, the most surprising fact was that interblock cooperation has a negative sign: that is, the stability of a block is higher when the neighboring block is melted.*

It is likely that the cooperative blocks in tropomyosin disturb each other, inducing mutual distortions.

5.3.5. Leucine Zipper

Leucine zippers are dimerization domains of the large family of bZIP transcription factors. Among them the dimerization domain of GCN4 transcription factor (the fragment from residue 249 to 281) attracted particular attention as an ideal example of an α-helical, double-stranded coiled-coil that folds and unfolds reversibly, rapidly, and, as believed, in a highly cooperative way. It was therefore considered as a perfect object for studying the energetics of formation of coiled-coils. The structure of this leucine zipper, composed of two identical 33–amino acid residue strands, has been determined crystallographically to high resolution (O'Shea et al., 1991; Alber, 1992). It was shown that the two strands in an α-helical conformation are held together by hydrophobic glue provided mostly by adjacent leucine residues at the interface. Many authors have investigated the unfolding of the GCN4 leucine zipper under increasing temperature, denaturant concentration, and pressure using various methods of observation, for example, circular dichroism (CD) (Krylov et al., 1994, 1998), calorimetry (Thompson et al., 1993; D'Aquino et al., 1996; Durr et al., 1999; Bosshard et al., 2001; Ibarra-Molero et al., 2001), stop-flow kinetics (Sosnick et al., 1996), and NMR (Lumb and Kim, 1995). This led to the conclusion that the unfolding–dissociation of the leucine zipper represents an almost perfect two-state transition.

The example of the GCN4 unfolding investigation is remarkable because it clearly shows how cautious an investigator should be in drawing conclusions, particularly those based on calorimetric information. Differential scanning calorimetry is indeed one of the most efficient methods for evaluating the cooperativity of temperature-induced processes in proteins. However, calorimetric analysis of such processes depends on the evaluation of the excess heat effect associated with the process being considered. In the case of coiled-coils, particularly of the short one, evaluation of this heat effect presents a problem, because their temperature-induced unfolding–dissociation proceeds over a broad temperature range. Moreover, heating

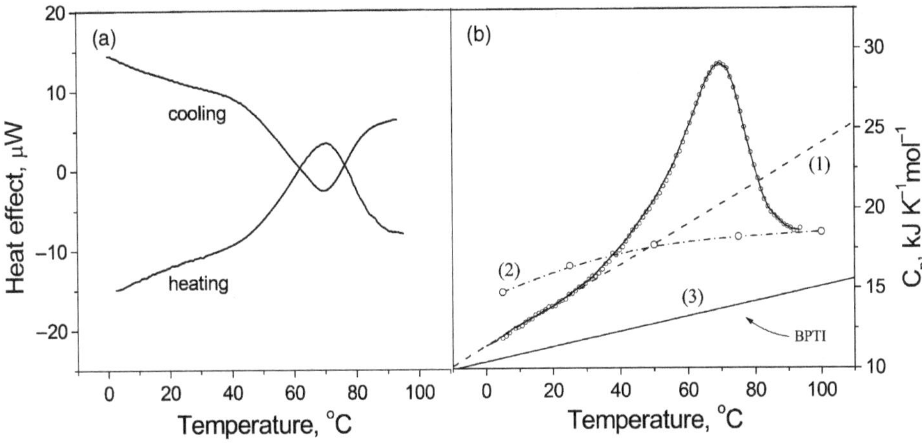

Figure 5.20. (a) The apparent heat effect as recorded by the Nano-DSC upon heating and subsequent cooling at 1 K/min of 136 mM GCN4 leucine zipper solution. (b) Temperature dependence of the partial molar heat capacity of GCN4 leucine zipper. The broken line (1) represents a linear extrapolation of the initial heat capacity into the zone of the heat absorption peak; the dashed-dotted curve (2) represents the calculated heat capacity of the fully unfolded polypeptides obtained by summing the heat capacities of the individual amino acid residues. The continuous line (3) represents the expected heat capacity of a compact globular protein having a stable rigid structure: it was obtained by recalculating the partial specific heat capacity of native BPTI for the molecular mass of the leucine zipper.

of the coiled-coils induces fraying of their ends. Their low-temperature apparent heat capacity function cannot therefore be regarded as an intrinsic heat capacity of the native protein and cannot simply be extrapolated into the transition zone in order to evaluate the excess heat effect of protein unfolding. Analysis of the heat effect of unfolding of the coiled-coils requires knowledge of their absolute partial heat capacity function over a broad temperature range, which can be determined only by using a DSC instrument having an extremely stable baseline. Using this instrument it was found that unfolding of a leucine zipper is not as simple as it was believed to be (Dragan and Privalov, 2002).

Figure 5.20a presents calorimetric recordings of the heat effect upon heating and subsequent cooling of 136-mM GCN4 leucine zipper. The extensive heat effects of scanning up and down appear as mirror images. This shows that the temperature-induced changes of this dimeric protein are reversible and proceed without noticeable hysteresis; that is, they are fast and at the heating–cooling rate used (1 K/min) are close to equilibrium. The deviation of the recordings from zero permits us to determine the absolute value of the partial heat capacity of the protein at any given temperature (see Section 3.2.3). Figure 5.20b presents the partial molar heat capacity function for the GCN4 leucine zipper. This heat capacity function does not differ much in shape from those reported previously (Thompson et al., 1993; D'Aquino et al., 1996; Durr et al., 1999; Bosshard et al., 2001; Ibarra-Molero et al., 2001).

However, in contrast to those, it represents not the apparent change of the heat capacity but the absolute molar heat capacity of the protein.

A remarkable feature of the absolute partial heat capacity function of the leucine zipper is that it increases rather steeply from the very beginning of heating, developing at the higher temperatures into a peak of heat absorption, after which it drops to the level expected for the fully unfolded polypeptide chain. The heat absorbed in a peak is usually associated with the cooperative unfolding of a protein and is evaluated by linear extrapolation of the initial and final heat capacities into the temperature zone of cooperative transition, assuming that these extrapolated functions represent the intrinsic heat capacities of the fully folded and fully unfolded states of the protein [lines (1) and (2) in Fig. 5.20b]. The unfolding enthalpy of the GCN4 leucine zipper determined in this way is very close to the calorimetric and van't Hoff enthalpies reported by other authors. However, in the case of the leucine zipper the use of such a procedure for evaluation of the unfolding enthalpy gives rise to serious concern for the following reasons.

(a) The heat capacity of a leucine zipper increases too steeply from the very beginning of heating: its temperature derivative (the slope) amounts to 0.125 kJ/K^2·mol, or calculated per gram to 15 mJ/K^2·g. This considerably exceeds the slope of the specific heat capacity function of any calorimetrically studied proteins with a stable, fully folded structure, which is usually of the order of 6 mJ/K^2·g (see Section 8.2.1).

(b) At 25°C, where the leucine zipper is supposed to be in the fully folded state, its partial molar heat capacity amounts to (14.5 ± 0.5) kJ/K·mol, that is, the partial specific heat capacity is (1.8 ± 0.06) J/K·g, whereas the partial specific heat capacity of fully folded proteins at this temperature never exceeds 1.4 J/K·g. It is only at 0°C, or even lower, perhaps at −10°C, that the partial specific heat capacity of the leucine zipper comes close to the value expected for the fully folded proteins at that temperature, (1.2 ± 0.1) J/K·g (Section 8.2.1).

(c) The linearly extrapolated initial heat capacity function of the leucine zipper crosses the heat capacity function of the unfolded protein at about 50°C, which means that the heat capacity effect of unfolding changes its sign at that temperature. This leads to the conclusion that the enthalpy and entropy of leucine zipper unfolding, being increasing functions of temperature below 50°C, become decreasing functions of temperature above 50°C, which is highly improbable.

All this leads to the conclusion that the heat capacity of the leucine zipper below the temperature zone of the extensive heat absorption peak does not represent the intrinsic heat capacity of a fully folded protein. It approaches the value expected for the fully folded state only below 0°C. It appears, thus, that the leucine zipper is in the fully folded state only at temperatures below 0°C and upon heating from that temperature some process is taking place that proceeds with significant heat absorption. This might be fluctuations of the protein structure that intensify with the

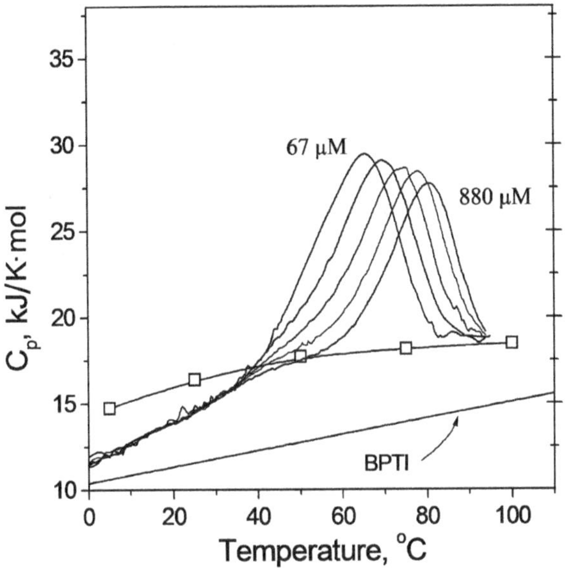

Figure 5.21. Temperature-dependence of the partial molar heat capacity of GCN4 leucine zipper measured at different concentrations of protein (67, 136, 260, 532, and 880 μM).

temperature, or some structural rearrangements that take place before the cooperative process of unfolding–dissociation of the two strands. Therefore, the apparent heat capacity of the leucine zipper before the temperature zone of the heat absorption peak cannot be used as a baseline for the evaluation of the excess heat effect of its conformational changes.

This becomes especially clear when comparing the heat capacity functions of the leucine zipper, obtained at different concentrations of protein, as shown in Figure 5.21. It appears that the increase of concentration results in significant change of the heat absorption peak: it shifts to a higher temperatures as one would expect for bimolecular reaction. However, the change of concentration does not affect the heat capacity function below 35°C, while the shift of the heat absorption peak to the higher temperatures reveals a shallow heat absorption peak with a maximum at 45°C. The presence of this peak explains why the height of the main heat absorption peak decreases with its shift to higher temperatures.

Thus, the temperature-induced unfolding of the GCN4 leucine zipper is clearly a complex, multistage process, which starts from a very low temperature, probably from 0°C or lower, and ends with the cooperative dissociation of strands above 80°C. The excess heat capacity in that case can be determined only using as a standard heat capacity of the fully folded state, that is, the heat capacity function of BPTI (see Section 3.2.5, Eq. 3.22). Using this function one can determine the excess heat effect of leucine zipper unfolding and analyze it thermodynamically (Fig. 5.22). Deconvolution analysis of the excess heat effect of a leucine zipper obtained at different concentrations shows that its unfolding proceeds in three cooperative stages,

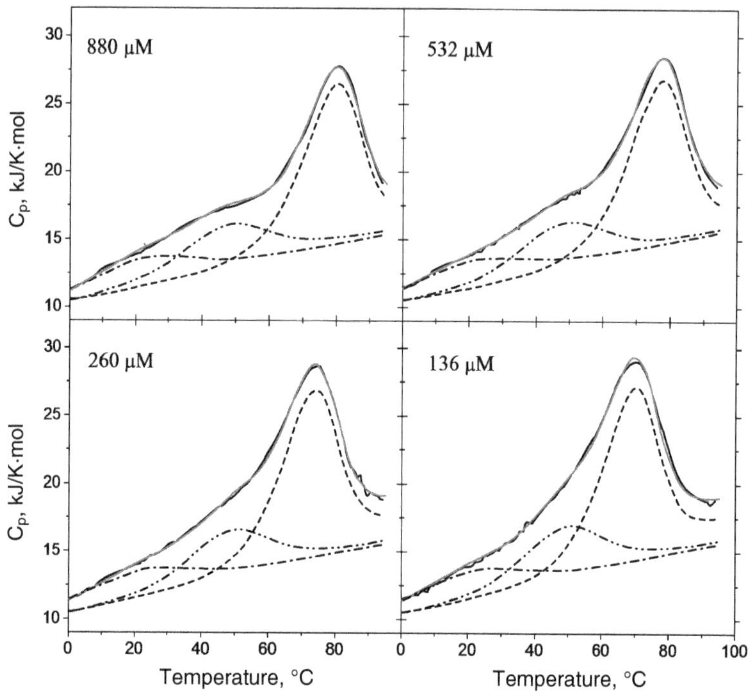

Figure 5.22. Deconvolution of the excess heat capacity functions of GCN4 leucine zipper at the concentrations indicated.

among which only the last one depends on concentration, that is, is associated with dissociation of the strands. The question is then what are the first two concentration-independent transitions?

The first transition starts at the very beginning of heating from 0°C and proceeds with significant heat absorption and decrease of ellipticity. It was found that this transition is sensitive to modification of the N-terminus (replacement of terminal arginine on valine). It is therefore associated with unfolding of this part of the leucine zipper. One could expect then that the second transition might be associated with unfolding of the C-terminal part of the leucine zipper. It was therefore surprising to find out that the replacement of Tyr17 by Trp at the middle part of the leucine zipper increased the temperature of the second transition. This transition is accompanied by a drop of helicity of about 15% and by change in the tryptophan fluorescence; that is, it involves Trp17, which is in the central part of the leucine zipper. It appears, therefore, that the second transition involves the whole of the leucine zipper. This transition is, however, unimolecular and thus not associated with the dissociation of strands. It must be therefore be associated with some rearrangements in the coiled-coil. In this connection, the observation made by Holtzer et al. (2001) is notable: over the temperature range below that of strand dissociation, the NMR resonances of Val19 and Leu13 split into two components, indicating some local repacking of the GCN4 leucine zipper. Thus, the N-terminal part of a leucine zipper does indeed unfold first, but it looks as if the unfolding of the C-terminal part cooperates with the repacking of the central part. The structural change of the leucine zipper at the second transition is certainly of great interest, particularly because it takes place at temperatures close to physiological.

5.3.6. Discreteness of the Coiled-Coils

The most impressive results of the studies of natural two-stranded α-helical coiled-coils is that the regular structure of these proteins is not continuous, but is subdivided into definite cooperative blocks at certain points that also determine the size of the proteolytic fragments. In such linear molecules the existence of singularities is rather unexpected. Even more unexpected is the fact that the neighboring cooperative regions (separated by the "singular points") destabilize each other.

It appears that, while the cooperation of the residues within the blocks is positive and leads to integration of all the residues in this region into a single structural unit, the cooperation of the residues in the space between the blocks is negative that is, at these points the cooperativity is converted to "anticooperativity." The mechanism of changing the sign of residue cooperation at the singular points along the linear molecules is obscure. It could be connected with accumulation of stresses in the deformed α-helixes wound into a supercoil which was predicted by Pauling and Corey (1953) and also Crick (1953a,b). It is obvious that the relief of a stress will take place at a weak point of the regular structure. The weak point might be some irregularity in the sequence of the polypeptide chain. However, because of the very different size of the cooperative blocks found in the coiled-coil proteins considered, an accumulation of stresses cannot alone be responsible for the discreteness of their structure.

Another aspect of the problem is the biological significance of the discrete character of the coiled-coil structure. In an analogy with multidomain proteins (see Chapter 10), it is tempting to assume that in the coiled-coils the observed discreteness reflects the evolutionary pathway of these molecules. As for the functional significance of the observed discreteness of coiled-coil structure, it is more or less evident. It is clear that discreteness can provide these rodlike molecules with the necessary flexibility by the joints between the rigid segments and also the ability to change length by fusion of the regular structure at some regions.

In myosin the segments without regular structures are located between the globular head and the rod (the "swivel") and between the HMM and LMM fragments (the "hinge"), that is, in the regions that should be flexible for the functioning of this molecule (Fig. 5.16). In α-tropomyosin the unstable segments are located in the middle of the molecule, and it is particularly interesting that their stability is regulated by changing the state of cysteine residues. This circumstance becomes especially attractive in that it is still unclear how tropomyosin can be shifted from the helical groove of muscle thin filament (as is believed to take place upon muscle functioning) without changing its length.

REFERENCES

Alber T. (1992). *Curr. Opin. Genet. Dev.*, **2**, 205–210.

Applequist J. (1963) *J. Chem. Phys.*, **38**, 934–941.

Baldwin R.L. (1986). *Proc. Natl. Acad. Sci. USA*, **83**, 8069–8072.

Bosshard H.R., Durr E., Hitz T., and Jelesarov I. (2001). *Biochemistry*, **40**, 3544–3552.

Braun V. (1975). *Biochim. Biophys. Acta*, **415**, 335–377.

Burke M., Himmelfarb S., and Harrington W.F. (1973). *Biochemistry*, **12**, 701–710.

Chen Y.H., Yang J.T., and Chau K.H. (1974). *Biochemistry*, **13**, 3350–3359.

Chou P.Y. and Scheraga H.A. (1971). *Biopolymers*, **10**, 657–680.

Crick F.H.C. (1952). *Nature*, **170**, 882–885.

Crick F.H.C. (1953a). *Acta Crystallogr.*, **6**, 685–688.

Crick F.H.C. (1953b). *Acta Crystallogr.*, **6**, 689–697.

D'Aquino J.A., Gomez J., Hisler V.J., Lee K.H., Amzel L.M., and Freire E. (1996). *Proteins: Struct. Funct. Genet.*, **25**, 143–156.

Dragan A.I. and Privalov P.L. (2002). *J. Mol. Biol.*, **321**, 891–908.

Dragan A.I., Potekhin S.A., Sivolob A., Lu M., and Privalov P.L. (2004). *Biochemistry*, **43**, 14891–14900.

Durr E., Jelesarov I., and Bosshard H.R. (1999). *Biochemistry*, **38**, 870–880.

Holtzer M.E., Bretthorst G.L., d'Avignon D.A., Angeletti R.H., Mints L., and Holtzer A. (2001). *Biophys. J.*, **80**, 939–951.

Ibarra-Molero B., Makhatadze G.I., and Matthews C.R. (2001). *Biochemistry*, **40**, 719–731.

Krishnan K.S., Brandts J.F., and Lehler S.S. (1978). *FEBS Lett.*, **91**, 206–208.

Krylov D., Barchi J., and Vinson C. (1998). *J. Mol. Biol.*, **279**, 959–972.

Krylov D., Mikhailenko I., and Vinson C. (1994). *EMBO J.*, **13**, 1849–1861.

Lifson S. and Roig G.D. (1961). *J. Chem. Phys.*, **34**, 1963–1974.

Liu J., Cao W., and Lu M. (2002). *J. Mol. Biol.*, **318**, 877–888.

Lowey S., Slayter H.S., Weeds A.G., and Baker H. (1969). *J. Mol. Biol.*, **42**, 1–29.

Lumb K.J. and Kim P.S. (1995). *Science*, **268**, 436–439.

Makhatadze G.I. and Privalov P.L. (1995). *Adv. Protein Chem.*, **47**, 307–425.

Nemethy G., Leach S.J., and Scheraga H.A. (1966). *J. Phys. Chem.*, **70**, 998–1004.

O'Shea E.K., Klemm J.D., Kim P.S., and Alber T. (1991). *Science*, **254**, 539–544.

Parry D.A.D. (1981). *J. Mol. Biol.*, **153**, 459–464.

Pauling L., Corey R.B., and Branson H.R. (1951). *Proc. Natl. Acad. Sci. USA*, **37**, 205–211.

Pauling L. and Corey R.B. (1953). *Nature*, **171**, 59–61.

Potekhin S.A. and Privalov P.L. (1978). *Biofizika (USSR)*, **23**, 219–223.

Potekhin S.A. and Privalov P.L. (1979). *Molek. Biol. (USSR)*, **13**, 666–671.

Potekhin S.A. and Privalov P.L. (1982). *J. Mol. Biol.*, **159**, 519–535.

Privalov P.L. (1982). *Adv. Protein Chem.*, **35**, 1–104.

Privalov P.L. and Gill S.J. (1988). *Adv. Protein Chem.*, **39**, 191–234.

Privalov P.L., Tiktopulo E.I., Venyaminov S.Yu., Griko Yu.V., Makhatadze G.I., and Khechinashvili N.N. (1989). *J. Mol. Biol.*, **205**, 737–750.

Ptitsyn O.B. (1972). *Pure Appl. Chem.*, **31**, 227–244.

Rialdi G. and Hermans H., Jr. (1966). *J. Am. Chem. Soc.*, **88**, 5719–5720.

Richardson J.M. and Makhatadze G.I. (2004). *J. Mol. Biol.*, **335**, 1029–1037.

Schellman J.A. (1955). *C.R. Trav. Lab. Carlsberg. Ser. Chim.*, **29**, 230–259.

Scholtz J.M., Marqusee S., Baldwin R.L., York E.J., Stewart J.M., Santoro M., and Bolen D.W. (1991). *Proc. Natl. Acad. Sci. USA*, **88**, 2854–2858.

Shu W., Liu J., Ji H., and Lu M. (2000). *J. Mol. Biol.*, **299**, 1101–1112.

Sosnick T.R., Jackson S., Wilk R.R., Englander S.W., and DeGrado W.F. (1996). *Proteins: Struct. Funct. Genet.*, **24**, 427–432.

Stone D., Sodek J., Johnson P., and Smillie L.B. (1975). *Proc. FEBS Meet.*, **31**, 125–136.

Taylor J.W., Greenfield, N.J., Wu B., and Privalov P.L. (1999). *J. Mol. Biol.*, **291**, 965–976.

Thompson K.S., Vinson C.R., and Freire E. (1993). *Biochemistry*, **32**, 5491–5496.

Woods E.F. (1976). *Aust. J. Biol. Sci.*, **29**, 405–418.

Yu Y., Monera O.D., Hodges R.S., and Privalov P.L. (1996a). *J. Mol. Biol.*, **255**, 367–372.

Yu Y., Monera O.D., Hodges R.S., and Privalov P.L. (1996b). *Biophys. Chem.*, **59**, 299–314.

Yu Y.B., Lavigne P., Kay C.M., Hodges R.S., and Privalov P.L. (1999). *J. Phys. Chem.*, **103**, 2270–2278.

Zimm B.H. and Bragg J.K. (1959). *J. Chem. Phys.*, **31**, 526–535.

6

POLYPROLINE-II COILED-COILS

6.1. COLLAGENS

6.1.1. Collagen Superhelix

An essential difference of the poly-L-proline-II helix from the α-helix is that it is not maintained by the intramolecular hydrogen bonds and it does not have a tight apolar core (Fig. 6.1). Therefore, this helical conformation is unstable. However, three such chains, wound together into a trimeric superhelix, form a fibrillar protein, **collagen**, forming a structural basis of skin, tendons, cartilage, and bones. This is a rather stable construction with unique mechanical and thermal properties (Brodsky and Persikov, 2008). The question is then: what are the forces that are stabilizing the collagen superhelix?

The conformation of the collagen molecule is dictated by the specificity of its primary structure, the sequence of amino acid residues in the polypeptide chain: every third residue in the part forming the superhelix is glycine, while the other two in the triplet could, with high probability, be imino acids, that is, proline or hydroxyproline:

Microcalorimetry of Macromolecules: The Physical Basis of Biological Structures, First Edition.
Peter L. Privalov.

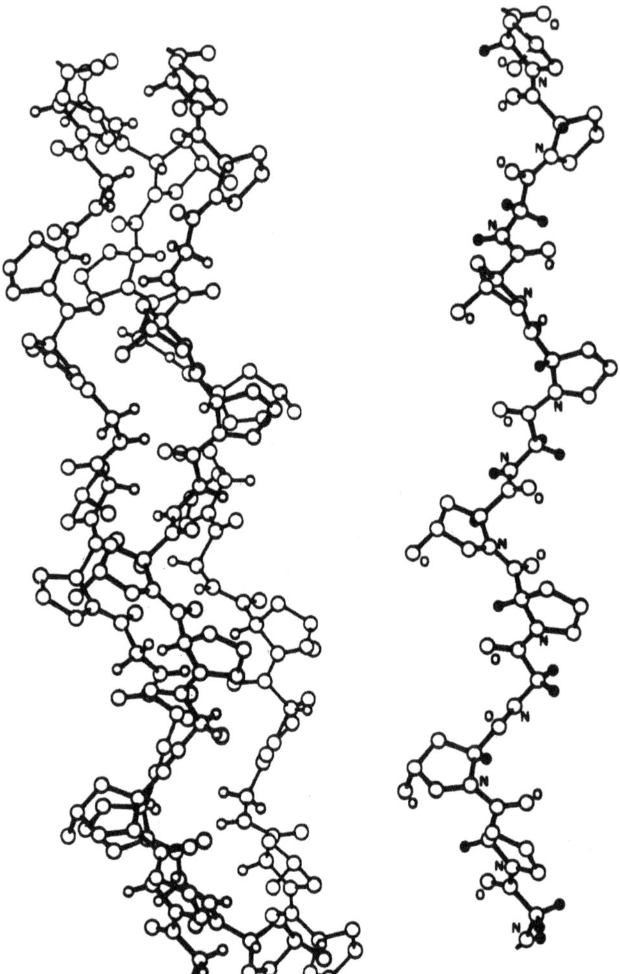

Figure 6.1. Model of the three-stranded collagen structure (left) and a single strand of the repeated sequence –Gly–Pro–Hyp– (right) (Fraser et al., 1979).

$$-\text{Gly}-\text{X}-\text{Y}-\text{Gly}-\text{X}-\text{Y}-\text{Gly}-\text{X}-\text{Y}-\text{Gly}-\text{X}-\text{Y}-\text{Gly}-\text{X}-\text{Y}-$$

Hydroxyproline appears as a result of post-translational modification of proline if it is in the Y position.

A theoretical consideration of possible conformations of polypeptides containing glycine in every third position and enriched by imino acid residues (fixing the ψ angle and essentially restricting the φ angle), led to two alternative coiled-coil models of collagen, with a somewhat different packing and different numbers of internal hydrogen bonds stabilizing the helical structure: the less tight one-bonded model suggested by Rich and Crick (1955) and tighter two-bonded model suggested

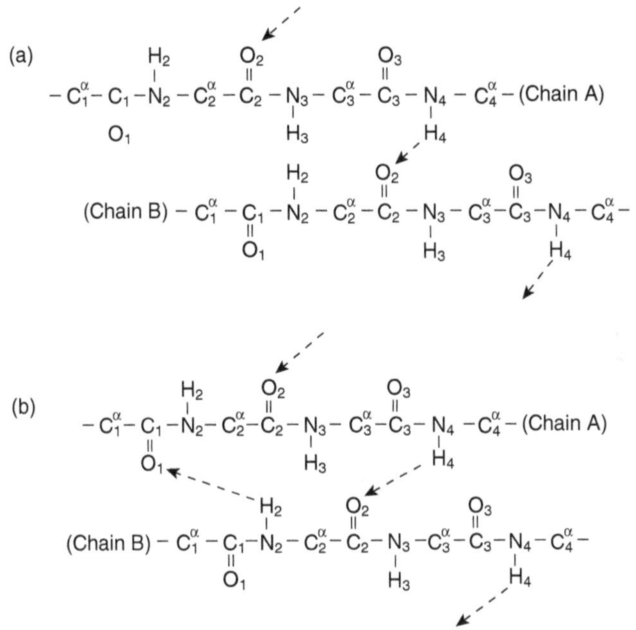

Figure 6.2. Schemes of hypothetical internal hydrogen bonds in collagen: (a) one-bonded model suggested by Rich and Crick (1955); (b) two-bonded model suggested by Ramachandran and Kartha (1955).

by Ramachandran and Kartha (1955) (Fig. 6.2). According to the one-bonded model, the collagen polypeptides in the poly-L-proline conformation are connected by one hydrogen bond per triplet, between the amide group of the glycyl, which is in the first position, and the carbonyl of a residue in the second position (X) of a triplet. According to the two-bonded model, there is one extra hydrogen bond between the carbonyl of the glycyl and the amide of a residue in the second position of a triplet, if this position is not occupied by an imino acid residue. As is evident, the quantity of these second bonds is less than one per triplet and decreases with the increase of the imino acid content in collagen, if the imino acid residues occupy with a somewhat equal probability the second (X) and third (Y) positions in a triplet.

6.1.2. Hydrogen Bonds in Collagen

There were many attempts to determine experimentally the quantity of hydrogen bonds in collagen macromolecule in aqueous solution by the hydrogen exchange method. For such experiments collagen seemed to be a perfect object, since in this linear and structurally homogeneous molecule should be a limited number of classes of exchangeable hydrogens and therefore simple kinetics of hydrogen exchange is *a priori* expected. Nevertheless, the results obtained by different authors were contradictory. According to some reports, there is one stable hydrogen bond per triplet

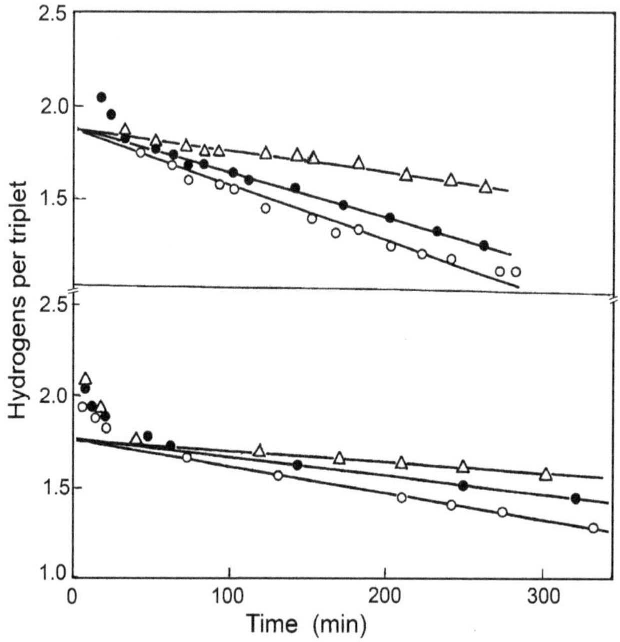

Figure 6.3. Hydrogen exchange kinetics at 5°C of cod (○), pile (●), and rat (△) skin collagens obtained by (a) deuterium exchange in solution at pD 3.5 and (b) tritium exchange in solution at pH 3.7 (Privalov et al., 1979).

in collagen (Katz, 1970; Traub and Piez, 1971; Esipova et al., 1972). According to others, there are two hydrogens per triplet (McBride and Harrington, 1967; Kingham and Brisbin, 1968). The cause of this discrepancy was clarified when it was shown that there are two classes, not one, of slowly and very slowly exchangeable hydrogens (Fig. 6.3) and that these classes are differently sensitive to temperature (Yee et al., 1974; Privalov et al., 1979). However, it appeared that the quantity of exchangeable hydrogens in each of these classes is the same for all studied collagens differing in their composition, stability, and melting enthalpy: it was found that all these collagens have (1.0 ± 0.1) very slowly and (0.7 ± 0.1) slowly exchangeable hydrogens per triplet. It seemed natural to attribute very slowly exchangeable hydrogen to the hydrogen bonds considered by the model of Rich and Crick (1955). As for the 0.7 slowly exchangeable hydrogens per triplet, it seemed reasonable to attribute it to the second bond in the model of Ramachandran and Kartha (1955), as the quantity of these bonds according to this model should be less than one and they were expected to be weaker.

Thus, at first sight it seemed that the results obtained by hydrogen exchange studies confirm the two-bonded model. But the model has one weak point that worried even its authors: the distance between the groups that are considered as a donor and acceptor for the hydrogen bond in the second position is too long (5.1 Å) for hydrogen bonding. To save the model, it was suggested that these groups are

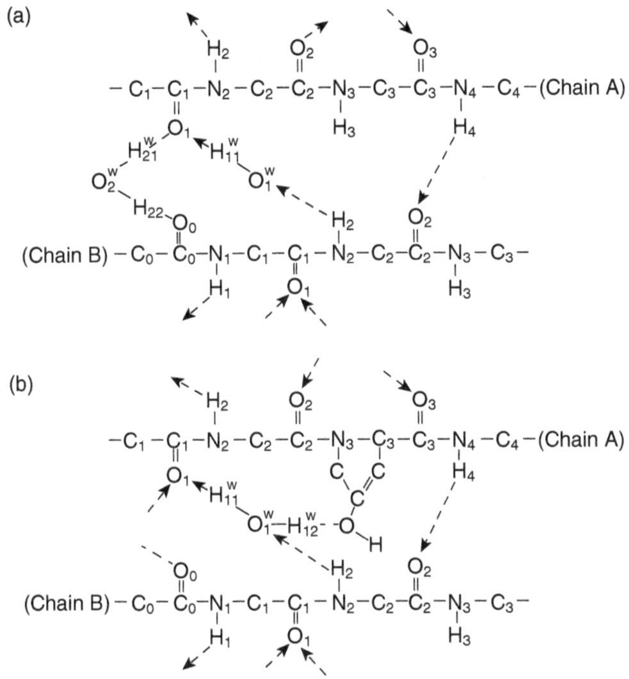

Figure 6.4. Diagram of water-bridged model of collagen: (a) not involving hydroxyprolil in binding; (b) involving hydroxyprolil in binding as suggested by Ramachandran and Ramakrishnan (1976).

connected by means of a water molecule, or even two water molecules (Ramachandran and Ramakrishnan, 1976). The new water-bridged model (Fig. 6.4) was much more acceptable from stereometrical point of view, but, as evident, it was no longer a two-bonded model.

Progress in collagen modeling by regular synthetic peptides and crystallographic studies of the formed three-peptides has confirmed the existence of the hydrogen bond between the NH of Gly in one chain and C=O of the residues in the X position of the neighboring chain, as predicted by the Rich-and-Crick model (Okuyama et al., 2009). However, in addition, peptides with a sequence where the X position is occupied by a residue other then Pro show a second interchain hydrogen bond between the amide group of the X-position residue and the C=O of the Gly residue, which is mediated by one water molecule (Emsley et al., 2004; Kramer et al., 2001) as was suggested by Ramachandran and Ramakrishnan (1976).

6.1.3. Stability of Collagens

A remarkable feature of collagens is that, at increased temperature above some critical level, their regular rigid structure breaks down within a rather short temperature range and the chains forming this superhelix separate and fold in three independent

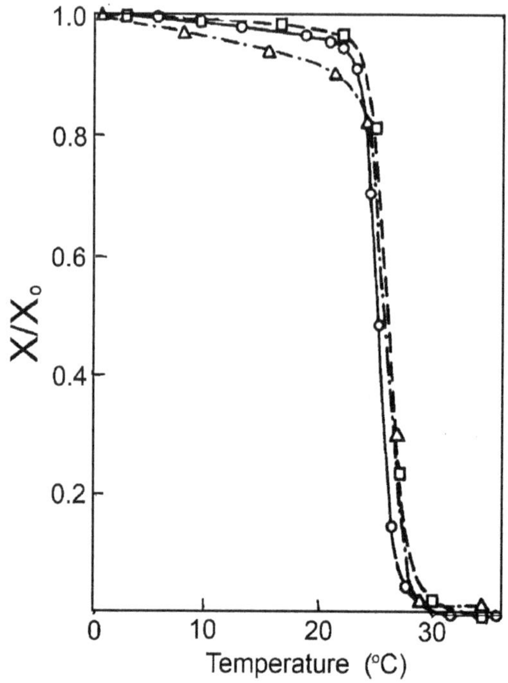

Figure 6.5. Melting profiles of collagen from carp swim bladder (ichtiocol) in 0.5 M $CaCl_2$, pH7.0 solution: (○) optical rotation, equilibrium curve; (□) optical rotation with 30 min incubation at each point; (△) viscosity with 30 min incubation at each point (von Hippel and Wong, 1963).

random coils that do not possess any residual structure (Boedtker and Doty, 1956; Engel, 1962; von Hippel, 1967; Privalov et al., 1971). This results in dramatic changes of all properties of collagen solution sensitive to shape, conformation, or molecular weight (Fig. 6.5). The process of denaturation or "melting" collagen structure is relatively slow compared with the denaturation of globular proteins. This might be connected with the slowness of *cis–trans* isomerization reaction of proline residues, which are present in large amounts in collagens (Bachinger et al., 1978). Therefore, the apparent transition temperature strongly depends on the heating rate. Correspondingly, only the value of the transition temperature obtained by extrapolation to zero heating rate can be used as a characteristic of this process. This value is usually called the **equilibrium melting temperature**.

The other problem in calorimetric studies of collagen is that upon reheating of the collagen solution its melting profile is not reproduced; that is, the melting of collagen appears to be an irreversible process and this raises the question of whether it can be treated thermodynamically. However, it was shown by comparative light scattering, viscosimetric, and optical rotation studies that the strand separation on collagen melting takes place only at the final stage of this process (Engel, 1962). It is likely that just this last stage is responsible for the observed irreversibility of collagen transition since the probability of proper alignment of three separated strands required for their assembling is rather low. Therefore, the main process of disruption of the helical conformation of the rather long collagen helix can be regarded as an equilibrium monomolecular reaction (von Hippel, 1967). Therefore, for a first approximation one can consider the melting profile of collagen as an "equilibrium" melting profile with transition temperature T_m.

The thermal stabilities of collagens from the various species of vertebrates are very different, but what is remarkable, all these collagens do not differ essentially in conformation, judging by their optical activity (i.e., helicity) and viscosity (von Hippel, 1967; Burjanadze, 1979). It is evident that there should be some correlation between the thermal stability and the chemical composition of collagens. According to Gustavson and Takahashi (1956), an important role in the stabilization of collagen structure is played by the hydroxyproline residue. Burge and Hyness (1959a,b) and Piez (1976) found that the thermostability of collagen correlates with the total imino acid content. Rigby (1967) noted the destabilizing role of the serine residue, and Neyfakh et al. (1978) assumed that the stability of collagen is determined by many different amino acids. Nevertheless, the most attention was concentrated on both imino acids, proline and hydroxyproline, which after glycine are present in the most substantial amounts specifically in collagens, and thus can be assumed to be responsible for its helical structure and stability (Fig. 6.6), This led to the assumption that for stabilization of the collagen helical structure the presence of rigid pyrrolidine rings in the polypeptide chain is essential because it reduces the number of conformations available to the random coil.

6.1.4. Role of Pyrrolidine Rings in Collagen Stabilization

According to Harrington (1964), if the entropy gain per residue on disruption of collagen structure (ΔS_{res}) for imino acid residues containing a rigid pyrrolidine ring is zero (whereas for amino acid residues it is about 17 J/K·mol), the entropy of melting of a collagen block consisting of 1000 residues should be

$$\Delta_m S_{1000} = (1000 - N_{1000})\Delta S_{res} \tag{6.1}$$

here N_{1000} is the number of hydroxyprolyl and prolyl residues per 1000 residues in collagen. As for the enthalpy of collagen melting, it should be different for the two existing models. For the one-bonded model it should be

$$\Delta_m H^{I}_{1000} = 333\,\Delta H^{H} \tag{6.2}$$

For the two-bonded model it should be

$$\Delta_m H^{II}_{1000} = (666 - N_{1000})\Delta H^{H} \tag{6.3}$$

where ΔH^{H} is the enthalpy of disruption of one peptide hydrogen bond.

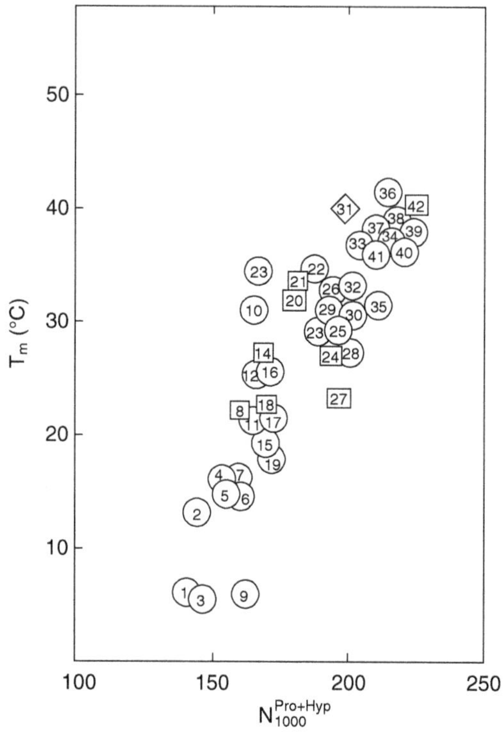

Figure 6.6. Plot of collagen melting temperature in salt-free solution with pH 3.7 versus the total imino acid content per 1000 residues. Circles, interstitial collagen of vertebrates; diamonds, noninterstitial collagens of vertebrates; squares, collagen of invertebrates: (1) icefish, *Trematomus eulepidotus*; (2) ray fish, *Raja*; (3) icefish, *Trematomus leonnbergi*; (4) hagfish, *Myxine glutinosa*; (5) cod skin, *Gadus morrhua*; (6) dogfish, *Squalus acanthias*; (7) cod swim bladder; (8) earthworm, *Lumbricus terrestris*; (9) icefish, *Racoutizia glacinalis*; (10) tadpole, *Rana catesbeiana*; (11) flounder, *Pleuronectus*; (12) frog, *Rana temporaria*; (13) frog, *Rana catasbeiana*; (14) sea anemone, *Actinia equine*; (15) hake swim bladder, *Merlatius*; (16) dogfish, *Mustelus*; (17) lamprey, *Petromyzom fluviatilis*; (18) earthworm *Lubricus terrestris*; (19) halibut; (20) blue crab, *Portunus trituberculatus*; (21) oyster, *Pinctada morterisii*; (22) newt, *Natophtalmus viridescens*; (23) shark, *Lamna nasus*; (24) octopus, *Octopus vulgaris*; (25) carp swim bladder; (26) carp skin; (27) *Nereis cuticle*; (28) pike, *Esox*; (29) perch, *Perska*; (30) maiger, *Sciaena diacanthus*; (31) sheep lens; (32) newt, *Natophtalmus viridescens*; (33) liver fluke, *Fasciola hepatica*; (34) rat skin; (35) dolphin; (36) chick skin; (37) rabbit skin; (38) pig kidney; (39) sheep skin; (40) rat tail; (41) calf skin (Privalov, 1982).

Assuming that at the midpoint of collagen melting $\Delta_m G = \Delta_m H - T_m \Delta_m S = 0$, and thus $T_m = \Delta_m H/\Delta_m S$, one finds the following for the temperature of collagen melting for the one-bonded model:

$$T_m = 333\,\Delta H^{\mathrm{H}}/[(1000 - N_{1000})\Delta S_{\mathrm{res}}] \tag{6.4}$$

For the two-bonded model one finds

$$T_m = (666 - N^*_{1000})\Delta H^{\mathrm{H}}/[(1000 - N_{1000})\Delta S_{\mathrm{res}}] \qquad (6.5)$$

where N^*_{1000} is the number of pyrrolidines occurring only in second position in the triplet, calculated per 1000 residues.

Thus, according to both models the melting temperature of collagen should increase with increase of the imino acids content. As for the enthalpy, by the one-bonded model it should not change with increase of the imino acids content; that is, it should not change with the increase of thermostability of collagen. By the two-bonded model it should decrease with increase of imino acids content; that is, the melting enthalpy should decrease with the increase of thermostability of collagens. Therefore, the calorimetric study of collagen melting became crucial for checking the existing concepts of collagen structure stabilization.

6.2. CALORIMETRIC STUDIES OF COLLAGENS

6.2.1. Enthalpy and Entropy of Collagen Melting

As is clear from the preceding discussion, the enthalpy and entropy of collagen unfolding have gained principal importance for understanding the mechanism of stabilization of collagen structure, and thus of the bases of this trimeric superhelix. To solve this problem and exclude speculations, we needed first of all to know what are indeed the enthalpy and entropy of unfolding–dissociation of this trimeric superhelix and how they depend on the content of imino acids in the collagens. Thus the calorimetry of collagen became a crucial experiment for solving the collagen problem. Therefore these experiments started immediately with the appearance of scanning microcalorimetry. Moreover, the collagen problem was one that stimulated design and development of this instrument.

Certain complications in the calorimetric studies of collagens should be mentioned. First is that the heat denaturation of collagen, that is, its intramolecular "melting," is a rather slow process, much slower than denaturation of globular proteins. Therefore, calorimetric study of temperature-induced collagen unfolding should be carried out at rather low heating rates (preferably at 0.2 K/min) at which the apparent melting profile is more or less close to the "equilibrium" melting profile. However, decreasing the scanning rate decreases the sensitivity of the calorimetric instrument for recording the temperature-induced heat effect. Therefore experiments with collagen require an extremely sensitive and precise instrument. The other complication is connected with the very high viscosity of collagen solution and drastic drop of this viscosity upon collagen denaturation (Fig. 6.5). This might induce a change of convection in the cylindrical calorimetric cell and redistribution of the temperature gradients, which results in artifacts in measuring the heat capacity change at collagen unfolding. Because of that, for studying collagen solutions, as well as some other highly viscous solutions (e.g., DNA), it is recommended to use a scanning calorimeter with capillary cells (see Section 3.2.2).

Melting of molecularly dispersed collagen on heating in salt-free aqueous solution proceeds in rather short temperature range with extremely extensive heat

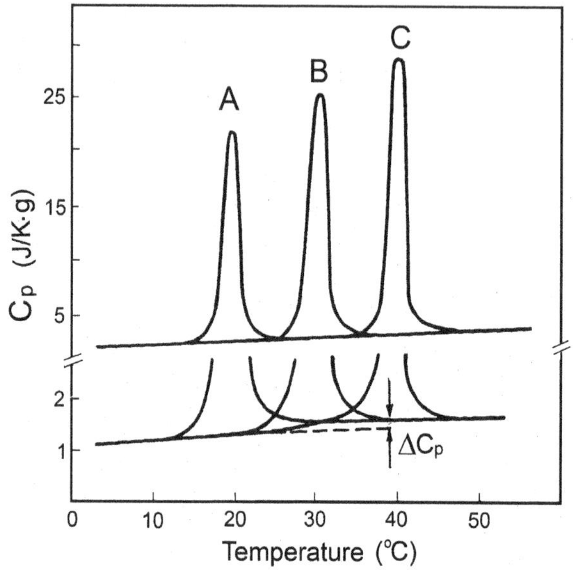

Figure 6.7. Temperature dependence of partial specific heat capacity of cod (A), pike (B), and rat (C) skin collagens in pH 3.5 salt-free solution. A fragment in magnified scale is presented under melting profiles to demonstrate the denaturational heat capacity increment (Privalov, 1982).

absorption (Privalov and Tiktopulo, 1970). There are two remarkable specificities in the heat effect of collagen melting: (a) the total heat of collagen melting is very large, much larger than the heat of denaturation of the α-helical coiled-coils and globular proteins. (b) the heat effect of collagen melting increases with increasing thermostability of the protein being considered (Fig. 6.7). The observed very steep dependence of the "melting" enthalpy on "melting" temperature largely exceeds the temperature dependence of the denaturation enthalpy of globular proteins (see Chapter 7). The question of principal importance is whether the enthalpy of collagen melting is a direct function of temperature or does it reflect some structural difference of the collagens isolated from different species?

If the enthalpy of collagen unfolding is a direct function of temperature, unfolding of collagen should result in a heat capacity increment, as $\partial\Delta H/\partial T = \Delta C_p$ (see Section 2.1.6). Measuring the heat capacity increment upon collagen unfolding is not simple because the heat absorption peak is very high. Nevertheless, using highly precise DSC with capillary cells it was found that the heat capacity increment at collagen unfolding is very small. To see it in Figure 6.7, a fragment of the partial heat capacity function is included in a fivefold magnified scale. It appears that the heat capacity increment at collagen unfolding is less than 0.2 J/K·g; that is, it is significantly lower than the heat capacity increment at unfolding of globular proteins and it is much lower than the observed dependence of collagen unfolding enthalpy on temperature, which amounts to 0.7 J/K·g. Thus, it appears that the enthalpy of collagen unfolding does not depend noticeably on temperature. It follows from this experimental fact that:

The observed difference in the enthalpies of unfolding of various collagens with different stabilities is not caused by the temperature dependence of the enthalpy function, but results from the structural differences of the considered collagens, which differ in their imino acid composition.

Calorimetric enthalpy of collagen melting amounts to 6000 kJ per mole of protein consisting of about 900 residues, which exceeds 4 times the van't Hoff enthalpy determined from the sharpness of transition. It appears, thus, that the collagen triple helix does not represent a single cooperative system.

The results of calorimetric studies of collagens reported by various groups are listed in Table 6.1. The table shows that the temperature and the enthalpy of collagen melting do not depend significantly on pH and salt content in the solution. At change of pH from 6 to 2.2, the melting temperature for rat skin collagen decreases by 7 K

TABLE 6.1. Calorimetric Studies of Acid-Soluble Collagen Melting

Source	pH	Salt	T_m (°C)	ΔH_m kJ/mol-res	References
Cod skin	3.5		20.0	3.17	Privalov and Tiktopulo (1970)
	3.9	0.1M NaCl	18.0	3.48	Burjanadze (1979)
	3.9	0.1 M NaCl	20.0	3.12	Monaselidze et al. (1973)
	2.2		14.0	3.13	Menashi et al. (1976)
	3.5	–	20	4.17	Privalov et al. (1979)
Ice fish skin	3.9	0.1 M NaCl	10	1.62	Burjanadze (1979)
	3.9	0.1 M NaCl	10.0	1.40	Monaselidze et al. (1973)
Whiting skin	3.5	–	21.3	3.87	Privalov and Tiktopulo (1970)
	3.9		25.0	4.10	Monaselidze et al. (1973)
Poke skin	3.5		30.6	5.27	Privalov and Tiktopulo (1970)
	2.2		21.0	4.91	Menashi et al. (1976)
	3.5		31.0	5.32	Privalov et al. (1979)
Halibut skin			13.0	4.24	Menashi et al. (1976)
Frog skin	3.9	0.1 M NaCl	320	5.27	Burjanadze (1979)
	3.9	0.1 M NaCl	35.0	5.36	Monaselidze et al. (1973)
	3.5	–	32.2	5.00	Privalov et al. (1979)
Carp swim bladder	3.5		36.3	5.20	Privalov et al. (1979)
Sheep skin	2.2	0.1 M NaCl	36.0	6.04	Menashi et al. (1976)
Calf skin			40.5	4.30	McClain and Wiley (1972)
Rat skin	6.0	–	42.0	6.7	Privalov (1968)
	2.2	–	35.0	6.4	Privalov (1968)
	3.5	0.1 M NaCl	36.5	6.12	Privalov and Tiktopulo (1970)
	3.5	0.1 M NaCl	36.5	6.12	Privalov and Tiktopulo (1970)
	3.9		41.0	6.38	Monaselidze et al. (1973)
	3.5		39.7	6.56	Privalov et al. (1979)
$(–Pro–Pro–Pro–)_{10}$	Acet. acid 4%			2.40	Engel et al. (1977)

TABLE 6.2. Structural Characteristics of Collagens and Thermodynamic Parameters of their Unfolding–Dissociation[a]

No.	Source	$(N_{1000})^{b}$ Pro	4-Hyp[c]	T_t (°C)	ΔH_t (J/mol-res)	ΔS_t (J/K·mol-res)
1	Cod skin	103	58	15	4100	14.0
2	Halibut	113	66	18	4490	15.4
3	Frog skin	106	68	31.2	4880	15.6
4	Pike skin	134	73	30.0	5240	17.3
5	Carp swim bladder	121	84	34.3	5150	16.7
6	Rat skin	112	115	39.7	6450	21.5
7	Sheep skin	133	97	39.0	6310	20.2

[a]Privalov (1982).
[b](N_{1000})-Imino acid content per 1000 residues.
[c]3-Hydroxyproline is regarded as Pro.

and the enthalpy by 0.6 kJ/mol, i.e. less than 5%. An increase of NaCl concentration from zero to 0.1 M leads to a decrease of melting temperature by 4 K and of the enthalpy by 0.5 kJ/mol. The observed decrease of enthalpy with variation of pH and salt content partly results from decrease in transition temperature. To exclude the temperature effect while considering thermodynamic characteristics of collagens, it is convenient to use the enthalpy and entropy reduced to some standard temperature. The thermodynamic parameters of the most thoroughly studied collagens reduced to 25°C, are listed in Table 6.2. They show that:

(a) *The enthalpy and entropy of collagen transition from the coiled-coil to the random-coil state are specific for characteristics of the given collagen molecule and vary greatly from species to species.*
(b) *The enthalpy and entropy of collagens unfolding correlate with their thermostability—both increase with increasing "melting temperature" and with increasing Gibbs energy of unfolding.*

The last fact is in drastic conflict with what was expected from the two models of collagen stabilization considered in the previous section. According to these models, with increasing melting temperature the enthalpy of collagen melting should not increase or even should decrease, but it increases (Fig. 6.8). Moreover, it was predicted that the entropy of collagen unfolding should decrease with increase of the imino acid content, but it increases.

6.2.2. Correlation between Thermodynamic and Structural Characteristics of Collagens

Figure 6.9 presents a plot of the reduced "melting" enthalpies of collagens versus the total imino acid content (a) and the content of hydroxyproline (b). It is seen

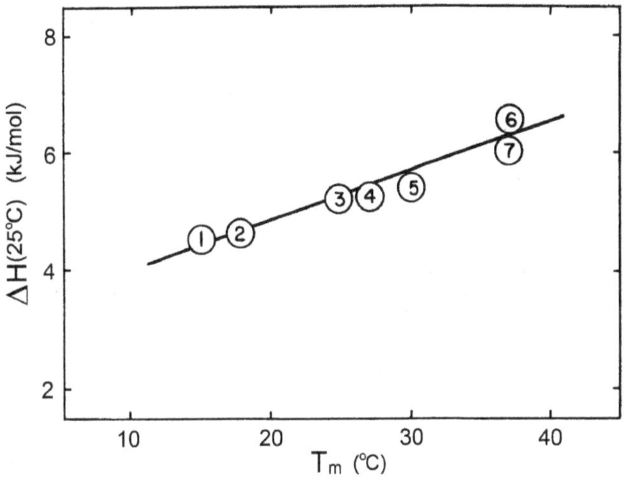

Figure 6.8. Plot of the reduced (to 25°C) melting enthalpy (per mole of residue values) versus melting temperature for collagens from different species listed in Table 6.2 (Privalov, 1982).

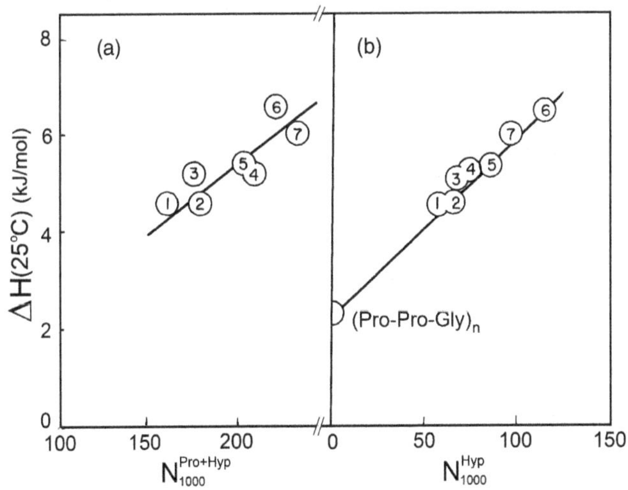

Figure 6.9. Plot of the reduced (to 25°C) melting enthalpy (per mole of residue values) of collagens versus (a) total prolyl and hydroxyprolyl content and (b) only hydroxyprolyl content in the helical parts of various species listed in Table 6.2 (Privalov, 1982).

that there is some correlation between the enthalpy and the total imino acid content, but a much better correlation is observed with the content of hydroxyproline alone.

Thus, calorimetric data on collagens "melting" showed unequivocally that both original models do not explain the thermal properties of collagens.

The fact that the melting enthalpy increases with increasing thermostability and its increase closely correlates with increasing hydroxyproline content clearly indicates that the main stabilizing effect comes not from the rigidity of the pyrrolidine ring but from hydroxyproline, that is, from the presence of a hydroxyl group on its ring. However, is unclear how this hydroxyl group could induce significant increase of the stability and the melting enthalpy of the collagen coiled-coil.

Nevertheless, it was found that with replacement of Pro on 4-Hyp in the synthetic polypeptide (Gly-Pro-Hyp)$_n$ the transition temperature increased by 35 K (Engel et al., 1977). It was also found that the transition temperature of the 26-residue fragment of αICB2 rat skin collagen on hydroxylation of its two proline residues (each in the third position in the triplets) increased by 8 K (Ward and Mason, 1973). An important discovery was also that the hydroxylation of prolines in collagen, which usually takes place at the post-translation stage, leads to an increase of the transition temperature by 15 K (Berg and Prockop, 1973; Rosenbloom et al., 1973; Fessler and Fessler, 1974). It appeared also that T_m correlates with the hydroxyproline only for collagens of vertebrates (Burjanadze, 1979). The absence of any correlation for invertebrates seemed to result from the 4-hydroxyproline being located in the third position in the triplets only in collagens of vertebrates, whereas in collagens of invertebrates 4-hydroxyproline is located with a rather equal probability in the second position too (Piez, 1976; Goldstein and Adams, 1970). Thus, for collagen stabilization presence of the hydroxyproline at the third position in the triplet is essential.

It is remarkable that at this position hydroxyproline is exposed to water; that is, it is unable to form a hydrogen bond within the triple helix—it can interact only with water.

Bearing in mind the well-known tendency of water molecules to cooperate with their neighbors, it does not seem improbable that hydroxyprolyl in these positions can serve as an initiator of an extensive cooperative network of hydrogen bonding enveloping the collagen that might be responsible for the exceptionally high enthalpy of collagen melting.

6.2.3. Role of Water in Maintaining the Collagen Structure

There is much experimental evidence of water participation in collagen stabilization. According to the early X-ray diffraction studies, collagen loses its regular structure with the removal of water (Luescher et al., 1973). This has been shown also by infrared spectroscopic studies of collagen films at various humidities (Susi et al., 1971). The total amount of water bound by collagen ("unfreezable water") is, according to various determinations, about 0.5 g per gram of collagen (Privalov and Mrevlishvili, 1967; Fung et al., 1974). Thus, hydration of collagen greatly exceeds that of other proteins (Privalov, 1968). From the water sorption isotherms of collagen it was concluded that there are two or three binding sites of water per triplet in collagen (Luescher et al., 1974; Grigera and Berendsen, 1979). According to infrared dichroism measurements on kangaroo tail tendon, strongly bound water molecules are preferentially oriented to the fiber axis (Suzuki and Fraser, 1974). The existence of water molecules oriented to the collagen axis has been shown also by nuclear magnetic resonance (NMR) studies on oriented collagen fibers (Fung and Siegel, 1972; Mighelsen and Berendsen, 1973) and by studies of dielectric properties of water absorbed on collagen (Hoeve and Lue, 1974).

It is remarkable that upon denaturation there is significant reduction of the water sorption capacity of collagens despite the much larger surface of the dissociated strands of collagen (Luescher et al., 1973). Therefore, the native triple-helical conformation is essential for strong binding of water. According to the calorimetric studies of collagen melting in the presence of a limited amount of water, when water concentration is decreased below the critical value of about three water molecules per triplet the enthalpy of collagen denaturation rapidly falls while the melting temperature increases (Luescher et al., 1974).

Although there were earlier reports of crystallization of collagen fragments, the collagen molecule itself has not proved amenable to investigation at the molecular level. The path to the molecular details of the collagen triple helix has been through collagen model peptides, which have yielded high-resolution X-ray structures (Brodsky and Persikov, 2008). They provide the first visualization of the elaborate water network that surrounds collagen molecules (Bella et al., 1994, 1995; Bella and Berman, 1996). Water molecules are seen to bridge C=O and Hyp–OH groups and repetitive networks of these water patterns are seen along the chain (Fig. 6.10). The increased number of high-resolution structures have confirmed that extended water networks are an inherent feature of all collagen triple-helix peptide crystal structures (Berisio et al., 2001, 2002; Kramer et al., 1999, 2001). NMR studies have indicated the kinetically labile nature of this collagen hydration shell (Melacini et al., 2000).

6.3. THERMODYNAMICS OF COLLAGENS

6.3.1. Cooperativity of Collagen Unfolding

Since separation of the collagen strands takes place only at the final stage of the melting process, the main process of disruption of the helical conformation of collagen can be regarded in principle as an equilibrium monomolecular reaction, which can be treated by the van't Hoff equation or its approximate expression (Eq. 3.24):

$$\Delta H^{\mathrm{vH}} \approx \frac{4RT_m^2}{\Delta T_{1/2}} \tag{6.6}$$

Thus, one can estimate the effective, or van't Hoff, enthalpy of collagen transition from the sharpness of the heat absorption peak, that is, from its half-width $\Delta T_{1/2}$—the temperature range in which the process progresses from one-fourth to three-fourths of completeness. The $\Delta T_{1/2}$ for collagens in a monodispersed state is about

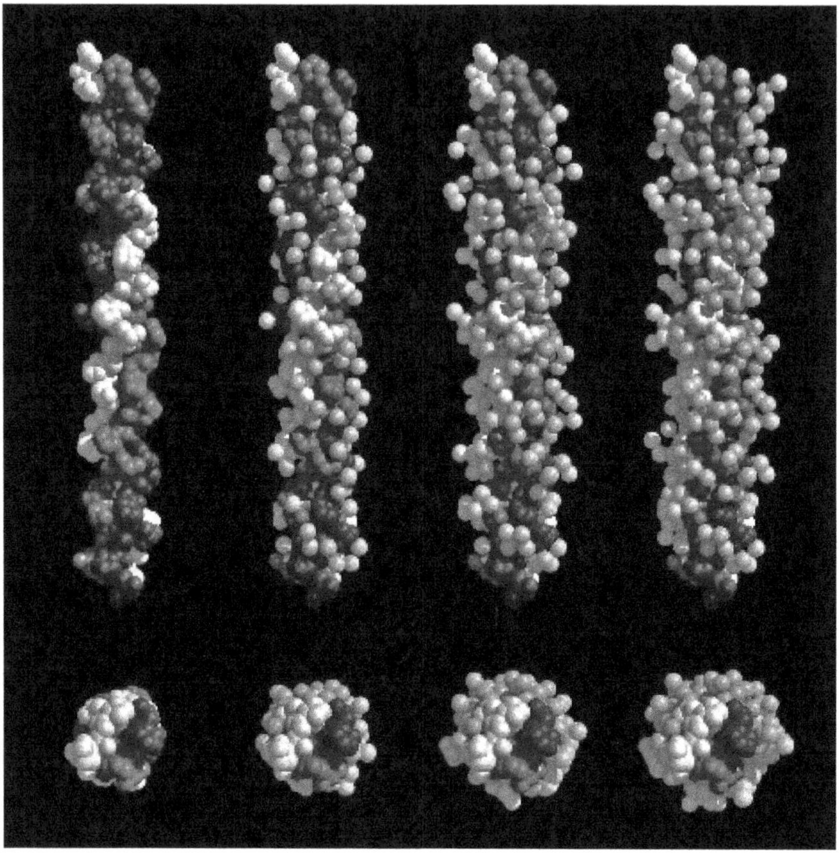

Figure 6.10. Layers of the fixed water molecules covering the collagen triple helix (Bella et al., 1995). (See insert for color representation of the figure.)

(2.1 ± 0.3) K. Therefore, for the effective van't Hoff enthalpies of cod skin, pike skin, and rat skin collagen transition one gets 1.20×10^3, 1.40×10^3, and 1.60×10^3 kJ/mol, respectively, which assumes rather large cooperative blocks in the collagen macromolecule. The number of residues forming a cooperative block, which melts as a single structural unit, can be found by a comparison of the effective enthalpy with the calorimetrically determined melting enthalpy (see Table 6.2). This gives that the cooperative block in collagen includes about 250 residues; that is, it is 80–85 residues long. The question is whether the cooperative regions in collagen are predetermined by some singularities in the sequence, as they are in the α-helical coiled-coils (see Section 5.4), or they have only statistical meaning, as in the single α-helix,

Examination of the known collagen sequences does not reveal any distinctive singularities in the regular arrangement of triplets in the polypeptide chain of collagen: in vertebrate collagens glycyl is always every third residue in the chain, and no long sequences, comparable in size with cooperative ones, have been found with low imino acid content (Piez, 1976; Fietzek et al., 1979; Allmann et al., 1979; Seyer and Kang, 1981). However, a thorough statistical analysis of the distribution of these residues in the sequence of vertebrate collagens demonstrates the existence of 78-residue repeat along the chain (Hofmann et al., 1984). The homologous 78-residue unit in collagen is a reminiscence of a primordial gene. The close correspondence of the size of this unit with the size of a cooperative region suggests that the cooperative region in collagen is predetermined, that is, that the collagen structure is discrete and consists of discrete structural blocks, just as does the structure of other large proteins. It is clear now that the sequence of most collagens includes sporadic interruptions in the repeating pattern (Kassner et al., 2004; Siebold et al., 1987; Brodsky and Persikov, 2008).

6.3.2. Factors Responsible for Maintaining the Collagen Coiled-Coil

As has been shown, the enthalpy of collagen melting strongly depends on the hydroxyproline content. The dependence between the enthalpy value, calculated per 1000 residues, and the hydroxyproline content n_{1000}^{Hyp} presented in Figure 6.9 can be expressed in the first approximation by the following linear equation:

$$\Delta H_{1000}(25°\mathrm{C}) = (2400 + 35.8 \times n_{1000}^{\mathrm{Hyp}})\ \mathrm{kJ/mol} \tag{6.7}$$

The first term in this equation corresponds to the transition enthalpy of a collagen-like structure that does not contain hydroxyproline, for example, (Pro–Pro–Pro)$_n$ polymer. If we assume that it results from water ordering by the 4-Hyp and assume that water ordering enthalpy in this case is similar to the water freezing enthalpy, which is about 6 kJ/mol, we find that each 4-Hyp in the third position in the triplet of the collagen chain stabilizes about six water molecules.

As for the entropy of collagen transition, it is *a priori* evident that it should depend on at least three variables: the content of the total imino acids, that of the hydroxyprolyl alone, and that of the proline residues alone. Earlier it was assumed that the entropy contribution of the residues containing pyrrolodine rings at collagen melting is zero, while the contribution of all other residues is about 17 J/K·mol (Harrington, 1964). However, now it is evident that the entropy contribution of hydroxyproline is far from being zero. As for the proline residues, it is also very unlikely that they do not contribute at all to the entropy of collagen chain in the random-coiled states. At the same time the assumption that the entropy contribution of the amino acids is 17 J/K·mol is also inconsistent with the entropy values obtained calorimetrically for collagens if the entropy dependence on the imino acid content is taken into account (see Table 6.2). Extrapolating the entropy values of collagens to zero imino acid content, one gets the entropy contribution of the amino acids at collagen melting as 11 J/K·mol. Since there is a large excess of glycine among the amino acids present in collagen, it can be assumed that its entropy contribution is close to this averaged entropy value. Then, using the entropy value for melting synthetic collagen-like structures containing proline, it is possible to estimate the entropy contribution of the proline residue at collagen melting. According to Engel et al. (1977) the entropy of melting (–Pro–Pro–Gly–)$_{10}$ is 21 J/K per mole of triplets. Excluding the contribution of glycine (11 J/K·mol) from this value, we get 5.0 J/K·mol for the entropy contribution of pyrrolidine-containing residues to the entropy of unwinding of the collagen chain upon denaturation.

The origin of the observed effect is hardly the flexibility of imino acid residue in the polypeptide chain when it is in a random-coil conformation. It is likely to be the effect of water ordering around collagen when it is in the triple-helical conformation if this conformation corresponds to the ice-lattice structure (Berendsen and Mighelsen, 1965; Lim, 1981). Using this value for the proline and amino acid residues, it is possible to estimate the entropy contribution of hydroxyproline from the known entropies of collagen melting (Table 6.2). This estimate gives the value of 103 J/K·mol. Thus, the entropy of collagen transition at 25°C can be approximated by the following empirical expression;

$$\Delta S_{1000} = (25°\text{C}) = [11 \times (1000 - n_{1000}^{\text{pro+Hyp}}) + 5 \times n_{1000}^{\text{pro}} + 103 n_{1000}^{\text{Hyp}}]\ \text{J/K·mol} \quad (6.8)$$

Here the impressive difference between the contribution of proline and hydroxyproline residues attracts the most attention. Assuming that this difference is connected with the ability of 4-hydroxyproline to interact with water, while the entropy contribution of the hydroxyproline residue itself is the same as that of the proline residue (5 J/K·mol), then the entropy contribution of water should be 98 J/K·mol. In this case, Equation (6.8) can be rewritten in the form:

$$\Delta S_{1000} = (25°\text{C}) = [11 \times 1000 - 6 n_{1000}^{\text{pro+Hyp}} + 5 \times n_{1000}^{\text{pro}} + 98 n_{1000}^{\text{Hyp.W}}]\ \text{J/K·mol} \quad (6.9)$$

Here the second term represents the entropy decrement caused by the rigidity of the pyrrolidine ring and the last term represents the entropy connected with the stabilization of water molecules around the collagen coiled-coil induced by hydroxyproline. Comparing it with the entropy of water freezing (22 J/K·mol) one can see that the hydroxyproline-induced entropy decrease is comparable with the effect of freezing about five water molecules.

Combining Equations (6.7) and (6.9), and having in mind that a cooperative block in collagen includes about 250 residues, one finds that the Gibbs energy of cooperative transition at 25°C can be expressed as follows:

$$\begin{aligned}\Delta G = (25°\text{C}) &= 0.25[\Delta H_{1000}(25°\text{C}) - 298 \times \Delta S_{1000}(25°\text{C})] \\ &= [-825 + 600 + 1.4 \times n_{1000}^{\text{Hyp.W}} + 0.45 \times n_{1000}^{p\text{Pro+Hyp}}]\ \text{kJ/mol} \quad (6.10)\end{aligned}$$

The descriptive capacity of this equation, as should be expected, is not high, but it helps to estimate the relative contribution of different factors in stabilization of the collagen structure. The first term in this equation presents a negative contribution of the entropy factor with the assumption that all residues are equally flexible. The second term represents the contribution of intramolecular bonding that is independent of the imino acid content in collagen. The third term represents the contribution of hydroxyproline interaction with water. The last term represents the contribution of rigid pyrrolidine rings. As can be seen, the most important contributors to stabilization of the collagen coiled-coil structure are the intramolecular bonds.

They alone, however, are absolutely insufficient to maintain this structure. The next contributor is water stabilized by hydroxyproline. Bearing in mind that in collagens of vertebrates the amount of hydroxyproline is about half of the total amount of imino acids, it follows that the interaction with water contributes to the stability of collagen structure about twice as much as the rigidity of pyrrolidines.

6.3.3. Flexibility of the Collagen Structure

Comparing the kinetics of hydrogen exchange of different collagens at a temperature where the classes of slowly and very slowly exchangeable hydrogens are not yet differentiated (Fig. 6.3) it becomes evident that the rates of exchange (i.e., the slopes of the curves) are different and that they correlate distinctly with the thermostability of the given collagens. Since the rate of hydrogen exchange can be considered as a measure of mobility or flexibility of molecular structures, one can conclude from this fact that there is some correspondence between flexibility of collagen structure and its thermostability.

Since the kinetics of hydrogen exchange in collagen is simple, the flexibility of its structure can be quantitatively determined from the observed rate of hydrogen exchange. For such simple kinetics of exchange the equilibrium constant of local micro-unfolding is expressed as follows:

$$K = \frac{1}{K_0} \cdot \frac{d(\mathrm{N} + \mathrm{U}^*)}{dT} \tag{6.11}$$

where N is the native and U is the unfolded state, the asterisk denotes the deuterated or tritiated state, and K_0 is the known rate constant for the exchange of peptide hydrogens exposed to water (Hvidt and Nielsen, 1966).

This type of analysis has been done for the classes of very slowly exchangeable hydrogens that can be clearly distinguished in a broad temperature range (Fig. 6.11). The results obtained are summarized in Figure 6.12 in the form of a van't Hoff plot of the equilibrium constant of micro-unfolding. As can be seen, the concentration of the local micro-unfoldings in the predenaturation temperature range is quite low, but it increases with an increase of temperature and then this dependence breaks abruptly at reaching the denaturation temperature range. Hence, one can conclude that there are two qualitatively different processes of collagen structure unfolding, the nondenaturational and denaturational.

Since the slope of the van't Hoff plot corresponds to the enthalpy of a process, it follows from the obtained results that the nondenaturational and the denaturational unfolding of collagen structure are characterized by qualitatively different enthalpies. The enthalpy of the nondenaturational process is small and is practically the same for all collagens studied, being about 25 kJ/mol (Table 6.3). Comparing it with the averaged calorimetric enthalpy of collagen melting calculated per mole of residue, 5.5 kJ/(mol-res), one can conclude that the nondenaturational

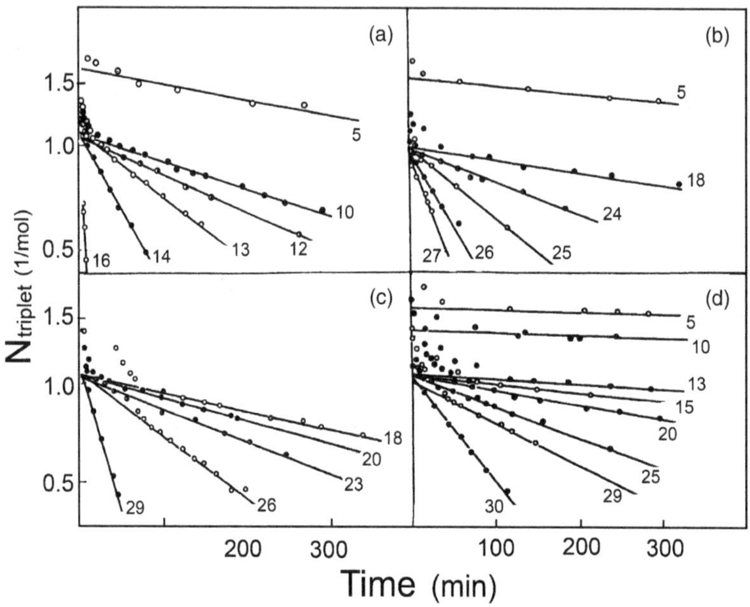

Figure 6.11. Temperature dependence of hydrogen exchange kinetics of collagens from (a) cod skin, (b) pike skin, (c) carp swim bladder, and (d) rat skin in solutions at pD 3.5 (Privalov et al., 1979).

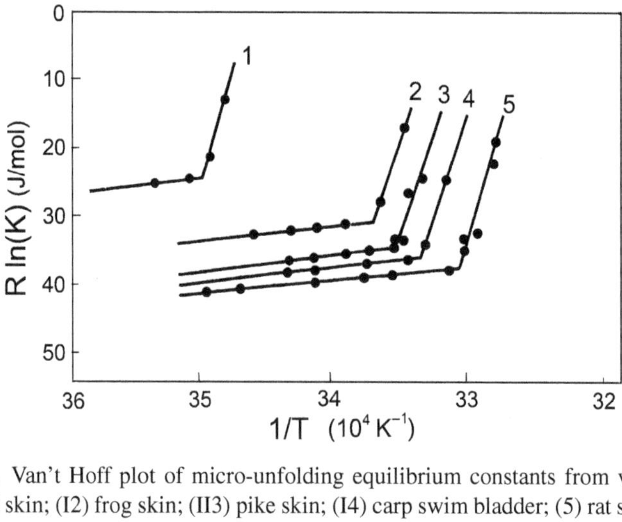

Figure 6.12. Van't Hoff plot of micro-unfolding equilibrium constants from various collagens: (1) cod skin; (I2) frog skin; (II3) pike skin; (I4) carp swim bladder; (5) rat skin (Privalov et al., 1979).

TABLE 6.3. Thermodynamic Parameters of Micro-unfolding in Collagens[a]

Organism	T (K)	ΔG^{mic} (kJ/mol)	ΔH^{mic} (kJ/mol)	ΔS^{mic} (kJ/Kmol)
Cod	282.9	7.2	25	63
	285.1	7.0	25	63
	286.3	6.1	≈600	≈200
	287.3	3.0	≈600	≈200
Frog	288.9	9.6	25	53
	291.3	9.4	25	53
	293.2	9.3	25	53
	295.1	9.3	25	53
	297.3	7.7	≈600	≈200
	299.1	5.1	≈600	≈200
Pike	291.1	10.7	25	49
	295.1	10.6	25	49
	296.6	10.5	25	49
	298.1	10.5	25	49
	299.1	7.9	≈600	≈200
	300.1	7.3	≈600	≈200
Carp	291.0	11.2	25	47
	293.1	11.1	25	47
	296.1	11.0	25	47
	299.1	10.9	25	47
	302.1	7.5	≈600	≈200
Rat	293.1	11.8	25	45
	296.1	11.7	25	45
	298.1	11.6	25	45
	301.9	11.4	25	45
	303.1	10.3	≈600	≈200
	305.1	8.8	≈600	≈200

[a]Privalov et al. (1979).

micro-unfolding of collagen structure involves about five residues, that is, one or two triplets. In contrast, the denaturational process involves two orders larger numbers of triplets, which unfold cooperatively.

The other fact that attracts attention in considering Figure 6.12 is the significant difference between the values of $R\ln K$ for various collagens if we compare them at the same temperature. Having in mind that the $RT\ln K$ is the Gibbs energy of the considered process, it follows that the Gibbs energies of micro-unfoldings, which can be denoted as ΔG^{mic}, are different for collagens from various species. The values of ΔG^{mic}, which can be regarded as a measure of microstability or of rigidity of collagen structure, are presented in Table 6.4. It can be seen that they correlate with the $\Delta G^{mac}(25°C)$ describing the macrostability of collagen structure obtained from its denaturation studies.

TABLE 6.4. Gibbs Energies of Macroscopic and Microscopic Unfolding of Collagens at Standard (25°C) and Physiological Temperatures[a]

Organism	Physiological Temperature T_{ph} (°C)	Macroscopic Unfolding ΔG^{mac} (J/mol)		Microscopic Unfolding ΔG^{mic} (kJ/mol)	
		At 25°C	At T_{ph}	At 25°C	At T_{ph}
Cod	5.0	−80	204	6.0	8.0
Pike	14.0	88	292	10.5	11.0
Frog	19.0	120	219	9.0	9.5
Carp	18.0	180	308	11.0	11.0
Rat	35.0	323	405	11.7	11.1
Relative spread (%)		125	36	24	13

[a]Privalov et al. (1979).

It appears that the microstability and macrostability of collagen structure correlate and both depend on the content of imino acids, and particularly hydroxyproline.

6.3.4. Biological Aspect of the Collagen Stability Problem

The question as to why collagens from different species have different stabilities has long been discussed. It has been found that the melting temperature of the collagen supermolecular structure (fibrils, tissues) correlates with the physiological temperature of donor species (Gustavson, 1956; Gustavson and Takahashi, 1956; Ushakov, 1964; Alexandrov and Andreeva, 1967). The same correlation is observed for monodispersed collagen molecules in solution, although their transition temperature is lower than that of the supermolecular collagen structures (Rigby, 1967; Rigby and Robinson, 1975) The difference between melting temperature of monodispersed collagen and collagen in fibrils is very similar for all collagens studied, being about 25 K. But the most remarkable fact is that the transition temperature of a monodispersed collagen is quite close to the upper limit of physiological temperature of the animal from which that collagen was obtained (Fig. 6.13). It is reasonable that the denaturational temperature of the collagen fibrils, which represent the functional state of collagen, is above the averaged physiological temperature. It is unclear, however, why the thermostability of all collagens is not significantly higher than physiological temperature, as it is in the case of globular proteins; moreover it decreases with decrease of the physiological temperature of the species.

According to Alexandrov (1977), the correlation between physiological temperature and collagen stability can be explained, assuming that some definite level of flexibility of protein structure is required for its efficient functioning in living systems and the stability of protein is a property linked with the flexibility of macromolecular structure. As Table 6.4 shows, the collagen's macrostabilities correlate indeed with its microstability, that is, rigidity. It is most interesting that:

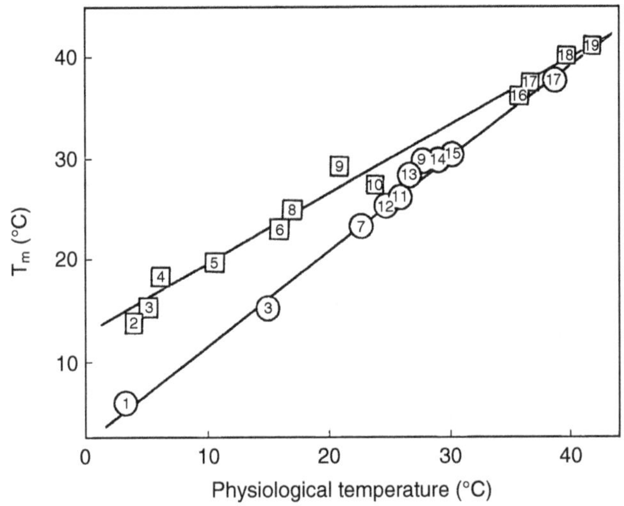

Figure 6.13. Plot of the melting temperature of collagens from various species versus the average physiological temperature (squares) of the species and the upper limit for physiological temperature (circles): (1) icefish; (2) antimora; (3) cod; (4) *Alepocephalus*; (5) whiting; (6) *Allolobaphora caliginosa*; (7) earthworm; (8) flatfish; (9) *Cyprinus carpio*; (10) butterfly fish; (11) tuna; (12) *Rana tempararia*; (13) *Aurelia coerula*; (14) *Rana ridibunda*; (15) *Helix aspersa*; (16) rat; (17) human; (18) hog; (19) chick (Privalov, 1982).

> *The Gibbs energies of micro-unfolding of different collagens are much the same if their values are compared at the corresponding physiological temperatures (Privalov et al., 1979).*

Thus, it follows that all collagens have indeed the same level of structure flexibility (rigidity) at physiological conditions, which is required for their proper functioning.

It still remains unclear why the transition temperature of monodispersed collagens is so close to the physiological one? This is especially evident in the case of collagens from warm-blooded animals that have a definite physiological temperature: while their physiological temperature is close to 37°C, their denaturation temperature is 40–42°C, that is, only 3–5 K higher (Table 6.3) It is also remarkable that this level of stability is achieved only as a result of post-translation modification of the collagen polypeptide chain by prolyl hydroxylase consisting in hydroxylation of prolyl residues in every third position in the triplet, while the newly synthesized polypeptide chains are forming a triple coiled-coil that is unstable at physiological temperature (Rosenbloom et al., 1973; Berg and Prockop, 1973; Fessler and Fessler, 1974; Prockop et al., 1976). Thus, it is very likely that the low stability of collagen structure is necessary for the easy transportation of synthesized polypeptide chains to where they are needed and for the technology of their assembly into the rigid triple coiled-coil.

REFERENCES

Alexandrov V.Ya. (1977). *Cells, Macromolecules and Temperature*. Springer-Verlag, Berlin and New York.

Alexandrov V.Ya. and Andreeva A.P. (1967). *Tsitologia*, **9**, 1288–1293.

Allmann H., Fietzek P.P., Glanville R.W., and Kühn K. (1979). *Hoppe Seyler's Z. Physiol. Chem.*, **360**, 861–868.

Bachinger H.P., Bruckner P., Timpl R., and Engel J. (1978). *Eur. J. Biochem.*, **90**, 605–613.

Bella J. and Berman H.M. (1996). *J. Mol. Biol.*, **264**, 734–742.

Bella J., Brodsky B., and Berman H.M. (1995). *Structure*, **3**, 893–906.

Bella J., Eaton M., Brodsky B., and Berman H.M. (1994). *Science*, **266**, 75–81.

Berendsen H.J.C. and Mighelsen C. (1965). *Ann. N.Y. Acad. Sci.*, **125**, 365–379.

Berg R.A. and Prockop D.J. (1973). *Biochem. Biophys. Res. Commun.*, **52**, 115–120.

Berisio R., Vitagliano L., Mazarella L., and Zagari A. (2001). *Biopolymers*, **56**, 8–13.

Berisio R., Vitagliano L., Mozzarella L., and Zagari A. (2002). *Protein Pept. Lett.*, **9**, 107–116.

Boedtker H. and Doty P. (1956). *J. Am. Chem. Soc.*, **78**, 4267–4280.

Brodsky B. and Persikov A. (2008). *Adv. Protein Chem.*, **70**, 301–339.

Burjanadze T.V. (1979). *Biopolymers*, **18**, 931–938.

Burge R.E. and Hyness R.D. (1959a). *Nature*, **184**, 1562–1565.

Burge R.E. and Hyness R.D. (1959b). *J. Mol. Biol.*, **1**, 155–164.

Emsley J., Knight C.G., Farndale R.W., and Barnes M.J. (2004). *J. Mol. Biol.*, **335**, 1019–1028.

Engel J. (1962). *Arch. Biochem. Biophys.*, **97**, 150–158.

Engel J., Chen H.T., and Prockop D.J. (1977). *Biopolymers*, **16**, 601–622.

Esipova N.G., Lazarev Yu.A., and Lazareva A.V. (1972). *Biofizika*, **17**, 949–953.

Fessler L.I. and Fessler J.H. (1974). *J. Biol. Chem.*, **249**, 7637–7646.

Fietzek P.P., Allman H., Rautenberg J., Henkel W., Wachter E., and Kuhn K. (1979). *Hoppe-Seyler's Z. Physiol. Chem.*, **360**, 809–820.

Fraser R.D., MacRae T.P., and Suzuki E. (1979). *J. Mol. Biol.*, **129**, 463–481.

Fung B.M. and Siegel M.M. (1972). *Biochim. Biophys. Acta*, **278**, 185–187.

Fung B.M., Witschel J., and McAmis L.L. (1974). *Biopolymers*, **13**, 1767–1776.

Goldstein A. and Adams E. (1970). *J. Biol. Chem.*, **245**, 5478–5483.

Grigera J.R. and Berendsen H.J.C. (1979). *Biopolymers*, **18**, 35–45.

Gustavson K.H., Ed. (1956). In: *The Chemistry and Reactivity of Collagen*. Academic Press, New York.

Gustavson K.H. and Takahashi J. (1956). In: *The Chemistry and Reactivity of Collagen* (K.H. Gustavson, Ed.). Academic Press, New York.

Harrington W.F. (1964). *J. Mol. Biol.*, **9**, 613–617.

Hoeve C.A.J. and Lue P.C. (1974). *Biopolymers*, **13**, 1661–1680.

Hofmann H., Voss T., Kuhn K., and Engel J. (1984). *J. Mol. Biol.*, **172**, 325–343.

Hvidt A. and Nielsen S.O. (1966). *Adv. Protein Chem.*, **21**, 287–386.

Kassner A., Tiedemann K., Notbohm H., Ludwig T., Morgelin M., Reinhardt D.P. et al. (2004). *J. Mol. Biol.*, **339**, 835–853.

Katz E.P. (1970). *Biopolymers*, **9**, 745–748.

Kingham D.J. and Brisbin D.A. (1968). *Can. J. Biochem.*, **46**, 1199–1203.

Kramer R.Z., Bella J., Brodsky B., and Berman H.M. (2001). *J. Mol. Biol.*, **311**, 131–147.

Kramer R.Z., Bella J., Mayville P., Brodsky B., and Berman H.M. (1999). *Nature Struct. Biol.*, **6**, 454–457.

Lim V.I. (1981). *FEBS Lett.*, **132**, 1–5.

Luescher M., Giovanoli R., and Hirter P. (1973). *Chimia*, **27**, 112–116.

Luescher M., Ruegg M., and Schindler P. (1974). *Biopolymers*, **13**, 2489–2503.

McBride O.W. and Harrington W.F. (1967). *Biochemistry*, **6**, 1484–1498.

McClain P.E. and Wiley E.R. (1972). *J. Biol. Chem.*, **247**, 692–697.

Melacini G., Bonvin A.M., Goodman M., Boelens R., and Kaptein R. (2000). *J. Mol. Biol.*, **300**, 1041–1049.

Menashi S., Finch A., Gardner A., and Ledward D.A. (1976). *Biochim. Biophys. Acta*, **444**, 623–625.

Mighelsen C. and Berendsen H.C.J. (1973). *J. Chem. Phys.*, **59**, 296–305.

Monaselidze D.R., Majagaladze G.V., Bakradze N.G., et al. (1973). In: *Conformational Changes of Biopolymers in Solution* (E.L. Andronikashvili, Ed.), pp. 176–180. Nauka, Moscow.

Neyfakh Yu.A., Tumanian V.G., and Esipova N.G. (1978). *Biofizika*, **23**, 158–159.

Okuyama K., Hongo C., Wu G., Mizuno K., and Nogushi K. (2009). *Biopolymers*, **91**, 361–372.

Piez K.A. (1976). In: *Biochemistry of Collagen* (G.N. Ramachandran and A.H. Reddi, Eds.), pp. 1–44. Plenum, New York.

Privalov P.L. (1968). *Biofizika*, **13**, 163–177.

Privalov P.L. (1982). *Adv. Protein Chem.*, **35**, 1–104.

Privalov P.L. and Mrevlishvili G. M. (1967), *Biofizika*, **12**, 22–29.

Privalov P.L. and Tiktopulo E.I. (1970). *Biopolymers*, **9**, 127–139.

Privalov P.L., Serdyuk I.N., and Tiktopulo E.I. (1971). *Biopolymers*, **10**, 1777–1794.

Privalov P.L., Tiktopulo E.I., and Tischenko V.M. (1979). *J. Mol. Biol.*, **127**, 203–216.

Prockop D.J., Berg R.A., Kivirikko K.I., and Uitto J. (1976). In: *Biochemistry of Collagen* (G.N. Ramachandran and A.H. Reddi, Eds.), pp. 163–273. Plenum, New York.

Ramachandran G.N. and Kartha G. (1955). *Nature*, **176**, 593–595.

Ramachandran G.N. and Ramakrishnan C. (1976). In: *Biochemistry of Collagen* (G.N. Ramachandran and A.H. Reddi, Eds.), pp. 45–84. Plenum, New York.

Rich A. and Crick F.H.C. (1955). *Nature*, **176**, 915–916.

Rigby B.J. (1967). *Nature*, **214**, 87–88.

Rigby B.J. and Robinson M.S. (1975). *Nature*, **253**, 277–279.

Rosenbloom J., Harsch M., and Jimenez S. (1973). *Arch. Biochem. Biophys.*, **158**, 478–484.

Seyer J.M. and Kang A.H. (1981). *Biochemistry*, **20**, 2621–2627.

Siebold B., Qian R.A., Glanville R.W., Hofmann H., Deitzmann R., and Kuhn K. (1987). *Eur. J. Biochem.* **168**, 569–575.

Susi H., Ard J.S., and Carroll R.J. (1971). *Biopolymers*, **10**, 1597–1604.

Suzuki E. and Fraser R.D.B. (1974). Proc. Rehovot Symp., 2nd, Israel, May 1974. *Peptides, Polypeptides,and Proteins* (E.R. Blout, Ed.), pp. 449–458. Wiley–Interscience, New York.

Traub W. and Piez K.A. (1971). *Adv. Protein Chem.*, **25**, 243–352.

Ushakov B.P. (1964). *Physiol. Rev.*, **44**, 518–560.

von Hippel, P.H. (1967). In: *Treatise on Collagen* (G.N. Ramachandran, Ed.), **1**, pp. 253–338. Academic Press, New York.

von Hippel P.H. and Wong K.Y. (1963). *Biochemistry*, **2**, 1399–1413.

Ward A.R. and Mason P. (1973). *J. Mol. Biol.*, **79**, 431–435.

Yee R.J., Englander S.W., and von Hippel P.H. (1974). *J. Mol. Biol.*, **83**, 1–16.

7

GLOBULAR PROTEINS

7.1. DENATURATION OF GLOBULAR PROTEINS

7.1.1. Proteins at Extremal Conditions

By "globular proteins" we mean compact, highly ordered proteins with molecular mass less than 30 kDa, that is, the proteins specified by low intrinsic viscosity and low hydrogen exchange rate. Thus, the polypeptide chain in globular proteins is tightly packed in a rather complicated unique conformation. Examples of atomic and backbone structures of two typical globular proteins, lysozyme (M_w = 14,300 Da) and barnase (M_w = 12,365 Da), are presented in Figure 4.12.

Transfer of globular proteins to the extreme conditions (e.g., high temperature, high pressure, high acidity, or high concentration of denaturants) leads to disappearance of their unique biological function, that is, to their denaturation. Denaturation of proteins also results in drastic changes of all their specific physical properties: decrease of solubility (judging by increasing tendency to aggregate), loss of compactness (judging by increase of their intrinsic viscosity), disordering of conformation (judging by decrease of their ellipticity), and increase of hydrogen exchange rate. It appears that denaturation of globular protein results in disruption of its native

Microcalorimetry of Macromolecules: The Physical Basis of Biological Structures, First Edition.
Peter L. Privalov.

compact structure, unfolding of its polypeptide chain. If so, investigation of this phenomenon might lead to understanding the essence of protein structure, its physical ground. This explains the great interest in the protein denaturation phenomenon. The first questions that have to be answered in solving this problem are "How stable is protein against denaturing action?" and "How far does protein unfold upon denaturation and how reversible is this process?"

The very first experiments with protein solutions showed that heating induces drastic decrease of their solubility over a narrow temperature range, resulting in intensive aggregation. Although the decrease in solubility in aqueous solutions is one of the most characteristic features of protein denaturation, it is the least studied quantitatively. Moreover, this decrease in solubility is the major complication in studying the state of denatured protein, since aggregation causes the greatest problem for all physical methods. Because of this tendency to aggregate, the process of denaturation could be studied only under conditions preventing protein aggregation, that is, far from the isoelectric point and in highly dilute solutions. This last requirement caused a great popularity of optical methods in studying denaturation of protein and was a stimulus for developing the supersensitive microcalorimetric techniques.

In the beginning most studies of protein denaturation were done using various active reagents, such as urea and guanidinium hydrochloride, as denaturing agents since the induced changes in proteins could be easily observed at room temperature by any optical parameter sensitive to changes in protein structure. However, with the appearance of information on protein structure and the increasing interest in its energetic base, investigation of temperature-induced denaturation of proteins has gained a special importance, particularly using scanning microcalorimetry, providing direct information on the relations between the two conjugate intensive and extensive basic parameters—temperature and enthalpy (Privalov, 1974, 1979).

7.1.2. The Main Problems of Protein Denaturation

Why are we interested in protein denaturation? Why does the process of native proteins losing their unique properties deserve attention?

There are several reasons for this interest. The practical one is clear: since proteins are the main components of our food, knowledge of conditions preserving them from degradation is certainly important.

Less clear is the academic interest for protein denaturation. It has risen along with the understanding that studying protein denaturation is the only possible approach to a fundamental problem of the formation of protein structure. We know that protein is synthesized as a linear polymer, which then folds into unique conformations. But how does that happen? All theories on the formation of protein structure are based on the assumption that this is a thermodynamically driven process of folding of the initially unfolded polypeptide chain into a compact, unique conformation. However, the only unfolded state of protein that could be obtained experimentally is its state at the extremal conditions, that is, the denatured state of protein. Therefore study of protein denaturation is the only practical approach for understanding the mechanism of formation of protein native structure, understanding the forces involved in this process and their cooperation. The questions that have to be answered in these studies are as follows:

1. How closely does the denatured protein approximate the ideal unfolded state of the polypeptide chain, which is taken as an initial state in all theories of protein folding and in all attempts to predict the protein spatial structure based on its chemical structure?
2. How reversible is the process of protein denaturation–renaturation?
3. Only then, if the answers to both questions 1 and 2 are positive, can one ask and get from denaturation studies the answer to the question: what are the forces involved in the formation of protein structure?

7.2. HEAT DENATURATION OF PROTEINS

7.2.1. DSC Studies of Protein Denaturation upon Heating

Differential scanning microcalorimetry (DSC) studies of the aqueous solutions of globular proteins showed that their partial molar heat capacity increases linearly with increasing temperature, up to some critical level, but then they denature with extensive heat absorption resulting in a significant increase of their heat capacity (Fig. 3.17). Thus the heat capacity of the thermally denatured protein is significantly higher than of the native one, $C_p^{\mathrm{D}} - C_p^{\mathrm{N}} = \Delta_{\mathrm{N}}^{\mathrm{D}} C_p \gg 0$. The heat capacity increment upon temperature-induced denaturation of proteins somewhat complicates determination of the enthalpy of this process, that is, the area under the excess heat Q. Usually the value of this heat, that is, the area of the heat absorption peak, is determined by extrapolating the initial and final heat capacities to the midtransition temperature, T_t (Fig. 3.18b). The area of the heat absorption peak above these extrapolated lines gives the calorimetric enthalpy of denaturation. More justified is the use of an explicit integral of the excess heat capacity to obtain a sigmoidal baseline that gradually increases from the initial to the final heat capacity value in proportion to the heat absorbed for the given temperature (Fig. 3.18c).

7.2.2. Reversibility of Heat Denaturation

Reversibility of a process is a prime requirement for its thermodynamic analysis. In the case of temperature-induced protein denaturation, this can be checked by scanning the studied protein both up and down the temperature scale. Upon heating, up to 80°C, many small proteins show good reversibility in solutions preventing their aggregation, for example, at pH values away from the isoelectric point of the protein (see Fig. 3.15). When the instrument used does not permit scanning down in temperature, the reversibility of denaturation can be checked by reheating the same solution after cooling.

Reversibility of protein unfolding decreases dramatically upon heating above 80°C for almost all mesophilic proteins because some degradation of their groups, particularly their oxidation by the dissolved oxygen. However, even if protein denaturation does not show perfect reversibility, it might still be analyzed thermodynamically because the temperature-induced cooperative unfolding of proteins is a much faster process than the aggregation of unfolded proteins and oxidation of their exposed groups. The effect of these concomitant factors can be reduced by decreasing protein concentration, increasing the heating–cooling rate, and removing dissolved oxygen from the protein solution being studied. To remove oxygen, nitrogen is bubbled through the solution, which is then placed for a short time under vacuum to get rid of the remnants of both gases.

7.2.3. Cooperativity of Denaturation

Temperature-induced denaturation of small globular proteins usually proceeds in a rather narrow temperature range: the half-width of the transition zone (i.e., the width of the denaturational heat absorption peak at its half-height, $\Delta T_{1/2}$) for these proteins is usually between 4 and 6 K. The sharpness of the temperature-induced denaturation of globular proteins suggests that this is not a gradual process of degradation of its highly ordered structure by the intensified thermal motion. It looks as if intensification of the thermal motions in the protein leads to cooperation of the micro-unfoldings, their synchronization, resulting in the overall unfolding of the whole macromolecule (Ikegami, 1977).

Considering cooperative unfolding of a single globular protein as a transition between two states, N $\Leftrightarrow$ D, one can specify this process by the equilibrium constant:

$$K = [\mathrm{D}]/[\mathrm{N}] \tag{7.1}$$

where [D] and [N] are the concentrations of the denatured and native states of protein, respectively. In that case the portion of heat absorbed up to a given temperature T, $\theta = Q(T)/Q_{\mathrm{tot}}$, can be regarded as a parameter indicating the progress of the reaction. Then the equilibrium constant can be expressed as $K = \theta/(1 - \theta)$, which can be used in the van't Hoff equation (see Section 3.2.6). Then, using Equation (3.23) or its simplified version, Equation (3.24), one can determine the effective van't Hoff enthalpy of protein denaturation from the observed sharpness of the heat absorption peak,—its relative height $\Delta C_p^{\max}/Q$ or relative width $\Delta T_{1/2}/Q$.

For many small compact globular proteins the van't Hoff enthalpy of temperature-induced denaturation (ΔH^{vH}) appeared to be in perfect correspondence with the calorimetric enthalpy (ΔH^{cal}) of this process determined from the area of the heat absorption peak: the deviation between ΔH^{vH} and ΔH^{cal} for various compact globular proteins was found to be less than 5% (Fig. 7.1). Since the van't Hoff equation assumes that the process being considered represents a two-state monomolecular reaction, validity of this equation for temperature-induced denaturation of globular proteins means that this process can be regarded indeed as a highly cooperative process, which is closely approximated by a two-state transition (Privalov and Khechinashvili, 1974; Privalov, 1979). This means that the population of the protein

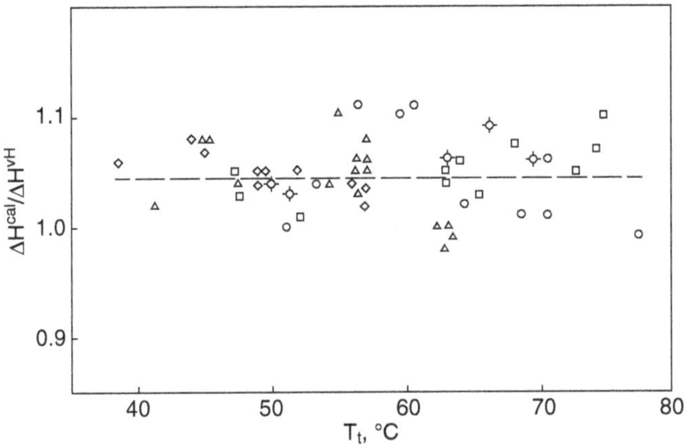

Figure 7.1. Ratio of the calorimetric and van't Hoff enthalpies of a number of globular proteins plotted against their transition temperature in solutions with various pH: (○) met-myoglobin; (Δ) ribonuclease A; (□) lysozyme; (◇) a-chymotrypsinogen; (☼) cytochrome c (Privalov and Khechinashvili, 1974).

states intermediate between the completely folded and unfolded states is very low; that is, it appears as if the protein jumps from the native (folded) state to the denatured (unfolded) state without noticeable intermediates. Moreover, the unfolded protein jumps back into the folded, far-from-simple conformation also without noticeable intermediates. This is not easy to imagine, but thus far all attempts to trace the intermediates upon unfolding or refolding of globular proteins by fast kinetic methods have failed: the compact globular proteins unfold and refold in less than 1 millisecond if the process is not complicated by the concomitant *cis–trans* slow conformational transition of prolines (Brandts et al., 1975, 1977; Baldwin, 1975).

It should be noted that the apparent sharpness of denaturation does not mean in itself that the temperature-induced process presents a two-state cooperative transition even if the various characteristics of protein are changing simultaneously in the narrow temperature range of protein denaturation. This is because, if a protein structure is subdivided into several subparts of identical stability, they might unfold in the same temperature range (see Chapter 10).

The only unequivocal criterion of a two-state transition of the considered temperature-induced process is the correspondence of its van't Hoff enthalpy and the real calorimetrically measured enthalpy.

In more detail one can analyze the temperature-induced denaturation of protein by computer modeling the heat capacity function assuming that it represents an ideal two-state transition (see Section 3.2.8). As shown in Figure 7.2, the deviation between the calorimetrically measured and the calculated best fit heat capacity profiles for barnase are very small. Such correspondence of the calorimetrically

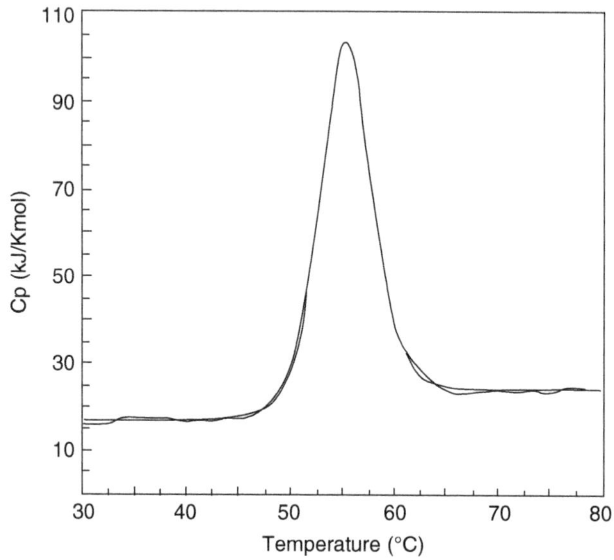

Figure 7.2. Partial heat capacity of barnase determined by DSC (blue line) and computer modeled assuming that the process represents a two-state transition (red line).

measured and the calculated ideal heat capacity functions unequivocally means that the population of intermediates is negligible. Therefore, unfolding of barnase is a highly cooperative process, which can be approximated by a two-state transition.

It should be emphasized that the possibility of performing such a detailed investigation of the cooperativity of a temperature-induced process is a privilege of DSC analysis—one cannot do the same with the optical or any other profiles. This is because this analysis is based on the fundamental relation between the conjugate extensive and intensive characteristics of the observed process, namely, the enthalpy and temperature. This functional dependence includes all information on the population of states that are realized in the observed temperature range (Freire and Biltonen, 1978; Privalov, 1979).

7.2.4. Heat Capacity of the Native and Denatured States

In general the heat capacities of all bodies increase with increasing temperature because of thermal motion intensification. The partial heat capacity of the native compact globular proteins in aqueous solutions is not an exception: it also increases with heating and in the accessible temperature range this increase appears as a linear function of temperature (Fig. 3.17). The heat capacity of denatured proteins is significantly higher than that of the compact native proteins and in the first calorimetric studies it also was assumed to be a linear function of temperature and to increase with temperature with the same slope as that of the native protein heat capacity. Thus, it was believed that the denaturational heat capacity increment, ΔC_p, does not depend on temperature but its magnitude is specific for any given protein (Privalov and Khechinashvili, 1974; Privalov, 1979). Further studies showed, however, that the situation is not so simple.

Determination of the complete heat capacity function of the unfolded protein is not straightforward because, if the temperature-induced protein denaturation is reversible, this state exists only above the transition temperature, that is, in a rather narrow temperature range. The heat capacity of the unfolded protein cannot be determined using the irreversibly denatured protein because its irreversibility assumes some degradation of its chemical structure, which might result in some changes of the thermal properties of the protein. This heat capacity cannot be measured in the presence of denaturant (e.g., urea or guanidinium hydrochloride) either, since all denaturants heavily solvate the polypeptide chain and change its partial heat capacity (see Section 7.5.2). One can irreversibly unfold protein only by disrupting the disulfide cross-links holding it. However, not all proteins have disulfide cross-links stabilizing their folded structures and it is not evident that an un-cross-linked protein is fully unfolded. However, the heat capacity of the ideally unfolded polypeptide chain can be modeled by summing the calorimetrically measured partial molar heat capacities of its components, the peptide group (CHCONH), the side chains (R), and two terminal groups (NH_2 and CHCOOH) (Makhatadze and Privalov, 1990, 1992, 1995; Privalov and Makhatadze, 1990, 1992):

$$C_p(\mathrm{pr})^{\mathrm{U}} = C_p(\mathrm{NH_2}) + (n-1)C_p(\mathrm{CHCONH}) + \sum_{i=1}^{n} C_p(\mathrm{R}) + C_p(\mathrm{CHCOOH}) \quad (7.2)$$

The calorimetrically determined partial heat capacities of the individual amino acid residues are listed in Table 7.1.

As shown in Figure 7.3, the heat capacities of the unfolded proteins are not linear functions of temperature but are slightly bent in such a way that their slope decreases with increasing temperature.

The situation with the native state of proteins appeared to be even more complicated than for the denatured state. Detailed calorimetric studies, using DSC instruments with highly stable baseline, showed that the partial specific heat capacities of the native compact globular proteins are linear functions of temperature with rather similar small slopes (Fig. 7.3). Assuming that they are linear functions of temperature above the denaturation temperature and extrapolating them to higher temperatures one finds that, since the heat capacity functions of the unfolded protein are not linear but curved functions of temperature, these two functions would cross at a temperature above 100°C. It appeared, then, that the difference between the heat capacities of the native and denatured proteins decreases with increasing temperature and even vanish above 100°C.

Comparison of different proteins also showed that the specific partial heat capacities of their folded native state might differ both in absolute value and in their temperature dependencies (Fig. 7.4). For the small compact and stable globular proteins this difference is rather small and is usually neglected.

TABLE 7.1. Temperature Dependence of the Partial Molar Heat Capacities of Peptide Units—of the N and C Terminal Groups, and of Side Chains R_i of Amino Acid Residues[a]

Group	5°C	25°C	50°C	75°C	100°C
–CO–CH–NH–	4	15	26	30	34
N + C termini	–158	–90	–21	–32	–150
R_i: side chains					
Ala	176	167	156	145	135
Arg	205	273	306	315	318
Asn	73	89	110	125	140
Asp	73	89	106	124	141
Cys	225	238	251	261	268
Gln	168	180	193	203	211
Glu	168	179	192	204	211
Gly	82	78	72	66	60
His	206	180	177	180	187
Ile	407	402	397	391	386
Leu	386	382	378	373	369
Lys	215	250	267	274	278
Met	197	176	158	150	148
Phe	396	383	370	358	348
Pro	214	178	152	143	136
Ser	76	81	86	91	97
Thr	194	184	182	186	199
Trp	471	458	446	434	424
Tyr	311	302	295	294	300
Val	325	314	305	295	286

[a]In J/K·mol.

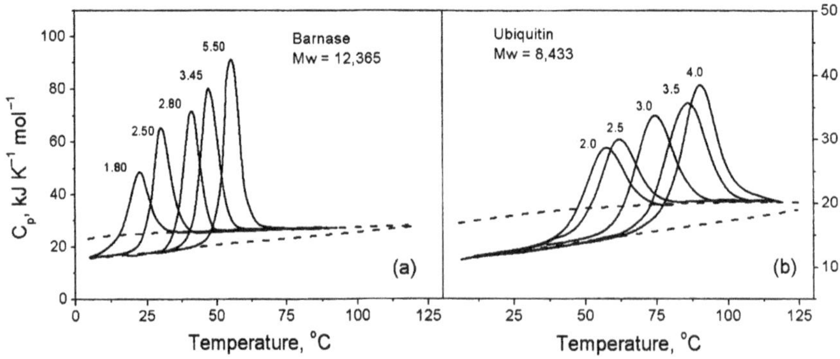

Figure 7.3. Partial molar heat capacity functions of barnase and ubiquitine at different pH values, showing the roughly linear dependence of C_p for native states. The heat capacity of unfolded states is approximated by the curved functions calculated by summing the heat capacities of amino acid residues forming the polypeptide chains of these proteins.

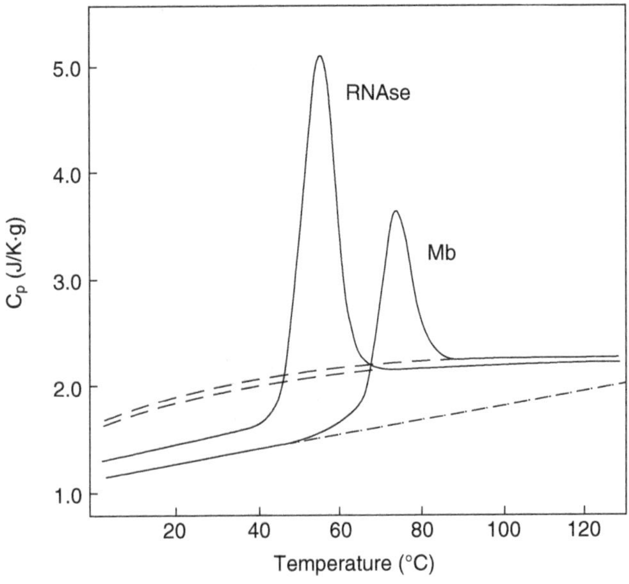

Figure 7.4. Partial specific heat capacity functions of ribonuclease A and myoglobin in pH 5.0 solutions providing maximum stability to proteins. The dashed lines show the heat capacity functions of ribonuclease with disrupted disulfide cross-links and apomyoglobin in an acidic solution (pH 2.5), that is, at conditions where they are unfolded. The dashed-and-dotted line is just a linear extrapolation of the heat capacity of the native myoglobin (Privalov, 1990).

7.2.5. Functions Specifying Protein Stability

When upon heating a protein unfolds cooperatively with a sharp heat absorption peak, from the area of the calorimetrically observed heat absorption peak one can determine the enthalpy of protein unfolding at the transition temperature. Also, if the temperature-induced process is highly cooperative (i.e., is approximated by a two-state transition) at the midtransition temperature, T_t, the equilibrium constant K equals 1 and the Gibbs energy of transition is zero:

$$\Delta_{\rm N}^{\rm D}G(T_t) = \Delta_{\rm N}^{\rm D}H(T_t) - T_t\Delta_{\rm N}^{\rm D}S(T_t) = 0.$$

Thus at that temperature the transition entropy is

$$\Delta_{\rm N}^{\rm D}S(T_t) = \Delta_{\rm N}^{\rm D}H(T_t)/T_t \qquad (7.3)$$

Moreover, the observed heat capacity difference between the native and denatured states, $\Delta_{\rm N}^{\rm D}C_p$, permits estimation of the expected enthalpy and entropy of protein unfolding at other temperatures, using Equations (2.11) and (2.17). As the temperature approaches 120°C the difference between the heat capacities of the unfolded and folded protein vanishes (Fig. 7.3). Correspondingly, the enthalpy and entropy of protein thermal denaturation asymptotically approach some constant level

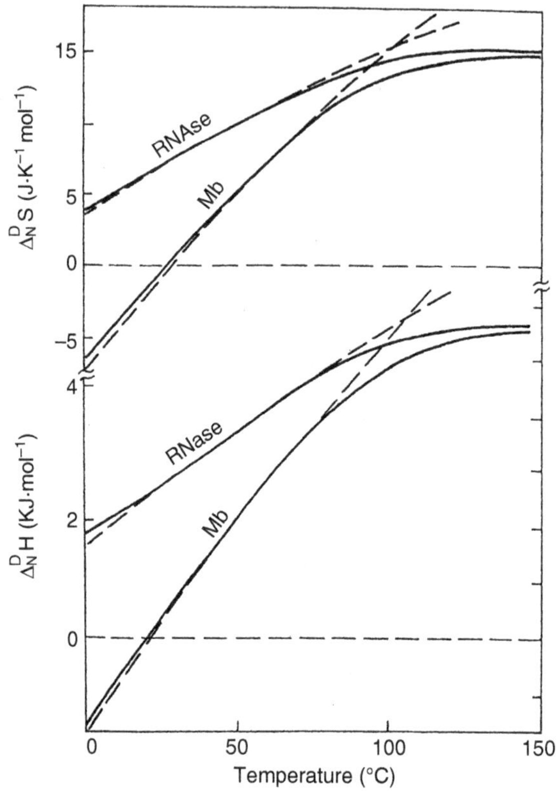

Figure 7.5. Temperature dependence of the specific enthalpy and entropy of denaturation of myoglobin and ribonuclease at the same conditions as in Figure 7.4, calculated per mole of amino acid residues from the assumption that their denaturation heat capacity increments are temperature independent (dashed lines) and temperature dependent (continuous lines) (Privalov, 1989).

at high temperatures (Fig. 7.5). For the various globular proteins this asymptotic level of the specific enthalpy, that is, the enthalpy calculated per gram or per mole of amino acid residues, might be different (Table 7.2; Fig. 7.6). However, the difference in maximal values of the specific enthalpies of thermal denaturation at 125°C does not exceed 15% of the averaged value of 50 J/g, or 5.5 kJ per mole of amino acid residue (kJ/(mol-res)). There is a similar situation with the specific entropies of thermal denaturation: their asymptotic values reached at about 125°C are somewhat different: they all are grouped between 12 and 17 J/K·(mol-res).

It is most remarkable that with decreasing temperature the enthalpy and entropy of protein denaturation decrease in magnitude, and one could expect that at some low enough temperature they become zero and then might even change sign. Ignoring the dependence of denaturation heat capacity on temperature, when enthalpy and entropy of protein denaturation can be approximated by Equations (2.14) and (2.21), one finds that enthalpy of denaturation reaches zero at temperature

TABLE 7.2. Thermodynamic Characteristics of Protein Unfolding[a]

Protein	Property	Temperature (°C) 5	25	50	75	100	125	References
BPTI								Makhatadze
$M_w = 6.565$, $N = 58$	$\Delta_N^U C_p$	2.8	2.7	2.6	2.1	1.3	0.3	et al.
$\Delta_N^U \text{WASA}_{\text{aliph}} = 4755$	$\Delta_N^U H$	72	130	200	259	303	323	(1993)
$\Delta_N^U \text{WASA}_{\text{arom}} = 756$	$\Delta_N^U S$	87	288	514	690	809	864	
$\Delta_N^U \text{WASA}_{\text{polar}} = 1728$	ΔG	47.8	44.2	34.0	18.9	1.2	−20.9	
Eglin C								Bae and
$M_w = 8.052$, $N = 70$	ΔC_p	4.2	3.9	3.5	2.5	1.6	0.2	Sturtevant
$\Delta \text{ASA}_{\text{aliph}} = 2145$	ΔH	33	115	208	283	335	358	(1995)
$\Delta \text{ASA}_{\text{arom}} = 769$	ΔS	−22	262	561	786	929	989	
$\Delta \text{ASA}_{\text{polar}} = 1.397$	ΔG	39.1	36.9	26.8	9.5	−11.5	−35.6	
G protein								Alexander
$M_w = 7.738$, $N = 70$	ΔC_p	3.6	3.6	3.3	2.6	2.0	1.0	et al.
$\Delta \text{ASA}_{\text{aliph}} = 1.834$	ΔH	−4	67	153	227	283	320	(1992)
$\Delta \text{ASA}_{\text{arom}} = 565$	ΔS	−103	145	422	743	801	897	
$\Delta \text{ASA}_{\text{polar}} = 1.364$	ΔG	24.6	23.8	16.7	3.2	−15.8	−37.0	
Tendamistat								Renner et al.
$M_w = 7.944$, $N = 75$	ΔC_p	3.5	3.6	3.6	2.9	2.3	1.4	(1992)
$\Delta \text{ASA}_{\text{aliph}} = 2644$	ΔH	−22	70	176	262	321	351	
$\Delta \text{ASA}_{\text{arom}} = 416$	ΔS	−213	109	452	711	877	955	
$\Delta \text{ASA}_{\text{polar}} = 2.141$	ΔG	37.2	37.5	30.0	14.6	−6.1	−29.1	
Ubiquitin								Wintrode
$M_w = 8.433$, $N = 76$	ΔC_p	5.8	5.7	5.1	3.8	2.5	0.8	et al.
$\Delta \text{ASA}_{\text{aliph}} = 3600$	ΔH	−88	27	162	273	351	393	(1994)
$\Delta \text{ASA}_{\text{arom}} = 270$	ΔS	−444	−44	393	727	959	1068	
$\Delta \text{ASA}_{\text{polar}} = 2.790$	ΔG	35.4	40.1	35.1	20.0	−6.7	−32.1	
RNAse T1								Yu et al.
$M_w = 11.071$, $N = 104$	ΔC_p	5.5	5.3	5.0	4.3	3.0	0.8	(1994)
$\Delta \text{ASA}_{\text{aliph}} = 3.375$	ΔH	173	281	410	528	621	672	
$\Delta \text{ASA}_{\text{arom}} = 1.482$	ΔS	444	817	1233	1584	1845	1976	
$\Delta \text{ASA}_{\text{polar}} = 2.791$	ΔG	49.6	37.5	11.7	−23.2	−67.2	−114	
Cytochrome c								Makhatadze
$M_w = 12.830$, $N = 104$	ΔC_p	6.9	6.8	6.1	4.4	2.8	0.6	and
$\Delta \text{ASA}_{\text{aliph}} = 3.922$	ΔH	−53	89	268	421	532	593	Privalov
$\Delta \text{ASA}_{\text{arom}} = 1020$	ΔS	−319	174	752	1210	1520	1681	(1993)
$\Delta \text{ASA}_{\text{polar}} = 3.358$	ΔG	35.7	37.1	25.1	−1.1	−35.0	−76	
Barnase								Griko et al.
$M_w = 12{,}365$, $N = 110$	ΔC_p	7.1	6.9	5.9	3.9	2.0	0.1	(1994)
$\Delta \text{ASA}_{\text{aliph}} = 4.317$	ΔH	167	307	467	590	664	690	
$\Delta \text{ASA}_{\text{arom}} = 1296$	ΔS	379	866	1384	1752	1959	2029	
$\Delta \text{ASA}_{\text{polar}} = 3.113$	ΔG	61.6	48.9	20.0	−19.7	−66.7	−117	
RNAse A								Makhatadze
$M_w = 13{,}600$, $N = 124$	ΔC_p	4.6	5.2	5.3	4.3	3.5	2.0	and
$\Delta \text{ASA}_{\text{aliph}} = 4.318$	ΔH	220	294	405	5.12	603	664	Privalov
$\Delta \text{ASA}_{\text{arom}} = 955$	ΔS	641	896	1254	1574	1826	1989	(1993)
$\Delta \text{ASA}_{\text{polar}} = 4.141$	ΔG	41.8	27.0	0.0	−35.8	−78.1	−127	

(Continued)

TABLE 7.2. (*Continued*)

Protein	Property	Temperature (°C) 5	25	50	75	100	125	References
Lysozyme								Makhatadze
M_w = 14,300, N = 129	ΔC_p	9.0	9.1	8.9	7.4	5.7	3.6	et al.
ΔASA_{aliph} = 5180	ΔH	111	242	408	562	683	7.53	(1993)
ΔASA_{arom} = 1185	ΔS	164	618	1153	1615	1954	2138	
ΔASA_{polar} = 4.231	ΔG	65.4	57.8	35.6	0.0	−45.8	−98	
Interleukin-1β								Makhatadze
M_w = 17.381, N = 153	ΔC_p	6.9	7.5	7.5	6.2	4.9	2.8	et al.
ΔASA_{aliph} = 6531	ΔH	7	151	330	501	640	736	(1994)
ΔASA_{arom} = 1374	ΔS	−99	401	106	1516	1903	2155	
ΔASA_{polar} = 4.141	ΔG	34.5	31.5	5.1	−26.6	−69.8	−128	
Myoglobin								Makhatadze
M_w = 17,800, N = 153	ΔC_p	14.4	14.0	12.8	10.03	7.7	4.4	et al.
ΔASA_{aliph} = 6842	ΔH	−231	6	291	555	774	920	(1993)
ΔASA_{arom} = 1565	ΔS	−919	−116	805	1595	2207	2588	
ΔASA_{polar} = 4.793	ΔG	24.5	40.6	31.0	−0.1	−49.2	−110	
T4 Lysozyme								Kitamura
M_w = 18,619, N = 163	ΔC_p	11.0	11.0	9.7	6.7	4.0	0.2	and
ΔASA_{aliph} = 7.306	ΔH	20	240	500	671	8.5	856	Sturtevant
ΔASA_{arom} = 1.337	ΔS	−190	576	1413	1928	2302	2439	(1989)
ΔASA_{polar} = 5.353	ΔG	72.8	68.4	42.6	0.1	−53.6	−115	

[a] $\Delta_N^D C_p$ in kJ/K·mol; $\Delta_N^D H$ and $\Delta_N^D G$ in kJ/(mol-res); $\Delta_N^D S$ in J/K·(mol-res).

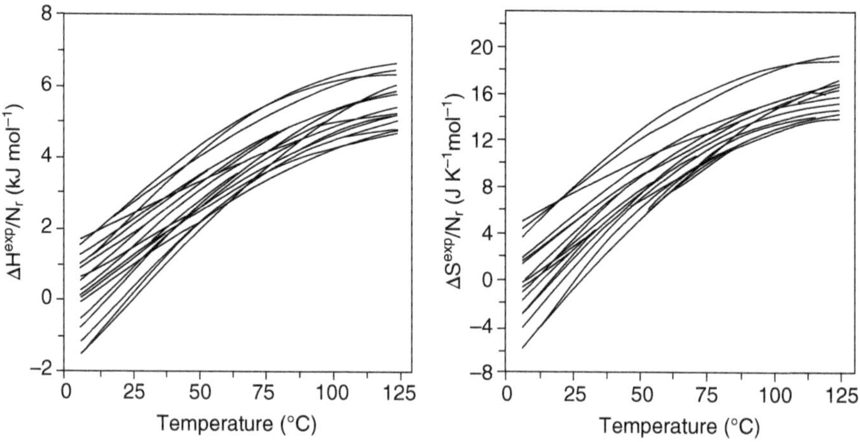

Figure 7.6. Temperature dependence of the specific enthalpies and entropies of denaturation of the globular proteins presented in Table 7.2 calculated per mole of amino acid residue, the averaged mass of which is assumed to be 115 Da (Makhatadze and Privalov, 1995).

$$T_H \approx T_t - \frac{\Delta_N^D H_t}{\Delta_N^D C_p} \tag{7.4}$$

The entropy of unfolding also decreases with temperature below T_t, but nonlinearly, and it reaches zero at a somewhat higher temperature, T_S, than does the enthalpy function:

$$T_S \approx T_t \exp\left(\frac{-\Delta_N^D H_t}{\Delta_N^D C_p T_t}\right) \tag{7.5}$$

Combining the enthalpy and entropy expressed by Equations (2.13) and (2.20), or their simplified forms in Equations (2.14) and (2.21), gives the Gibbs energy difference between the unfolded (denatured) and folded (native) states of a protein as a function of temperature (Fig. 7.7).

For the simplified cases, when the temperature dependence of the denaturational heat capacity increment could be ignored, we will have

$$\Delta_N^D G(T) = \Delta_N^D H(T) - T\Delta_N^D S(T) \approx \Delta_N^D H_t[T_t - T]/T_t - \Delta_N^D[(T_t - T) + T\ln(T/T_t)] \tag{7.6}$$

Expanding the logarithmic term

$$\ln(T_t/T) = \ln(1 + \Delta T/T) \approx \Delta T/T - (\Delta T)2/2T^2 + (\Delta T)^3/3T^3 + \cdots$$

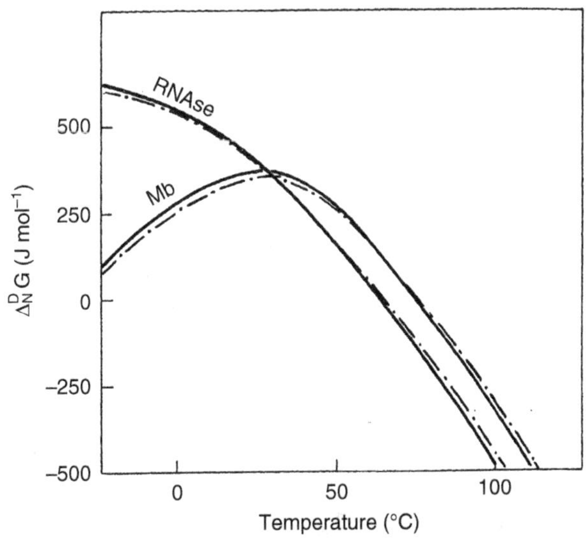

Figure 7.7. The ΔG function for ribonuclease A and myoglobin for the same conditions as in Figure 7.4 calculated under the assumptions that their denaturation heat capacity increments are temperature independent (dashed lines) and that they are temperature dependent (continuous lines) (Privalov, 1989).

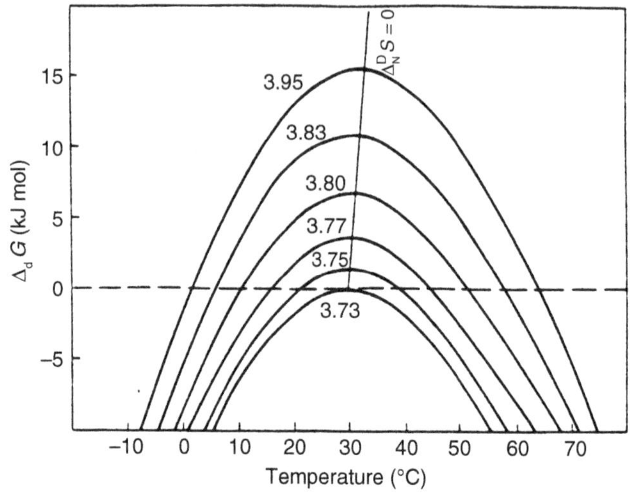

Figure 7.8. The Gibbs energies of the metmyoglobin in solutions of different pH in the presence of acetate buffer (Privalov et al., 1986).

one gets

$$\Delta_N^D G(T) \approx \frac{\Delta_N^D H_t [T_t - T]}{T_t} - \Delta_N^D C_p \frac{[T_t - T]^2}{2T} \quad (7.7)$$

Here the first term is positive and increases linearly with decreasing temperature, while the second term is negative and increases rapidly with decreasing temperature. Therefore, in contrast to the enthalpy and entropy functions, which are the monotonously increasing functions of temperature, the Gibbs energy difference between the unfolded (denatured) and folded (native) states of protein is a function with an extremum (Fig. 7.8). It is notable that, in contrast to the enthalpy and entropy functions, the shape of the Gibbs energy function is almost identical for the two cases being considered, when the denaturation heat capacity increment does and does not depend on temperature. It appears that the effects in enthalpy and entropy, resulting from changes in the heat capacity increment, efficiently compensate each other.

The maximum of the Gibbs energy is reached at the temperature $T_{\max}$, where the temperature derivative of this function is zero:

$$\frac{\partial \Delta_N^D G}{\partial T} = \frac{\partial(\Delta_N^D H - T\Delta_N^D S)}{\partial T} = \Delta_N^D C_p - \Delta_N^D S - T\frac{\Delta_N^D C_p}{T} = -\Delta_N^D S(T_s) = 0 \quad (7.8)$$

At temperatures above and below $T_{\max}$ the Gibbs energy difference decreases and drops to zero at temperatures T_t and T_t^*:

$$T_t^* \approx \frac{\Delta_N^D C_p T_t^2}{2\Delta_N^D H_t - \Delta_N^D C_p T_t} \quad (7.9)$$

Thus, the ΔG has a maximum at the temperature where the entropy of unfolding is zero, that is, $T_{\max} = T_S$. At temperatures above and below $T_{\max}$ the Gibbs energy difference decreases. Since ΔG represents the work required to transfer a protein from the folded to the unfolded state, it determines the stability of the protein structure. It follows that the zero stability of the native state is reached at two different temperatures: at a high temperature, T_t, at which the heat denaturation of protein takes place; and at a low temperature, T_t^*, at which one can expect the cold denaturation of protein.

7.3. COLD DENATURATION

7.3.1. Proteins at Low Temperatures

The denaturation of proteins upon heating has never been considered as surprising. It always seemed to be a natural phenomenon, even when nothing was known about protein structure; it was only supposed that it should be rather sophisticated to fulfill various very delicate functions. According to Le Chatellier's principle, any process that is induced by increasing temperature should proceed with heat absorption and thus with increase in enthalpy and entropy, that is, with disordering of the considered system. Therefore, disruption of the native protein structure upon heating, the heat denaturation of protein, appeared as an obvious effect: it should unfold upon heating because of increasing entropic dissipative forces and that should proceed with heat absorption, that is, the enthalpy and entropy increase. By the same argument, decreasing temperature should only induce processes leading to increasing order and, since a protein in its native state is believed to be the most ordered, one cannot expect that cooling could lead to any significant change of protein structure and, more so, to its denaturation. Nevertheless, this is just what was found experimentally with proteins. Cold denaturation appeared as a surprising, but very general, property of globular proteins. This paradox attracted the greatest attention since it was considered as a most important test for reliability of protein thermodynamics, which was very unusual and therefore doubtful. However, without thermodynamics one could not hope to progress in understanding these macromolecular systems, particularly in understanding the mechanism of folding and stabilization of the native protein structure. This interest is also enhanced by the practical importance of cold denaturation as low temperatures are widely used by biochemists in their usual practice to keep proteins in the native state. Moreover, understanding the effect of cold on food was practically important for food storage.

The interest in the influence of low temperature on the stability of native proteins began with a paper published by Hopkins (1930). Studying precipitation of ovalbumin, denatured by concentrated urea at various temperatures, he noticed that the rate of denaturation was higher at 0°C than at 23°C. The negative influence of temperature upon the denaturation rate was surprising since it had already been established that the rate of heat denaturation of protein in aqueous solution is characterized by a positive and large temperature coefficient. This finding of Hopkins was confirmed by others on other globular proteins (Clark, 1945; Simpson and Kauzmann, 1953; Christensen, 1951; Schellman, 1958; Chen and Schellman, 1989).

Meanwhile, the facts were accumulating which showed that many enzymes in aqueous solution, without urea or GuHCl, were inactivated at temperatures close to 0°C, but their activity was restored at room temperature (for a review see Privalov, 1990). The simplest explanation of the observed cold lability of enzymes in aqueous solutions, not containing denaturants, was based on the assumption that it is caused by the ionization of some essential groups of protein molecules with decreasing temperature (Bock et al., 1975; Bock and Frieden, 1974). Since the enthalpy of protonation of the protein groups is negative and for some of the groups is rather large (e.g., for histidine it is –28.8 kJ/mol and for α-amino group, –43.8 kJ/mol), one can expect that at decreasing temperatures the protonation of these groups increases. This, in principle, could switch off the active site in an enzyme. However, using the van't Hoff equation, one can calculate that even –43.8 kJ/mol is not a sufficient value to explain the observed sharpness of cold-induced denaturation of protein, which takes place in the interval of a few degrees. It was evident that there should be a much more general reason for the cold denaturation of protein.

7.3.2. Experimental Observation of Cold Denaturation

Direct observation of cold denaturation of proteins is not easy since for most proteins it is expected at temperatures below freezing of aqueous solutions, while cold denaturation is specific only for the proteins in aqueous environment. As mentioned previously (see Section 2.3.4), dust-free aqueous solution can be supercooled without freezing down to −15°C. On the other hand, the expected temperature of cold denaturation of a protein can be raised by the appropriate choice of the solution conditions, particularly of the solution pH and the buffer used. This is because denaturation of protein usually results in unmasking of internal titratable groups and their protonation or deprotonation, depending on the pH at which the experiment is carried out. Enthalpies of deprotonation are positive and for the acidic groups are rather small, while for the alkaline groups they are quite considerable (see Table 2.1 in Chapter 2). Moreover, if protein unfolding is studied in a buffered solution, the protonation–deprotonation of the protein induces compensating deprotonation–protonation of the buffer that is in the solution. Therefore, the calorimetrically measured heat of protein unfolding also includes the heat of the protein protonation–deprotonation and, correspondingly, the heat of the buffer deprotonation–deprotonation:

$$\mathrm{Pr{-}H + Buf^{-} \leftrightarrow Pr^{-} + Buf{-}H}$$

Therefore presence of the buffer can change considerably the overall enthalpy of protein unfolding and shift the T_H value (i.e., the temperature at which denaturation enthalpy becomes zero) to the higher temperatures. Also, decrease of pH results in decrease of the stability of protein and, correspondingly, decrease of its heat denaturation temperature, but increase of its cold denaturation temperature (Fig. 7.8).

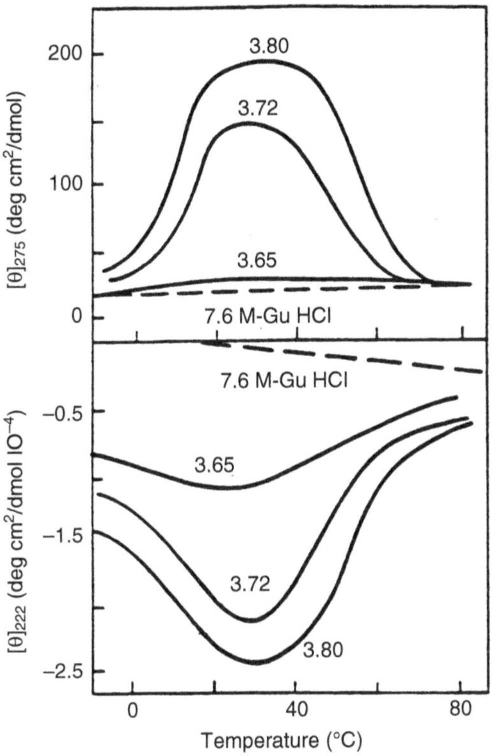

Figure 7.9. Temperature dependence of the ellipticity in the near and far ultraviolet for metmyoglobin in sodium acetate solutions with various pH values (Privalov, 1990).

One can observe the denaturation of protein by changing its ellipticity upon cooling, showing disappearance of the helical conformation of this highly helical protein both upon cooling and upon heating myoglobin solution from 40°C (Fig. 7.9).

Cold denaturation of protein results in considerable increase of its intrinsic viscosity, which reaches the value specific for the completely unfolded protein at low pH (Fig. 7.10). It is remarkable that the intrinsic viscosity of the unfolded polypeptide chain is very high at low temperature and decreases with increasing temperature. This is because at low temperature the unfolded polypeptide chain is less flexible and occupies much larger hydrodynamic volume than at high temperatures. Nuclear magnetic resonance (NMR) studies of myoglobin cooling and heating showed disappearance of resonances in the range of 1.3 to −1.5 ppm, indicating the breakdown of the compact and ordered protein structure (Privalov et al., 1986).

According to DSC experiments, the changes in metmyoglobin and apomyoglobin (i.e., myoglobin without a heme group) in these two temperature regions proceed with significant heat effect, which is positive if the solution is heated and negative if it is cooled (Fig. 7.11). It is remarkable that the heat effects observed upon cooling and subsequent heating of apomyoglobin are much more symmetrical than in the

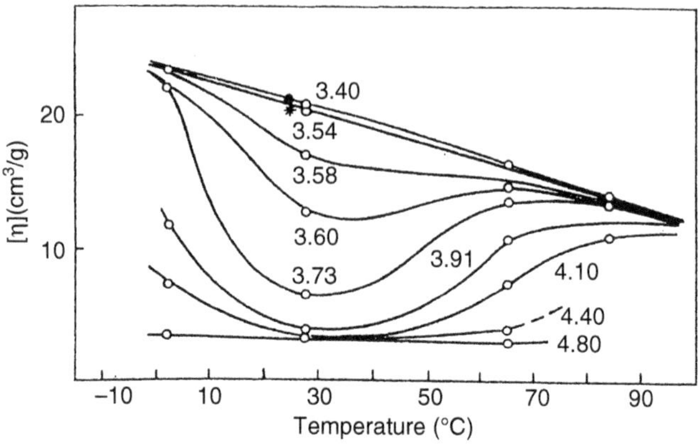

Figure 7.10. Temperature dependence of the intrinsic viscosity of metmyoglobin in sodium acetate solutions with various pH values (Privalov et al., 1986).

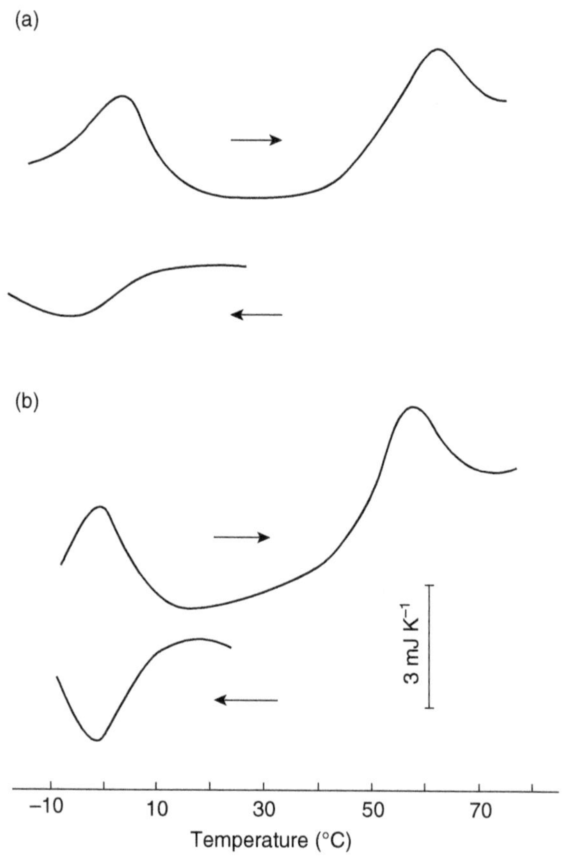

Figure 7.11. Original microcalorimetric recordings of the heat effects upon cooling and subsequent heating with the scan rate 0.5 K/min of (a) metmyoglobin and (b) apomyoglobin in sodium acetate solution (Griko et al., 1988b).

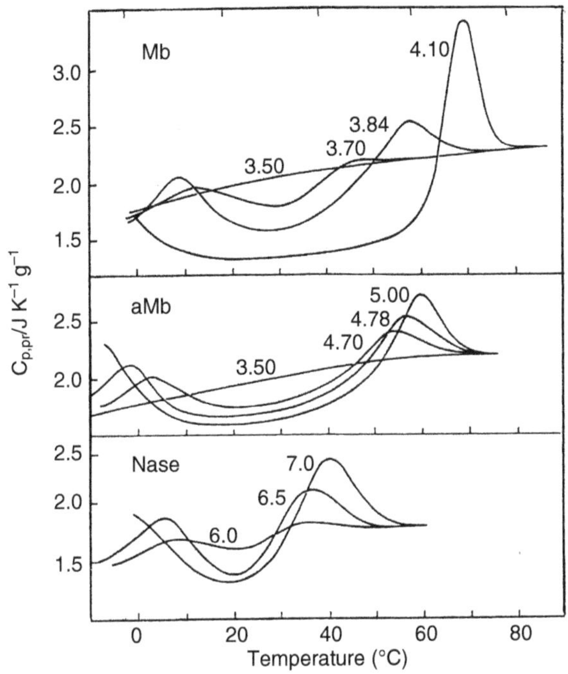

Figure 7.12. Partial molar heat capacity of metmyoglobin (Mb), apomyoglobin (aMb), and staphylococcal nuclease (Nase) in sodium acetate solutions with various pH values (Privalov, 1990).

case of metmyoglobin. Since the DSC experiment proceeds with the same cooling and heating rates (usually 1 K/min), the difference in the observed profiles shows that kinetics of unfolding and subsequent refolding of apomyoglobin, which lacks heme, are significantly faster than that of metmyoglobin, which loses the heme upon unfolding at cooling and then has to regain it upon subsequent folding at heating.

After pioneering works on myoglobin (Privalov et al., 1986), the cold denaturation phenomenon was demonstrated on a number of proteins, particularly apomyoglobin (Griko et al., 1988b), staphylococcal nuclease (Griko et al., 1988a), phosphoglycerate kinase (PGK) (Griko et al., 1989), and lactoglobulin (Griko and Privalov, 1992) (Fig. 7.12). Unlike myoglobin, staphylococcal nuclease has not many buried histidines. Therefore, its total denaturation enthalpy in acetate buffer is much higher than that of metmyoglobin. Correspondingly, according to Equation (7.9), it should have a much lower T_H value, about 15°C, and, consequently, its cold denaturation takes place at considerably lower temperatures than that of metmyoglobin and apomyoglobin.

The other example presents PGK, which is a much larger protein than the usual globular proteins (M_w = 48,000), for example, myoglobin (M_w = 17,800) or staphylococcal nuclease (M_w = 16,000). According to the crystallographic studies, the single polypeptide chain of PGK forms two well-defined domains (Blake and Rice,

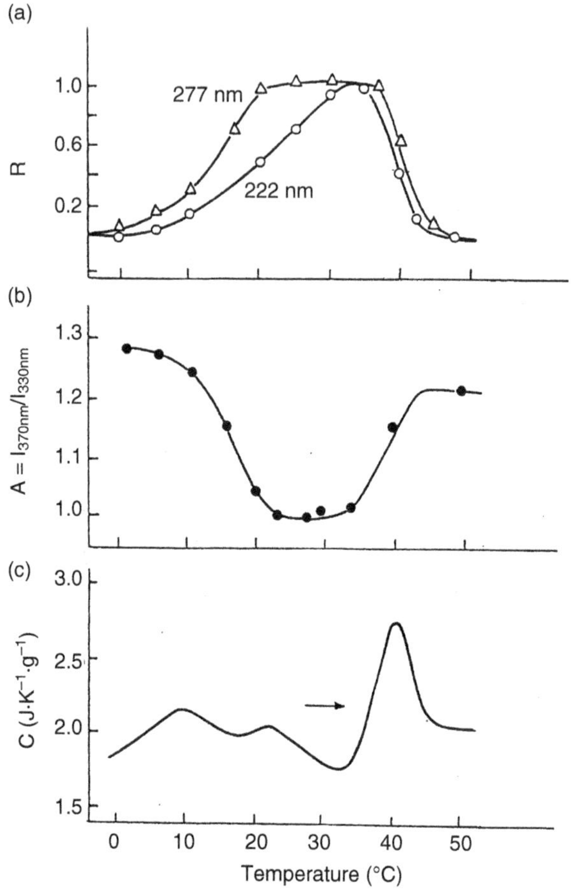

Figure 7.13. Temperature-induced denaturation of phosphoglycerate kinase in solutions containing 0.7 M GuHCl observed in heating experiments by (a) the relative changes of ellipticity at 222 and 277 nm, (b) the tryptophan emission spectrum, and (c) the partial specific heat capacity (Griko et al., 1989).

1981). Nevertheless, calorimetric studies of the heat denaturation of this enzyme showed that this process represents a two-state transition (Hu and Sturtevant, 1987); that is, in this process PGK figures as a single cooperative unit. Calorimetric studies of this protein in a broad temperature range revealed that at low temperatures it undergoes cold denaturation (Griko et al., 1989). Most interesting is that, while the heat denaturation of this enzyme is a two-state process (judging by the simultaneous change of various optical properties and the sharp heat absorption peak with equal van't Hoff and calorimetric enthalpies), the cold denaturation of this protein proceeds in two distinct and rather independent stages (Fig. 7.13). Since a yeast PGK molecule has two tryptophans, which are located in the C-terminal domain, it became possible to identify the observed stages, analyzing the change of fluorescence spectra of tryptophan in the studied temperature range. This showed that the two stages in the cold denaturation of PGK correspond to independent disruption of two domains.

It follows that the interaction of two domains in PGK is temperature dependent: it decreases as the temperature decreases, and increases as temperature increases (Freire et al., 1992).

Thus, at high temperature the two domains are merged into a single cooperative block, whereas at a low temperature they figure as independent cooperative units.

Studies of cold denaturation of globular proteins demonstrated that the thermodynamic formalism correctly describes their thermal properties. These studies showed that:

Proteins are indeed thermodynamically controlled systems and cold denaturation is a very general property of globular proteins.

The most intriguing specificity of cold denaturation is that it proceeds with heat release, that is, decrease of entropy and increase of order in contrast to heat denaturation. Formal explanation of this fact is that enthalpy and entropy functions specifying protein unfolding in aqueous solutions change sign at temperatures between T_t and T_t^*. The problem is, however, why are these functions changing sign? Why do they depend so steeply on temperature? The formal answer is: because unfolding of globular proteins proceeds with a heat capacity increment. The question then is: why does denaturation of globular proteins occur with significant heat capacity increment? This problem is discussed in Chapter 8.

7.4. pH-INDUCED PROTEIN DENATURATION

7.4.1. Isothermal pH Titration of Globular Proteins

Using temperature as a variable showed that the thermal properties of globular proteins strongly depend on the pH of the solution: the temperature and the heat effect of protein denaturation both decrease with decrease or increase of pH from neutral. The interdependence of denaturation temperature and pH could be also observed when using pH as a variable parameter. As shown in Figure 7.14, lysozyme unfolds upon decreasing pH, and the pH range at which it denatures and the heat effect both depend on temperature. At 25°C protein does not unfold; therefore the titration curve at this temperature shows protonation of the acidic groups, which are on the surface of the native protein. However, at higher temperatures protein unfolds with extensive heat absorption. Therefore, the difference between the titration curves at these temperatures and that at 25°C shows the quantity of proton uptake upon protein unfolding.

Protonation–deprotonation of the titratable groups of protein proceeds with certain heat effects. The enthalpy, entropy, and Gibbs energy of deprotonation, as well as the p*K* values of these groups are listed in Table 2.1 in Chapter 2. One can see that the enthalpy of deprotonation is very small for the acidic groups but is quite

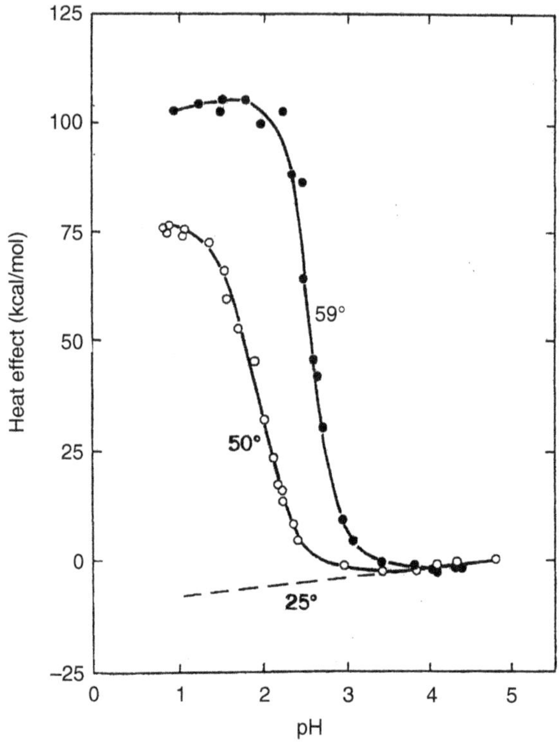

Figure 7.14. Calorimetric titration of the lysozyme at three different temperatures (Pfeil and Privalov, 1976a).

significant for the alkaline groups. It is also large for histidine, which is titrated at the neutral pH range.

The number of protons released or taken up by protein upon denaturation can be determined from the dependence of the denaturation temperature T_d on pH (Privalov et al., 1969):

$$\Delta_N^D \nu = \frac{\Delta_N^D H}{2.3RT_t^2} \frac{\partial T_t}{\partial \text{pH}} \tag{7.10}$$

In the pH range where denaturation temperature does not depend on pH, that is, $dT_d/d\text{pH} = 0$ (which for most proteins is a pH in the range from 5 to 9) the ionization effects at protein denaturation are negligible.

It should be noted that Equation (7.10) assumes that denaturation of the protein being considered is a cooperative two-state transition. An assumption on the cooperativity of the acid-induced denaturation is usually based on the observation that many different properties of globular proteins are changing simultaneously in a rather narrow range of pH. This, however, does not mean that pH-induced denaturation represents a two-state transition. Indeed, as in the case of temperature-induced denaturation (see Section 7.2), if a protein consists of several more-or-less independent subunits (domains) of identical stability, they all will denature in the same narrow pH range, although the process as a whole does not represent a single two-state transition. One can check the correctness of the assumption that pH-induced denaturation presents a two-state transition by comparing the effective value of $\Delta\nu$ calculated by Equation (7.10) with the real value determined by a direct potentiometric experiment (Pfeil and Privalov, 1976a).

7.5. DENATURANT-INDUCED PROTEIN UNFOLDING

7.5.1. Use of Denaturants for Estimating Protein Stability

By "denaturants" we mean the active reagents that denature proteins, for example, urea or guanidinium hydrochloride (GuHCl). Protein denaturation upon increase of the denaturant concentration can be observed by changes in various characteristics of protein, particularly the optical characteristics such as ellipticity or fluorescence. The changes observed in various properties of compact globular proteins usually occur in a rather narrow range of denaturant concentration, suggesting that it is a cooperative process. Assuming that this process represents a two-state transition, one can determine the apparent equilibrium constant and the apparent Gibbs energy of protein denaturation in the whole range of the denaturant-induced changes in protein (see Section 2.2.1).

Plotting Gibbs energy against denaturant concentration and extrapolating the plot to zero denaturant concentration (Fig. 7.15) one gets the apparent Gibbs energy of

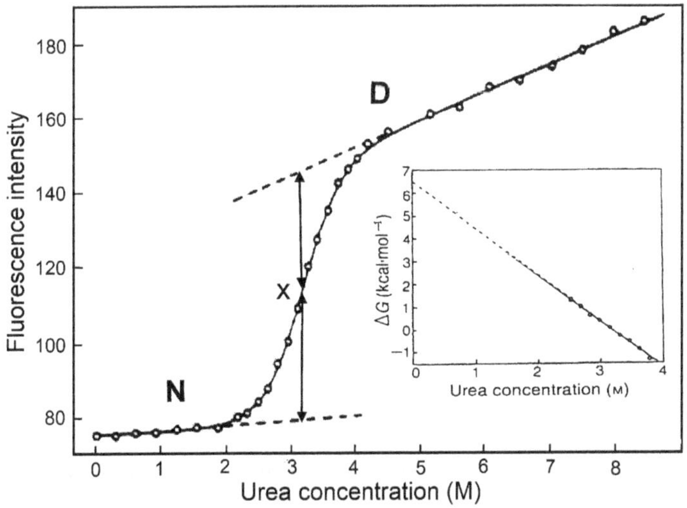

Figure 7.15. Protein denaturation by urea observed by change in fluorescence intensity. Inset: Gibbs energies determined from the progress of reaction in the transition zone and its extrapolation to zero concentration of denaturant (Pace, 1990).

stabilization of the native protein structure in the absence of denaturant (Pace, 1990; Pace et al., 1990). Moreover, the dependence of the Gibbs energy of protein unfolding on denaturant concentration, that is, the slope of the of ΔG(denaturant) function, which is usually called the m factor, correlates with the amount of protein surface exposed to solvent upon unfolding. Thus, such an analysis provides information on the stability of a protein and on the extent of its unfolding. It does not require any special instrument and could be performed using any characteristics of protein sensitive to its conformational changes, for example, circular dichroism (CD), optical rotatory dispersion (ORD), or fluorescence spectrum. The simplicity of this method of estimating the stability of protein structure led to its great popularity.

However, there are several things that are raising doubt about the reliability of this method in estimating stability of proteins. First, this method is based on the assumption that the denaturant-induced unfolding of protein represents a cooperative two-state transition, but there is no direct and reliable proof of that. The only argument that the denaturant-induced denaturation of protein is a cooperative process is that it takes place in a rather narrow range of denaturant concentration, and many different characteristics of protein change in this region. However, as discussed for thermal and acid denaturation of proteins, this certainly is not a strong argument since protein consisting of several domains of similar stability will also behave in this way but its unfolding would not represent a two-state transition.

The other concern is raised by too far a linear extrapolation of the Gibbs energy of unfolding, which is estimated in the range of denaturant concentrations where protein denaturation is observed (i.e., usually from 2 to 5 M of denaturant) to zero concentration of denaturant (Fig. 7.15, inset). Linearity of this dependence has no experimental justification. This is particularly doubtful in the case of GuHCl-induced denaturation, since this compound is a salt and increasing the concentration of this salt to 6 M results in a significant increase of the ionic strength of the solution—and the effect of this on protein stability might not be linear. Therefore, it was suggested to compensate this change in ionic strength by adding some other salt in solution upon decreasing the GuHCl concentration (Yao and Bolen, 1992). However, this suggestion has not found wide acceptance: not only because its realization is not easy but also because if one gets the value of Gibbs energy characterizing protein stability in a 6 M salt solution, it is too far from physiological conditions that present a real interest.

7.5.2. Calorimetric Studies of Protein Unfolding by Denaturants

The calorimetric investigation of denaturant-induced protein denaturation was done using an LKB flow-mix calorimeter (Pfeil and Privalov, 1976b; Makhatadze and Privalov, 1992). Results of calorimetric titration of lysozyme with GuHCl is shown in Figure 7.16. One can clearly see three phases in the denaturant-induced effect. At relatively low GuHCl concentrations, exothermal effects with nearly linear concentration dependence are observed. At a higher concentration of GuHCl a sigmoidal curvature is seen, suggesting appearance of an endothermic process. At the subsequent increase in concentration, the curve again becomes almost linear but its slope

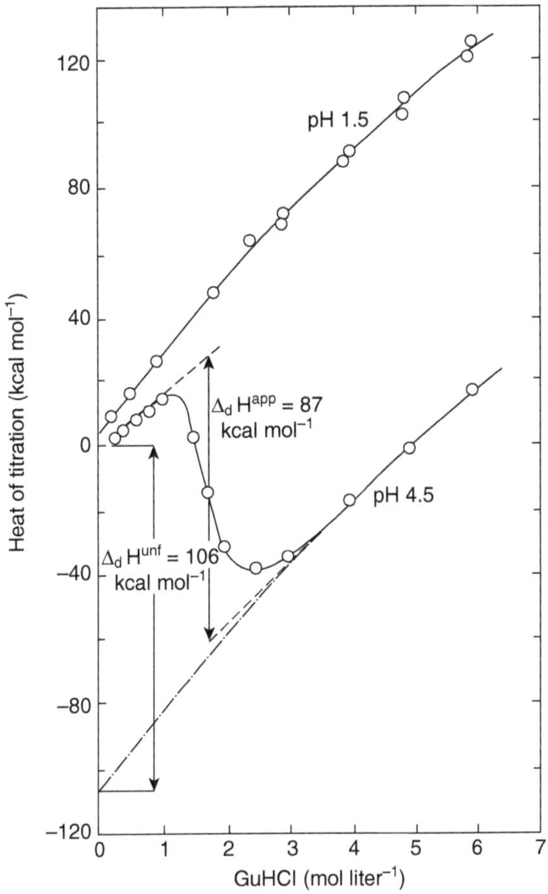

Figure 7.16. ITC titration of lysozyme by GuHCl at 59.1°C and indicated pH of solutions (Pfeil and Privalov, 1976b).

is steeper than what is observed at the initial stage. The slopes before and after the sigmoidal part are manifestations of the heat of progressive solvation by the native and GuHCl denatured protein with an increase of its concentration. This shows that:

The native protein, as well as the denatured one, is able to interact with GuHCl.

As one would expect these interactions are more extended in the case of unfolded protein, which has more exposed groups for binding denaturant. To exclude these heat effects of denaturant binding by the native and denatured protein and to determine the net heat effect of unfolding, it is necessary to extrapolate the enthalpies of the native and denatured states to zero concentration of denaturant. This is simple for the native state; for the denatured state one has to use the dependence obtained for the unfolded protein, for example, either a protein with reduced disulfide cross-links or a protein

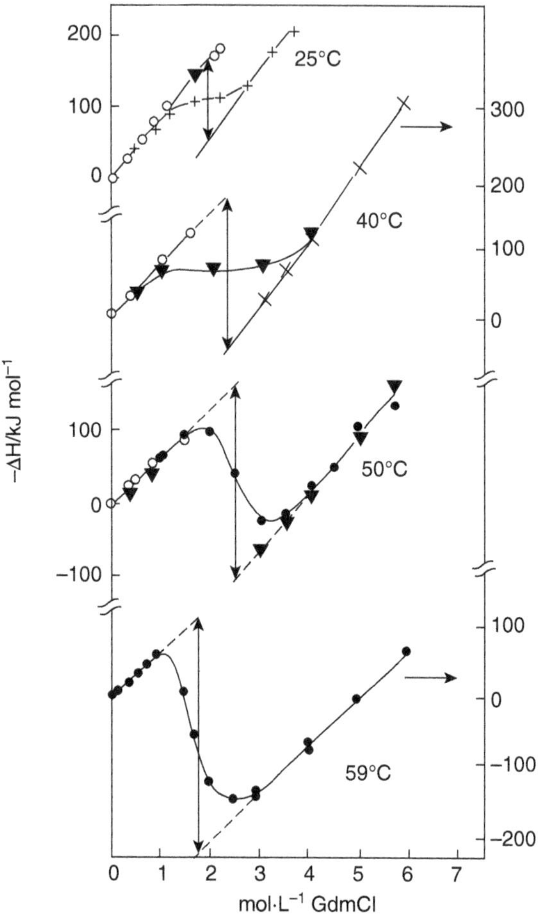

Figure 7.17. Calorimetric titration of lysozyme at indicated temperatures of solution (Pfeil and Privalov, 1976b).

at low pH (e.g., pH 1.5), at which the protein is fully unfolded. As shown in Figure 7.16, extrapolation to zero concentration of GuHCl gives, for the net enthalpy of lysozyme unfolding, 106 kcal/mol = 443 kJ/mol; and for the value largely exceeding the apparent heat effect of denaturation, 87 kcal/mol = 364 kJ/mol. It appears thus that the enthalpy of preferential binding of GuHCl is negative. Studies of this process at different fixed temperatures showed that the enthalpy of protein denaturation by GuHCl depends on temperature (Fig. 7.17): increasing the temperature increases the heat effect of the denaturant-induced unfolding of protein. Thus, it appears that the denaturant-induced unfolding results in the heat capacity increment.

Direct DSC studies of protein unfolding in the presence of denaturants showed that with increasing denaturant concentration protein thermostability decreases, and the protein denatures at lower temperatures with lower heat effect (Fig. 7.18). Presence of GuHCl and urea also leads to a noticeable increase of the partial heat

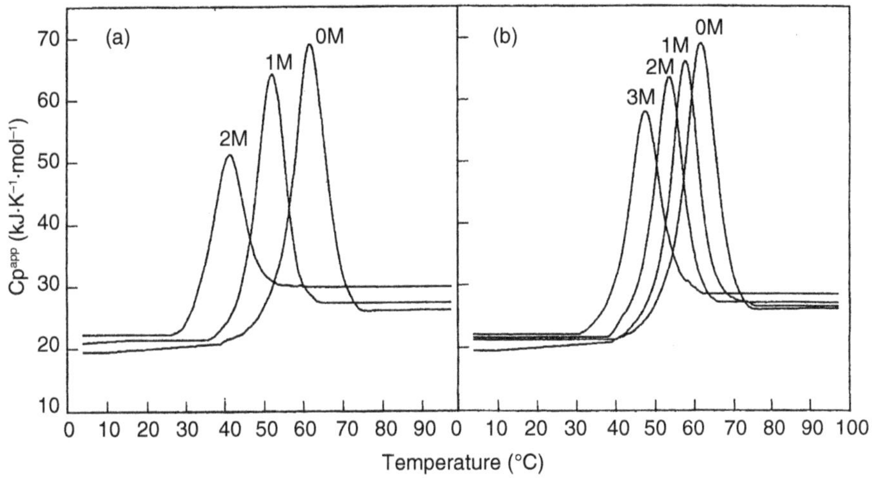

Figure 7.18. Temperature dependence of the partial molar heat capacity of RNAse in aqueous solutions at pH 5.5 containing various concentrations of (a) GuHCl and (b) urea (Makhatadze and Privalov, 1992).

capacity of proteins, especially in the unfolded state. The partial specific heat capacity of the unfolded polypeptide chain of RNAse at 25°C is (1.90 ± 0.01) J/K·g, but in the presence of 3 M urea it increases to (2.10 ± 0.01) J/K·g (Makhatadze and Privalov, 1992).

7.5.3. Urea and GuHCl Interactions with Protein

Figure 7.19 shows the results of calorimetric titration of RNAse with disrupted disulfide cross-links at pH 2.5 and different temperatures by urea and GuHCl. This un-cross-linked protein is in the completely unfolded random coiled state. The titration curves, especially in the case of urea, are far from saturation at the practical limit of the denaturant concentration (8 M). Nevertheless, the heat effect of titration at this concentration is quite large, about 50 J per gram of protein (i.e. several hundred kilojoules per mole), for both reagents.

If we assume independent binding sites for a number of denaturant molecules (X) bound by the protein molecule in solution with denaturant activity a, then

$$\mathrm{X}(a) = \sum n_i \frac{k_i \times a}{1 + k_i a} \tag{7.11}$$

where n_i is the number of i-type binding sites and k_i is the binding constant. If Δh_i is the binding enthalpy of denaturant by the i-type site, then the heat effect of protein titration by the denaturant is

$$Q(a) = \sum \Delta h_i n_i \frac{k_i \times a}{1 + k_i a} \tag{7.12}$$

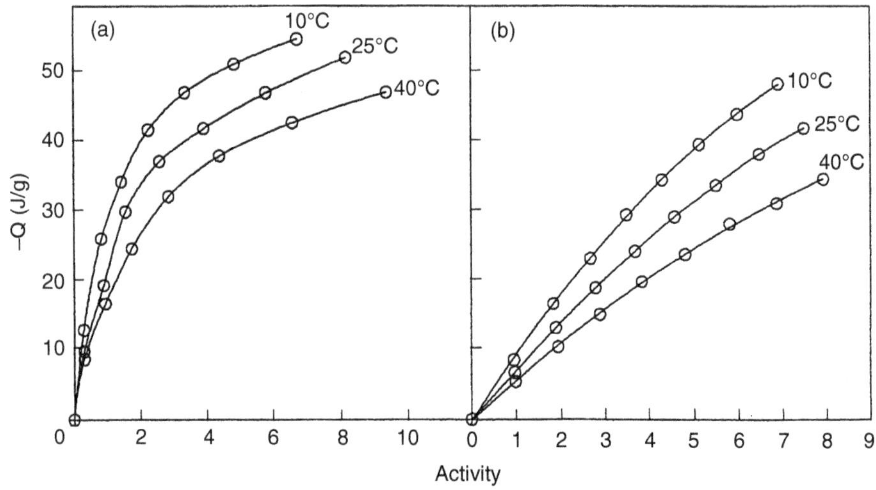

Figure 7.19. Specific heats of isothermal calorimetric titration of the unfolded RNAse by (a) GuHCl and (b) urea at three different temperatures (Makhatadze and Privalov, 1992).

Assuming that all binding sites are similar, this equation boils down to

$$Q(a) = n\Delta h \frac{k \times a}{1 + k \times a} \tag{7.13}$$

Upon rearrangement we get Scatchard's expression:

$$Q(a)/a = (n\,\Delta h) \times k - k \times Q(a) \tag{7.14}$$

A plot of $Q\,(a)/a$ against $Q(a)$ should give a straight line if the binding sites are independent and similar. The slope of this line corresponds to the binding constant **k** and the abscissa intercept gives $(n\Delta h)$ if Q is calculated per mole of protein. Analysis of the calorimetrically measured heat effects of titration of the unfolded polypeptide chains by GuHCl and urea show that their Scatchard plots are close to linear (Fig. 7.20), suggesting that the binding of GuHCl and urea molecules occur at independent and similar binding sites on the protein.

From the binding constant and its temperature dependence one can estimate the following:

The intrinsic Gibbs energy of binding:

$$\Delta \mathbf{g} = -RT \ln \mathbf{k} \tag{7.15}$$

The enthalpy of binding:

$$\Delta h = \partial(\Delta \mathbf{g}/T)/\partial(1/T) = RT^2(\partial \ln \mathbf{k}/\partial T) \tag{7.16}$$

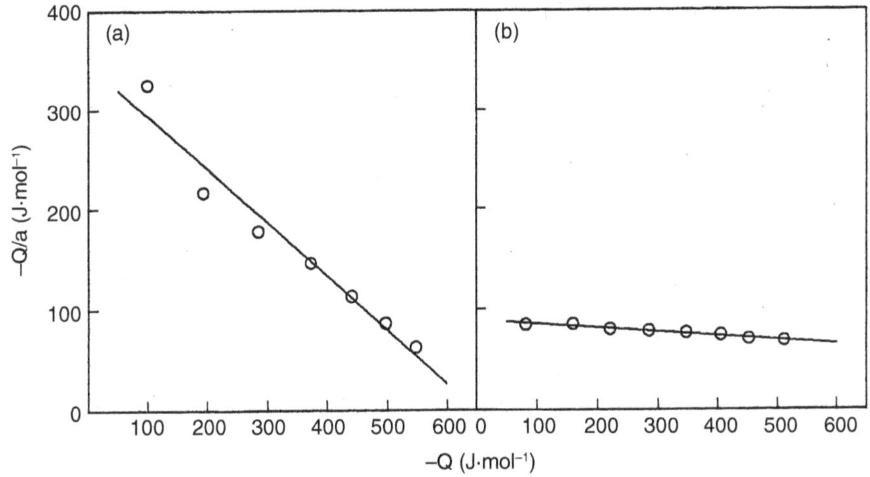

Figure 7.20. Scatchard plots of the calorimetric titration data at 25°C for apo-cytochrome c in (a) GuHCl and (b) urea (Makhatadze and Privalov, 1992).

TABLE 7.3. Intrinsic Binding Characteristics of GuHCl and Urea Molecules to Proteins at 25°C[a]

Parameter	Urea	GuHCl
k	0.061 ± 0.010	0.60 ± 0.09
Δh (kJ/mol)	−9 ± 2	−11 ± 2
Δg (kJ/mol)	6.9 ± 1	1.3 ± 1
Δs (J/K·mol)	−53 ± 8	−41 ± 8
n (Apo–Cyt)	142 ± 19	56 ± 2
n ($RNAse^{-SS}$)	240 ± 28	74 ± 8
n (Lys^{-SS})	229 ± 16	82 ± 6

[a]Makhatadze and Privalov (1992).

The entropy of binding:

$$\Delta s = -\partial \Delta g / \partial T = R \ln \mathbf{k} + RT(\partial \ln \mathbf{k} / \partial T) \tag{7.17}$$

When $(n\Delta h)$ and Δh are determined, one can estimate the number of binding sites:

$$n = (n\ \Delta h)/\Delta h \tag{7.18}$$

Analysis of the intrinsic binding characteristics shows both differences and similarities between the interaction of urea and GuHCl with proteins (Table 7.3). They differ, first of all, in the binding constants: while for GuHCl it is 0.60, for urea it is 0.06 at 25°C—that is, they differ by an order of magnitude (Makhatadze and

Privalov, 1992). On the other hand, the enthalpies of binding of these two compounds to the polypeptide chains do not differ much. Therefore, the difference in the binding constants is mainly caused by the entropies of binding. The entropies of binding are negative and significantly larger in magnitude for urea. However, the overall denaturing effects of these two compounds are not dramatically different because they differ in the number of binding sites: the unfolded polypeptide chain of RNAse has 74 binding sites for the GuHCl and 240 sites for the urea. Therefore, the overall numbers of GuHCl and urea molecules actually bound by the polypeptide chains are not too different for these two compounds.

To understand the mechanism of denaturant action on protein, it is necessary to find out what structural parameters of the protein molecule determine the number of binding sites of these two denaturants. Such correlation analyses, carried on three globular proteins (cytochrome c, ribonuclease A, and lysozyme), showed that the best correlation is observed between the numbers of binding sites and the total number of exposed polar groups. It appears as if one urea molecule is bound by two hydrogen-bond-forming groups and the binding site of GuHCl molecule is formed by four or five hydrogen-bond-forming groups (Makhatadze and Privalov, 1992).

The correlation of the estimated number of binding sites with the exposed nonpolar surface was found to be rather poor. Therefore, one can hardly assume that the urea and GuHCl interact with nonpolar groups of protein. This does not mean, however, that they cannot influence the protein stability by changing the hydrophobic interactions, as was supposed by many authors (see, e.g., Breslow and Gui, 1990; Creighton, 1991). Denaturants in high concentration could affect properties of water determining hydrophobic interactions; however, this has not been yet shown by direct experiments.

7.6. UNFOLDED STATE OF PROTEIN

7.6.1. Completeness of Protein Unfolding at Denaturation

In theoretical modeling of protein folding, the completely unfolded polypeptide chain is usually considered as a standard, initial state. However, the only state of protein that can be obtained practically and occurs unfolded is the denatured state of protein. The question is, therefore, how closely the denatured state of protein approximates the ideal unfolded state: are there residual interactions between the groups in the denatured protein, the residual elements of the structure? This question is of principal importance for the experimental investigation of the protein folding mechanism.

It is widely assumed that a protein in aqueous solution undergoes incomplete unfolding upon heat denaturation, in contrast to its denaturation by concentrated GuHCl or urea (Tanford, 1968). The evidence for this was provided by a smaller intrinsic viscosity specific to the heat-denatured proteins (Hamaguchi and Sakai, 1965) and by the additional changes in the optical properties of heat-denatured proteins with the addition of GuHCl (Aune et al., 1967). Analyzing equilibrium profiles upon heat and GuHCl denaturation of lysozyme it was concluded that the state of heat-denatured protein corresponds to 70% of the fully unfolded state, which is achieved in concentrated GuHCl (Tanford and Aune, 1970; Tanford, 1970). Additional evidence for this conclusion was found in the apparent heat capacity change that, according to the indirect estimates of these authors, appeared to be 0.95 kcal/K·mol = 3.93 kJ/K·mol for the heat denaturation of lysozyme and 1.38 kcal/K·mol = 5.74 kJ/K·mol for the unfolding in GuHCl. This difference was interpreted as direct evidence that the presence of GuHCl results in exposure of more internal groups of protein, that is, that denaturant induces more complete unfolding. Thus, the state of protein in the concentrated GuHCl and urea solutions was regarded as corresponding to the ideal random-coiled state. As a result, most studies on protein unfolding concentrated on the investigation of protein denaturation by GuHCl and urea, more so that the denaturation progress with increasing concentration of the denaturant observed by any method sensitive to protein state suggested a simple procedure for the determination of protein stability in terms of Gibbs energy of unfolding (see Section 7.5). However, detailed study of proteins denatured by various means showed that the situation with the state of proteins unfolded by denaturants is not as simple—while the states of protein unfolded by temperature or pH are not as complicated—as had been supposed.

First, the direct calorimetric studies of lysozyme unfolding showed that the heat capacity increment of its temperature-induced unfolding amounts to (6.7 ± 0.8) kJ/K·mol, that is, is significantly larger than what was indirectly estimated for unfolding by GuHCl (5.74 kJ/K·mol). It was also found that the heat capacity increment for unfolding of lysozyme by GuHCl, if corrected for the preferential binding, is very close in magnitude to that for temperature-induced unfolding, (6.3 ± 0.8) kJ/K·mol (Pfeil and Privalov, 1976b; Makhatadze and Privalov, 1992).

As for the viscosimetric studies of protein solutions, they usually were carried out by the Oswald or rotational viscosimeter. The protein intrinsic viscosity $[\eta]$ is determined by extrapolating the reduced viscosity, determined from the solution flow-time τ (or of the period of rotation in the case of rotational viscosimeter), to zero protein concentration, ω:

$$[\eta] = \lim_{\omega \to 0}[(\tau - \tau_0)]/\omega\tau_0 \tag{7.19}$$

The intrinsic viscosity of globular proteins is low, on the order of 3 cm^3/g, and is almost independent of pH through the pH region in which the given protein is in the native state, but it increases dramatically upon heat denaturation of the protein (Fig. 7.21). The intrinsic viscosity of proteins, the polypeptide chain of which is not cross-linked by disulfide bonds, in solutions with low pH is about 20 cm^3/g at room temperature, which is close to that expected for the polypeptide chain in the random coiled conformation calculated by the equation $[\eta] = 77 \cdot n^{0.666}/M_{ar}$ (Table 7.4). However, an increase of temperature leads to significant decrease of the intrinsic viscosity of the unfolded polypeptide chain (Fig. 7.10). The effect of temperature on the intrinsic viscosity of unfolded polypeptide chain results from increase of its flexibility caused by the increase of rotational freedom of torsions with temperature. Increasing temperature also leads to decreasing intrinsic viscosity of protein in concentrated GuHCl solution: at 80°C it does not differ from the intrinsic viscosity

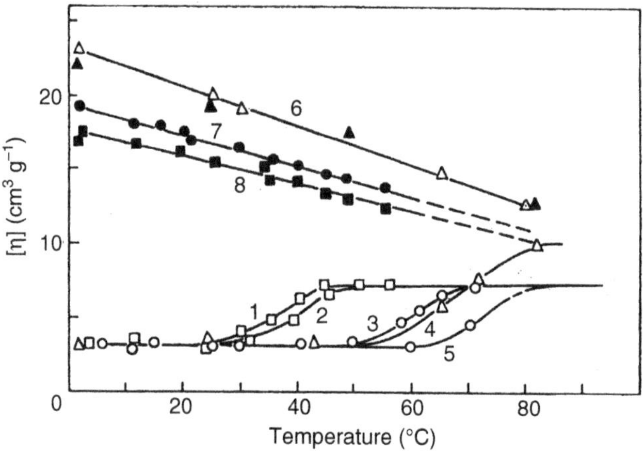

Figure 7.21. Temperature dependence of the intrinsic viscosity of proteins. Proteins folded at room temperature: (1) ribonuclease A (RNAse), pH 3.0; (2) RNAse, A, pH 4.0; (3) lysozyme (Lys), pH 3.0; (4) myoglobin (Mb), pH 5.0; (5) Lys, pH 4.0. Unfolded proteins: (6) apo myoglobin and myoglobin, pH 3.0; (7) lysozyme with disrupted S–S, pH 2.2; (8) ribonuclease with disrupted S–S, pH 2.2 (Privalov et al., 1989).

TABLE 7.4. Intrinsic Viscosity of Unfolded Proteins at 25°C in Solutions with pH 2.2 and 6 M GuHCl[a]

Protein	M_{ar}	n	$[\eta]^b$ pH 2.2	$[\eta]^b$ 6 M GuHCl	$[\eta]^{calc\,b}$
aCyt	11,702	104	15.0	–	14.8
RNAse^{-SS}	13,683	124	15.6	16.1	16.6
Lys^{-SS}	14,300	129	17.0	17.6	17.0
aMb	17,170	153	20.0	20.1	19.1

[a]Privalov et al. (1989).
[b]$[\eta]$ in cm^3/g.
aCyt, apo cytochrome c; RNAse^{-SS}, ribonuclease A with disrupted S–S cross-links; Lys^{-SS}, lysozyme with disrupted S–S cross-links; aMb, apo myoglobin.

of the heat-denatured protein (Fig. 7.21). Therefore, if the GuHCl-denatured protein is considered as completely unfolded, the conformation of the heat-denatured protein should also be regarded as completely unfolded.

A similar situation holds for the ellipticity of proteins at 222 nm, which is usually considered as an index of its secondary structure (Chen et al., 1972). It was found that the ellipticity at 222 nm of the heat-denatured protein and that of protein in the concentrated GuHCl solutions differ considerably. If one accepts that the polypeptide chain in a concentrated solution of GuHCl is in a completely random-coiled conformation, then it follows that the heat-denatured protein has considerable residual secondary structure. However, the question is: can one use the CD spectra of protein in concentrated solution of GuHCl as a reference for the random-coil conformation of the polypeptide chain? Detailed investigation of this subject showed that the ellipticities of the heat-denatured proteins are rather close to those of the unfolded polypeptide chain without disulfide cross-links at acidic pH, but they deviate significantly from the spectra of proteins in 6 M GuHCl, especially at 222 nm and temperature 10°C. It appeared, however, that with temperature increase this difference decreases and is likely to disappear above 80°C (Privalov et al., 1989). It is notable that this difference in ellipticity disappears due to a considerable temperature dependence of the ellipticity of protein in the presence of GuHCl!

The experiments with short synthetic peptides, which could hardly have any helical conformation in aqueous solutions, showed that the addition of GuHCl induces significant changes in their ellipticity (Hvidt et al., 1985; Makhatadze and Privalov, 1992). This, however, is not unexpected, as it is known that GuHCl interacts enthalpically with the peptide groups (Pfeil and Privalov, 1976b). As discussed in Section 7.5.3, one molecule of GuHCl links with four or five hydrogen-bond-forming groups of protein. Then, one should expect that GuHCl can affect the conformation of the peptide group and the optical properties of the individual amino acid residues. According to Tiffany and Krimm (1972), the high concentration of GuHCl and urea favors the locally extended 3_{10}-helix (EH conformation) of polypeptides. If so, the effect of temperature on the ellipticity of proteins in concentrated GuHCl solutions could be explained by the decrease of content of the EH conformation with temperature increase. This is because increasing temperature should lead to desolvation of GuHCl bound to polypeptide since their binding enthalpy is negative. It appears thus that:

Proteins in concentrated solutions of GuHCl attain a random coiled conformation only at a rather high temperature.

Figure 7.22 summarizes the calorimetrically measured enthalpies of lysozyme unfolding by temperature, pH, and GuHCl and corrected, as described previously, for the protonation and denaturant solvation effects. One can see that they all are expressed by the same function of temperature, that is, they are identical. Since the slope of this function corresponds to the heat capacity effect of unfolding, it follows that the heat capacity effects of protein unfolding by any of the means considered are identical. This is just what was found when the measured heat capacity effects of unfolding were corrected for the solvation effects of the denaturant.

This conclusion was confirmed by comparison of the Gibbs energies of protein unfolding by various means. It was shown that the ΔG of the ribonuclease T1 unfolding at 25°C, determined by DSC studies of its heat denaturation, is 5.60 kcal/mol = 23.3 kJ/mol and by the equilibrium studies of urea denaturation is 5.55 kcal/mol = 23.1 kJ/mol (Hu et al., 1992). Similarly, the ΔG of ribonuclease A unfolding at 25°C, pH 7.0, determined by DSC studies of heat denaturation was found to be 9.0 kcal/mol = 37.4 kJ/mol and by the urea denaturation studies, 9.1 kcal/mol = 37.8 (Pace et al., 1999). One can conclude then that:

If a protein is completely unfolded in concentrated GuHCl, it should be unfolded to the same extent upon acid and heat denaturation.

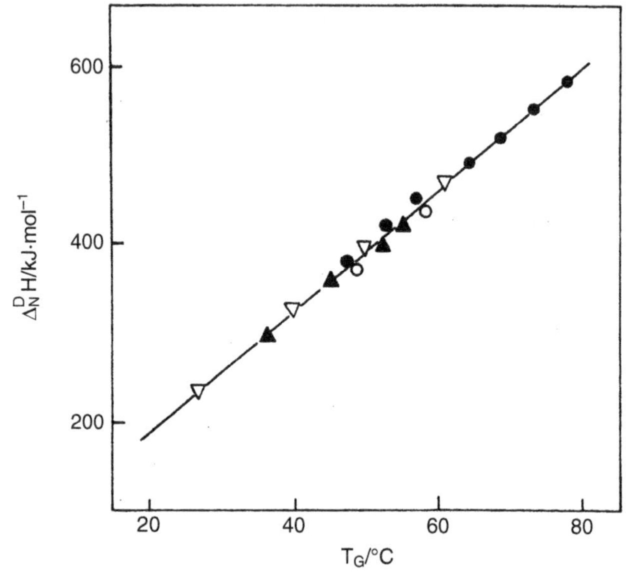

Figure 7.22. Enthalpies of lysozyme denaturation by various methods under different conditions plotted against the temperature of denaturation. *Circles* indicate solutions without GuHCl: *open*, denaturation by temperature at fixed pH; *filled*, denaturation by pH at fixed temperatures. *Triangles* indicate solutions with GuHCl: *filled*, denaturation by temperature at fixed concentration of GuHCl; *open*, denaturation by GuHCl at fixed temperature (Pfeil and Privalov, 1976b).

Thus, in contrast to a rather widespread opinion, one can conclude that:

The residual interactions between the groups of a temperature-unfolded nonaggregating protein are insignificant and can be neglected, at least to a first approximation.

This is a very important conclusion because in the theoretical modeling of protein folding, the energetics of this process on the unfolded polypeptide chain is considered as a standard reference state. We see now that this ideal unfolded state of protein can be approximated by the temperature- and acid-denatured state, which needs minimal corrections for the residual interactions between the groups.

7.6.2. Thermodynamic Functions Describing Protein States

It is clear from the preceding discussion that unfolding of proteins depends on the environmental conditions and that external variables affect the folded and unfolded states of proteins. For thermodynamic specification of these states one has to choose the standard state from which the changes would be counted. A standard that could be chosen for proteins is the state at neutral pH, room temperature, and normal pressure, that is, $pH_0 = 7.0$, $T_0 = 25°C = 398.16$ K, and $P = 1$ atm. When the variables are temperature and pH, for an infinitesimal change of some function F describing a protein state we will have

$$dF = \left(\frac{\partial F}{\partial T}\right)_{pH} dT + \left(\frac{\partial F}{\partial pH}\right)_T dpH \tag{7.20}$$

In particular, departing from this standard condition we will have the following for the enthalpy and entropy of the protein:

$$\begin{aligned} H^N(T, pH) &= \int_{T_{ot}}^{T} \left(\frac{\partial H}{\partial T}\right)_{pH} dT + \int_{pH_0}^{pH} \left(\frac{\partial H}{\partial pH}\right)_T dpH \\ &= \int_{T_0}^{T} C_{p,pH}^N dT + H^N(pH)_{T_0} \end{aligned} \tag{7.21}$$

$$S^N(T, pH) = \int_{T_0}^{T} \left(\frac{\partial S}{\partial T}\right)_{pH} dT + \int_{pH_0}^{pH} \left(\frac{\partial S}{\partial pH}\right)_T dpH = \int_{T_0}^{T} \frac{C_{p,pH}^N}{T} dT + S^N(pH)_{T_0} \tag{7.22}$$

The temperature-dependent terms contain the partial heat capacity of the native protein in solution at given pH, $C_p^N(T)_{pH}$, which is determinable by scanning microcalorimetry. The pH-dependent term $H^N(pH)_T$, which describes the molar heat of titration of native protein with initial pH 7.0 at T_0, can be determined by isothermal microcalorimetry. The pH-dependent term of standard entropy cannot be determined directly, but it can be calculated if the standard enthalpy $H^N(pH)$ and the Gibbs function $G^N(pH)$ are known for the given standard temperature.

Determination of $G^N(pH)$ is possible by the direct use of titration curves. From the general equation

$$G(x) = G(x_0) + \int_{x_0}^{x} \left(\frac{\partial G}{\partial x}\right) dx \tag{7.23}$$

Equation (7.24) follows for the multiple equilibrium of protein ionization:

$$G(\alpha) = G(\alpha_0) - 2.3RT \int_{\alpha_0}^{\alpha} pH(\alpha) d\alpha \tag{7.24}$$

Introducing instead of the degree of ionization (α) the relative number of protons (ν), measured potentiometrically for a standard condition, where the integration constant $G^0(\nu_0)$ is zero, we have the following for the Gibbs function:

$$G^N(\nu) = -2.3RT \int_{\nu_0}^{\nu} pH(\nu) d\nu \tag{7.25}$$

which can be easily transformed into G^N(pH) on the basis of the experimental titration curve (Pfeil and Privalov, 1976a). Having G^N(pH), we will have for the pH-dependent entropy term:

$$S_{pH}^{N} = \frac{H_{pH}^{N} - G_{pH}^{N}}{T_0} \tag{7.26}$$

For the Gibbs energy function of the native protein we have

$$G^{N}(T, \text{pH}) = H^{N}(T, \text{pH}) - TS^{N}(T, \text{pH}) \tag{7.27}$$

Corresponding functions for the denatured state can be obtained from the standard functions of the native state and denaturational changes in enthalpy ($\Delta_d H$), entropy ($\Delta_d S$), and Gibbs energy ($\Delta_d G$).

The validity of the standard functions can be tested using a cyclic pathway of protein transfer from State I to State II (Fig. 7.23). The defined function will be potential if its value does not depend on the pathway and for any closed cycle the total change will be zero. Detailed investigation, using lysozyme as an example, showed that its transfer from State I ($\text{pH}_{\text{I}} = 4.8$, $T_{\text{I}} = 25°\text{C}$) to State II ($\text{pH}_{\text{II}} = 1.5$, $T_{\text{II}} = 50°\text{C}$) by two pathways, by (Step1–Step2) or (Step3–Step4), leads to very similar overall results (Pfeil and Privalov, 1976a).

Thus using the scanning and isothermal titration calorimetric data one can construct the standard enthalpy and entropy functions of the folded and unfolded states of protein in temperature and pH coordinates (Fig. 7.24). The difference between these two surfaces at each point gives the value of the enthalpy, or entropy, at protein unfolding at the corresponding condition. The examples of the Gibbs energy differences between the native and unfolded states for several globular proteins are shown in Figure 7.25.

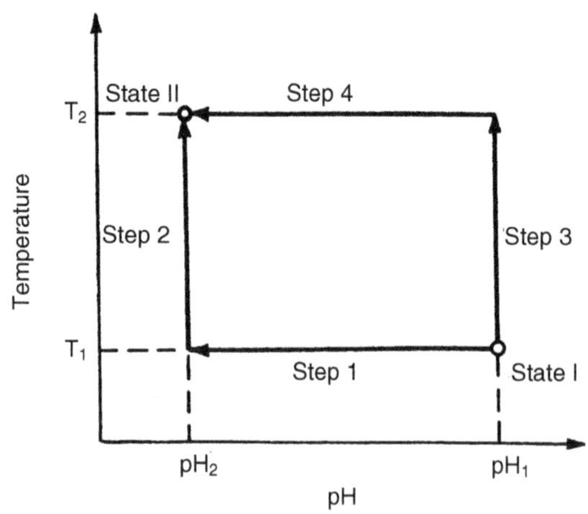

Figure 7.23. Two different pathways to reach from state I to state II.

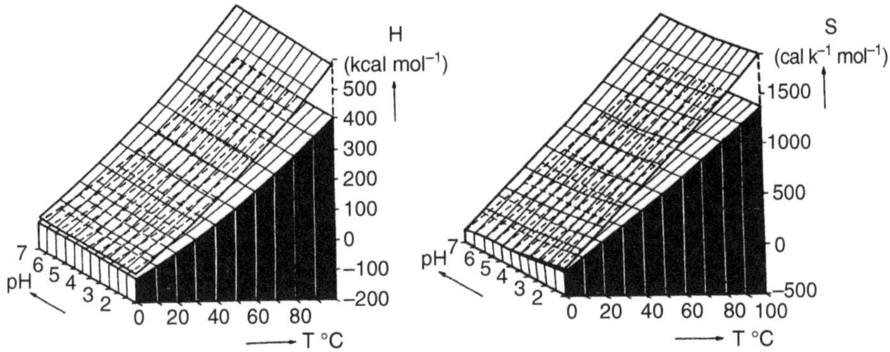

Figure 7.24. The standard enthalpy and entropy of the folded (lower surface) and unfolded (upper surface) states of lysozyme in the pH and temperature coordinates. The difference between the surfaces at any (pH, *T*) point gives the enthalpy and entropy of protein unfolding at this point (Pfeil and Privalov, 1976c).

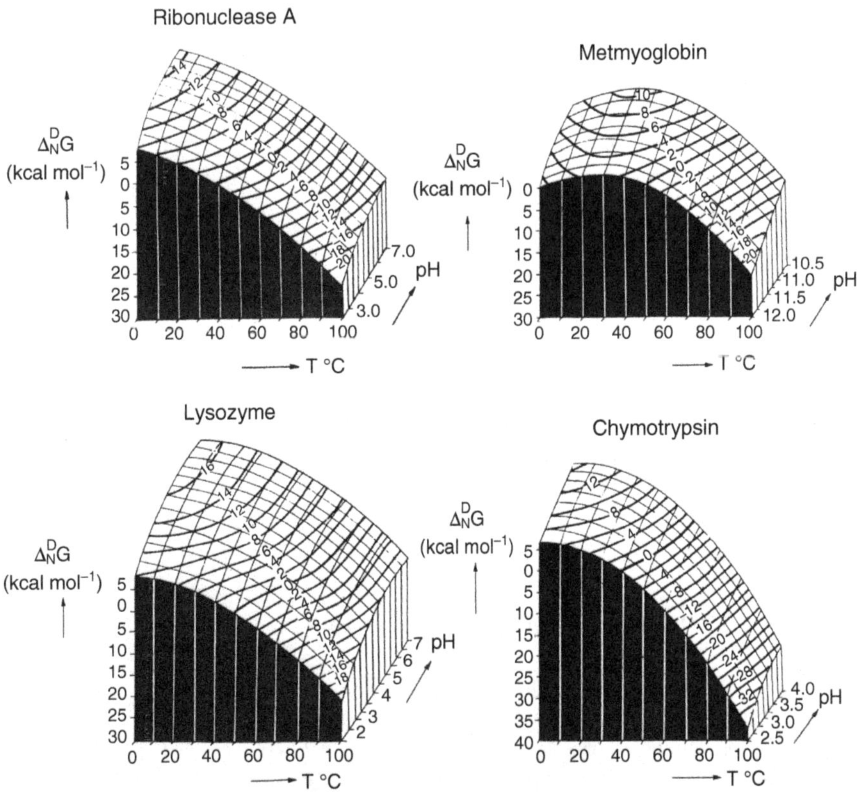

Figure 7.25. Gibbs energies of unfolding of various globular proteins determining their stabilities at any (*T*, pH) values (Pfeil and Privalov, 1976c).

The main conclusion that follows from the previously described thermodynamic studies of globular proteins, is that:

The native and denatured states of these proteins represent their qualitatively different macroscopic states, and transition between them closely resembles a first-order phase transition that proceeds with drastic change in all thermodynamic characteristics: enthalpy, entropy, Gibbs energy, and heat capacity.

REFERENCES

Alexander P., Fahnestock C.S., Lee T., Orban J., and Bryan P. (1992). *Biochemistry*, **31**, 3597–3603.

Aune K.C., Salahuddin, A., Zarlenyo M.H., and Tanford C. (1967). *J. Biol. Chem.*, **242**, 4486–4489.

Bae S.J. and Sturtevant J.M. (1995). *Biophys. Chem.*, **55**, 247–252.

Baldwin R.L. (1975). *Annu. Rev. Biochem.*, **44**, 453–475.

Blake C.C. and Rice D.W. (1981). *Philos. Trans. Roy. Soc. London A, Biol. Sci.*, **293**, 93–104.

Bock P.E. and Frieden C. (1974). *Biochemistry*, **13**, 4191–4196.

Bock P.E., Gilbert H.R., and Frieden C. (1975). *Biochem. Biophys. Res. Commun.*, **66**, 564–569.

Brandts J.F., Brennan M., and Lin L.N. (1977). *Proc. Natl. Acad. Sci. USA*, **72**, 4178–4181.

Brandts J.F., Halvorson H.R., and Brennan M. (1975). *Biochemistry*, **14**, 4953–4963.

Breslow R. and Gui T. (1990). *Proc. Natl. Acad. Sci. USA*, **87**, 167–169.

Chen B.-L. and Schellman J.A. (1989). *Biochemistry*, **28**, 685–691.

Chen Y.-H., Yang J.T., and Martinez H.M. (1972). *Biochemistry*, **11**, 4120–4131.

Christensen L.K. (1951). *C.R. Lab. Carlsberg. Ser. Chim.*, **28**, 39–46.

Clark J.H. (1945). *J. Gen. Physiol.*, **28**, 539–545.

Creighton T.E. (1991). *Curr. Opin. Struct. Biol.*, **1**, 5–16.

Freire E. and Biltonen R.L. (1978). *Biopolymers*, **17**, 463–479.

Freire E., Murphy K.P., Sanchez-Ruiz J., Galisteo M.L., and Privalov P.L. (1992). *Biochemistry*, **31**, 250–256.

Griko Yu.V. and Privalov P.L. (1992). *Biochemistry*, **31**, 8810–8815.

Griko Yu.V., Makhatadze G.I., Privalov P.L., and Hartley R.W. (1994). *Protein Sci.*, **3**, 669–676.

Griko Yu.V., Privalov P.L., Sturtevant J.M., and Venyaminov S.Yu. (1988a). *Proc. Natl. Acad. Sci. USA*, **85**, 3343–3347.

Griko Yu.V., Privalov P.L., Venyaminov S.Yu., and Kutyshenko V.P. (1988b). *J. Mol. Biol.*, **202**, 127–138.

Griko Yu.V., Venyaminov S.Yu., and Privalov P.L. (1989). *FEBS Lett.*, **244**, 276–278.

Hamaguchi K. and Sakai H. (1965). J. Biochem. *(*Tokyo*)*, **57**, 103–114.

Hopkins P.G. (1930). *Nature*, **126**, 383–386.

Hu C.Q. and Sturtevant J.M. (1987). *Biochemistry*, **26**, 178–182.

Hu C.Q., Sturtevant J.M., Thomson J.A., Erickson R.E., and Pace C.N. (1992). *Biochemistry*, **31**, 4876–4882.

Hvidt S., Rodgers M.E., and Harrington W.F. (1985). *Biopolymers*, **24**, 1647–1662.

Ikegami A. (1977). *Biophys. Chem.*, **6**, 117–130.

Kitamura S. and Sturtevant J.M. (1989). *Biochemistry*, **28**, 3788–3792.

Makhatadze G.I. and Privalov P.L. (1990). *J. Mol. Biol.*, **213**, 375–384.

Makhatadze G.I. and Privalov P.L. (1992). *J. Mol. Biol.*, **226**, 491–505.

Makhatadze G.I. and Privalov P.L. (1993). *J. Mol. Biol.*, **232**, 639–659.

Makhatadze G.I. and Privalov P.L. (1995). *Adv. Protein Chem.*, **47**, 307–425.

Makhatadze G.I., Clore G.M., Gronenborn A.M., and Privalov P.L. (1994). *Biochemistry*, **33**, 9327–9332.

Makhatadze G.I., Kim K.S., Woodward C., and Privalov P.L. (1993). *Protein Sci.*, **2**, 2028–2036.

Pace C.N. (1990). *Trends Biotech.*, **8**, 93–98.

Pace C.N., Grimsley G.R., Thomas S.T., and Makhatadze G.I. (1999). *Protein Sci.*, **8**, 1500–1504.

Pace C.N., Laurents D.V., and Thomson J.A. (1990). *Biochemistry*, **29**, 2564–2572.

Pfeil W. and Privalov P.L. (1976a). *Biophys. Chem.*, **4**, 23–32.

Pfeil W. and Privalov P.L. (1976b). *Biophys. Chem.*, **4**, 33–40.

Pfeil W. and Privalov, P.L. (1976c). *Biophys. Chem.*, **4**, 41–55.

Privalov P.L. (1974). *FEBS Lett.*, **40**, S140-S153.

Privalov P.L. (1979). *Adv. Protein Chem.* **33**, 167–241.

Privalov P.L. (1989). *Annu. Rev. Biophys. Biophys. Chem.*, **18**, 47–69.

Privalov P.L. (1990). *CRC Crit. Rev. Biochem. Mol. Biol.*, **25**, 281–305.

Privalov P.L. and Khechinashvili N.N. (1974). *J. Mol. Biol.*, **86**, 665–684.

Privalov P.L. and Makhatadze G.I. (1990). *J. Mol. Biol.*, **213**, 385–391.

Privalov P.L. and Makhatadze G.I. (1992). *J. Mol. Biol.*, **224**, 715–723.

Privalov P.L., Griko Yu.V., Venyaminov S.Yu., and Kutyshenko V.P. (1986). *J. Mol. Biol.*, **190**, 487–498.

Privalov P.L., Ptitsyn O.B., and Birshtein, T.M. (1969). *Biopolymers*, **8**, 559–571.

Privalov P.L., Tiktopulo E.I., Venjaminov S.Yu., Griko Yu.V., Makhatadze G.I., and Khechinashvili N.N. (1989). *J. Mol. Biol.*, **205**, 737–750.

Renner M., Hinz H.-J., Scharf M., and Engels J.W. (1992). *J. Mol. Biol.*, **233**, 769–779.

Schellman, J.A. (1958). *C.R. Lab. Carlsberg. Ser. Chim.*, **30**, 395–401.

Simpson R.B. and Kauzmann W. (1953). *J. Am. Chem. Soc.*, **75**, 5139–5143.

Tanford C. (1968). *Adv. Protein Chem.*, **23**, 121–275.

Tanford C. (1970). *Adv. Protein Chem.*, **24**, 1–95.

Tanford C. and Aune, K.C. (1970). *Biochemistry*, **9**, 206–211.

Tiffany M.L. and Krimm S. (1972). *Biopolymers*, **11**, 2309–2316.

Wintrode P.L., Makhatadze G.I., and Privalov P.L. (1994). *Proteins: Struct. Funct. Genet.*, **18**, 246–253.

Yao M. and Bolen D.W. (1992). ASBMB/Biophysical Society Joint Meeting, Houston, Texas.

Yu Y., Makhatadze G.I., Pace C.N., and Privalov P.L. (1994). *Biochemistry*, **33**, 3312–3319.

8

ENERGETIC BASIS OF PROTEIN STRUCTURE

8.1. HYDRATION EFFECTS

8.1.1. Proteins in an Aqueous Environment

One of the most remarkable features of protein unfolding is that its enthalpy and entropy are asymptotically increasing functions of temperature: they are positive at room temperature and increase with increasing temperature to some limit reached above 120°C (Fig. 7.6). Correspondingly, they decreasing with decreasing temperature and can even change sign at low temperatures, as shown by cold denaturation studies (see Section 7.3). It follows that the temperature derivative of enthalpy, that is, the heat capacity effect of protein unfolding, is positive but decreases in magnitude with increasing temperature. These specificities of proteins raise certain questions:

(a) Why does the heat capacity of native protein increase on transfer into aqueous solution? Does the water somehow change the thermal properties of the protein? Or is the protein changing the properties of the surrounding water?

Microcalorimetry of Macromolecules: The Physical Basis of Biological Structures, First Edition.
Peter L. Privalov.

(b) Why does the heat capacity of protein increase on unfolding? It was first supposed that the increase in the heat capacity of a protein results from the increase of conformational freedom upon its unfolding. However, the observed heat capacity increment is at least one order of magnitude larger than what one could expect from the increase of conformational freedom at protein unfolding (Sturtevant, 1977).

(c) Why does the denaturation heat capacity increment decrease in magnitude with increasing temperature? If the heat capacity increment of unfolding resulted from increasing conformational freedom of the protein upon unfolding, this effect would only increase with rising temperature since the conformational freedom of the polypeptide backbone and of the side chains increases with increasing temperature.

(d) Why do enthalpy and entropy of protein unfolding decrease with decreasing temperature and even change sign at low temperatures? The negative entropy of unfolding means an increase of order. However, protein unfolding upon cooling, cold denaturation, assumes disruption of the highly ordered native structure; that is, it results in conformational disorder.

The main cause of all these unusual thermal properties of protein appears to be the hydration of its groups exposed to water in the folded and unfolded state (Privalov, 1979; Makhatadze and Privalov, 1995). Hydration of protein groups proceeds with significant enthalpy and entropy effects resulting from the formation of new interactions between the water molecules and the protein groups and also between the water molecules themselves. These rearrangements of interactions with water and in water should certainly depend on temperature, and this dependence should not necessarily be simple. Therefore, hydration of protein groups should result in some heat capacity effect, which also might depend on temperature. Thus, understanding the protein unfolding–refolding energetics needs an understanding of the concomitant effects of protein hydration–dehydration. This, however, has required accumulation of experimental data on transfer of various low-molecular-weight organic compounds into water and, subsequently, development of methodology for the evaluation of these data to derive the thermodynamic characteristics of protein hydration. To apply these results to proteins it was necessary to find out if the hydration effects of individual chemical groups are additive, and if the overall hydration effect on the molecule can be estimated, at least for a first approximation, by summing the hydration contributions of individual groups.

8.1.2. Hydration of Protein Groups

There have been many attempts to evaluate hydration effects on protein unfolding using results on transfer of various low-molecular-weight compounds into water from the liquid phase (Baldwin, 1986), solid phase (Murphy and Gill, 1991), or gaseous phase (Wolfenden, 1978; Wolfenden et al., 1981; Ben-Naim, 1987; Makhatadze and Privalov, 1993, 1995; Privalov and Makhatadze, 1993). One can consider the solid phase as the most appropriate reference state of the densely packed interior of a native protein in analyzing transfer of its internal groups into water upon unfolding because, if the liquid phase were used as a reference, it would be impossible to determine the energies of interaction between these groups in the native protein. Also, one cannot determine the conformational entropy of a polypeptide chain unfolding using transfer from the liquid phase because in the native protein the groups are fixed, unlike in the liquid. However, to evaluate the internal interactions in proteins and conformational entropy of polypeptide chain unfolding, we need to know the net hydration effects (i.e., the enthalpy, entropy, and heat capacity effect as functions of temperature), which do not include the interactions between the considered groups and the effects associated with thermal motion of these groups. These net hydration effects can be determined only considering transfer of the model compounds from a fixed position in the gaseous phase to a fixed position in water (see Section 2.4).

The surface normalized values of hydration effects for all protein groups are listed in Table 8.1. It is notable that $\delta C_{p,i}^{\mathrm{hyd}}$ values for all aliphatic groups are positive and similar. They are positive also for aromatic groups, but smaller in magnitude than for aliphatic groups (Makhatadze and Privalov, 1987, 1988, 1989, 1990; Murphy and Gill, 1991; Livingstone et al., 1991; Spolar et al., 1989, 1992). It is incorrect therefore to assume that the aliphatic and aromatic groups are similar in their hydration properties as is often assumed in the literature.

The temperature dependence of the surface normalized values of the hydration effects for aliphatic and polar groups are shown in Figures 8.1 and 8.2. It appears that the hydration enthalpies and entropies of the polar and aliphatic groups are negative at room temperature; however, although for nonpolar aliphatic groups they decrease in magnitude with increasing temperature, for polar groups they increase. Thus, the heat capacity change at hydration of polar and nonpolar groups is of opposite sign: positive for the nonpolar groups and negative for the charged and polar groups. The situation with aromatic groups is in between: the hydration enthalpy and entropy both are negative and decrease in magnitude with increasing temperature, as in the case of aliphatic groups, but the change is not so steep (Fig. 8.3).

TABLE 8.1. Surface Normalized Values of Heat Capacities, Enthalpies, and Entropies of Hydration for Various Parts of Amino Acid Residues[a]

		Temperature				
Surface	Property	5°C	25°C	50°C	75°C	100°C
Aliphatic	$\delta C_p^{\mathrm{hyd}}$	2.24	2.14	2.03	1.91	1.80
	δH^{hyd}	−166	−122	−70	−21	26
	δS^{hyd}	−730	−578	−409	−263	−134
Aromatic	$\delta C_p^{\mathrm{hyd}}$	1.65	1.55	1.41	1.29	1.19
	δH^{hyd}	−180	−148	−111	−77	−46
	δS^{hyd}	−430	−319	−199	−98	−12

(Continued)

TABLE 8.1. (*Continued*)

Surface	Property	Temperature				
		5°C	25°C	50°C	75°C	100°C
Polar parts of: Arg	δC_p^{hyd}	−0.38	−0.20	−0.12	−0.04	0.01
	δH^{hyd}	−821	−827	−831	−833	−834
	δS^{hyd}	−458	−478	−492	−497	−498
Asn	δC_p^{hyd}	−1.27	−1.01	−0.67	−0.41	−0.16
	δH^{hyd}	−871	−894	−915	−928	−936
	δS^{hyd}	−575	−654	−723	−763	−783
Asp	δC_p^{hyd}	−1.72	−1.40	−1.07	−0.71	−0.40
	δH^{hyd}	−684	−715	−746	−768	−782
	δS^{hyd}	−360	−469	−569	−636	−675
Cys	δC_p^{hyd}	1.80	2.01	2.23	2.42	2.54
	δH^{hyd}	−309	−271	−218	−160	−98
	δS^{hyd}	−535	−402	−232	−59	113
Gln	δC_p^{hyd}	−0.38	−0.22	−0.06	0.07	0.17
	δH^{hyd}	−697	−703	−706	−706	−703
	δS^{hyd}	−571	−591	−604	−603	−594
Glu	δC_p^{hyd}	−0.71	−0.55	−0.35	−0.17	−0.05
	δH^{hyd}	−459	−562	−573	−580	−583
	δS^{hyd}	−392	−436	−473	−492	−500
His	δC_p^{hyd}	−1.96	−2.43	−2.38	−2.26	−2.07
	δH^{hyd}	−1084	−1128	−1188	−1247	−1301
	δS^{hyd}	−542	−693	−888	−1060	−1211
Lys	δC_p^{hyd}	−1.31	−1.53	−1.59	−1.36	−1.15
	δH^{hyd}	−685	−714	−753	−789	−821
	δS^{hyd}	−394	−482	−609	−716	−804
Met	δC_p^{hyd}	−3.51	−3.83	−4.07	−4.04	−3.91
	δH^{hyd}	−399	−473	−572	−672	−774
	δS^{hyd}	−158	−412	−732	−1031	−1308
Ser	δC_p^{hyd}	−1.62	−1.40	−1.20	−096	−072
	δH^{hyd}	−1015	−1045	−1078	−1104	−1126
	δS^{hyd}	−878	−983	−1089	−1168	−1227
Thr	δC_p^{hyd}	−1.09	−1.29	−1.22	−0.89	−0.29
	δH^{hyd}	−1262	−1287	−1318	−1343	−1359
	δS^{hyd}	−971	−1053	−1156	−1232	−1274
	δC_p^{hyd}	1.05	0.96	1.07	1.08	1.03
	δH^{hyd}	−1181	−1161	−1135	−1110	−1084
	δS^{hyd}	−766	−693	−615	−534	−460
Trp	δC_p^{hyd}	−1.46	−1.48	−1.36	−1.15	−0.86
	δH^{hyd}	−824	−854	−889	−921	−946
	δS^{hyd}	−314	−415	−531	−625	−695
CONH	δC_p^{hyd}	−2.08	−1.81	−1.56	−1.53	−1.49
	δH^{hyd}	−1662	−1702	−1745	−1785	−1823
	δS^{hyd}	−890	−1026	−1162	−1278	−1383

[a]Makhatadze and Privalov (1995, 1996).
δC_p^{hyd} in J/K·mol·Å^2; δH^{hyd} in J/mol·Å^2; δS^{hyd} in J/K·mol·Å^2.

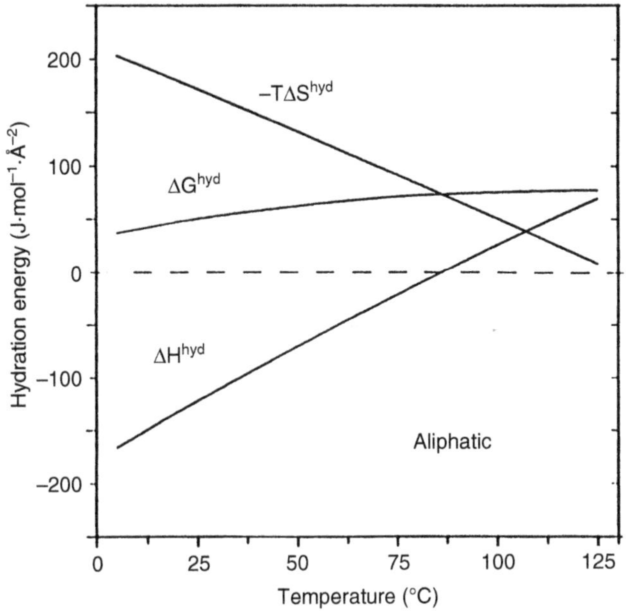

Figure 8.1. Temperature dependence of the hydration enthalpy, Gibbs energy, and entropy factor for aliphatic groups (Makhatadze and Privalov, 1995).

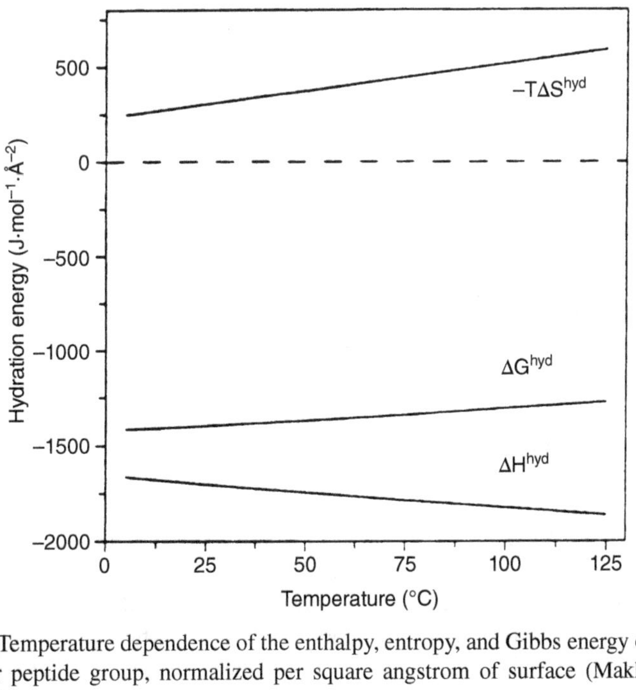

Figure 8.2. Temperature dependence of the enthalpy, entropy, and Gibbs energy of hydration for the polar peptide group, normalized per square angstrom of surface (Makhatadze and Privalov, 1995).

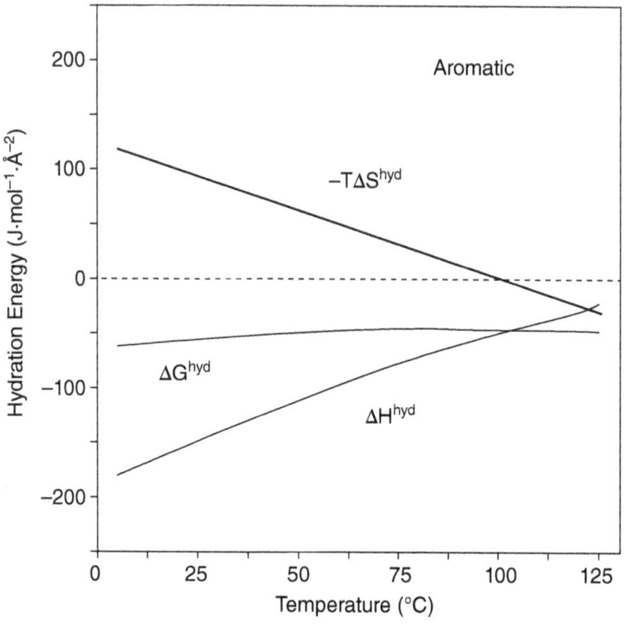

Figure 8.3. Temperature dependence of the hydration enthalpy, Gibbs energy, and entropy factor for aromatic groups (Makhatadze and Privalov, 1995).

The main consequence of the differences in the temperature dependence of the hydration enthalpy and entropy functions and, correspondingly, in the temperatures where these parameters become zero (T_H and T_S values) is that the Gibbs energies of hydration of aromatic and aliphatic groups differ qualitatively and they differ from that for polar groups. The Gibbs energy of hydration of polar groups is large and negative, and its magnitude decreases with increasing temperature. Therefore, these groups are clearly **hydrophilic**, and their hydrophilicity decreases with increasing temperature. In contrast, the Gibbs energy of hydration of aliphatic groups is positive and it increases with increasing temperature. These groups are, therefore, **hydrophobic**, and their hydrophobicity increases with increasing temperature. The Gibbs energy of hydration of aromatic groups, which have been traditionally regarded as typical hydrophobic groups, is negative but much smaller in magnitude than it is for polar groups (Privalov and Makhatadze, 1993; Makhatadze and Privalov, 1994). Therefore, we cannot classify these groups as hydrophobic. Moreover, these groups can hardly be classified as nonpolar, as is usually assumed. Indeed, the presence of double bonds between ring carbon atoms in the aromatic groups results in segregation of partial electron charges to a π-electron cloud covering the face of the aromatic ring—thus leaving hydrogen atoms on the edge of the ring, which makes this ring positively charged (Burley and Petsko, 1988). Therefore, these compounds are in fact polar and their interaction with other molecules, particularly with water, is at least partly polar (Makhatadze and Privalov, 1994). Judging by the magnitude of hydration effects, interaction of aromatic rings with water is weaker than that of the conventional polar groups. Consequently, one can call them **weak polar interactions** (Burley and Petsko, 1988). On the other hand, the temperature dependence of hydration effects of aromatic rings differs qualitatively from that of the polar groups: their hydration enthalpy and entropy decrease in magnitude with increasing temperature, as these functions do for aliphatic compounds. The only difference is that in the hydration of aromatic compounds the enthalpy dominates over the entropy term in the Gibbs energy.

The negative enthalpy and entropy of hydration of polar groups are evidently caused by their interaction with polar water molecules, mostly by hydrogen bonding. The hydration enthalpy and entropy of polar groups, however, are too large to be explained by formation of a single hydrogen bond with water. Bearing in mind that water is a cooperative liquid (see Section 2.3.1), one could assume that the presence of polar groups leads to some reorganization in water structure with reinforcement of the hydrogen-bonding network between the water molecules. A similar situation exists with the hydration of nonpolar groups: their enthalpy of hydration is too large to be explained by additional van der Waals interactions with water molecules. This suggests that these groups lead to some reorganization in the hydrogen-bonding network in the surrounding water. This is usually explained by dumping fluctuations of the flickering clusters of water upon insertion of the neutral molecule into the cavity of the transparent water structure.

If the presence of polar or nonpolar solute increases order in water, judging by their negative entropy of hydration, why then is the temperature dependence of the hydration enthalpy and entropy of polar groups and apolar aliphatic groups different? In other words, why do the heat capacity effects of hydration of these groups have opposite signs? A possible explanation might be that the presence of aliphatic groups in water specifically intensifies the icelike structure, increases the probability of formation of the "flickering" clusters of the icelike structure, whereas polar groups essentially reorganize the water structure and this is achieved more readily at high temperatures because reorganization of the less-structured water is easier. Unfortunately we still do not have a quantitative theoretical model of liquid water and can only speculate on the effects of solutes on its structure and thermodynamic properties.

8.1.3. Hydration of the Folded and Unfolded Protein

The total hydration effect for the folded and unfolded protein can be calculated if the effects caused by the individual groups are additive and are proportional to the surface of the group that is exposed to water:

$$F_k^{\mathrm{hyd}}(T) = \sum_i \Delta_n^{\mathrm{U}} \mathrm{WASA}_{k,j} \times \Delta_n^{\mathrm{U}} f_{k,j}^{\mathrm{hyd}}(T) \tag{8.1}$$

Here, $\Delta_{\mathrm{N}}^{\mathrm{U}}\mathrm{WASA}_{k,j}$ is the change of water-accessible surface area (WASA) of type k of a group j in the protein upon protein unfolding, and $\Delta_{\mathrm{n}}^{\mathrm{U}} f_{k,j}^{\mathrm{hyd}}(T)$ is the hydration enthalpy, entropy, or heat capacity of this type of group, normalized per square angstrom. The additivity of hydration effects was examined with various model compounds and short model polypeptides (Privalov and Gill, 1988). This holds for nonpolar groups with high accuracy. The applicability of this principle for the polar groups seems less evident because the polar groups' influence on water should be of a longer range. Nevertheless, by using various model compounds it was shown that additivity holds for polar compounds at least within 7% accuracy (Makhatadze and Privalov, 1995).

The hydration effects for the charged and polar groups, in contrast to the aliphatic ones, are negative and very different for different groups (Table 8.1). For a first approximation, however, one can use an averaged value of the hydration effects, which takes into account the typical distribution of various groups in globular proteins. For 25°C this averaged hydration effect is approximated by the following expressions:

$$\delta H^{\text{hyd}}(25°\text{C}) = [-122 \times \text{WASA}_{\text{aliph}} - 148 \times \text{WASA}_{\text{arom}} - 1240 \times \text{WASA}_{\text{pol}}] \quad \text{J/mol} \tag{8.2}$$

$$\delta G^{\text{hyd}}(25°\text{C}) = [50 \times \text{WASA}_{\text{aliph}} - 53 \times \text{WASA}_{\text{arom}} - 1000 \times \text{WASA}_{\text{pol}}] \quad \text{J/mol} \tag{8.3}$$

$$\delta S^{\text{hyd}}(25°\text{C}) = [-0.58 \times \text{WASA}_{\text{aliph}} - 0.30 \times \text{WASA}_{\text{arom}} - 0.81 \times \text{WASA}_{\text{pol}}] \quad \text{J/K·mol} \tag{8.4}$$

$$\delta C_p(25°\text{C})^{\text{hyd}} = [2.14 \times \text{WASA}_{\text{aliph}} + 1.55 \times \text{WASA}_{\text{arom}} - 1.27 \times \text{WASA}_{\text{pol}}] \quad \text{J/K·mol} \tag{8.5}$$

For any other temperature in the Celsius scale the components of ΔC_p^{hyd} can be calculated by the following expressions (Makhatadze and Privalov, 1995):

$$\delta C_p(T)^{\text{hyd}}_{\text{aliph}} = [2.14 - 4.48 \times 10^{-3} \times (T-25) - 2.5 \times 10^{-6} \times (T-25)^2] \quad \text{J/K·mol} \tag{8.6}$$

$$\delta C_p(T)^{\text{hyd}}_{\text{arom}} = [1.55 - 5.47 \times 10^{-3} \times (T-25) - 8.6 \times 10^{-6} \times (T-25)^2] \quad \text{J/K·mol} \tag{8.7}$$

$$\delta C_p(T)^{\text{hyd}}_{\text{pol}} = [-1.27 + 8.04 \times 10^{-3} \times (T-25) - 31.1 \times 10^{-6} \times (T-25)^2] \quad \text{J/K·mol} \tag{8.8}$$

The water accessible surface area of proteins with known structure is determined by rolling a probe of radius 1.4 Å (approximating a water molecule) over the surface being studied (Fig. 8.4). There are several commercial programs that can be used for this purpose (Shrake and Rupley, 1973; Miller et al., 1987).

Determination of the WASA for the unfolded protein is not as straightforward as for the folded protein because the unfolded polypeptide does not have a definite structure and even the most powerful computers cannot efficiently model its rapidly changing random-coiled conformation. In earlier studies, therefore, the WASA of the unfolded protein was estimated by summing WASA_i of each amino acid i as determined in the extended tripeptides Gly–X–Gly or Ala–X–Ala. However, since Gly does not have a side chain and Ala has a very small one, this method gives WASA values of the side chain X without screening by neighboring side chains, which certainly takes place in the random-coiled conformation of the polypeptide

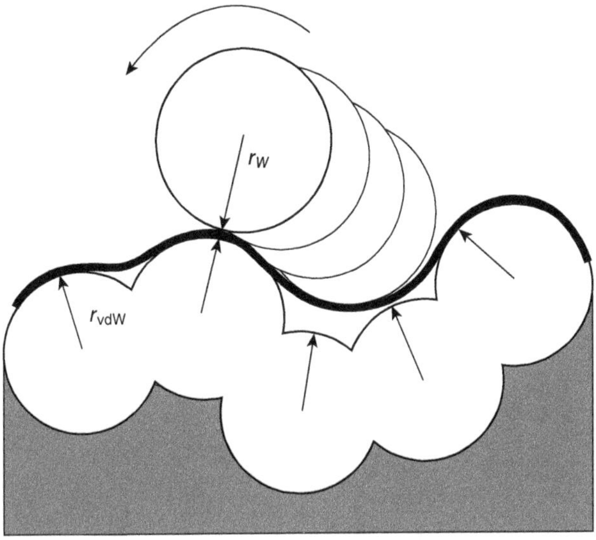

Figure 8.4. Water-accessible surface is defined by the path of the contact made between the ball and the van der Waals surface.

chain. It is more justified, therefore, to determine $WASA_i$ by approximating the unfolded polypeptide chain with an extended conformation or β-conformation. In these conformations, as a result of screening effects from neighboring longer side chains, the $WASA_i$ values of groups determined by rolling the probe are smaller by about 20% than those calculated by simple summation of the surfaces of amino acid residues determined in Ala–X–Ala tripeptide (Makhatadze and Privalov, 1995).

Comparison of the WASA of the unfolded and the native, folded protein gives the change in WASA upon protein unfolding, $WASA^U - WASA^N = \Delta_N^U WASA$. Analysis of the structure of the globular proteins in Table 7.2 shows that the $\Delta_N^U WASA$ is largest for aliphatic groups and smallest for aromatic groups: the nonpolar buried surface in globular proteins amounts to $(62 \pm 1)\%$ of the total surface and the rest accounts for polar and charged surface (Fig. 8.5). It is remarkable that $\Delta_N^U WASA$ for all three types of groups (aliphatic, polar, and aromatic) increases linearly with increasing number of residues in the polypeptide chain, N_r; that is, their ratio does not depend on the molecular weight of the protein (Makhatadze and Privalov, 1995). Thus, the $(\Delta_N^U WASA) / N_r$ of polar groups and nonpolar groups is not complementary as one might expect. Their simultaneous change could be associated with change in the shape of the protein molecule, that is, with change in the ratio of the outer surface to the volume with increase of protein size. It appears that:

> *The deviation of protein shape from globularity increases with increase of protein size, and this seems to be a very general principle of protein architecture, the essence of which needs explanation: is it a technological requirement for protein folding or for its functioning?*

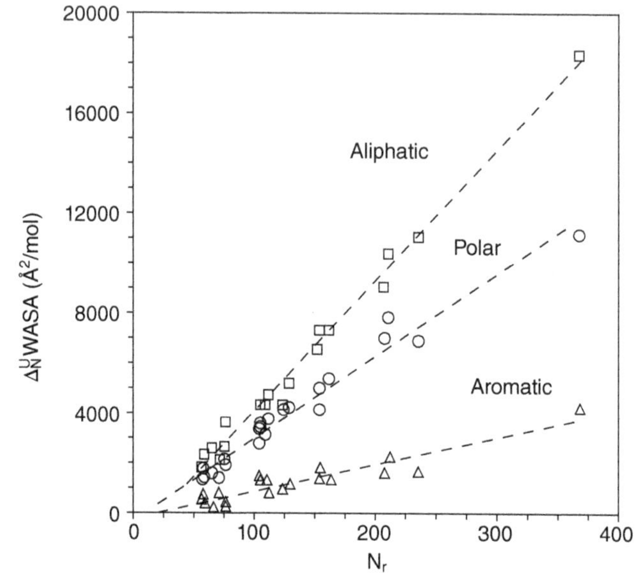

Figure 8.5. A plot of Δ_N^UWASA for the aliphatic, polar, and aromatic groups in the globular proteins listed in Table 8.4 versus the number of residues in the polypeptide chain, N_r (Makhatadze and Privalov, 1995).

8.2. PROTEIN IN VACUUM

8.2.1. Heat Capacity of Globular Proteins

The partial specific heat capacities of globular proteins in the native state depend on temperature and they increase with increasing temperature, as happens with the heat capacities of all macroscopic bodies because of the thermal motion intensification upon heating (Table 8.2). It is interesting, however, that as temperature decreases the specific heat capacities converge to some point at −20°C. Their values at −20°C are very close to (125 ± 3) J/K·(mol-res) = (1.09 ± 0.03) J/Kg (Fig. 8.6). Among all proteins calorimetrically studied so far, the smallest slope is shown by bovine pancreatic trypsin inhibitor (BPTI), a highly stable protein with rigid structure cross-linked by three S–S bridges, which unfolds upon heating above 100°C (Makhatadze et al., 1993). Therefore, the partial specific heat capacity of this protein in aqueous solution is usually considered as a standard partial specific heat capacity of fully folded globular proteins (see Section 3.2.5). The partial specific heat capacity of this protein is expressed by Equation (3.22).

It is remarkable that the partial specific heat capacities of globular proteins in aqueous solution, as determined from differential scanning calorimetry (DSC) data,

TABLE 8.2. Temperature Dependence of the Partial Specific Heat Capacity of Native Globular Proteins in Aqueous Solution and the Specific Heat Capacity of the Anhydrous Proteins all Calculated per Mole of Amino Acid Residues[a]

Protein, number of residues	$C_p(0°C)$ J/K·(mol-res)	$\partial C_p/\partial T$ J/K^2·(mol-res)	N	References
RNAse A, 124	144	1.10	11	Makhatadze and Privalov (1993)
RNAse T1, 104	140	1.00	10	Yu et al. (1994)
T4 lysozyme, 163	145	0.95	9	Kitamura and Sturtevant (1989)
Cytochrome c, 104	143	0.90	8	Makhatadze and Privalov (1993)
Ubiquitin, 76	143	0.87	7	Wintrode et al. (1994)
Lysozyme, 129	140	0.82	6	Makhatadze and Privalov (1993)
Myoglobin, 153	138	0.78	5	Makhatadze and Privalov (1993)
Barnase , 110	140	0.70	4	Griko et al. (1994)
BPTI, 58	141	0.65	3	Makhatadze et al. (1993)
Anhydrous protein	133	0.47	2	Hutchens et al. (1969)
Anhydrous polypeptide	132	0.42	1	Wen (1996)

[a]Averaged molecular mass of residue is assumed to be 115.

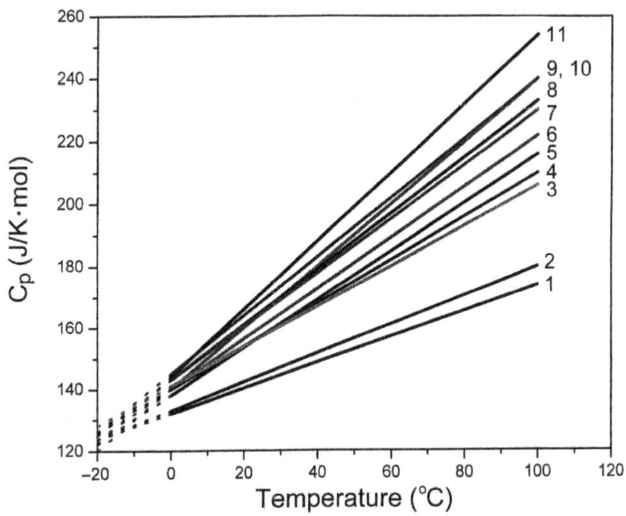

Figure 8.6. Extrapolations of partial specific heat capacities of the native globular proteins listed in Table 8.2.

are significantly higher than the specific heat capacities of the anhydrous proteins, measured by absolute calorimeter, and they increase much more steeply with increasing temperature (Table 8.2). Therefore, it is clear that the presence of water somehow affects the protein, but this influence is very selective and is likely to depend on the compactness and rigidity of the globular protein (see Chapter 9).

Excluding the calculated heat capacity effects of hydration from the heat capacity of folded and unfolded protein, one can get the heat capacity of the folded and unfolded protein in the absence of hydration effects, that is, as if they are placed in vacuum (Privalov and Makhatadze, 1992). Corrected for hydration the specific heat capacities of folded and unfolded proteins are very similar at lower temperatures (Table 8.3). This is rather unexpected since it was believed that the heat capacity of a protein in the unfolded state should be higher than of the folded one because of increasing conformational freedom of the polypeptide chain (Sturtevant, 1977). It is remarkable also that the vacuum specific heat capacities of the native proteins are rather similar and at 5°C and 25°C are very close in value to the specific heat capacity of the anhydrous (dry) chymotrypsinogen measured with an absolute heat capacity calorimeter by Hutchens et al. (1969). Thus, it appears that:

The heat capacity increment of a protein upon its transfer into water and the denaturational heat capacity increment of a protein both are provided by the hydrated water and not the protein.

The other remarkable and quite unexpected fact is that with increasing temperature the heat capacities of the unfolded and folded protein deviate and the heat capacity of the "dry" unfolded protein becomes lower than the heat capacity of "dry" folded protein (Fig. 8.7). This, however, is not as surprising as it seems: it is known that the heat capacities of substances in the condensed state are significantly higher than the heat capacities of these substances in the gaseous state. This is because with heating of the condensed phase the energy is accumulated in the fluctuating interactions between the components of the substance, which are absent in the gaseous phase.

8.2.2. Enthalpy of Protein Unfolding in Vacuum

The calculated hydration enthalpies of aliphatic, aromatic, and polar groups upon unfolding of the proteins considered above are presented in Table 8.4. Excluding them from the calorimetrically measured enthalpy of protein unfolding (Table 7.2), one can determine the enthalpy of internal interactions maintaining the compact folded state of protein in vacuum, $\Delta_{\mathrm{N}}^{\mathrm{U}}H^{\mathrm{int}}$. The specific values of these enthalpies of internal interaction, calculated per mole of amino acid residues for the calorimetrically studied globular proteins, are grouped between 37 and 52 kJ/mol (Fig. 8.8). It is remarkable that these enthalpies do not depend much on temperature: they just decrease slightly with increasing temperature due to the thermal expansion of the folded protein structure. The question is: what kind of interactions in the protein interior are providing the enthalpies for protein folding in vacuum?

TABLE 8.3. Specific Heat Capacities of Proteins in Vacuum (Corrected for Hydration Effects)[a,b]

Protein M_w	Property	Temperature (°C) 5	25	50	75	100	125
Ubiquitin	$C_p(\mathrm{N})^{\mathrm{vac}}$	1.12	1.29	1.48	1.67*	1.87*	2.07*
8.433	$C_p(\mathrm{U})^{\mathrm{vac}}$	1.17	1.35	1.45	1.52	1.58	1.63
	$\Delta_{\mathrm{N}}^{\mathrm{U}}C_p^{\mathrm{vac}}$	**0.05**	**0.06**	**−0.03**	**−0.15**	**−0.29**	**−0.44**
RNAse T1	$C_p(\mathrm{N})^{\mathrm{vac}}$	1.58	1.78	1.98	2.20*	2.40*	2.60*
7,944	$C_p(\mathrm{U})^{\mathrm{vac}}$	1.50	1.66	1.83	1.99	2.07	2.12
	$\Delta_{\mathrm{N}}^{\mathrm{U}}C_p^{\mathrm{vac}}$	**−0.08**	**−0.12**	**−0.15**	**−0.21**	**−0.33**	**−0.47**
Cytochrome c	$C_p(\mathrm{N})^{\mathrm{vac}}$	1.23	1.36	1.53	1.71*	1.87*	2.04*
12,830	$C_p(\mathrm{U})^{\mathrm{vac}}$	1.16	1.27	1.40	1.46	1.52	1.57
	$\Delta_{\mathrm{N}}^{\mathrm{U}}C_p^{\mathrm{vac}}$	**−0.07**	**−0.09**	**−0.13**	**−0.25**	**−0.35**	**−0.47**
Barnase	$C_p(\mathrm{N})^{\mathrm{vac}}$	1.05	1.23	1.35	1.63*	1.83*	2.03*
12,365	$C_p(\mathrm{U})^{\mathrm{vac}}$	0.98	1.14	1.26	1.31	1.38	1.45
	$\Delta_{\mathrm{N}}^{\mathrm{U}}C_p^{\mathrm{vac}}$	**−0.07**	**−0.09**	**−0.09**	**−0.32**	**−0.45**	**−0.58**
RNAse A	$C_p(\mathrm{N})^{\mathrm{vac}}$	1.43	1.53	1.65	1.78*	1.90*	2.02*
13,600	$C_p(\mathrm{U})^{\mathrm{vac}}$	1.32	1.43	1.54	1.59	1.65	1.69
	$\Delta_{\mathrm{N}}^{\mathrm{U}}C_p^{\mathrm{vac}}$	**−0.11**	**−0.10**	**−0.11**	**−0.19**	**−0.25**	**−0.33**
Lysozyme	$C_p(\mathrm{N})^{\mathrm{vac}}$	1.27	1.40	1.55	1.71*	1.87*	2.02*
14,300	$C_p(\mathrm{U})^{\mathrm{vac}}$	1.29	1.34	1.47	1.54	1.59	1.65
	$\Delta_{\mathrm{N}}^{\mathrm{U}}C_p^{\mathrm{vac}}$	**0.02**	**−0.06**	**−0.08**	**−0.17**	**−0.28**	**−0.37**
Myoglobin	$C_p(\mathrm{N})^{\mathrm{vac}}$	1.22	1.36	1.53	1.70*	1.88*	2.05*
17,800	$C_p(\mathrm{U})^{\mathrm{vac}}$	1.23	1.34	1.45	1.52	1.58	1.64
	$\Delta_{\mathrm{N}}^{\mathrm{U}}C_p^{\mathrm{vac}}$	**0.01**	**−0.02**	**−0.08**	**−0.18**	**−0.30**	**−0.41**
Averaged	$C_p(\mathrm{N})^{\mathrm{vac}}$	**1.27 ± 0.10**	**1.42 ± 0.10**	**1.58 ± 0.10**	**1.77 ± 0.10**	**1.94 ± 0.10**	**2.12 ± 0.10**
Anhydrous chymotrypsinogen (Hutchens et al., 1969)	$C_p(\mathrm{N})^{\mathrm{dry}}$	**1.11 ± 0.05**	**1.21 ± 0.05**				

[a]Makhatadze and Privalov (1995, 1996).

[b]In kJ/K·g; * indicates extrapolated values.

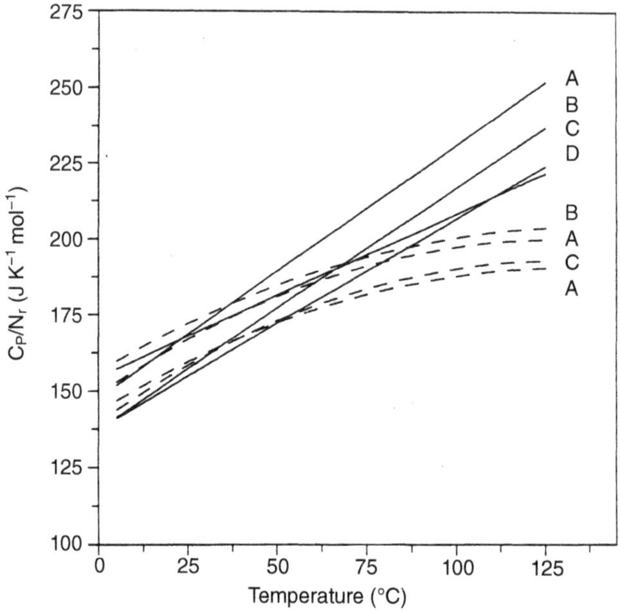

Figure 8.7. Vacuum specific heat capacity of native (continuous line) and unfolded (broken line) proteins: (A) myoglobin; (B) cytochrome c; (C) lysozyme,; (D) RNAse A (Privalov and Makhatadze, 1992).

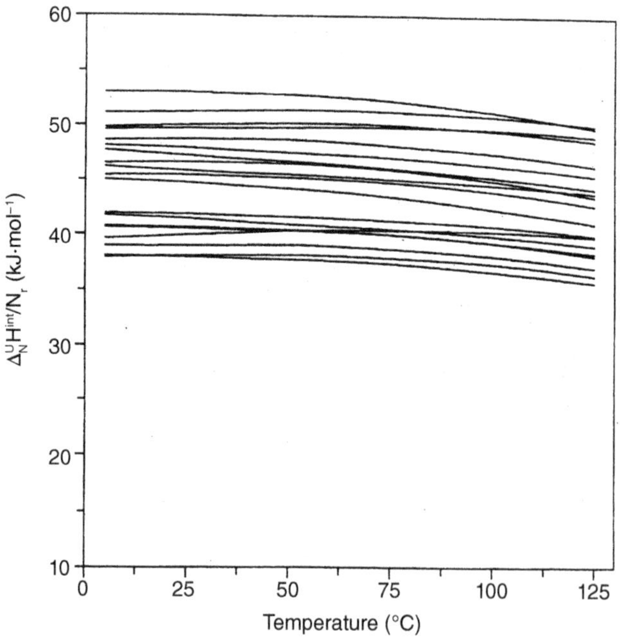

Figure 8.8. Temperature dependence of the enthalpies of internal interactions in the considered proteins, calculated per mole of amino acid residues for a number of calorimetrically studied globular proteins (Makhatadze and Privalov, 1995).

TABLE 8.4. Enthalpies of Hydration of Aliphatic, Aromatic, and Polar Groups on Unfolding of Proteins[a,b]

		Temperature (°C)					
Protein	Property	5	25	50	75	100	125
BPTI	Aliphatic	−274	−202	−116	−35	43	114
	Aromatic	−136	−112	−84	−58	−35	−14
	Polar	−1879	−1912	−1942	−1967	−1986	−2006
Eglin C	Aliphatic	−356	−262	−150	−45	56	148
	Aromatic	−138	−114	−85	−59	−35	−14
	Polar	−1670	−1707	−1748	−1784	−1814	−1842
G protein	Aliphatic	−304	−224	−128	−39	48	127
	Aromatic	−102	−84	−65	−44	−26	−10
	Polar	−1814	−1860	−1910	−1955	−1995	−2031
Tendamistat	Aliphatic	−439	−323	−185	−56	69	183
	Aromatic	−75	−62	−46	−32	−10	−7
	Polar	−2469	−2520	−2570	−2612	−2647	−2676
Ubiquitin	Aliphatic	−597	−439	−252	−76	94	248
	Aromatic	−48	−40	−30	−21	−12	−5
	Polar	−2395	−2454	−2517	−2574	−2625	−2671
RNAse T1	Aliphatic	−560	−412	−236	−71	88	233
	Aromatic	−267	−219	−165	−114	−68	−27
	Polar	−3362	−3437	−3516	−3584	−3644	−3696
Cytochrome c	Aliphatic	−720	−529	−304	−91	113	300
	Aromatic	−229	−189	−141	−98	−59	−23
	Polar	−4365	−4466	−4578	−4677	−4764	−4842
Barnase	Aliphatic	−717	−527	−302	−91	112	298
	Aromatic	−233	−192	−144	−100	−60	−23
	Polar	−3829	−3921	−4018	−4105	−4181	−4249
RNAse A	Aliphatic	−717	−527	−302	−91	112	298
	Aromatic	−172	−141	−106	−74	−44	−17
	Polar	−4614	−4716	−4823	−4914	−4992	−5058
Lysozyme	Aliphatic	−860	−632	−363	−109	135	357
	Aromatic	−213	−175	−132	−91	−55	−21
	Polar	−5023	−5124	−5226	−5313	−5387	−5451
Interleukin 1β	Aliphatic	−1084	−797	−457	−137	170	451
	Aromatic	−247	−203	−153	−106	−63	−25
	Polar	−5048	−5178	−5320	−5449	−5567	−5675
Myoglobin	Aliphatic	−1205	−885	−508	−152	189	501
	Aromatic	−327	−269	−202	−140	−84	−33
	Polar	−6292	−6450	−6625	−6784	−6928	−7062
T4 lysozyme	Aliphatic	−1213	−891	−511	−153	190	504
	Aromatic	−241	−198	−148	−103	−62	−24
	Polar	−6497	−6653	−6822	−6974	−7111	−7236

[a]Makhatadze and Privalov (1995, 1996).
[b]In kJ/mol.

Let us first consider enthalpy, which might be provided by packing the nonpolar and aromatic groups in the protein interior. The contribution of van der Waals interactions between aliphatic groups and between aromatic groups to the energy of protein stabilization can be approximated by the interactions in the crystals of aliphatic or aromatic molecules, which have almost the same packing density as the averaged globular proteins (Richards, 1974, 1977; Richards and Lim, 1994). The energy of interaction between molecules packed in a crystal can be estimated from the enthalpy of sublimation of this crystal, corrected for the volume difference between the condensed and gaseous phases. At 25°C these enthalpies, calculated per square angstrom, are about 130 J/mol·Å^2 for the aliphatic and 180 J/mol·Å^2 for the aromatic groups and decrease slightly with temperature increase due to thermal expansion of the crystals (Makhatadze and Privalov, 1993; Gavezotti, 1989; Sneddon and Tobias, 1992). The overall contribution of the expected van der Waals interactions to the stabilization of the compact state of the considered proteins calculated per mole of amino acid residues are shown in Figure 8.9. At low temperatures they are grouped between 6 and 8 kJ/(mol-res) and decrease slightly with increasing temperature due to thermal expansion. It appears that the van der Waals contribution amounts to only 10% of the total enthalpy of maintaining the compact state of protein, while the rest, about 40 kJ/(mol-res), is provided mostly by the interactions between polar groups, that is, by the hydrogen bonding of these groups.

The theoretical evaluation of the energy of internal interactions in protein is based on the computer summation of all secondary interactions between the atoms of the

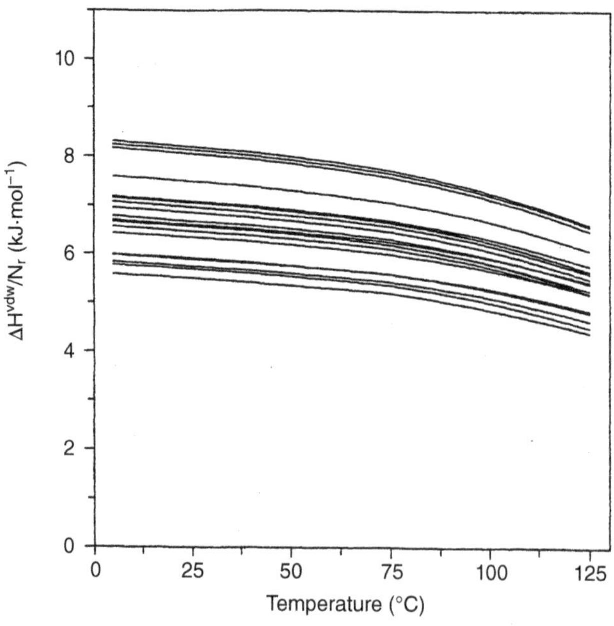

Figure 8.9. Contribution of van der Waals interactions to the stabilization of protein structure (Makhatadze and Privalov, 1995).

polypeptide chain in the folded and unfolded conformation. According to the current understanding of protein structure, all groups that are packed in the protein interior are charged to some extent. In the computer analysis of the energy of internal interactions, a certain distribution of charges between the groups in protein is postulated, and interactions between these charges are calculated using the Poisson–Boltzmann equation. However, one of the basic parameters in this equation is the dielectric constant, which has no meaning in heterogeneous systems—such as protein interiors. Therefore, the value of dielectric constant was just a fitting parameter. One of the most detailed computer analyses of the energy of internal interactions in proteins was done using the program CHARMM by Lazaridis et al. (1995). The energies of internal interactions calculated by this program for the four proteins in the folded and unfolded states are listed in Table 8.5. It shows that the total specific enthalpies of interactions in the native proteins in vacuum, calculated per mole of amino acid residues for four different globular proteins, are rather similar—varying between 36 and 41 kJ/(mol-res). According to the empirical estimates these enthalpies vary between 46 and 49 kJ/mol; that is, they are 25% higher. However, much more serious deviation was found in estimations of the van der Waals interactions and interactions between polar groups. According to the theoretical estimates, the enthalpic contribution of van der Waals interaction between the nonpolar groups largely dominates over the contribution of polar groups, while according to the empirical information, the situation is opposite: interactions between polar groups (i.e., hydrogen bonding and salt links) dominate over the van der Waals interactions. The possible reasons for this discrepancy might be that, according to the theory used, the interactions between polar groups include not only an electrostatic component but also a component associated with the induced polarizability of these groups (Lazaridis et al., 1995). In theory the enthalpic contribution of this induced polarizability is assigned to the van der Waals interaction. However, no one has yet succeeded experimentally in separating these two components, the electrostatic and

TABLE 8.5. Vacuum Total Enthalpies and Their van der Waals Components of Protein Unfolding[a]

	Theoretical Estimates (Lazaridis et al., 1995)		Empirical Estimates (Makhatadze and Privalov, 1995)	
Protein	ΔH_{vac}^{total}	ΔH_{vac}^{tvdW}	ΔH_{vac}^{total}	ΔH_{vac}^{vdW}
Cytochrome c	3946	883	5213	771
RNAse A	4469	619	5678	712
Lysozyme	4669	858	6173	861
Myoglobin	6243	1280	7611	1235

[a]Estimated theoretically using program CHARMM (Lazaridis et al., 1995) and empirically by excluding the hydration effect from the calorimetrically measured enthalpy of protein unfolding (Makhatadze and Privalov, 1995).
ΔH in kJ/mol.

polarizability, of the interactions between polar groups. Traditionally in experimental sciences, and particularly in protein chemistry, the overall interactions between polar groups are regarded as hydrogen bonding.

8.2.3. Entropy of Protein Unfolding in Vacuum

The calorimetrically measured total entropy of protein unfolding is associated with disordering of the polypeptide backbone and side chains and ordering of the surrounding water by the groups exposed upon unfolding, that is, the entropy of hydration of these groups. Therefore, by subtracting the entropies of hydration of polar and nonpolar groups (Table 8.6) from the calorimetrically determined entropy of protein unfolding (Table 7.2), one can get the conformational entropy of protein unfolding.

The hydration entropies of polar and nonpolar groups upon unfolding of various globular proteins are shown in the lower part of Figure 8.10. Since the contribution of weakly polar aromatic groups to hydration effects is rather low, it is considered together with that of the nonpolar aliphatic groups. One could see that the hydration effects of polar and nonpolar groups are very different: the hydration entropies of nonpolar groups are negative at room temperature, decreasing in magnitude with increasing temperature and vanishing at about 125°C. In contrast, the hydration entropies of polar groups are also negative at room temperature, but become even more negative with increasing temperature. Thus, at 120°C hydration entropies of polar groups are quite considerable, varying between –25 and –37 J/K·(mol-res). If these two hydration effects of polar and nonpolar groups are subtracted from the calorimetrically determined entropy of protein unfolding, one gets the conformational entropy of protein unfolding (Fig. 8.10, upper part). For various globular proteins these vacuum entropies appear to be between 45 and 62 J/K·(mol-res) at 5°C and decrease slightly with increasing temperature. This decrease of conformational entropy of protein unfolding with increasing temperature results from the dominating increase of the entropy of folded state with heating. The fact that the entropic contribution of a single amino acid residue to the entropy of protein unfolding is almost the same for quite different proteins, differing by less than 10% from the mean value of 54 J/K·(mol-res), is very remarkable. It actually shows that the contribution of amino acid residues to the conformational entropy of a polypeptide chain and the contribution of the groups of this chain to the hydration entropy are additive to a good approximation.

There might be several reasons for the observed spread in the conformation entropy contribution of a single residue. First, this might be the S–S cross-links in the polypeptide chain in the proteins being considered, which restrict its conformational space (see Section 4.2.3). Since the considered proteins differ in the number of S–S cross-links, their specific entropies calculated per amino acid residue should be different. Figure 8.11a presents the plot of the conformational entropy factors of protein unfolding for 25°C (calculated per mole of residue) versus the number of S–S cross-links in the polypeptide chains, N_{SS}. It does not show a simple correlation: the entropy even increases with increase of the number of cross-links in the protein, which might mean that the cross-links are increasing the order of the native state,

TABLE 8.6. Entropies of Hydration of Aliphatic, Aromatic, and Polar Groups upon Unfolding of Proteins[a,b]

		Temperature (°C)					
Protein	Property	5	25	50	75	100	125
BPTI	Aliphatic	−1206	−955	−676	−434	−221	−36
	Aromatic	−325	−241	−150	−74	−9	47
	Polar	−1151	−1262	−1356	−1428	−1483	−1523
Eglin C	Aliphatic	−1566	−1240	−877	−564	−287	−47
	Aromatic	−331	−245	−155	−75	−9	48
	Polar	−946	−1074	−1204	−1310	−1391	−1465
G protein	Aliphatic	−1339	−1060	−750	−482	−246	−40
	Aromatic	−243	−180	−112	−55	−7	35
	Polar	−1011	−1164	−1325	−1457	−1568	−1661
Tendamistat	Aliphatic	−1932	−1529	−1082	−696	−355	−58
	Aromatic	−179	−133	−83	−41	−5	26
	Polar	−1471	−1648	−1805	−1928	−2027	−2098
Ubiquitin	Aliphatic	−2627	−2080	−1472	−947	−482	−79
	Aromatic	−116	−86	−54	−26	−3	17
	Polar	−1378	−1581	−1783	−1945	−2088	−2206
RNAse T1	Aliphatic	−2464	−1951	−1380	−885	−452	−74
	Aromatic	−637	−473	−295	−135	−18	92
	Polar	−1910	−2164	−2415	−2615	−2783	−2915
Cytochrome c	Aliphatic	−3167	−2507	−1774	−1141	−581	−95
	Aromatic	−548	−406	−254	−125	−15	79
	Polar	−2489	−2836	−3192	−3483	−3727	−3925
Barnase	Aliphatic	−3151	−2495	−1766	−1135	−578	−95
	Aromatic	−557	−413	−258	−127	−16	80
	Polar	−2115	−2433	−2740	−2994	−3209	−3382
RNAse A	Aliphatic	−3152	−2496	−1766	−1136	−579	−95
	Aromatic	−411	−305	−190	−94	−11	59
	Polar	−2802	−3154	−3495	−3759	−3981	−4148
Lysozyme	Aliphatic	−3781	−2994	−2119	−1362	−694	−114
	Aromatic	−510	−378	−236	−116	−14	73
	Polar	−3004	−3349	−3675	−3925	−4134	−4296
Interleukin 1β	Aliphatic	−4768	−3775	−2671	−1718	−875	−144
	Aromatic	−591	−438	−273	−135	−16	85
	Polar	−2924	−3372	−3827	−4201	−4534	−4809
Myoglobin	Aliphatic	−5298	−4195	−2969	−1909	−973	−160
	Aromatic	−782	−580	−362	−178	−22	113
	Polar	−3514	−4060	−4616	−5080	−5483	−5824
T4 lysozyme	Aliphatic	−5333	−4223	−2988	−1921	−979	−161
	Aromatic	−575	−427	−266	−131	−16	83
	Polar	−3665	−4205	−4743	−5187	−5571	−5889

[a]Makhatadze and Privalov (1995).
[b]In J/K·mol.

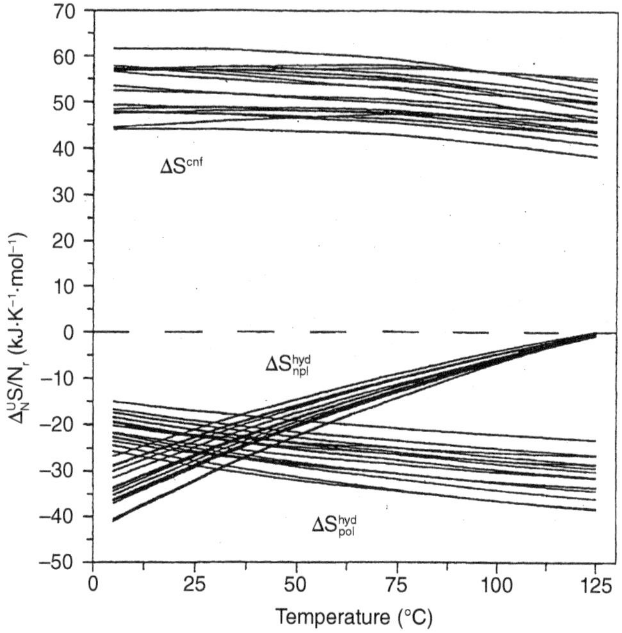

Figure 8.10. Temperature dependence of the entropies of hydration for polar and nonpolar groups, and the conformational entropies of unfolding of a number of small globular proteins with compact native structure, calculated per mole of amino acid residues (Makhatadze and Privalov, 1995).

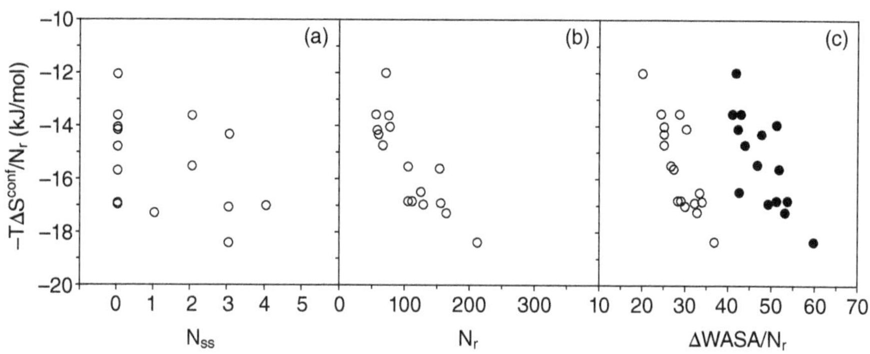

Figure 8.11. Plot of the conformational entropy factors at 25°C versus: (a) the number of S–S cross-links in the polypeptide chains, N_{ss}; (b) the number of amino acid residues, N_r, in protein; (c) the buried polar (○) and nonpolar (●) surface area per amino acid residue (Makhatadze and Privalov, 1995).

that is, its rigidity. The per-residue conformational entropy increases with decreasing size of a single-domain protein (number of residues less than 200) but does not change in large proteins consisting of several domains (Fig. 8.11b). This can be explained by the increase in the proportion of looser residues with decreasing single-domain size. It is remarkable that the conformational entropy of protein unfolding correlates with the amount of buried polar and nonpolar surface area—in fact, with the number of hydrogen bonds and van der Waals contacts in the native protein (Fig. 8.11c). This is understandable, since the internal secondary interactions are increasing the rigidity and order of the native state, that is, decreasing its entropy. Thus, it appears that the conformational entropy increment on protein unfolding is larger for the more compact, more globular proteins with maximal secondary interactions and minimal surface-to-volume ratio. Since proteins are considered as ideal globules in theoretical modeling, the theoretical estimations of the conformational entropy of protein unfolding should be compared not with the averaged entropy estimated for the number of experimentally studied proteins, but with the maximum values found, which for the conformational entropy factor is $\Delta S^{conf} = 60$ J/K·(mol-res) at temperature 25°C = 298.15 K. At this temperature the empirically determined conformational entropy factor is $T\Delta S^{conf} = 18$ kJ/(mol-res).

As discussed in Section 4.2.2, the averaged theoretical value calculated by various authors for the side-chain entropy factor is $T\Delta S^{sch} = (4 \pm 1)$ kJ/(mol-res); the backbone entropy factor is about $T\Delta S^{bb} = 5.5$ kJ/(mol-res) at 25°C. Thus, the overall theoretical conformational entropy factor, $T\Delta S^{conf}$, is expected to be about 10 kJ/(mol-res) at this temperature. This theoretically calculated value is almost 2 times smaller than the empirically estimated value of the conformational entropy factor. What is the reason for such discrepancy between the values of the conformational entropies estimated by two different approaches, the empirical and theoretical? Is it caused by incorrectness in evaluating the hydration effects (which figures in the empirical approach) or by incorrectness in counting the available conformations (in theoretical modeling)?

The accuracy of the estimation of entropies of hydration of aliphatic and aromatic groups does not induce much concern because these entropies approach zero at 100°C. Thus, they cannot influence much the magnitude of the conformational entropy at the elevated temperatures at which the deviation between the experimental and theoretical values of ΔS^{conf} is as large as that at lower temperatures. Therefore, this deviation may be caused only by the hydration of polar groups. However, to decrease the estimated value of the conformational entropy 2 times, one should suppose that the hydration of polar groups does not contribute anything to the entropy of protein unfolding at 25°C (see Fig. 8.10), which is absolutely improbable. As for the theoretical estimates, one cannot exclude that some sources of entropy gain on protein unfolding are missed, such as changes in the vibrational modes, which are hotly discussed in literature (Karplus and Kushick, 1981; Doig and Williams, 1992; Williams, 1991; Karplus et al., 1987). It might be also the force field used in calculations of the entropies of side chains of amino acid residues, namely, the electrostatic interactions in media with dielectric constant 78, that is, in aqueous media. If so, the calculated entropy of the polypeptide chain certainly includes the entropies of hydration of protein groups, which is negative.

8.3. BACK INTO THE WATER

8.3.1. Enthalpies of Protein Unfolding in Water

Protein unfolding in vacuum is an abstraction, which is needed only for a very special purpose—to evaluate the internal interactions stabilizing protein structure and the conformational entropy of the polypeptide chain. But the real proteins are hydrated. They interact with water, and it is the delicate balance of these interactions with water and the internal interactions that determines the native protein structure. Therefore, since the enthalpies of the nonpolar and polar interactions are determined and the enthalpies of hydration for polar, aromatic, and aliphatic groups are known, one can estimate contribution of these interactions to the overall enthalpy of protein unfolding in water:

$$\Delta_N^U H = \left(\Delta_N^U H^{\text{pol}} + \Delta_N^U H^{\text{hyd}}\right)_{\text{pol}} + \left(\Delta_N^U H^{\text{npl}} + \Delta_N^U H^{\text{hyd}}\right)_{\text{npl}} \tag{8.9}$$

These enthalpy effects of unfolding are presented in Figure 8.12 as functions of temperature. It appears that these enthalpy effects depend strongly on temperature, but those associated with polar groups and those associated with nonpolar (aliphatic + aromatic) groups change in the opposite direction with increasing temperature. They are also of opposite signs through almost all of the temperature range of

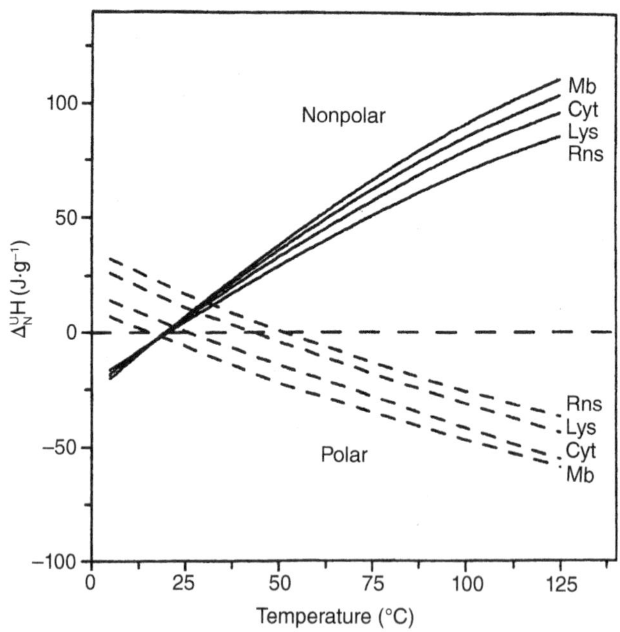

Figure 8.12. Temperature dependence of the enthalpy of transfer of polar groups (dashed lines) and nonpolar groups (continuous lines) from the interior of the protein molecule into water on protein unfolding for 20 different globular proteins (Makhatadze and Privalov, 1993).

aqueous solution in the liquid state. Thus, the enthalpic effects associated with disruption of contacts between nonpolar groups and hydrations of these groups differ qualitatively from the effects associated with disruption of bonds between polar groups and their hydration.

The specific enthalpies of transfer of nonpolar groups into water converge to zero at about 20°C and diverge with increasing temperature. The enthalpy convergence temperature, T_H, is about 10 degrees lower than T_H for transfer of nonpolar molecules from the pure liquid phase into water (see Section 2.4.1). This is because the enthalpy of interaction of nonpolar groups in the crystalline state is larger than in the liquid state on the enthalpy of fusion (Nicholls et al., 1991). This difference in T_H demonstrates that the approximation of a protein interior by a nonpolar liquid phase is an unjustified simplification. Even worse is the approximation of a protein interior by crystalline cyclic dipeptides (diketopiperazins). The T_H for transfer of these dipeptides from the crystalline phase into water is 71°C, indicating poorer interactions in the initial phase (Murphy and Freire, 1992). This is because the packing densities of these crystals are exceptionally low (Harpaz et al., 1994), significantly lower than it in the protein interior. Therefore, these crystals can hardly serve as appropriate models of proteins.

Zero enthalpy of transfer of nonpolar groups into water means that the enthalpy of van der Waals interactions between nonpolar groups packed in the protein interior and the enthalpy of hydration of these groups are very similar at room temperature. As temperature increases, the van der Waals interactions between nonpolar groups do not change much, but the enthalpy of hydration of these groups decreases. Therefore, at high temperatures, the enthalpy effect of interactions between nonpolar groups in the interior of a native compact protein overbalances the hydration enthalpy of these groups on protein unfolding.

Distinct from nonpolar groups, the enthalpy of transfer of polar groups from the interior of protein into water changes almost in parallel with increasing temperature. It is positive at low temperatures, then changes sign at some elevated temperature, and continues to decrease almost linearly with decreasing temperature. This is because the positive enthalpy of interaction between polar groups in a protein does not depend significantly on temperature, while the negative hydration enthalpy of these groups increases in magnitude linearly with increasing temperature (see Fig. 8.2). Therefore, above some temperature, the hydration enthalpy of polar groups overcompensates the enthalpy of hydrogen bonding between polar groups.

There were some attempts to estimate theoretically the hydration effects. However, water is still a mysterious substance: we do not yet have a theory describing even pure liquid water and more so its solutions. Therefore, it is not surprising that so far all attempts at theoretical determination of water's interaction with proteins, and of its contribution to various features of proteins, have failed. The difference in the energies of hydration of folded and unfolded proteins calculated by using the Poisson–Boltzmann equation, that is, the hydration enthalpy of protein unfolding (Lazaridis et al., 1995), appeared to be 4 times smaller than the empirical estimates obtained by summing hydration effects of the exposed polar groups. The sum of the computer-calculated hydration effects with the energy of protein unfolding calculated by CHARMM, which should represent the energy of protein unfolding in water, appeared to be an order magnitude greater than the calorimetrically measured enthalpy of protein unfolding (see the Epilogue in Makhatadze and Privalov, 1995). Thus, the continuum electrostatic calculations do not work with water.—Water still presents a major obstacle for the theoretical modeling of proteins.

8.3.2. Hydrogen Bonds

As discussed previously (see Section 8.2.2), although part of the interactions between polar groups might be attributed to van der Waals interactions, separating this part from the experimentally measured enthalpy of polar group interactions is improbable. Therefore, traditionally, in considering proteins the whole contribution of polar groups to the stabilization of protein structure is attributed to hydrogen bonding.

There is some controversy regarding the estimates of the number of hydrogen bonds in native protein structures. Assuming that every donor–acceptor pair forms a single hydrogen bond, there should be about 0.62–0.72 hydrogen bonds per residue in globular proteins (Privalov, 1979; Stickle et al., 1992; Makhatadze and Privalov, 1995). These bonds are usually called the **reduced hydrogen bonds**. However, in proteins each donor (acceptor) can interact with several acceptors (donors). Thus, there are more hydrogen bonds than donor–acceptor pairs in a protein (Burley and Petsko, 1988; McDonald and Thornton, 1994). The number of these **network hydrogen bonds** in a large group of analyzed proteins was found to be about 1.08 per residue. In fact, the number of network hydrogen bonds depends on the conformation of a protein, particularly on its helicity, and ranging from 0.8 to 1.6 per residue (Stickle et al., 1992). This is because in an α-helix a hydrogen bond is formed between the amide group of an amino acid residue and the carboxyl groups of the third and also fourth residues along the chain; that is, this bond is bifurcated. Therefore, an increase of helicity of a protein increases the average number of hydrogen bonds per residue.

Dividing the integral enthalpic contribution of polar groups in the protein by the number of hydrogen bonds, one obtains the enthalpy of a hydrogen bond in proteins. For a reduced hydrogen bond, enthalpy varies between 50 and 70 kJ/mol. For a network hydrogen bond the enthalpy values are lower, between 25 and 50 kJ/mol, with a mean of 38 kJ/mol. The enthalpy values of hydrogen bonding are smaller for proteins with a more extensive network of hydrogen bonds, with longer averaged length of the bonds, and with larger content of α-helices.

The enthalpy of hydrogen bonding determined from calorimetric analysis of globular protein unfolding is close to values of these bonds between various organic molecules. According to these estimates, the enthalpy of hydrogen bond formation between various polar compounds in nonpolar organic solvents or in the gas phase amounts to 60 kJ/mol (Joesten and Schaad, 1974) or 60–120 kJ/mol for groups carrying a single positive or negative charge (Meot-Ner and Sieck, 1986; Del Bene, 1994).

In the presence of water the enthalpy of formation of a hydrogen bond between polar groups, $\Delta H_{\mathrm{HB}}^{\mathrm{hyd}}$, becomes almost zero, or even becomes positive (Fig. 8.12). In

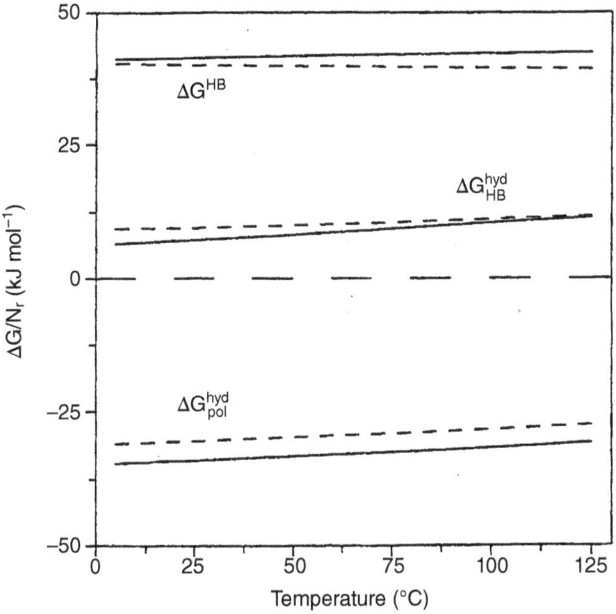

Figure 8.13. Temperature dependence of the contributions of the mean Gibbs energy of hydrogen bonding in the interior of a protein, ΔG^{HB}, of hydration of the polar residues exposed to water on unfolding, ΔG_{pol}^{hyd}, and the resultant Gibbs energy, ΔG_{HB}^{hyd}, which contributes to the stability of protein in aqueous media. The solid lines represent myoglobin; the dashed lines represent ribonuclease A (Makhatadze and Privalov, 1995).

the α-helix it is about 2.7 kJ/mol at 25°C (see Section 5.1.2). However, in the Gibbs energy of formation of a hydrogen bond this small enthalpic contribution is overbalanced by the positive entropic term of dehydration of the polar groups, as is shown for the example of myoglobin and ribonuclease A in Figure 8.13:

$$\Delta G_{HB}^{hyd} = \Delta G_{HB} + \Delta G_{pol}^{hyd} = \Delta H_{HB} + \Delta H_{pol}^{hyd} - T\Delta S_{pol}^{hyd} = \Delta H_{HB}^{hyd} - T\Delta S_{pol}^{hyd} \quad (8.10)$$

At 25°C, the mean value of the contribution of polar interactions to the Gibbs energy of stabilizing these two proteins in the native state amounts to 13 kJ/mol. Recalculating it per mole of network hydrogen bonds, one finds that it varies between 4 and 11 kJ/mol, depending on helicity of the protein, because the relative number of network hydrogen bonds depends on the α-helicity of the protein. These estimates are close to that for the isolated α-helix (see Section 5.1.2). Thus, despite many doubts (Dill, 1990; Williams, 1991):

The hydrogen bond stabilizes a protein structure in aqueous solution.

It should be emphasized, however, that the stabilizing effect of hydrogen bonding in proteins is mostly entropic but not enthalpic, as was widely believed.

It is notable that the enthalpy of hydrogen bonding between the polar groups of protein in aqueous media is close to zero or is even negative at temperatures 25–50°C. However, as we find out, the enthalpy of melting of the α-helix is positive at that temperature (see Section 5.1.2). This positive enthalpy might come only from the van der Waals interactions between the apolar groups forming the core of the helix.

8.3.3. Hydrophobic Effect

After Kauzmann (1959) the low solubility of nonpolar groups in water, their hydrophobicity, was regarded as a main cause of polypeptide folding into the compact globular structure with nonpolar core. The most peculiar aspect of this "hydrophobic force," which repels the nonpolar groups from water, was that it seemed to be totally entropic, as the enthalpy of transfer of nonpolar solutes into water at room temperature was found to be negligibly small, while the entropy of transfer was large and negative (see Section 2.4.2). It appeared that the thermodynamically unfavorable entropy decrease associated with water ordering in the presence of nonpolar molecules was responsible for the hydrophobicity of nonpolar substances. Application of this entropic concept of hydrophobicity of nonpolar groups to proteins seemed especially attractive since it led to the conclusion that the negative entropy of polypeptide folding is balanced by the positive entropy of removing nonpolar groups from water, and thus protein folding does not cost much in entropy. However, water ordering actually means an increase of hydrogen bonding between water molecules, which should proceed with heat release, that is, with negative enthalpy effect. If an enthalpy decrease is not observed in this process, that means that it is compensated by some other process, which might be only disruption of the van der Waals interactions between the nonpolar groups. Therefore, the effect, which we can regard as hydrophobic interactions in proteins, should include the enthalpy of the van der Waals interactions between nonpolar groups and the Gibbs energy of hydration of these groups:

$$\Delta G^{\text{HPH}} = \Delta H^{\text{vdW}} + \Delta G^{\text{hyd}} = \Delta_S^g H - \Delta_S^W H + T\Delta_S^w S \qquad (8.11)$$

The enthalpy of van der Waals interactions between nonpolar groups, $\Delta_S^g H$, is positive and decreases slightly with increasing temperature. The enthalpy of hydration of nonpolar groups is negative and strongly temperature dependent, increasing in magnitude as the temperature decreases. At room temperature the enthalpy of hydration of nonpolar groups becomes equal in magnitude and opposite in sign to the enthalpy of van der Waals interactions between these groups. Therefore, at room temperature (20°C) these two enthalpy effects compensate each other and it is only the entropy of hydration of nonpolar groups that contributes to the Gibbs energy of their transfer into water, that is, to the magnitude of the hydrophobic effect at this temperature. At high temperature the entropy of hydration of nonpolar groups decreases to zero and the Gibbs energy of the hydrophobic effect becomes completely enthalpic (see Fig. 2.10a). Thus, we cannot say that the hydrophobic effect is entropic in nature, as had been supposed. This effect is entropic at low temperatures, where the hydration effects are dominant, but it appears to be enthalpic at high temperatures, where the hydration effects are smaller (Baldwin, 1986; Privalov and Gill, 1988). It is clear that this change in the nature of the hydrophobic effect results from the heat capacity increment specific for the hydration of nonpolar groups. However:

> *The hydrophobicity of these groups is provided not by their hydration, as had been believed, but by the van der Waals interactions between themselves (see Section 2.4.2).*

Using the van der Waals interactions between nonpolar groups of protein and their hydration effects, calculated as discussed previously, one can estimate the contribution of hydrophobic interactions to the stabilization of the native compact protein structure. The hydrophobic interactions and their components are shown in Figure 8.14 for two proteins, myoglobin and ribonuclease A. These two examples represent extreme cases among the proteins yet studied, with the largest and smallest number of nonpolar contacts.

8.3.4. Balance of Forces Stabilizing and Destabilizing Protein Structure

All thermodynamic factors responsible for the stabilization and destabilization of the compact native structure of globular proteins are presented in Figure 8.15. It shows that the main destabilizing action of the compact highly ordered state of protein comes from a thermal dissipative force, which is proportional to the gain in

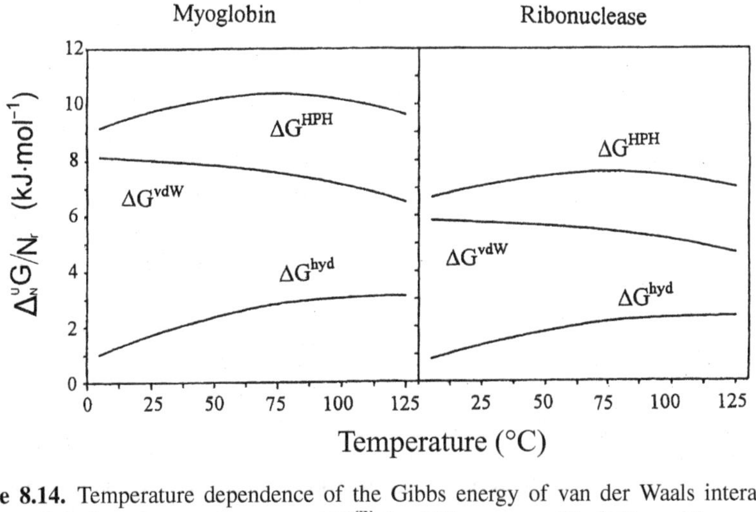

Figure 8.14. Temperature dependence of the Gibbs energy of van der Waals interactions between aliphatic and aromatic groups, ΔG^{vdW}, the Gibbs energy of hydration of these groups, ΔG^{hyd}, and the resultant Gibbs energy, which specifies hydrophobic interactions, ΔG^{HPH}, in (a) myoglobin and (b) ribonuclease A (Makhatadze and Privalov, 1995).

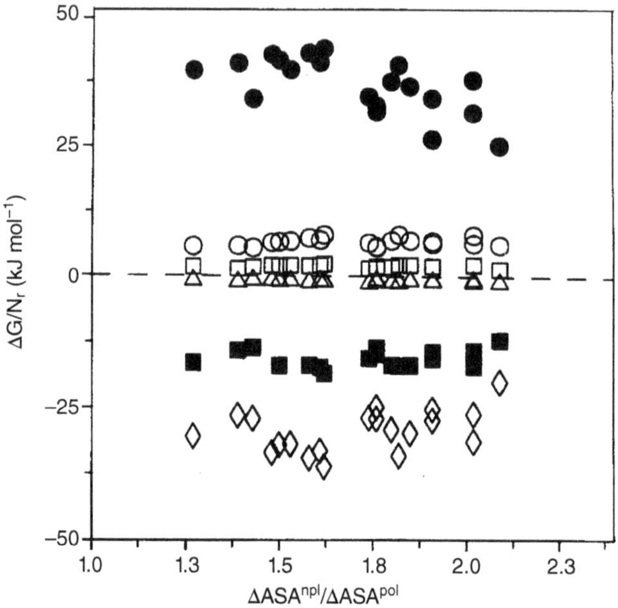

Figure 8.15. The relative contribution of van der Waals interaction (○), hydrogen bonding (●), hydration effects of aliphatic groups (□), aromatic groups (△), polar groups (■), and dissipative force, $-T\Delta S^{conf}$ (◇), to the stabilization of the folded conformation (calculated per mole of amino acid residue) at 25°C versus the ratio of $\Delta ASA_{npl}/\Delta ASA_{pol}$ in the given protein (Makhatadze and Privalov, 1995).

conformational entropy on protein unfolding and the absolute temperature, $-T\Delta S^{conf}$. Thus, this negative force increases with increasing temperature. The destructive action of this dissipative force is supported by the hydration of polar groups. The joint destructive effect of these two factors reaches almost 45 kJ per mole of residue. These negative factors are counterbalanced first of all by the hydrogen bonding of polar groups, which contributes about 40 kJ per residue to the Gibbs energy of protein stabilization. The contribution of van der Waals interactions between nonpolar and aromatic groups and of the hydration of these groups is rather modest. Their combined effect, which includes van der Waals interactions between these groups and their hydration effects, and is usually regarded as hydrophobic interactions, is almost of the same order of magnitude as the combined effect of polar groups, that is, of hydrogen bonding of these groups and their hydration. The relative contribution of all these enthalpic interactions varies depending on the relative extent of the polar and nonpolar groups in the protein. In myoglobin, which has the largest nonpolar core, van der Waals interactions per amino acid residue are larger than those in ribonuclease A, which has fewer nonpolar contacts. On the other hand, in ribonuclease A, which is not too rich in regular α-helical conformation, the hydrogen bonds contribute more to the stabilization of the folded conformation than those in myoglobin, which has a high α-helical content.

Because the energy value of a square angstrom of a buried polar surface (about 275 J/mol·Å^2) is higher than that of a nonpolar surface (about 170 J/mol·Å^2), one cannot expect that increasing the extent of hydrophobic contacts could increase the protein stability. The stability of a globular protein is hardly changed by the proportion of buried polar and nonpolar surfaces since in all examined globular proteins with compact structure this proportion is almost the same (see Fig. 8.5). However, the stability of a protein can certainly be increased without changing the proportion of buried surfaces just by increasing the strength of van der Waals interactions and hydrogen bonding. Because the energies of these interactions, and particularly of the van der Waals interactions, are very sensitive to distance, a small repacking of groups, even below the resolution of X-ray crystallography, can provide a significant change in stability. This might explain the mysterious thermostability of proteins from thermophiles. Unfortunately, the packing density of the thermophiles' interior has not yet been investigated.

Temperature dependence of the joint efforts of protein stabilizing and destabilizing factors is especially interesting. As was shown, the enthalpy and entropy of protein unfolding are strongly increasing functions of temperature and this is because protein unfolding proceeds with significant heat capacity increment. This positive heat capacity increment is provided by the hydration of exposed nonpolar groups; the heat capacity effect of hydration of polar groups is negative but is significantly smaller in magnitude than the effect of hydration of nonpolar groups. Both these heat capacity effects of hydration depend on temperature: that of the nonpolar groups decreases and of the polar groups increases with increasing temperature. As a result, the overall heat capacity increment of protein unfolding vanishes at some temperature T_x, which is at about 120°C. Correspondingly, the enthalpy and entropy of protein unfolding approach their limits asymptotically at that temperature (Fig. 7.6). Surprisingly, in the enthalpy and entropy of protein unfolding the changes associated with the decrease of the heat capacity increment with increasing temperature efficiently compensate each other; as a result, the Gibbs energy of protein unfolding, which specifies the stability of protein structure, is insensitive to the temperature-induced changes in the denaturational heat capacity increment (Fig. 7.7). Therefore, taking as a standard the state of protein at this temperature T_x, the enthalpy and entropy of protein denaturation can be presented in the first approximation as

$$\Delta_{\rm N}^{\rm U}H(T) = \Delta_{\rm N}^{\rm U}H(T_x) - \Delta_{\rm N}^{\rm U}C_p(T_x - T) \tag{8.12}$$

$$\begin{aligned}\Delta_{\rm N}^{\rm U}S(T) &= \Delta_{\rm N}^{\rm U}S(T_x) - \Delta_{\rm N}^{\rm U}C_p \ln(T_x/T) \\ &\approx \Delta_{\rm N}^{\rm U}S(T_x) - \Delta_{\rm N}^{\rm U}C_p\left(\frac{T_x - T}{T}\right) + \Delta_{\rm N}^{\rm U}C_p\frac{1}{2}\left(\frac{T_x - T}{T}\right)^2\end{aligned} \tag{8.13}$$

where $\Delta_{\rm N}^{\rm U}H(T_x)$ is the enthalpy of protein unfolding without hydration effects, that is, protein unfolding in vacuum, and $\Delta_{\rm N}^{\rm U}S(T_x)$ is the conformational entropy of protein unfolding in vacuum. Thus, to the first approximation, we have for the Gibbs energy of stabilization of the native protein state:

$$\begin{aligned}\Delta_{\mathrm{N}}^{\mathrm{U}}G(T) &= \Delta_{\mathrm{N}}^{\mathrm{U}}H(T) - T\Delta_{\mathrm{N}}^{\mathrm{U}}S(T) \\ &\approx \Delta_{\mathrm{N}}^{\mathrm{U}}H(T_x) - T\Delta_{\mathrm{N}}^{\mathrm{U}}S(T_x) - \Delta_{\mathrm{N}}^{\mathrm{D}}C_p\left[\frac{(T_x - T)^2}{2T}\right]\end{aligned} \tag{8.14}$$

It is only the first term in Equation (8.14), representing the total enthalpy of hydrogen bonding and van der Waals interactions in protein, which is positive. The second term, which represents the action of dissipative forces, is negative and increases in absolute value as the temperature increases. The third term in this equation is the only one that represents the effect of the nonpolar groups' hydration. This term is always negative. It is zero at T_x but rapidly increases with decreasing temperature (Fig. 8.16).

Thus, it appears that the native compact protein structure is stabilized only by intramolecular hydrogen and van der Waals bonding, while the effect of water solvation by nonpolar groups has only a destabilizing action: at physiological temperature it decreases the stability of globular protein 6 times, from about 300 kJ/mol to about 50 kJ/mol.

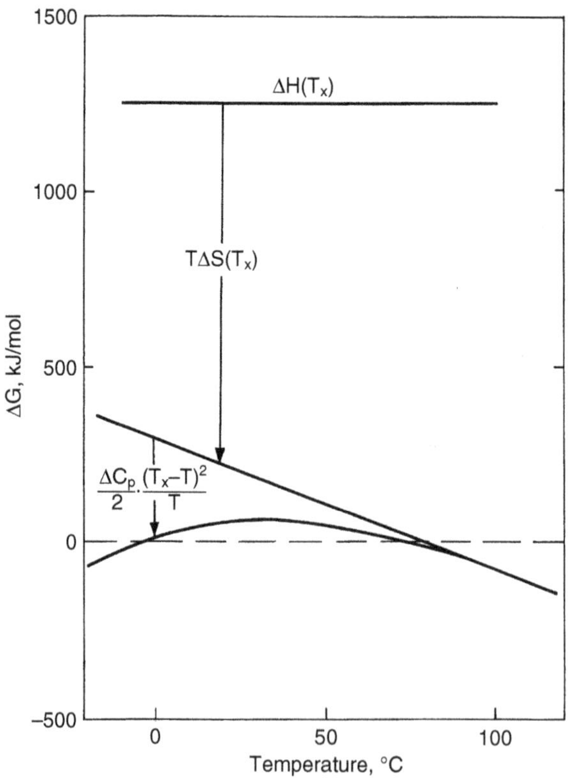

Figure 8.16. Contribution of dissipative force [$T\Delta S(T_x)$] and water solvation effect [$(\Delta C_p/2)(T_x/T - 1)^2$] to the stabilization of an abstract globular protein (Privalov and Gill, 1988).

It is remarkable that this is the maximal stability of globular protein and it is reached at temperatures close to physiological. At high temperatures, this structure breaks down due to an increase of dissipative forces, whereas at low temperatures it breaks as a result of an increase of the solvation tendency caused by hydration of nonpolar groups, leading to cold denaturation of protein (see Section 2.4.2). It looks as though the stability close to 50 kJ/mol is optimal for compact globular proteins. The question is: why is it optimal?

REFERENCES

Baldwin R.L. (1986). *Proc. Natl. Acad. Sci. USA*, **83**, 8069–8072.

Ben-Naim A. (1987). *Solvation Thermodynamics.* Plenum, New York and London.

Burley S.K. and Petsko G.A. (1988). *Adv. Protein Chem.*, **39**, 125–189.

Del Bene J.E. (1994). *J. Phys. Chem.*, **98**, 5902–5905.

Dill K.A. (1990). *Science*, **250**, 297–298.

Doig A.J. and Williams D.H. (1992). *J. Am. Chem. Soc.*, **114**, 338–343.

Gavezotti A. (1989). *J. Am. Chem. Soc.*, **111**, 1835–1843.

Griko Yu.V., Freire E., and Privalov P.L. (1994). *Biochemistry*, **33**, 1889–1899.

Harpaz Y., Gerstein M., and Chothia M.C. (1994). *Structure*, **2**, 641–649.

Hutchens J.O., Cole A.G., and Stout J.W. (1969). *J. Biol. Chem.*, **244**, 26–32.

Joesten M.D. and Schaad L.J. (1974). *Hydrogen Bonding.* Dekker, New York.

Karplus M. and Kushick J.N. (1981). *Macromolecules*, **14**, 325–332.

Karplus M., Ichiye T., and Pettitt B.M. (1987). *Biophys. J.*, **52**, 1083–1085.

Kauzmann W. (1959). *Adv. Protein Chem.*, **14**, 1–63.

Kitamura S. and Sturtevant J.M. (1989). *Biochemistry*, **28**, 3788–3792.

Lazaridis T., Archontis G., and Karplus M. (1995). *Adv. Protein Chem.*, **47**, 231–306.

Livingstone J.R., Spolar R.S., and Record M.T., Jr. (1991). *Biochemistry*, **30**, 4237–4244.

Makhatadze G.I. and Privalov P.L. (1987). *Proc. USSR Acad. Sci.*, **296**, 746–749.

Makhatadze G.I. and Privalov P.L. (1988). *J. Chem. Thermodynam.*, **20**, 405–412.

Makhatadze G.I. and Privalov P.L. (1989). *J. Solution Chem.*, **18**, 927–936.

Makhatadze G.I. and Privalov P.L. (1990). *J. Mol. Biol.*, **213**, 375–384.

Makhatadze G.I. and Privalov P.L. (1993). *J. Mol. Biol.*, **232**, 639–659.

Makhatadze G.I. and Privalov P.L. (1994). *Biophys. Chem.*, **50**, 285–291.

Makhatadze G.I. and Privalov P.L. (1995). *Adv. Protein Chem.*, **47**, 307–425.

Makhatadze G.I. and Privalov P.L. (1996). In: *Physical Properties of Polymers Handbook* (J.M. Mark, Ed.), pp. 91–99. American Institute of Physics, New York.

Makhatadze G.I., Kim K.S., Woodward C., and Privalov P.L. (1993). *Protein Sci.*, **2**, 2028–2036.

McDonald I.K. and Thornton J.M. (1994). *J. Mol. Biol.*, **238**, 777–793.

Meot-Ner M. and Sieck L.W. (1986). *J. Am. Chem. Soc.*, **108**, 7525–7529.

Miller S., Janin J., Lesk A.M., and Chothia C. (1987). *J. Mol. Biol.*, **196**, 641–656.

Murphy K.P. and Freire E. (1992). *Adv. Protein Chem.*, **43**, 313–361.

Murphy K.P. and Gill S.J. (1991). *J. Mol. Biol.*, **222**, 699–709.

Nicholls A., Sharp K.A., and Honig B. (1991). *Proteins: Struct. Funct. Genet.*, **11**, 281–296.

Privalov P.L. (1979). *Adv. Protein Chem.*, **33**, 167–241.

Privalov P.L. and Gill S.J. (1988). *Adv. Protein Chem.*, **39**, 191–234.

Privalov P.L. and Makhatadze G.I. (1992). *J. Mol. Biol.*, **224**, 715–723.

Privalov P.L. and Makhatadze G.I. (1993). *J. Mol. Biol.*, **232**, 660–679.

Richards F.M. (1974). *J. Mol. Biol.*, **82**, 1–14.

Richards F.M. (1977). *Annu. Rev. Biophys. Bioenerg.*, **6**, 151–176.

Richards F.M. and Lim W.A. (1994). *Quart. Rev. Biophys.*, **26**, 423–498.

Shrake A. and Rupley J.A. (1973). *J. Mol. Biol.*, **79**, 351–371.

Sneddon S.F. and Tobias D.J. (1992). *Biochemistry*, **31**, 2842–2846.

Spolar R.S., Ha J.-H., and Record M.T., Jr. (1989). *Proc. Natl. Acad. Sci. USA*, **86**, 8382–8385.

Spolar R.S., Livingstone J.R., and Record M.T., Jr. (1992). *Biochemistry*, **31**, 3947–3955.

Stickle D.F., Presta L.G., Dill K.A., and Rose G.D. (1992). *J. Mol. Biol.*, **226**, 1143–1159.

Sturtevant J.M. (1977). *Proc. Natl. Acad. Sci. USA*, **74**, 2236–2240.

Wen J. (1996). In: *Physical Properties of Polymers Handbook* (J.M. Mark, Ed.), pp. 101–111. American Institute of Physics, New York.

Williams D.H. (1991). *Aldrichimica Acta*, **24**, 71–80.

Wintrode P.L., Makhatadze G.I., and Privalov P.L. (1994). In: *Proteins: Struct. Funct. Genet.*, **18**, 246–253.

Wolfenden R. (1978). *Biochemistry*, **17**, 201–204.

Wolfenden R., Anderson L., Cullis P.M., and Southgate C.C.B. (1981). *Biochemistry*, **20**, 849–855.

Yu Y., Makhatadze G.I., Pace C.N., and Privalov P.L. (1994). *Biochemistry*, **33**, 3312–3319.

9

PROTEIN FOLDING

9.1. MACROSTABILITIES AND MICROSTABILITIES OF PROTEIN STRUCTURE

9.1.1. Macrostability of Proteins

With the appearance of crystallographic information on proteins, we became increasingly convinced that protein structures are extremely sophisticated and highly ordered. However, the structures of real proteins cannot be as fixed as are their models since at the molecular level fluctuations are not negligible at all. The question arises then: how do we define the stability of these fluctuating structures? The assumption that protein denaturation represents an "all-or-none" process has opened the possibility of its qualitative treatment in physical terms. The equilibrium studies led to the conclusion that protein denaturation proceeds with enormous increase in enthalpy. This was regarded as an indication that protein denaturation is a highly cooperative process involving the whole macromolecule and results in unfolding of the polypeptide chain forming it. In fact, the conclusion concerning extreme cooperativity of protein denaturation was achieved only because it was *a priori* assumed that this is a cooperative process. There was only one way to escape from this

Microcalorimetry of Macromolecules: The Physical Basis of Biological Structures, First Edition.
Peter L. Privalov.

circulus vitiosus: to measure directly the thermodynamic functions describing a protein unfolding, such as enthalpy, volume change, and ligand binding. Only by comparing the directly measured thermodynamic values specifying protein unfolding with the values derived from equilibrium analysis of this process, could one decide whether protein can be treated thermodynamically and be considered as a single cooperative system.

The very first calorimetric measurements of the protein unfolding showed that enthalpy of temperature-induced unfolding is indeed very large and in the case of small globular proteins it is in a very close correspondence with the enthalpy derived from equilibrium analysis of the observed process using the van't Hoff equation, which assumes its all-or-none character. Thus, it becomes clear that the native structure of a small compact globular protein breaks down in an all-or-none way; that is, such a protein indeed represents a highly cooperative quasi-macroscopic system.

Extrapolating the calorimetrically obtained protein unfolding characteristics to some conditions one can specify protein stability in terms of the Gibbs energy, that is, the work required to unfold this whole macroscopic system at the considered condition (see Section 7.6.2). For small compact globular proteins consisting of about 100 amino acid residues, this Gibbs energy at standard temperature (25°C = 298.15 K) amounts to about 50 kJ/mol (see Table 7.2). Thus, it appears that each amino acid residue contributes about 0.5 kJ/mol to the stabilization of protein structure. On the other hand, at room temperature the dissipative force of thermal motion is $RT = 8.36$ J/K·mol × 298.15 K ≈ 2.5 kJ/mol. This is 5 times larger than the energy contribution of a single residue to the stabilization of protein structure. It appears that protein structure is stable at room temperature only because the amino acid residues forming it strongly cooperate; that is, they cannot change their state individually but only in the ensemble, all together. The total work for breaking this cooperative system exceeds 20 times the energy of thermal motion at room temperature. Therefore the protein structure is quite stable at this standard temperature and, as well, at the physiological temperature.

The need for cooperation of all groups of globular proteins is evident: to fulfill its delicate function, protein structure should withstand the dissipative action of thermal motion, so its macroscopic stability should be high enough. It should be emphasized, however, that:

> *The secret of the stability of the native structure of protein is not in the magnitude of intramolecular interactions, which are always too weak to withstand individually the dissipative action of thermal motion, but in the effective cooperation of these interactions.*

9.1.2. Microstability of Proteins

The local unfoldings of a compact protein structure fluctuating in thermal motion, which result in occasional exposure of its internal residues to the solvent, are of some interest. The work required for this local unfolding specifies the "micro" stability of a protein structure. This "micro-unfolding" of a protein structure can be registered by the rate of hydrogen exchange between the protein and surrounding water enriched with deuterium or tritium.

In considering microstability and macrostability of proteins we are dealing with a somewhat similar process—the unfolding of a compact protein structure. Therefore, one might expect that these properties of protein structure are interrelated and, moreover, that formally they are expressed in the same dimensions, in kilojoules per mole (kJ/mol). However, in the case of macrostability the "mole" applies to the cooperative unit, which might be the whole protein, whereas in the case of microstability "mole" is assumed to be the single amino acid residue. Thus, these two stabilities differ in scale, that is, quantitatively, but not only that: the macrostability describes the stability of the whole cooperative structure, whereas the microstability describes the local stability of the macromolecule, or—when averaged over the whole protein—expresses the rigidity of the protein structure, while its inverse value expresses the flexibility of the protein structure. Therefore, the averaged Gibbs energy of micro-unfoldings, $\langle\Delta_N^U G^{mic}\rangle$, does not correlate with the specific macroscopic Gibbs energy of unfolding, calculated per gram or per mole of amino acid residue (Table 9.1). However, surprisingly it correlates with the specific enthalpy and entropy of protein unfolding calculated per gram of protein, increasing with their decrease. But these inverse correlations of the $\langle\Delta_N^U G^{mic}\rangle$ with specific enthalpy and entropy of protein unfolding seem to be in themselves the result of correlation with the specific heat capacity increment upon protein unfolding, $\Delta_N^U C_p^{mac}$, and this correlation is especially simple for the inverse specific microstability, $(\Delta_N^U G^{mmic})^{-1}$, expressing the flexibility of protein structure (Fig. 9.1). Since the specific value of the unfolding heat capacity increment $\Delta_N^U C_p^{msc}$ actually expresses the saturation of the protein globule with apolar contacts (see Chapter 8), one can conclude that the flexibility of protein structure decreases with the increase in concentration of the apolar contacts in protein.

TABLE 9.1. Micro and Macro Characteristics of Globular Proteins at 25°C[a]

Proteins	$\langle\Delta G^{mic}\rangle$ (kcal/mol)	$\Delta_N^U G^{mac}$ (cal/g)	$\Delta_N^U H^{mac}$ (cal/g)	$\Delta_N^U S^{mac}$ (cal/K·g)	$\Delta_N^U C_p^{msc}$ (cal/K·g)
Ribonuclease A	2.5	0.8	5.0	1.42	0.090
Lysozyme	2.7	1.0	4.4	1.14	0.100
Trypsin	3.1	0.5	2.6	0.64	0.120
Chimotrypsin	3.3	0.5	2.7	0.72	0.120
Cytochrome c	3.4	0.7	1.7	0.34	0.130
BPTI	3.7	0.8	0.0	−0.26	0.110
Chymotrypsinogen	4.7		2.0		0.130
Carbonic anhydrase	6.6	0.4	0.1	−0.10	0.140
Metmyoglobin	7.1	0.7	0.0	−0.22	0.150

[a]P. Privalov and Tsalkova (1979).

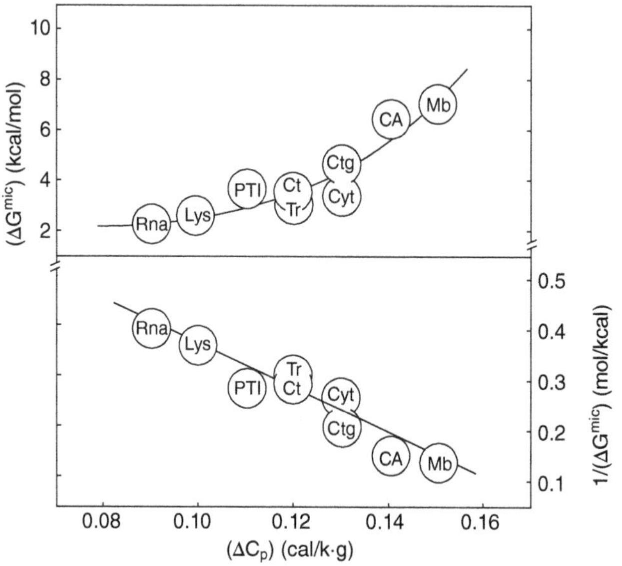

Figure 9.1. Plot of rigidity $\langle \Delta G^{\text{mic}} \rangle$ and flexibility $\langle \Delta G^{\text{mic}} \rangle^{-1}$ versus the partial specific heat capacity increment upon protein unfolding, $\Delta_{\text{N}}^{\text{U}} C_p^{\text{msc}}$ for various globular proteins (P. Privalov and Tsalkova, 1979).

> *Thus, it appears that the degree of rigidity of protein structure is determined by the clusters of nonpolar groups in the protein interior (P. Privalov and Tsalkova, 1979).*

9.1.3. Packing in Protein Interior

When hydrogen exchange is observed by nuclear magnetic resonance (NMR), one can trace the hydrogen exchange of a certain (*i*th) residue and from its rate determine the equilibrium constant of the local, micro-unfolding of this residue, K_i^{mic}, and the Gibbs energy of this micro-unfolding: $\Delta G_i^{\text{mic}} = -RT\ln(K_i^{\text{mic}})$. Figure 9.2 demonstrates such a hydrogen exchange experiment carried out by Englander and colleagues on cytochrome c in the presence of denaturant, GuHCl, in solution (Englander and Mayne, 1992; Krishna et al., 2006; Maity et al., 2005; Jeng and Englander, 1991). With decreased denaturant concentration, the rate of hydrogen exchange decreases; that is, the Gibbs energy of micro-unfolding increases. Extrapolating its value to zero concentration of denaturant, one can estimate the microstability of a particular element of protein structure in aqueous solution. In the case of cytochrome c, the most unstable appeared to be the segment shown in red, followed by the segment in yellow; the most stable are the helices shown in dark blue. It appears that the various parts of a protein molecule differ in flexibility, that is, rigidity. Therefore, one can conclude that the amino acid residues forming these parts should differ in the number of contacts with neighbors; that is, these parts should differ in packing densities.

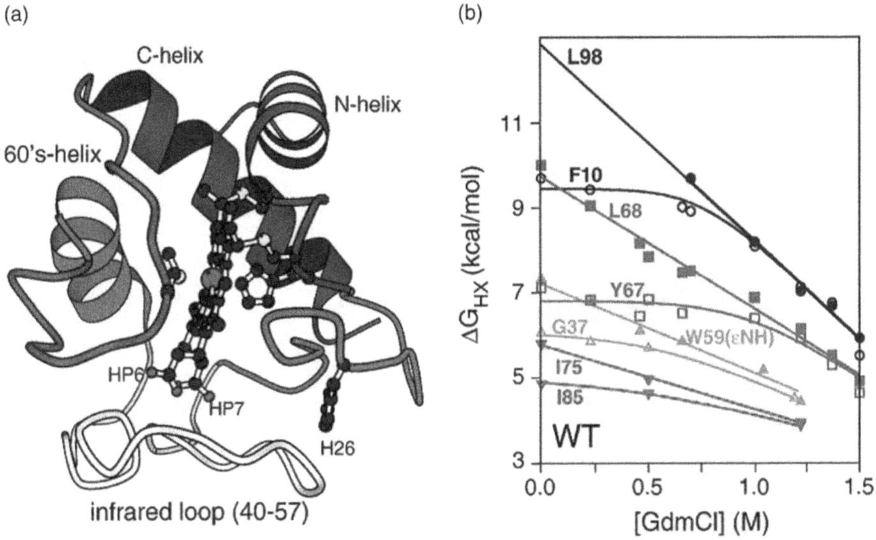

Figure 9.2. Hydrogen exchange by the different structural elements of cytochrome c. (Reproduced from Krishna et al., 2006.) (See insert for color representation of the figure.)

In view of the preceding discussion, the packing of groups in the protein interior is of particular interest. Macroscopically, packing density is defined as the ratio between the van der Waals envelope of the molecule and the total volume the molecule occupies in the context of surrounding molecules (occupancy volume). However, at the atomic level the definition of the volume that any given atom occupies is less obvious (Richards, 1974). The most widely used definition of volume occupancy is based on a geometric composition known as a Voronoi diagram (Harpaz et al., 1994; Janin and Chothia, 1990; Richards, 1985; Finney, 1975). In the case of atoms within protein, the area of proximity for any given atom can be constructed by connecting this atom with every reasonably close atom, then bisecting these interatomic vectors by planes normal to these vectors (Fig. 9.3).

Analysis of the packing density of the protein interior showed that it is not uniform: there are clusters of tightly packed groups and clusters of rather loosely packed groups. Comparison of the cytochrome c interior packing (Fig. 9.4) with the results of hydrogen exchange studies (Fig. 9.2) shows that the segment with the low packing density is most easily exchanging its hydrogens, while the structural elements that are specified by slow hydrogen exchange are packed the most tightly. Thus, the tightly packed clusters are indeed more stable than the loosely packed clusters. It is remarkable also that the tightly packed clusters do not form a single apolar core in the interior of the globular protein, as was believed assuming that the apolar groups are removed from the entropically unfavorable contacts with water, but these clusters are spread in the protein interior. Such uneven distribution of packing densities appears to be a general feature of all globular proteins (Fig. 9.5).

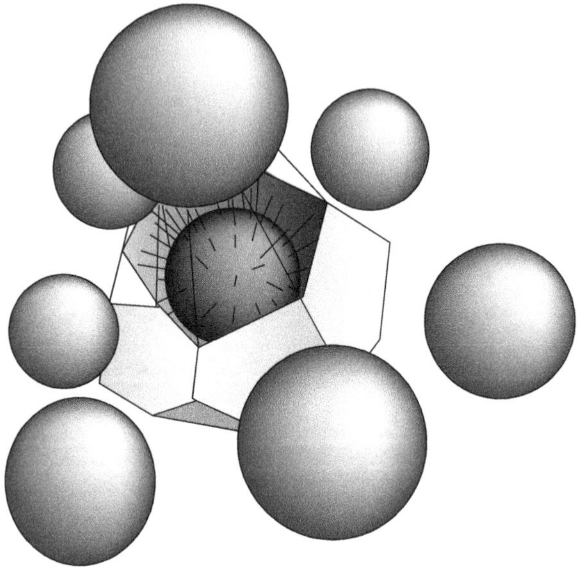

Figure 9.3. Voronoi polyhedron for the atom shown inside. One facet of the polyhedron was left transparent. The space around the atom in the center is probed by random beams. The points where beams cross the vertices of the polyhedron determine the length of the needle. Volume of the partition is calculated by summing up the lengths of the needles (G. Privalov, 1995).

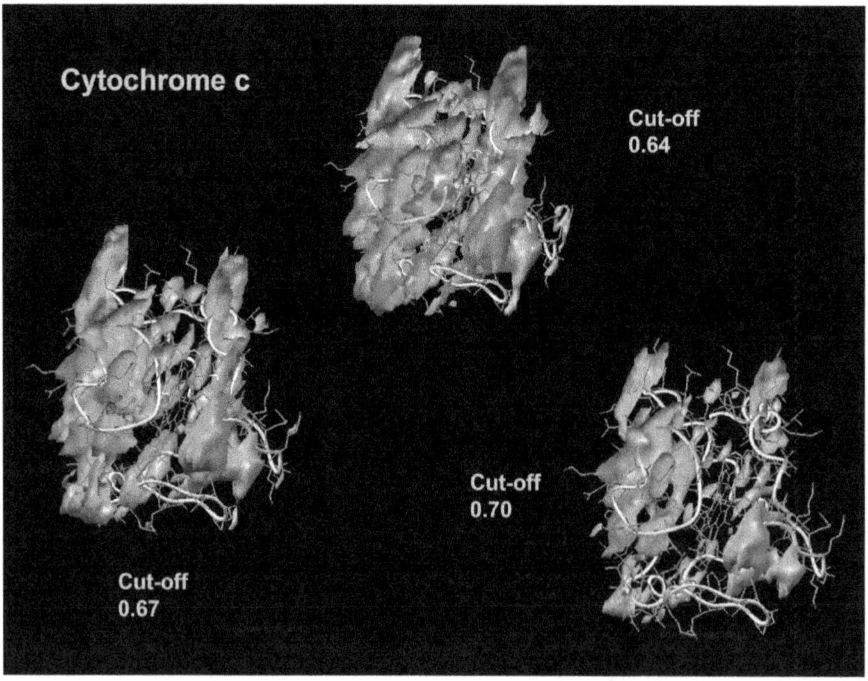

Figure 9.4. Packing density maps of cytochrome c at different cutoffs. (See insert for color representation of the figure.)

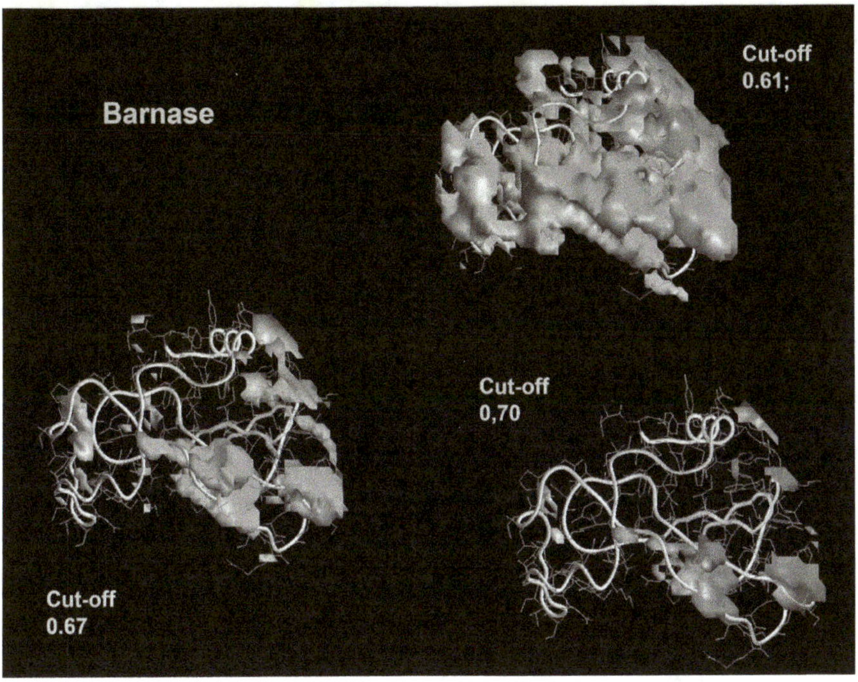

Figure 9.5. Packing density maps of barnase at different cutoffs. (See insert for color representation of the figure.)

It has been found that **the densely packed clusters are highly conservative** (Richards and Lim, 1994; G. Privalov, 1995). Thus, it is the loosely packed positions in proteins that allow the variation of amino acids upon evolution: the replacement of amino acids in these loose positions does not cost much in energy and the protein can easily accommodate these replacements. The case of cytochrome c, considered previously, presents an impressive example; this extremely important protein has existed in all living species. The primary structures of this 104-residue protein from baker's yeast and horse differ in 45 positions; nevertheless, the conformations of these two proteins are almost identical (Fig. 9.6). Thus the tertiary structure of protein is more conservative than its primary structure and this certainly was needed not only to preserve its function in evolution but also to preserve the folding technology that was perfected during a billion of years of evolution. This, however, assumes grouping of certain residues into stable formats that then might direct folding of the rest of the parts of the polypeptide chain. Such stable formats might be the tightly packed clusters in protein, which most probably formed first in the protein folding process.

One can assume that upon protein heating the energy is accumulated mainly in the more easily fluctuating loosely packed clusters. This might be the reason for the observed increase of the heat capacity of the native protein upon heating (see Section 8.2.1 and Fig. 8.6) and, corresponding, steep increase of its excess enthalpy (Fig. 9.7). Indeed, it appears that the initial slope of the heat capacity function is minimal for the stable proteins with rigid structure, such as bovine pancreatic trypsin inhibitor

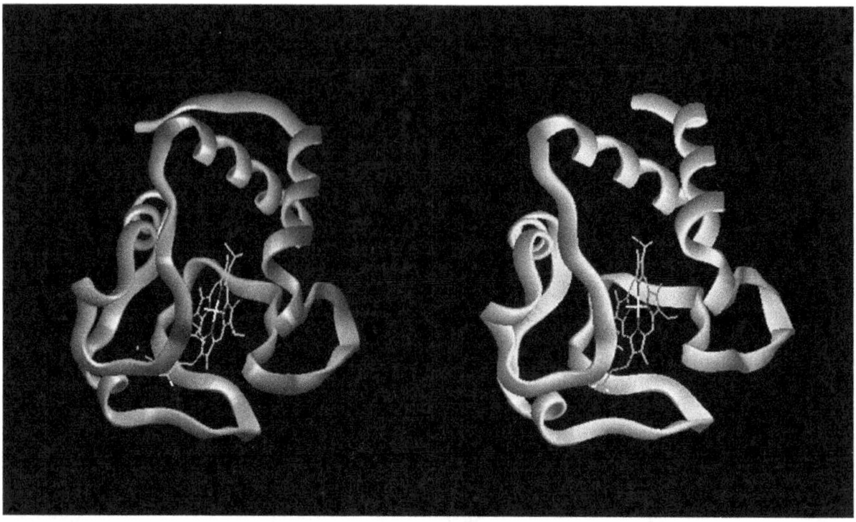

Figure 9.6. Structures of cytochrome c from baker's yeast (green) and horse (yellow) have rather similar tertiary structures despite significant (45%[!]) difference in their primary structures. (See insert for color representation of the figure.)

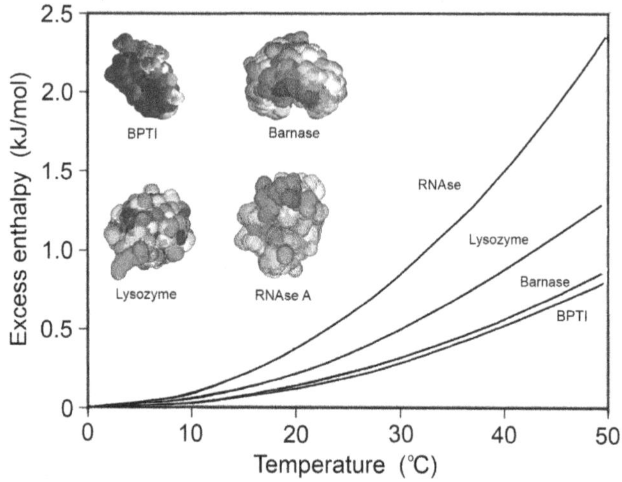

Figure 9.7. The B-factors of four different proteins and their excess enthalpy accumulation with temperature rise. The blue corresponds to low B-factor and red to high B-factor. (See insert for color representation of the figure.)

(BPTI), and is larger for proteins with flexible structure, as one can judge by the B-factor of the considered proteins, which shows directly the extent of thermal fluctuations in the protein. The steep increase of the excess enthalpy of protein with increasing temperature raises the question why proteins do not unfold gradually upon heating. Is it because the tightly packed clusters still keep their structure?

9.2. PROTEIN FOLDING TECHNOLOGY

9.2.1. Intermediate States in Protein Folding

The mechanism of protein folding is one of the most puzzling problems in protein science. Protein is synthesized as a linear heterogeneous polymer, a polypeptide, and then, if it is designed to function in the folded state, it folds into a functionally active unique native conformation utilizing the information included in its primary structure. All attempts to trace the intermediate states realized upon unfolding and refolding of small compact globular proteins have failed so far. Protein unfolding proceeds extremely fast; however, what is most surprising, folding of its polypeptide chain into rather complicated, unique conformations also proceeds very fast (Sosnick et al., 1994; Bhuyan and Udgaonkar, 2001). Its experimental investigation required development of special techniques for the fast initiation of the protein unfolding–refolding process: stop-flow, temperature-jump, and pressure-jump instruments. However, all that was found using these sophisticated devices is that protein unfolding and refolding processes proceed in milliseconds (Sosnick et al., 1994). This rules out the possibility that folding is a simple stochastic process. Indeed, if it were a stochastic process of random search of the proper arrangement of all residues, bearing in mind that a polypeptide chain of 100 amino acid residues displays about 10^{100} conformations (see Section 4.2), and in one second each bond might change 10^{12} times, stochastic search of the proper conformations would require 10^{100} $/2 \times 10^{14}$ sec = 5×10^{85} sec = 10^{78} years, the paradox noticed first by Levinthal (1968). This complicated process is certainly somehow directed; that is, it proceeds through a definite sequence of intermediate states along decreasing Gibbs energies (Anfinsen, 1972, 1973). The nature of these intermediates induces much debate due to a lack of reliable information, largely because these states are too unstable, that is, too short-lived, to be observed directly.

However, with some proteins under certain conditions states have been observed that do not appear to be the native state or completely unfolded (Tanford, 1968; Ptitsyn, 1992). These cases have attracted a special attention, since they might be a key for solving the problem of protein folding. What are these states? Are they incorrectly refolded proteins, or are they partly folded polypeptide chains, that is, the intermediates in its folding to the compact unique conformation? If they are intermediates, do they represent some global state of the whole polypeptide chain of the protein or a state in which only part of the chain is folded into a nativelike structure? If this part of the chain is stable and unfolds cooperatively, more or less independently of the rest of the molecule, then it could be regarded as a folding domain. It is clear that folding of large protein consisting of several cooperative domains should proceed in discrete steps corresponding to the folding of individual domains (see Chapter 10). The main problem, however, is the folding of a small protein without a subdomain organization: does it fold and unfold through an intermediate state that is stable under some conditions and does not represent an artifact of folding, that is, is not a misfolded state of a polypeptide chain trapped in a potential well?

9.2.2. Molten Globule Concept

One could imagine that folding of a polypeptide chain starts from collapse into a compact, disordered liquidlike globule from which it later acquires the native conformation by the proper adjustment of all groups. One could expect that the collapse of a polypeptide into a disordered compact globule could be a very fast process. However, the search for proper adjustment of all the groups in the highly viscous condensed phase cannot be as fast and should take quite a long time. Nevertheless, it was believed that collapse of the polypeptide chain, which reduces significantly the configurational phase space, can solve Levinthal's paradox (Ptitsyn, 1972, 1995a,b). The hypothesis of a "molten globule intermediate" in protein folding was therefore met with enthusiasm; its popularity had rapidly increased with the number of new observations that seemed supportive for this new idea and the idea itself was also evolving under the pressure of the new facts.

Initially, the molten globule was assumed to be a collapsed, mobile state of the polypeptide chain without fixed, specific interactions. This collapsed state was supposed to be maintained by hydrophobic force, which was regarded as an external pressure of water on the apolar groups condensing them into a compact, liquidlike globule (droplet), but not penetrating inside the globule (Ptitsyn, 1992). Thus, it was a dry molten globule. Then it was proposed that the compaction promotes a helical conformation of the polypeptide chain (Dill, 1990; Chan and Dill, 1991). Thus, the molten globule appeared to be a highly dynamic compact state of the polypeptide chain, without tertiary structure but with some helicity not necessarily related to the secondary structure of the native protein. Therefore, this concept missed the point that the "hydrophobic force" is not just the entropic pressure of water that does not want to be ordered by apolar groups. As discussed previously (see Sections 2.4.2 and Section 8.3.3), the main contributors to the hydrophobic effect are the van der Waals interactions between the apolar groups. These van der Waals interactions will dump mobility of the dry molten globule. The concept also missed the point that the polar groups of protein, which would appear inside the compact globule, will form hydrogen bonds and the energy of a single hydrogen bond in the absence of polar water molecules amounts to 50 kJ/mol. Thus, in the absence of water the occasionally formed hydrogen bonds will never rearrange and search for their proper arrangement. Therefore, later the "dry molten globule" was replaced by the "wet molten globule" and "swollen molten globule," which assumed the presence of some water inside the globule (Finkelstein and Shakhnovich, 1989). It was then suggested that a polypeptide chain in a molten globule has a "nativelike fold" and could be a "highly ordered molten globule" (Ptitsyn, 1995a,b; Peng and Kim, 1994; Redfield et al., 1994). This idea was argued for based on the observation that proteins in the molten globule state are separated chromatographically from the unfolded proteins (Ptitsyn and Uverski, 1994). However, if the isomeric forms of protein move in two different peaks in a chromatographic experiment, this means that their rate of interconversion is very low. If so, the state that is considered cannot be a dynamic state with a highly mobile conformation and thus cannot be regarded as a "molten globule." Could it then serve as an intermediate

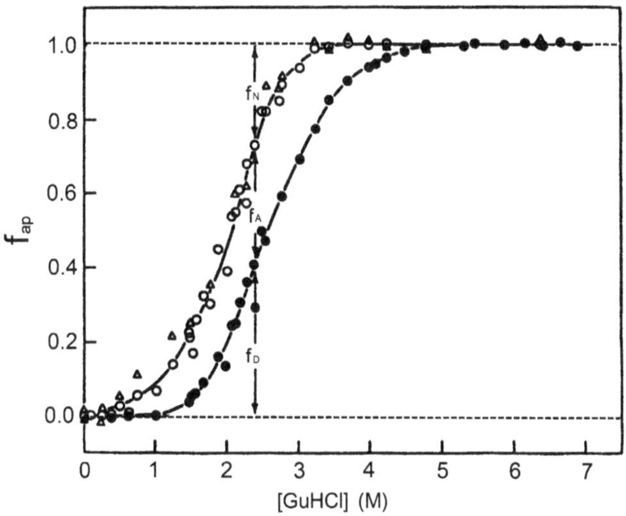

Figure 9.8. Urea-induced changes of ellipticity of bovine α-lactalbumin at 270 nm (○) and 222 nm (●) (Kuwajima et al., 1976).

in folding providing a fast search for the unique native conformation that is required to resolve Leventhal's paradox?

There were many reports of the experimental facts demonstrating the existence of stable, compact, partly disordered intermediate states in protein folding, but most studies concentrated on the unfolding of four proteins that were regarded as demonstrating a molten globule state under mild denaturing conditions: α-lactalbumin, equine lysozyme, staphylococcal nuclease, and apomyoglobin.

It was noticed that the ellipticities of α-lactalbumin in the near and far ultraviolet (UV) do not change in parallel upon denaturation (Fig. 9.8): the change in the far UV proceeded over broader pH and denaturant concentration ranges, suggesting that the protein first loses its tertiary structure, but retains almost all of its initial helicity, which vanishes only at the second stage (Kuwajima et al., 1976). A similar biphasic change of α-lactalbumin structure was observed calorimetrically upon heating (Fig. 9.9): the first stage of α-lactalbumin unfolding is accompanied by extensive heat absorption and the calorimetric enthalpy determined from the area of the heat absorption peak is in good correspondence with the van't Hoff enthalpy determined from the sharpness of this process, assuming it to be a two-state transition, that is, a cooperative process, which proceeds with significant change in enthalpy and entropy. Thus, it can be regarded as a first-order-like phase transition. This transition is accompanied by a significant increase in heat capacity, but its value does not reach the level expected for the completely unfolded polypeptide chain with all groups exposed to water. The heat capacity level of the fully unfolded polypeptide chain is reached with a further increase in temperature, which proceeds with shallow heat absorption peak.

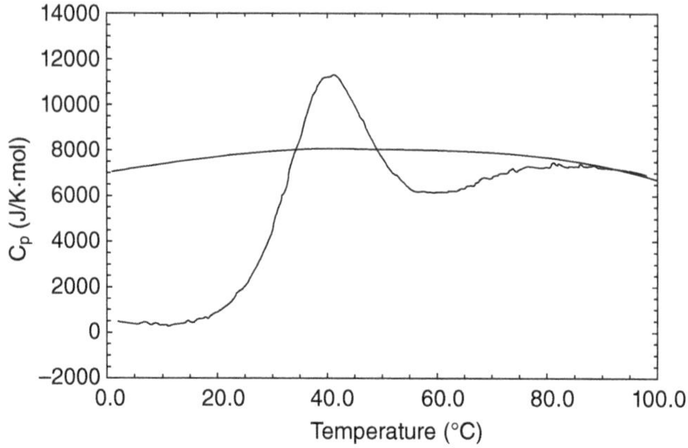

Figure 9.9. Temperature dependence of the excess heat capacity of α-lactalbumin at pH 8.00, 500 mM Tris, and the expected heat capacity of fully unfolded polypeptide chain of lactalbumin (Griko et al., 1994a).

A very similar situation was found with equine lysozyme, the structure of which closely resembles that of α-lactalbumin. Unlike the hen-egg-white lysozyme (HEWL), which unfolds upon temperature increase in one cooperative stage with sharp heat absorption (see Fig. 3.17), the equine lysozyme unfolds in two distinct stages (Fig. 9.10), each proceeding with significant excess heat absorption and significant increase of heat capacity (Morozova et al., 1991, 1995; Van Dael et al., 1993; Griko et al., 1995). Thermodynamic analysis of the excess heat observed upon heating equine lysozyme shows that both stages represent highly cooperative two-state transitions. At the end of the second stage the heat capacity of equine lysozyme reaches the value specific for the completely unfolded polypeptide chain. Therefore, it was assumed that the first stage represents protein transition into the molten globule state.

The structure of equine lysozyme and all c-type lysozymes are very similar to that of α-lactalbumin: they all have two domains separated by a cleft and one of these domains is rich in α-helical conformation, and the other in β-conformation. However, analysis of packing densities in these proteins shows that in equine lysozyme there are two separated clusters of groups with packing density higher than 0.70, while in the hen-egg-white lysozyme these two densely packed clusters merge together forming a single densely packed core (Fig. 9.11). Thus, the intermediate state that is observed upon denaturation of α-lactalbumin and equine lysozyme is the state in these two-domain proteins with one domain unfolded and the other one preserved.

In the case of staphylococcal nuclease there were numerous publications devoted to the thermal and acid denaturation of various mutants of this protein reporting that they retain a considerable amount of residual structure in the denatured state, as well as some compactness (see, e.g., Dill and Shortle, 1991; Shortle, 1993, 1996). This state was believed to represent a molten globule although its sensitivity to mutations

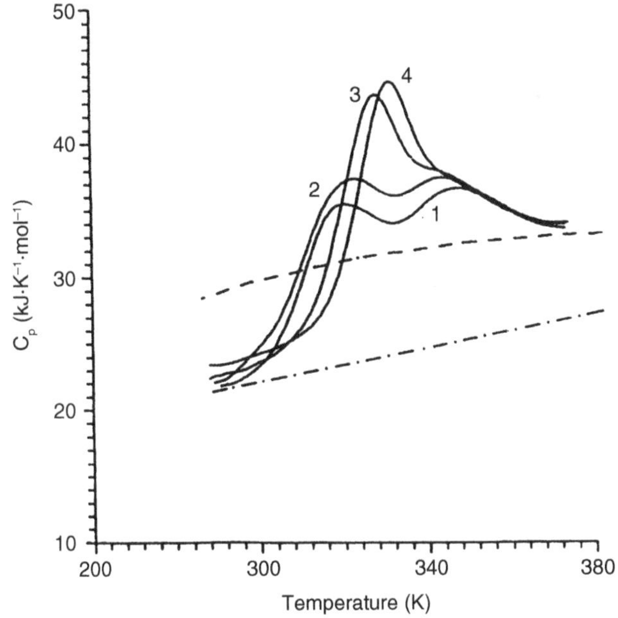

Figure 9.10. Temperature dependence of the partial heat capacity of equine lysozyme in the presence of various concentrations of Ca^{2+} in 10 mM Na acetate, pH 4.5: (1) 0.0 mM $CaCl_2$; (2) 0.10 mM $CaCl_2$; (3) 0.75 mM $CaCl_2$; (4) 1.5 mM $CaCl_2$ (Griko et al., 1995).

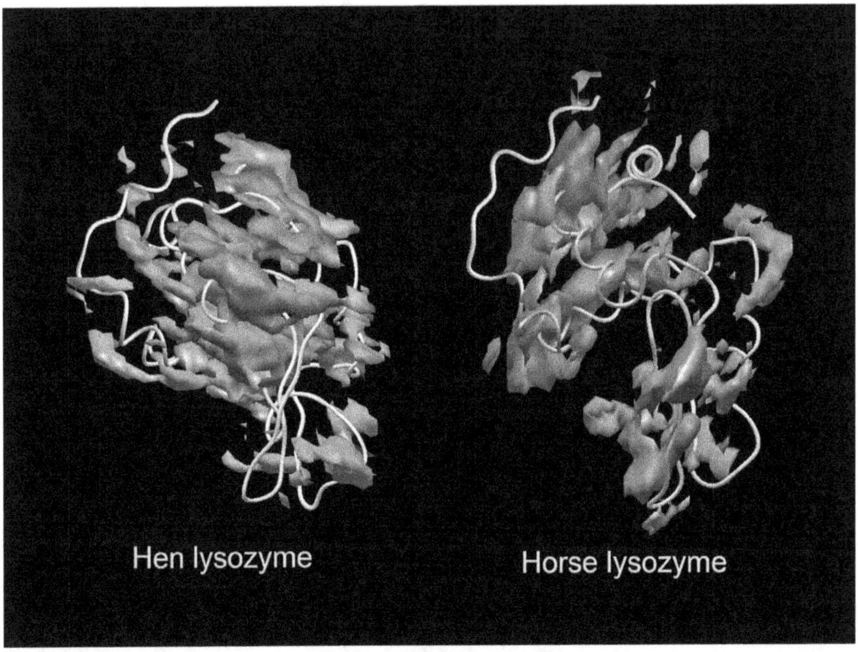

Figure 9.11. Packing density maps of the hen-egg-white and horse (equine) lysozymes at cutoff 0.7, showing that two domains in the hen-egg-white lysozyme merge into a single dense cluster; in equine lysozyme the two domains are separated by a cleft (Griko et al., 1995). (See insert for color representation of the figure.)

TABLE 9.2. Transitions in the Staphylococcal Nuclease Mutants[a,b]

Protein	ΔH_{tot}	T_i (°C)	ΔH_1 (kJ/mol)	T_2 (°C)	ΔH_2 (kJ/mol)	ΔG_2(25°C) (kJ/mol)
Wild-type	278	41.1	186	69.7	92	5.9
V66L	233	38.9	117	70.9	116	9.1
V66W	219	43.5	145	66.3	74	3.8
D77A	286	45.3	198	72.6	88	5.2
E75V	220	40.1	152	65.5	68	3.1
G88V	239	45.2	144	58.6	95	6.1
1–136				66.1	128	5.0

[a]According to Carra and Privalov (1996).
[b]pH 4.0, 400 mM NaCl.

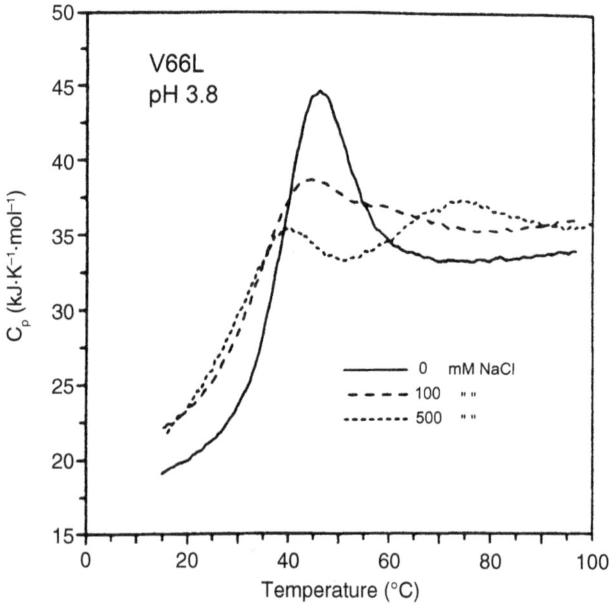

Figure 9.12. Temperature dependence of the partial heat capacity of the V66L mutant of staphylococcal nuclease in solutions at pH 3.8 with various concentrations of salt (Carra et al., 1994).

was totally inconsistent with the idea of a state lacking fixed long-range interactions (Flanagan et al., 1993). Calorimetric study of these mutants (Table 9.2) indeed showed that unfolding of most of these mutants is not a single two-state transition, but proceeds in two more or less independent cooperative stages, depending on the mutant (Carra et al., 1994; Carra and Privalov, 1995, 1996). These stages can be clearly distinguished by varying the solvent conditions, pH, and concentration of salts (Fig. 9.12). Each of these stages represents a two-state transition and is

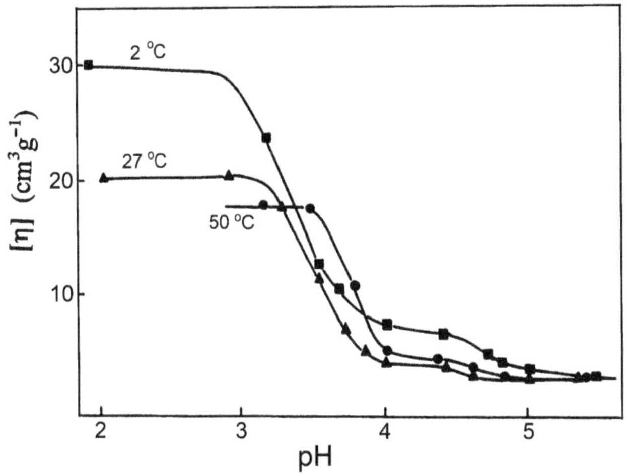

Figure 9.13. pH dependence of the intrinsic viscosity of apomyoglobin (Griko et al.,1988).

associated with an increase in the heat capacity. They therefore correspond to the unfolding of some more or less independent compact cooperative structure, that is, two domains of this protein. Studies of various mutants showed that replacements of amino acid residues in the part of the molecule containing the β-barrel leads to a shift of the second endotherm, while replacements of amino acid residues in the α-helical part induces a shift of the first endotherm. The fragment of staphylococcal nuclease that lacks 13 amino acid residues from the C-terminal end shows a simple heat absorption peak in the temperature range of the second endotherm (Griko et al., 1994b).

For apomyoglobin it was found that upon decreasing the pH in solutions containing a low concentration of salts apomyoglobin expands, but expansion of its hydrodynamic volume measured by intrinsic viscosity occurs in two distinct stages (Fig. 9.13). The first stage takes place in the pH region from 5.0 to 4.5; the second, from 4.0 to 3.5. The first is the region of protonation of histidine residues, of which there are many in myoglobin; the second is the region of protonation of carboxyl groups.

With decreasing pH, not only does the hydrodynamic volume of apomyoglobin change, but also its ellipticity. It is particularly interesting that the ellipticity in the near UV drops to zero in the first stage of the volume expansion, that is, in the pH range 5.0 to 4.5, while the ellipticity in the far UV decreases only partly at this stage (Fig. 9.14). Thus, in the pH region from 4.5 to 4.0 apomyoglobin is in the state that was classified as a molten globule: it is compact, it retains some helicity, and it lacks tertiary structure according to the near-UV circular dichroism (CD) criterion. Further decrease of pH leads to complete unfolding of apomyoglobin. It unfolds also upon increase in temperature, and its intrinsic viscosity and ellipticity reach values that are characteristic of polypeptide chain in 8 M urea solution.

The problem is that, although apomyoglobin in the pH range 4.5 to 4.0 does not show ellipticity in the near UV, according to NMR studies, it has elements of ordered

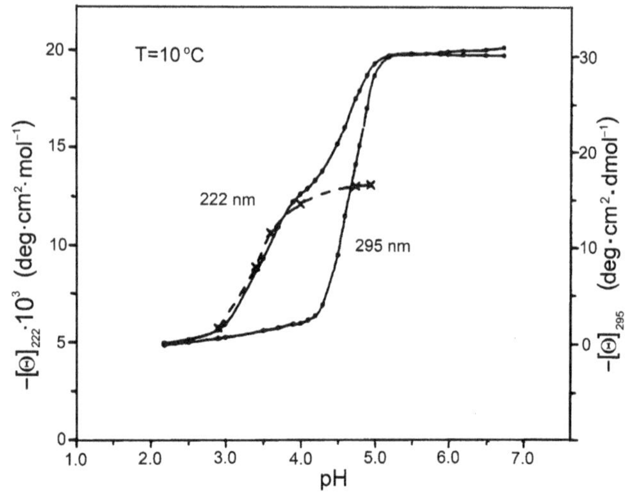

Figure 9.14. pH dependence of the ellipticity of apomyoglobin in the near and far UV (Griko and Privalov, 1994).

structure (Hugheson et al., 1990). In this state some of the helices (A, G, H, and partly F) of the native myoglobin are still preserved and do not exchange their NH protons. It is notable that these helices are also the most packed in the native myoglobin, forming a distinct compact subdomain. When isolated as fragments, these helices are unstable (Hugheson et al., 1991; Barrick and Baldwin, 1993; Waltho et al., 1993). Therefore, one can assume that in the compact intermediate state of apomyoglobin at acidic pH these helices are stabilized by some long-range interactions. The part of apomyoglobin containing A, G, and H helices has been isolated by limited proteolysis and was shown to unfold upon acid titration in exactly the same pH region as the molten globule apomyoglobin (De Sanctis et al., 1994). According to Jennings and Wright (1993) the amide protons of residues from the A, G, and H helices are fully protected within 6 msec after exposure to refolding conditions. Since these helices in isolation are only marginally stable under folding conditions, their stabilization in the compact intermediate could be due to some specific long-range interactions. Thus, in the part of apomyoglobin corresponding to the densely packed subdomain on the native myoglobin, the secondary and tertiary structures appear simultaneously at the very first stage of folding. What occurs later is not the rearrangement of the formed structure but folding of the rest of the polypeptide chain, namely the helix, C–D loop, and E helix, which takes several seconds, perhaps because it proceeds in a compact state that does not prevent inappropriate interactions between protein groups.

It appears thus that these four examples, which were considered as the most supportive of the molten globule concept, in fact demonstrate folding–unfolding of two-domain proteins proceeding in two distinct cooperative stages. Unfolding of the multidomain protein is considered in Chapter 10.

9.3. FORMATION OF PROTEIN STRUCTURE

9.3.1. Transient State in Protein Folding

In view of the preceding discussion, the folding of small proteins is of special interest—proteins smaller than apomyoglobin, or even equine lysozyme and staphylococcal nuclease; that is, the proteins that do not contain subparts which might unfold–refold as independent domains. Detailed calorimetric studies of equilibrium unfolding of such proteins show that they unfold without noticeable stable intermediates. Examples are SH3 domain (M_w 6.6×10^3, Viguera et al., 1994), BPTI (M_w 6.5×10^3, Makhatadze et al., 1993), chymotrypsin inhibitor (M_w 9.2×10^3, Jackson and Fersht, 1991), eglin-c (M_w 8.0×10^3, Bae and Sturtevant, 1995), Protein G (M_w 6.6×10^3, Alexander et al., 1992), tendamistat (M_w 7.9×10^3, Renner et al., 1992), ubiquitin (M_w 8.4×10^3, Wintrode et al., 1994), ribonuclease T_1 (M_w 11.0×10^3, Yu et al., 1994), barnase (M_w 12.3×10^3, Griko et al., 1994c). In all these cases the temperature-induced unfolding is approximated almost perfectly by a two-state transition.

Udgaonkar and Baldwin (1995) concluded that an early folding intermediate of ribonuclease A has a well-developed tertiary structure with the same protection against hydrogen exchange as the native proteins and specifically binds 2′ CMP. Furthermore, a single rate-limiting step breaks the entire network of peptide hydrogen bonds and causes overall unfolding of ribonuclease A (Kiefhaber and Baldwin, 1995). According to Matouschek et al. (1992), on folding of barnase the major hydrophobic core begins to form in the early stage together with some β-sheet, which is an element of tertiary structure. Kinetic experiments of Fersht and colleagues (Otzen et al., 1994) on the folding of chymotrypsin inhibitor led to the conclusion that tertiary structure coalesced consistently with the rearrangement of a "globally collapsed" structure. According to these authors, the transient state of barnase appeared to be an expanded form of the native state in which secondary and tertiary elements have been concurrently partly formed.

It thus appears that the native structure of small single-domain proteins folds very rapidly if the polypeptide chain is not trapped in a misfolded form at some inappropriate folding conditions. This process does not consist of two independent stages, of fast collapse and slow rearrangement, but proceeds as a single process of condensation of the polypeptide chain into a native format which then grows. As discussed earlier, a tightly packed cluster of groups might serve as a folding nucleus, which is the most stable format in the protein interior and most probably folds first (see Section 9.1.3). If condensation of a protein leads to a near-native arrangement of its major structural elements, subsequent folding to the native state may occur within milliseconds. All intermediates in this process are extremely unstable, transient. At some stage they might appear as a highly dynamic molten globule, but it is certainly a transient situation and not a specific thermodynamic state of the protein macromolecule.

> *What is therefore important for folding is not the collapse of the polypeptide chain into a globule, but condensation of the chain into a nativelike structure (Creighton, 1994; Creighton et al., 1996).*

9.3.2. Mechanism of Cooperation

As follows from the above, the most important factor in maintaining the native protein structure is not the value of individual secondary interactions in the protein interior (which all are much lower than the energy of thermal motion) but their extensive cooperation. According to current understanding, the extreme cooperativity, which is specific for the native proteins, proceeds from a tight and unique packing of groups in the protein interior providing specific long-range interactions between the groups of the polypeptide chain (Ikegami, 1977; Abkevich et al., 1995; P. Privalov, 1979, 1989, 1992; Shakhnovich and Finkelstein, 1989; Karplus and Shakhnovich, 1992; Bromberg and Dill, 1994).

After Kauzmann (1959) it was believed that the essential role in folding of polypeptide chain into a compact globule is played by the hydrophobic interactions expelling the apolar groups from the entropically unfavorable contacts with water and thus forming the apolar core of globular protein. However, as we discussed (see Sections 2.4.2 and 8.3.3), the hydrophobic effect, which is responsible for separation of apolar groups from water, is provided not by the unwillingness of water to be ordered in the presence of apolar groups, but by the enthalpy of van der Waals interactions between apolar groups, while the hydration of these groups has just the opposite action. However, these two components of the hydrophobic effect operate at different ranges: the van der Waals interactions are short range, while the hydration effect, water ordering, manifests at a range exceeding the size of water molecule. As a result, what is regarded as a "hydrophobic effect" responsible for separation of apolar groups from water, is a biphasic action that is attractive at the short distances between apolar groups and is repulsive at the longer distances. It looks as though just this biphasic character of the "hydrophobic effect" might be one of the major factors responsible for the extreme cooperativity of this molecular system.

Thus, protein folding should be regarded not as expelling apolar groups from the contacts with water, but as expelling water from the apolar protein interior (P. Privalov and Gill, 1988). The consequence of expelling water from the protein interior is an increase of the probability of hydrogen bonding between the polar groups. It is essential that the polar pair held by a hydrogen bond figures actually as a nonpolar unit in the protein interior, which can be surrounded by apolar groups isolating it from contacts with the polar water molecules. Screening of the polar pair from water by the nonpolar groups, in fact placing it in surroundings with low dielectric constant, increases the strength of hydrogen bonding and this enhances the cooperativity of all this process of polypeptide chain folding. On the other hand, when there is a decrease of tight packing of apolar groups, water penetrates into the protein interior, where it competes with the polar groups for hydrogen bonding and thus destabilizes the compact protein structure.

As is clear, this extreme cooperativity could be realized only when all groups of the polypeptide chain are in a certain configuration. All other configurations of these groups must be far less stable and hence will be rapidly replaced, until only the correct configuration is reached, which corresponds to the native state. This is especially clear with the example of the polar groups along the polypeptide chain, which, if not correctly placed, cannot form hydrogen bonds in the nonpolar surroundings. However small the positive contribution of a hydrogen bond might be in the maintenance of protein structure, the negative contribution of unpaired polar groups in the protein interior would be very large, about 50 kJ/mol; that is, it amounts to the entire energy of the protein structure stabilization.

As follows from this discussion, the lower limit on the size of the cooperative domain is determined by the requirement that it should be stable. To be sufficiently stable at physiological temperatures, the Gibbs energy of protein stabilization should be at least one order magnitude higher than the energy of thermal motion at that temperature; that is, a cooperative domain should include at least 50 amino acid residues. The upper size limit of the cooperative unit is likely to be determined by difficulties in the formation of a completely integrated cooperative aperiodic structure, which increase rapidly with increasing number of amino acid residues that should cooperate. Therefore, the cooperative domain usually does not include more than 200 amino acid residues; that is, 20 kDa is the upper limit for the mass of a globular protein (P. Privalov, 1982).

Thus, aperiodicity and discreteness are closely interrelated properties of a protein: the highly ordered aperiodic structure must be discrete, while a discrete structure cannot be regular—it must be aperiodic (P. Privalov, 1989).

These two interrelated properties of protein proceed from the ability of weak intramolecular forces to cooperate. Studying the detailed mechanism of their cooperation is, perhaps, the most pressing physical problem in protein science. The main problem that arises in modeling such systems is that:

The contributions of various factors to protein stabilization are not additive: it is the nonadditivity of various factors participating in protein stabilization that turns these macromolecules into cooperative systems.

9.3.3. Thermodynamic States of Proteins

A single protein molecule consisting of thousands of atoms actually represents a quasi-macroscopic system and thus its various states could be qualified as its various thermodynamic states. The term "thermodynamic state of a macroscopic system" is usually understood to mean the state that differs significantly from all other states of this system in its extensive thermodynamic parameters, such as the enthalpy, entropy, volume—that is, the first derivatives of the thermodynamic potential. Transition from one thermodynamic state into another thermodynamic state therefore should proceed with a drastic change of these parameters. In other words, it is associated with a discontinuity of the first derivatives of thermodynamic potential. After Ehrenfest such a change of a macroscopic state is called a **first-order phase transition**. Changes of the macroscopic system that proceed with a change in the second derivative of the thermodynamic potential, that is, the heat capacity or compressibility, are considered as **second-order phase transitions**. The first-order phase transition usually proceeds with change of the symmetry of the macroscopic system (Landau and Lifshits, 1958). An obvious example of a first-order phase transition is the melting of ice, which presents transformation of a crystal into liquid and proceeds with drastic change of enthalpy, entropy, and volume. An example of a second-order phase transition is the transformation of a liquid into a gas at the critical point. For water this critical point is at high temperature and pressure. At the critical point the difference between the expanded liquid and the highly compressed gas disappears. Because of its high compressibility, a macroscopic system in the critical state fluctuates dramatically.

Considering the protein macromolecule as a macroscopic system, one can define its definite thermodynamic states as states that are separated by cooperative transitions. The native state of a small compact globular protein is a thermodynamic state since it is separated from the denatured state by a highly cooperative transition with drastic change in enthalpy and volume, that is, the first derivatives of thermodynamic potential. Thus, it can be regarded as a first-order phase transition (P. Privalov, 1979, 1989). The native and denatured states of protein represent different phases that differ in symmetry, that is, in order. Therefore, one phase cannot be transformed into the other gradually; these phases can only be transformed abruptly and entirely. From this point of view, the folded cooperative structure of a globular protein resembles a crystal, but a crystal of a critical size, a nucleus, which exists or does not exist as a whole, but cannot melt gradually, as all its intermediate states are highly unstable.

The native and denatured states of globular protein are sensitive to environmental conditions and can change significantly upon their variation. However, these changes of the native protein or of the denatured protein cannot be considered as transitions between the different thermodynamic states. Consequently, they cannot be classified as first-order phase transitions; correspondingly, these different forms of protein cannot be considered as different phases, that is, as different thermodynamic states. In particular, the different forms of the denatured protein cannot be regarded as different phases even if they differ significantly in some characteristics such as hydrodynamic volume or ellipticity, because these are not the derivatives of thermodynamic potential. From a thermodynamic point of view, all such forms represent the same macroscopic state and variation of the external conditions just results in redistribution of the population of microscopic states within the same macroscopic state. This redistribution of population of the microscopic states manifests itself in a gradual change in the protein characteristics, in contrast to the qualitative change that occurs at a phase transition.

According to the statistical mechanics of polymers, the collapse of a polypeptide chain into a compact liquidlike globule should not represent a first-order phase transition, but a second-order phase transition, which proceeds with a change in the second derivative of the thermodynamic potential, that is, the heat capacity and compressibility (Lifshits et al., 1978; Bryngelson and Wolynes, 1987; Chen and Prohofsky, 1994).

REFERENCES

Abkevich V.L., Gutin A.M., and Shakhnovich E.L. (1995). *J. Mol. Biol.*, **252**, 460–471.

Alexander P., Fahnestock S., Lee T., Orban J., and Bryan P. (1992). *Biochemistry*, **31**, 3597–3603.

Anfinsen C.B. (1972). *Biochem. J.*, **128**, 737–749.

Anfinsen, C.B. (1973). Principles that govern the folding of protein chain (Nobel Lecture). *Science*, **181**, 223–230.

Bae S.-J. and Sturtevant J.M. (1995). *Biophys. Chem.*, **55**, 247–252.

Barrick D. and Baldwin R.L. (1993). *Protein Sci.*, **2**, 869–876.

Bhuyan A.K. and Udgaonkar J.B. (2001). *J. Mol. Biol.*, **312**, 1135–1160.

Bromberg S. and Dill K.A. (1994). *Protein Sci.*, **3**, 997–1009.

Bryngelson J.D. and Wolynes P.G. (1987). *Proc Natl. Acad. Sci. USA*, **84**, 7524–7528.

Carra J.H. and Privalov P.L. (1995). *Biochemistry*, **34**, 2034–2041.

Carra J.H. and Privalov P.L. (1996). *FASEB J.*, **10**, 67–74.

Carra J.H., Anderson E.A., and Privalov P.L. (1994). *Biochemistry*, **33**, 10842–10850.

Chan H.S. and Dill K.A. (1991). *Annu. Rev. Biophys. Biophys. Chem.*, **20**, 447–490.

Chen Y.Z. and Prohofsky E.W. (1994). *Biophys. J.*, **66**, 202–206.

Creighton T.E. (1994). *Nature Struct. Biol.*, **1**, 135–138.

Creighton T.E., Darby N.J., and Kemmink J. (1996). *FASEB J.*, **10**, 110–118.

De Sanctis G., Ascoli F., and Brunori M. (1994). *Proc. Natl. Acad. Sci. USA*, **91**, 11507–11511.

Dill K.A. (1990). *Biochemistry*, **29**, 7133–7155.

Dill K.A. and Shortle D. (1991). *Annu. Rev. Biochem.*, **60**, 795–825.

Englander S.W. and Mayne L. (1992). *Annu. Rev. Biophys. Biomol. Struct.*, **21**, 243–265.

Finkelstein A.V. and Shakhnovich E.I. (1989). *Biopolymers*, **28**, 1681–1689.

Finney J.L. (1975). *J. Mol. Biol.*, **96**, 721–732.

Flanagan J.H., Kataoka M., Fujisawa T., and Engelman D.M. (1993). *Biochemistry*, **32**, 10359–10370.

Griko Yu.V. and Privalov P.L. (1994). *J. Mol. Biol.*, **235**, 1318–1325.

Griko Yu.V., Freire E., and Privalov P.L. (1994a). *Biochemistry*, **33**, 1889–1899.

Griko Yu.V., Freire E., Privalov, G.P., Van Dael H., and Privalov P.L. (1995). *J. Mol. Biol.*, **252**, 447–459.

Griko Yu.V., Gittis A., Lattman E.E., and Privalov P.L. (1994b). *J. Mol. Biol.*, **243**, 93–99.

Griko Yu.V., Makhatadze G.I., Privalov P.L., and Hartley R.W. (1994c). *Protein Sci.*, **3**, 669–676.

Griko Yu.V., Privalov P.L., Venyaminov S.Yu., and Kutyshenko V.P. (1988). *J. Mol. Biol.*, **202**, 127–138.

Harpaz Y., Gerstein M., and Chothia C. (1994). *Structure*, **2**, 641–649.

Hugheson F.M., Barrick D., and Baldwin R.L. (1991). *Biochemistry*, **30**, 4113–4118.

Hugheson F.M., Wright P.E., and Baldwin R.L. (1990). *Science*, **249**, 1544–1548.

Ikegami A. (1977). *Biophys. Chem.*, **6**, 117–130.

Jackson S.E. and Fersht, A.R. (1991). *Biochemistry*, **30**, 10428–10435.

Janin J. and Chothia C. (1990). *J. Biol. Chem.*, **265**, 16027–16030.

Jeng M.F. and Englander S.W. (1991). *J. Mol. Biol.*, **221**, 1045–1061.

Jennings P.A. and Wright P.E. (1993). *Science*, **265**, 892–896.

Karplus M. and Shakhnovich E.I. (1992). In: *Protein Folding* (T.E. Creighton, Ed.), pp. 127–195. Freeman, New York.

Kauzmann W. (1959). *Adv. Protein Chem.*, **14**, 1–63.

Kiefhaber T. and Baldwin R.L. (1995). *Proc. Natl. Acad. Sci. USA*, **92**, 2657–2661.

Krishna M.M., Maity H., Rumbley J.N., Lin Y., and Englander S.W. (2006). *J. Mol. Biol.*, **359**, 1410–1419.

Kuwajima K., Nitta K., Yoneyama M., and Sugai S. (1976). *J. Mol. Biol.*, **106**, 359–373.

Landau L.D. and Lifshits E.M. (1958). *Statistical Physics*. Addison-Wesley, Reading, MA.

Levinthal C. (1968). *J. Chem. Phys.*, **65**, 44–45.

Lifshits E.M., Grosberg A.Ya., and Khokhlov A.R. (1978). *Rev. Mod. Phys.*, **60**, 683–713.

Maity M., Krishna M.M., Mayne L., and Englander S.W. (2005). *Proc. Natl. Acad. Sci. USA*, **102**, 4741–4746.

Makhatadze G.I., Kim K.S., Woodward C., and Privalov P.L. (1993). *Protein Sci.*, **2**, 2028–2036.

Matouschek A., Serrano I., and Fersht A.R. (1992). *J. Mol. Biol.*, **224**, 819–835.

Morozova L.A., Haezebrouck P., and Van Cauwelaert F. (1991). *Biophys. Chem.*, **41**, 185–191.

Morozova L.A., Haynie D.T., Arico-Mandel C., Van Dael H., and Dobson C.M. (1995). *Nature Struct. Biol.*, **2,** 871–875.

Otzen D.E., Itzhaki L.S., El Masry N.F., Jeckson S.E., and Fersht A.R. (1994). *Proc. Natl. Acad. Sci. USA*, **91**, 10422–10425.

Peng Z. and Kim P.S. (1994). *Biochemistry*, **33**, 2136–2141.

Privalov G.P. (1995). Packing of protein interior: structure and distribution of packing density. PhD dissertation, The Johns Hopkins University, Baltimore, MD.

Privalov P.L. (1979). *Adv. Protein Chem.*, **33**, 167–241.

Privalov P.L. (1982). *Adv. Protein Chem.*, **35**, 1–104.

Privalov P.L. (1989). *Annu. Rev. Biophys. Biophys. Chem.*, **18**, 47–69.

Privalov P.L. (1992). In: *Protein Folding* (T.E. Creighton, Ed.), pp. 83–126. Freeman, New York.

Privalov P.L. and Gill S.J. (1988). *Adv. Protein Chem.*, **39**, 191–234.

Privalov P.L. and Tsalkova T.N. (1979). *Nature*, **280**, 693–696.

Ptitsyn O.B. (1972). *Pure Appl. Chem.*, **31**, 227–244.

Ptitsyn O.B. (1992). In: *Protein Folding* (T.E. Creighton, Ed.), pp. 243–300. Freeman, New York.

Ptitsyn O.B. (1995a). *Adv. Protein Chem.*, **47**, 83–229.

Ptitsyn O.B. (1995b). *Curr. Opin. Struct. Biol.*, **5**, 74–78.

Ptitsyn O.B. and Uverski V.N. (1994). *FEBS Lett.*, **341**, 15–18.

Redfield C., Smith R.A., and Dobson C.M. (1994). *Nature Struct. Biol.*, **1**, 23–29.

Renner M., Hinz H.-J., Scharf M., and Engels S.J.W. (1992). *J. Mol. Biol.*, **233**, 769–779.

Richards F.M. (1974). *J. Mol. Biol.*, **82**, 1–14.

Richards F.M. (1985). *Methods Enzymol.*, **115**, 440–464.

Richards F.M. and Lim W.A. (1994). *Quart. Rev. Biophys.*, **26**, 423–498.

Shakhnovich E.I. and Finkelstein A.V. (1989). *Biopolymers*, **28**, 1667–1680.

Shortle D. (1993). *Curr. Opin. Struct. Biol.*, **3**, 66–74.

Shortle D. (1996). *FASEB J.*, **10**, 27–34.

Sosnick T.R., Mayne L., Hiller R., and Englander S.W. (1994). *Nature Struct. Biol.*, **3**, 149–156.

Tanford C. (1968). *Adv. Protein Chem.*, **23**, 121–282.

Udgaonkar J.B. and Baldwin R.L. (1995). *Biochemistry*, **34**, 4088–4096.

Van Dael H., Haezebrouck P., Morozova L., Arico-Muandel C., and Dobson C.M. (1993). *Biochemistry*, **32**, 1186–1194.

Viguera A.R., Martinez J.C., Filimonov V.V., Mateo P.L., and Serrano L. (1994). *Biochemistry*, **33**, 2142–2150.

Waltho J.P., Feher V.A., Merutka G., Dyson H.J., and Wright P.E. (1993). *Biochemistry*, **32**, 6337–6355.

Wintrode P.L., Makhatadze G.I., and Privalov P.L. (1994). *Proteins: Struct. Funct. Genet.*, **18**, 246–253.

Yu Y.H., Makhatadze G.I., Pace C.N., and Privalov P.L. (1994). *Biochemistry*, **33**, 3312–3319.

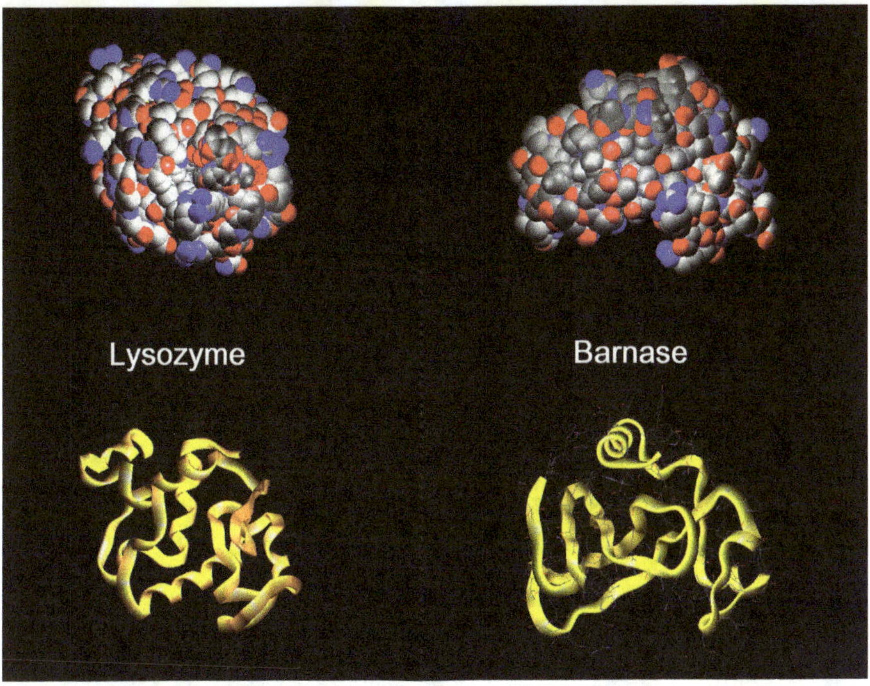

Figure 4.12. The atomic model and the backbone conformation of the typical globular proteins: hen egg white lysozyme and barnase.

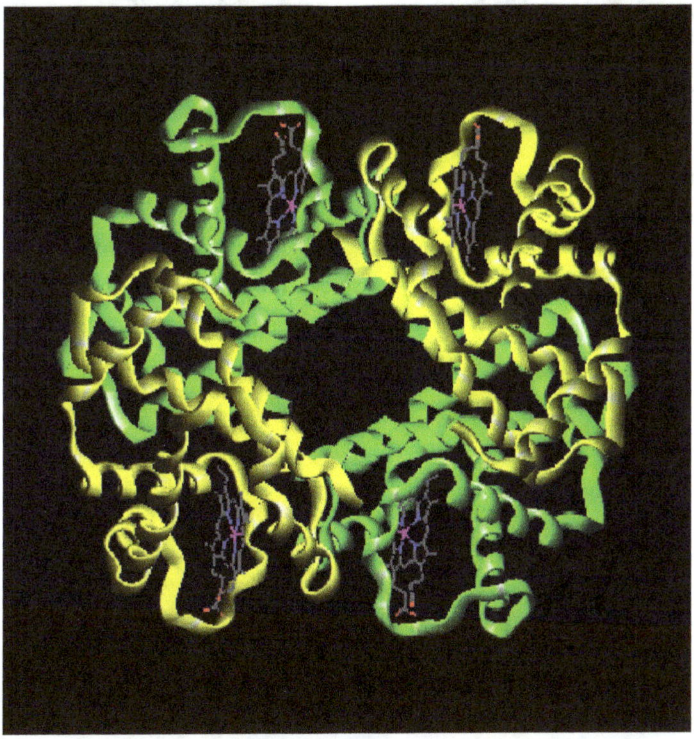

Figure 4.13. Hemoglobin represents a complex of four noncovalently held subunits. Note that these subunits also represent the complexes of protein with the low-molecular-weight heme group.

Microcalorimetry of Macromolecules: The Physical Basis of Biological Structures, First Edition.
Peter L. Privalov.

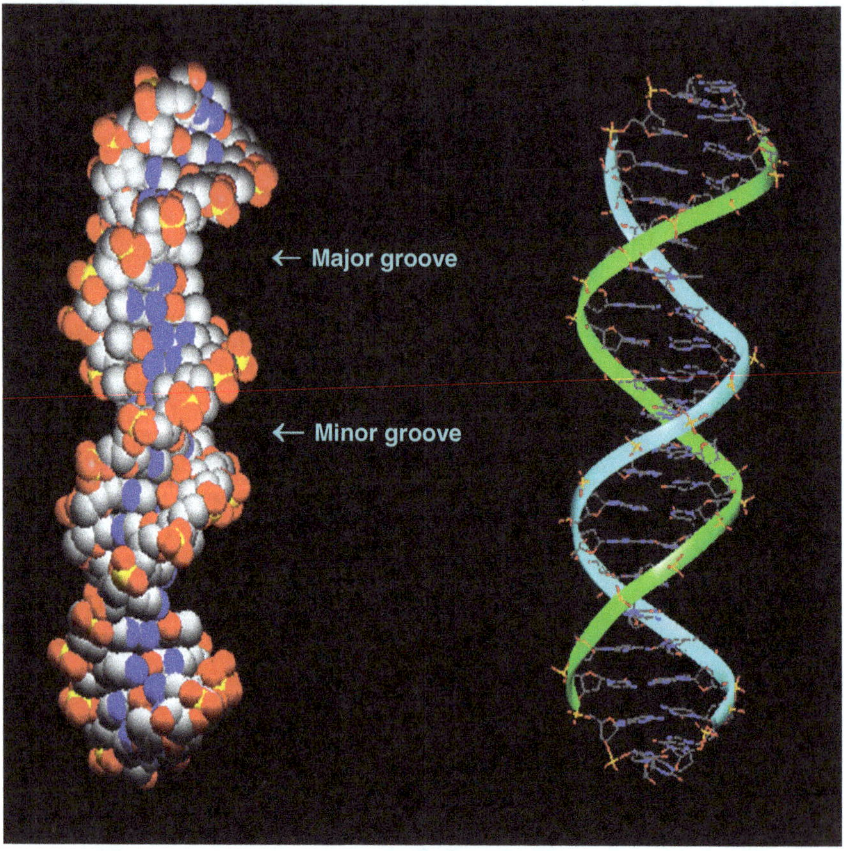

Figure 4.17. The B-DNA double helix.

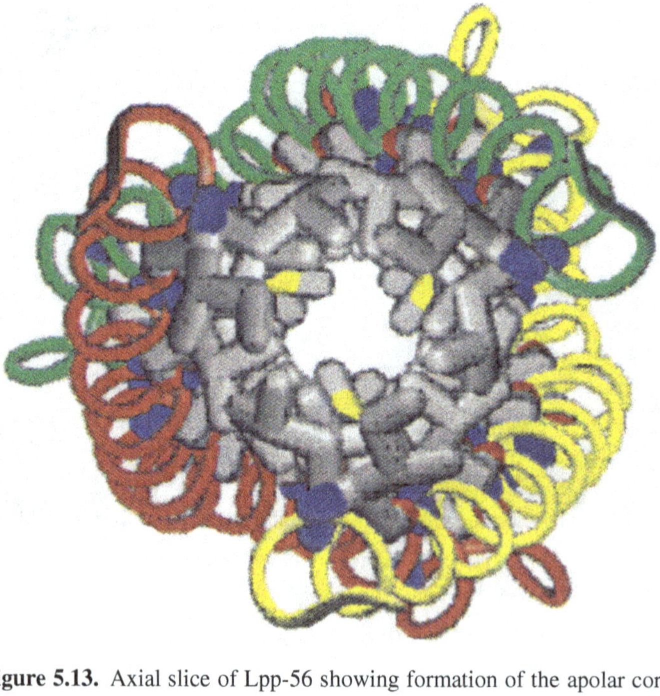

Figure 5.13. Axial slice of Lpp-56 showing formation of the apolar core.

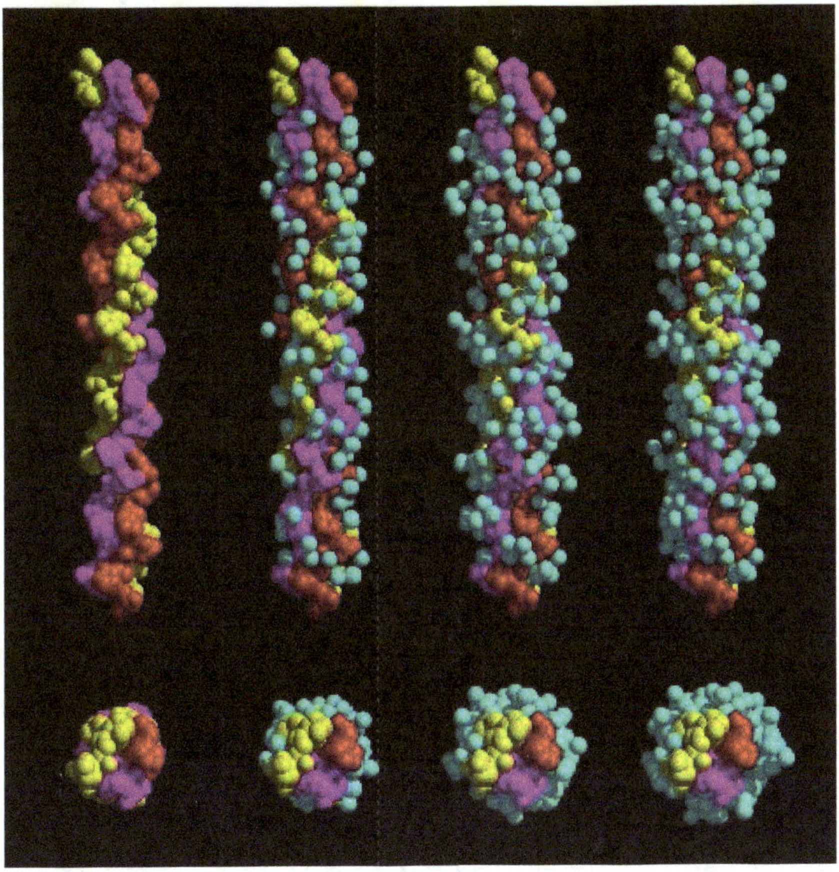

Figure 6.10. Layers of the fixed water molecules covering the collagen triple helix (Bella et al., 1995).

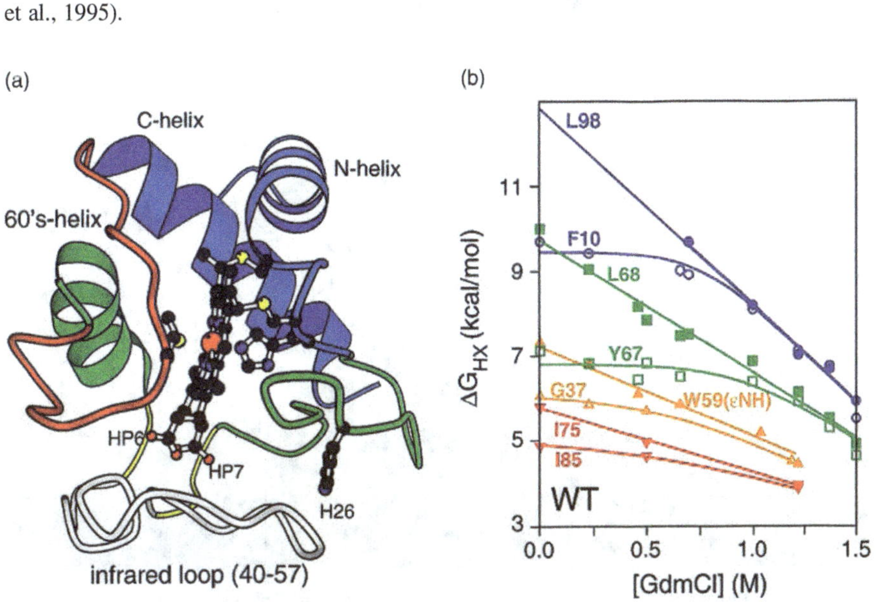

Figure 9.2. Hydrogen exchange by the different structural elements of cytochrome c. (Reproduced from Krishna et al., 2006.)

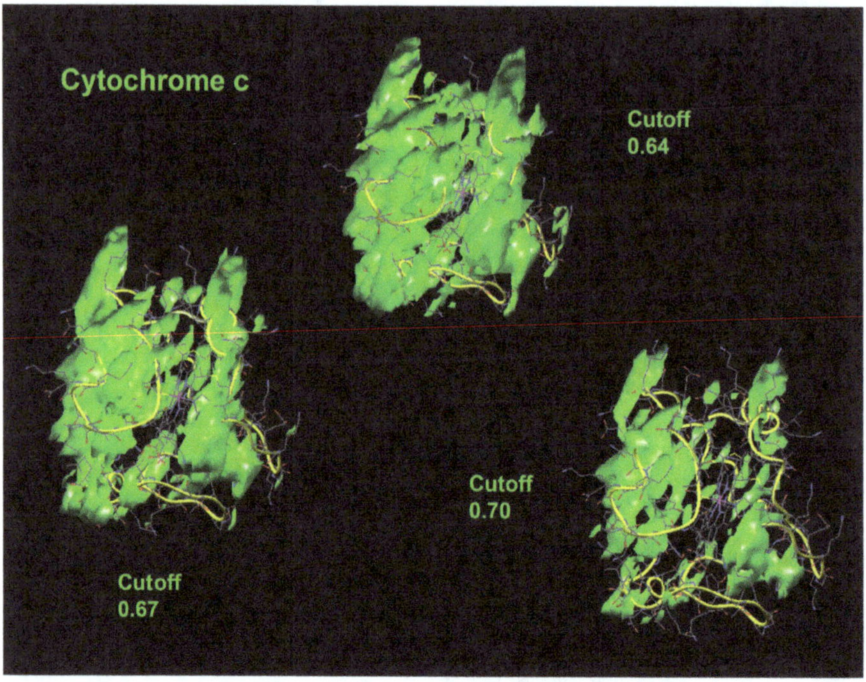

Figure 9.4. Packing density maps of cytochrome c at different cutoffs.

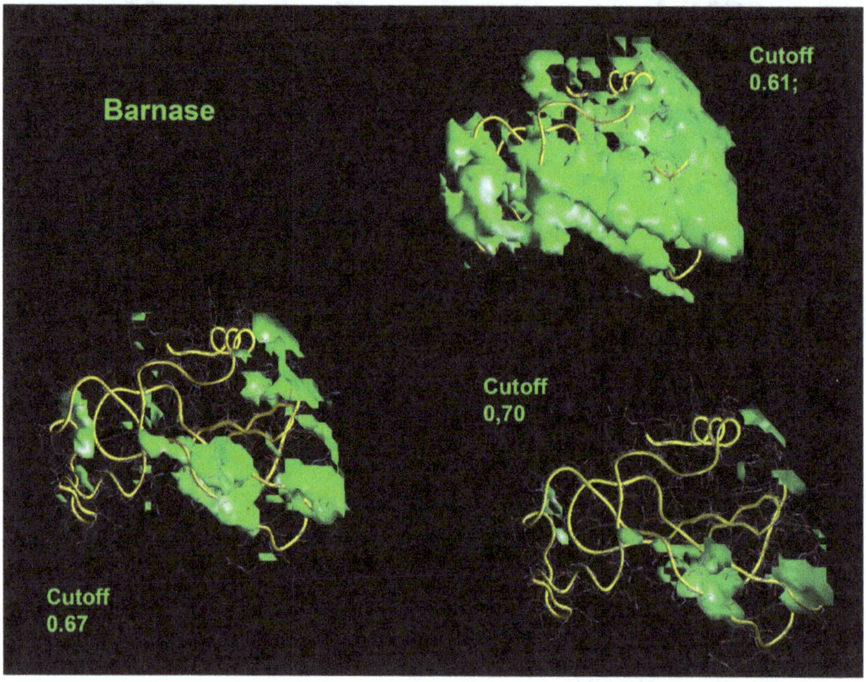

Figure 9.5. Packing density maps of barnase at different cutoffs.

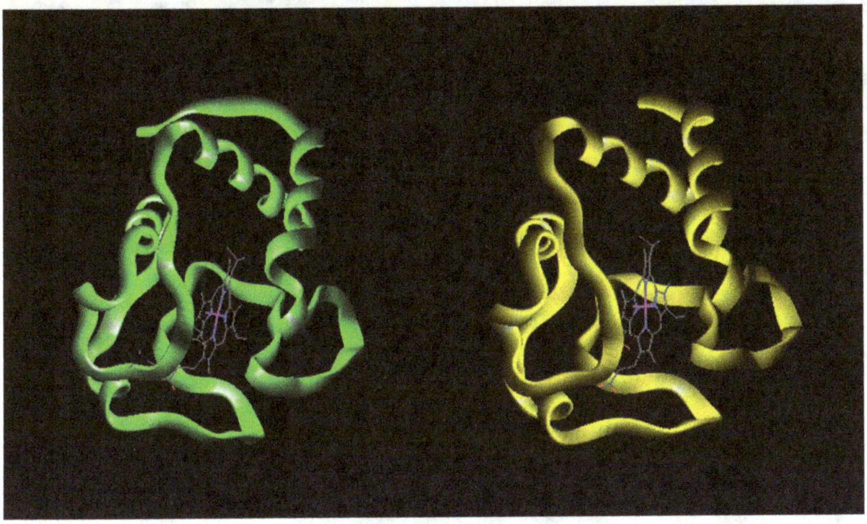

Figure 9.6. Structures of cytochrome c from baker's yeast (green) and horse (yellow) have rather similar tertiary structures despite significant (45%[!]) difference in their primary structures.

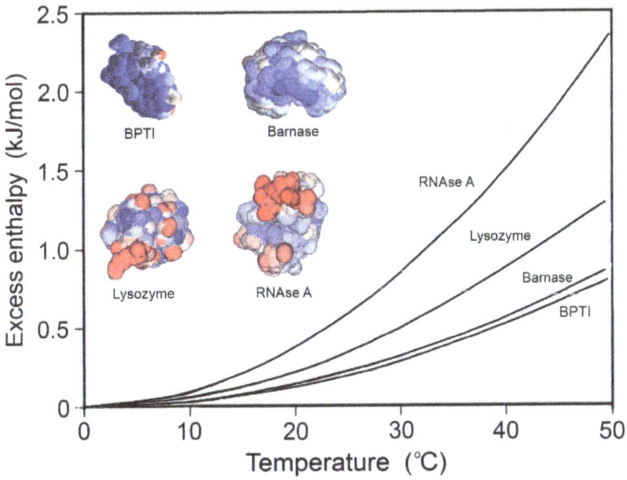

Figure 9.7. The B-factors of four different proteins and their excess enthalpy accumulation with temperature rise. The blue corresponds to low B-factor and red to high B-factor.

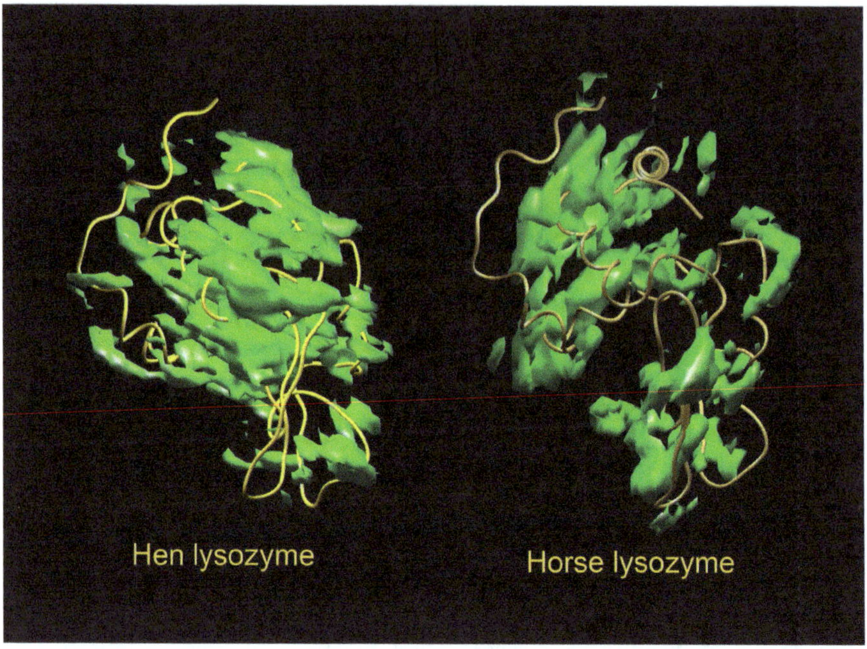

Figure 9.11. Packing density maps of the hen-egg-white and horse (equine) lysozymes at cutoff 0.7, showing that two domains in the hen-egg-white lysozyme merge into a single dense cluster; in equine lysozyme the two domains are separated by a cleft (Griko et al., 1995).

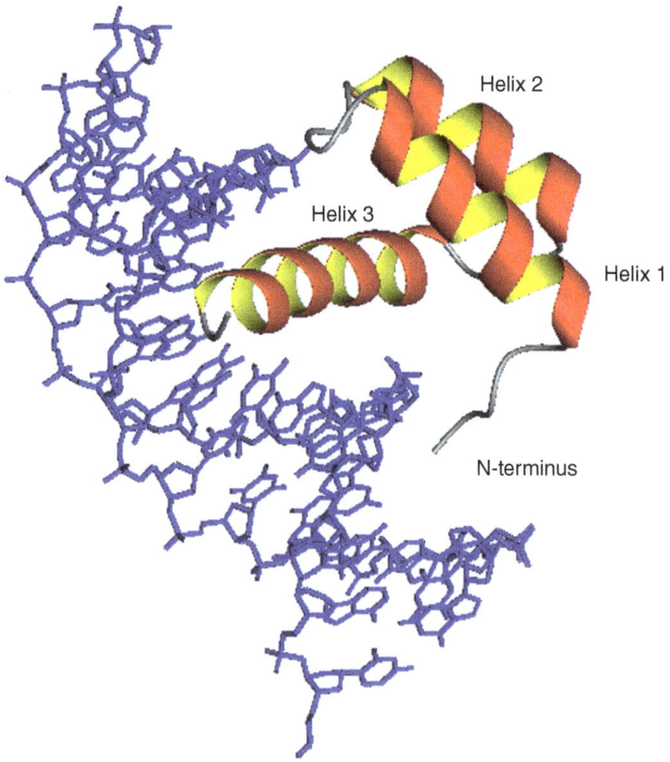

Figure 12.1. Association of the Antennapedia homeodomain with its target DNA. The DNA-recognition helix sits in the major groove of DNA without its essential distortion, while the N-terminal tail enters into the minor groove (PDB 9ANT).

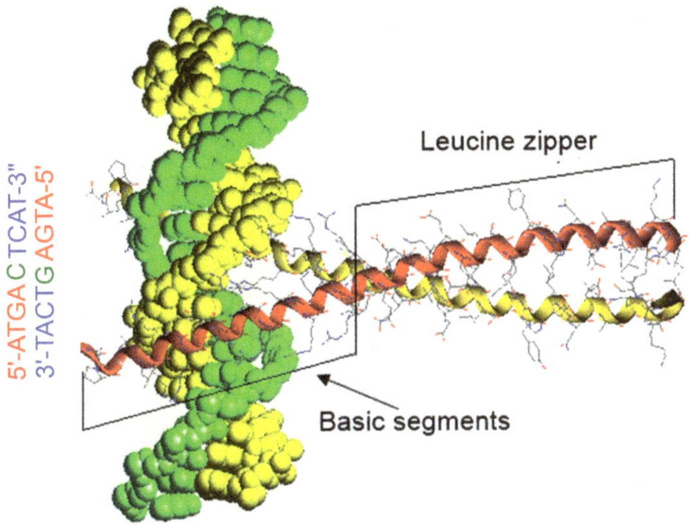

Figure 12.5. The model of bZIP-GCN4 complex with the pseudopalindromic AP-1 DNA showing the DNA-recognizing basic segments entering into the major grooves of DNA.

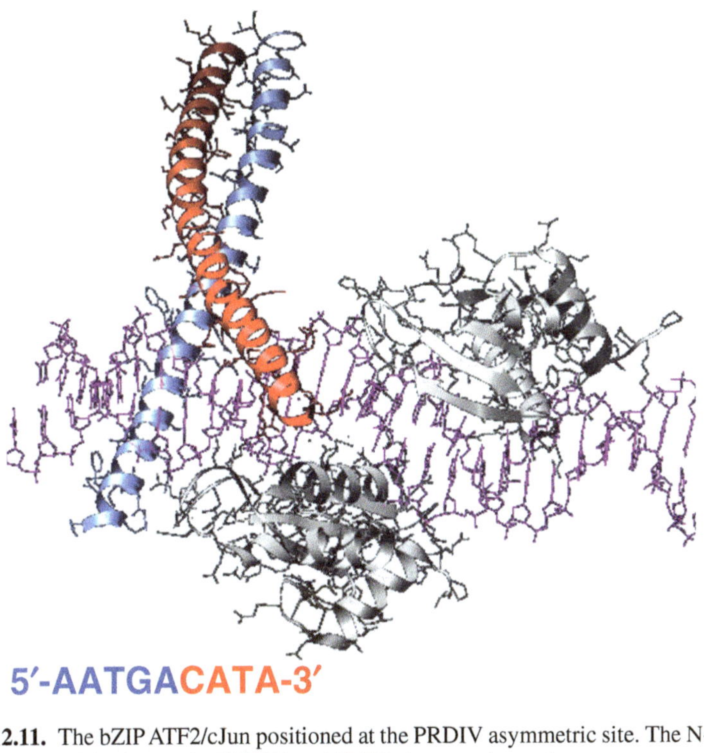

Figure 12.11. The bZIP ATF2/cJun positioned at the PRDIV asymmetric site. The N-terminal domains of IRF-3 bound to the PRDIII and PRDI sites are shown in gray.

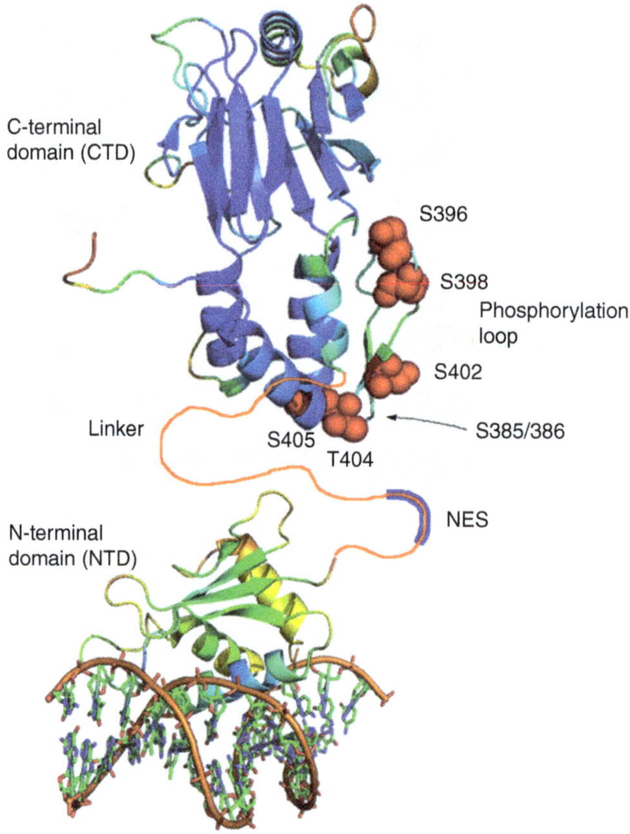

Figure 12.12. Structures of the DNA-binding N-terminal domain and C-terminal dimerization domains of IRF-3 connected by the hypothetical 75-residue linker. The Ser/Thr residues corresponding to the *in vivo* specific sites of phosphorylation are shown in creatinine phosphokinase (CPK) (Dragan et al., 2008).

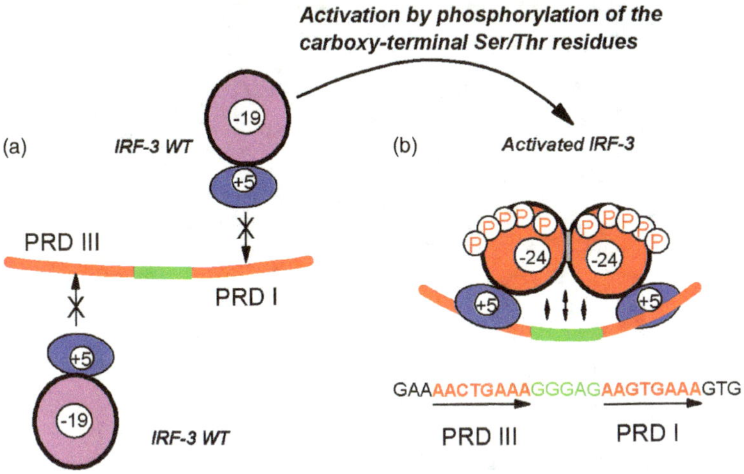

Figure 12.13. Cartoons showing: (a) the independent binding of two monomeric IRF-3 to the PRDI and PRDIII binding sites located on the opposite faces of the IFN-β enhancer DNA; (b) phosphorylation of the IRF-3 resulting in its dimerization and binding as a dimer to the PRDI and PRDIII sites which should result in bending of the shown in green spacer.

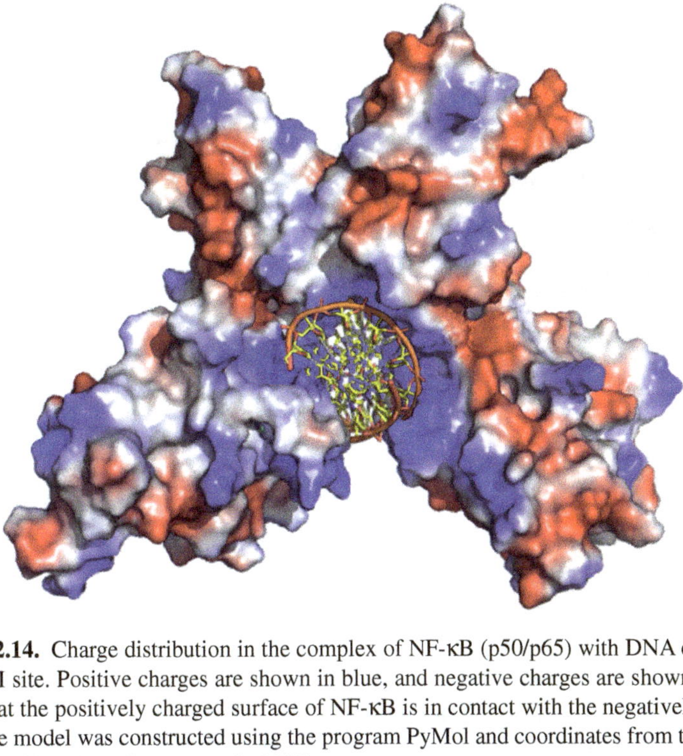

Figure 12.14. Charge distribution in the complex of NF-κB (p50/p65) with DNA containing the PRDII site. Positive charges are shown in blue, and negative charges are shown in red. It is seen that the positively charged surface of NF-κB is in contact with the negatively charged DNA. The model was constructed using the program PyMol and coordinates from the Protein Data Bank file 1VKx (Dragan et al., 2008).

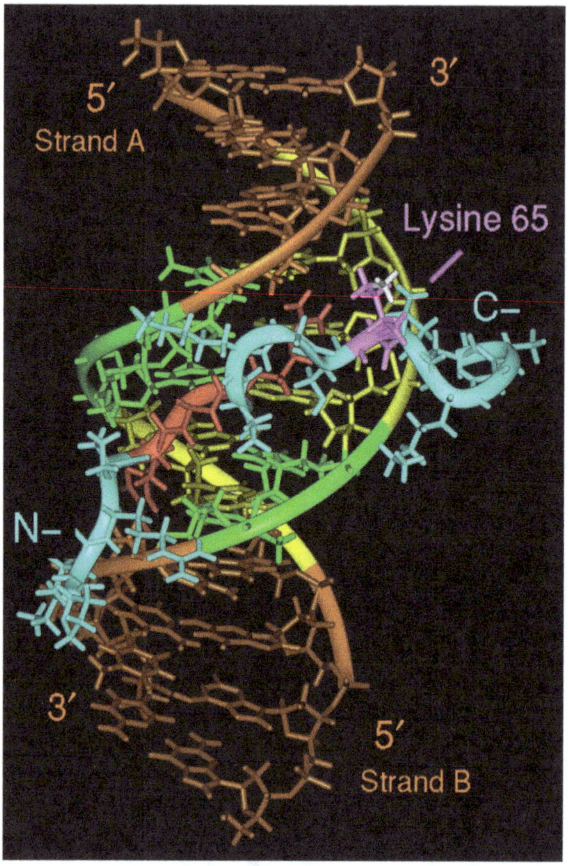

Figure 12.16. Ribbon drawing of the DBD2/DNA complex. The five AT base pairs, which make up the binding site, are shown in yellow (A) and green (T); the rest of the DNA duplex is shown in orange. The central RGR motif, which penetrates deep into the minor groove of DNA, is shown in red; the other amino acids in the peptide are shown in blue (Dragan, 2003b).

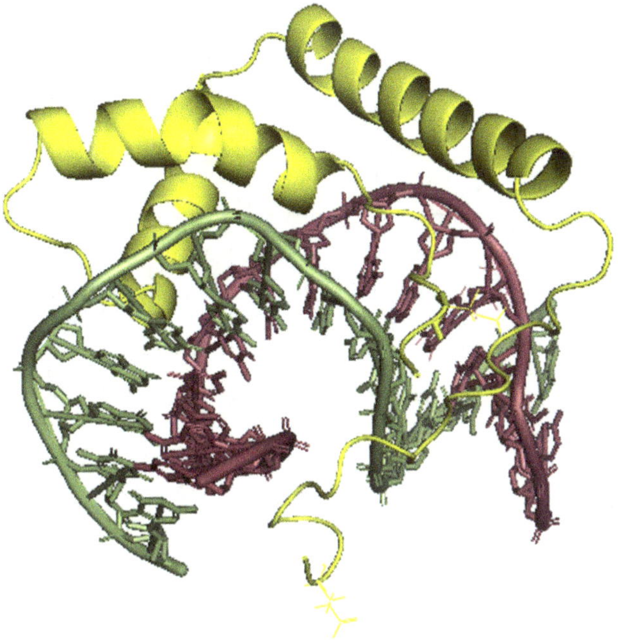

Figure 12.19. HMG DBD sits in the minor groove of DNA and bends it considerably, while its C-terminal basic tail enters into the major groove of DNA.

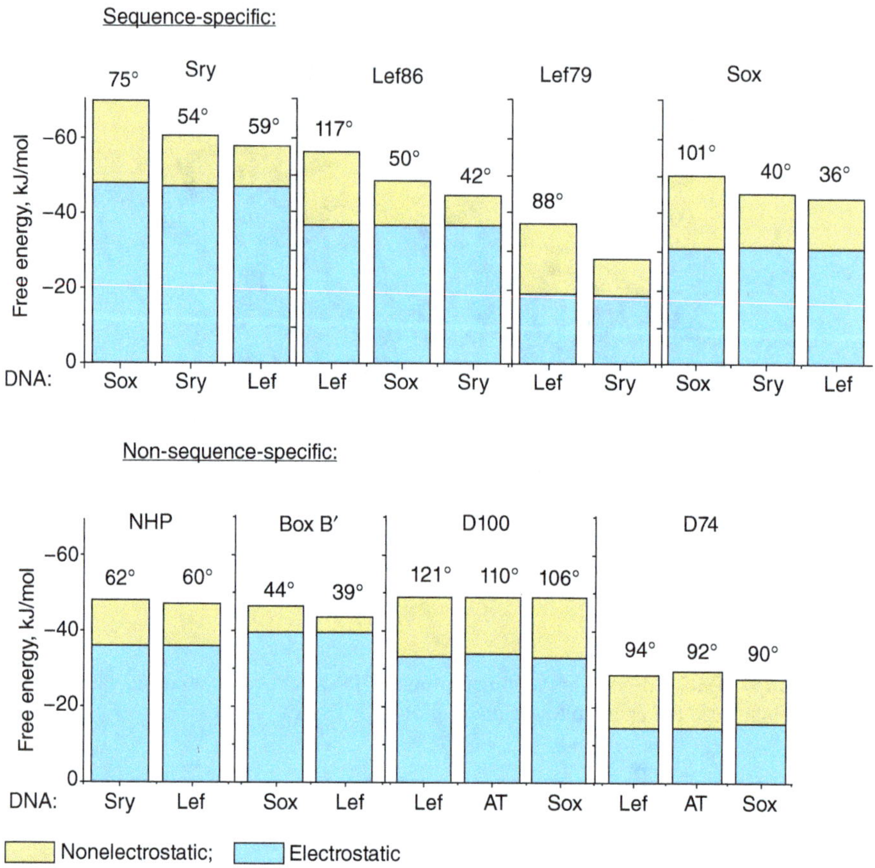

Figure 12.25. Electrostatic (blue) and nonelectrostatic (yellow) components of the total Gibbs free energy of binding the DBDs with various DNAs. The numbers above the bars indicate the induced DNA bend angles measured in the standard buffer (Dragan et al., 2004b).

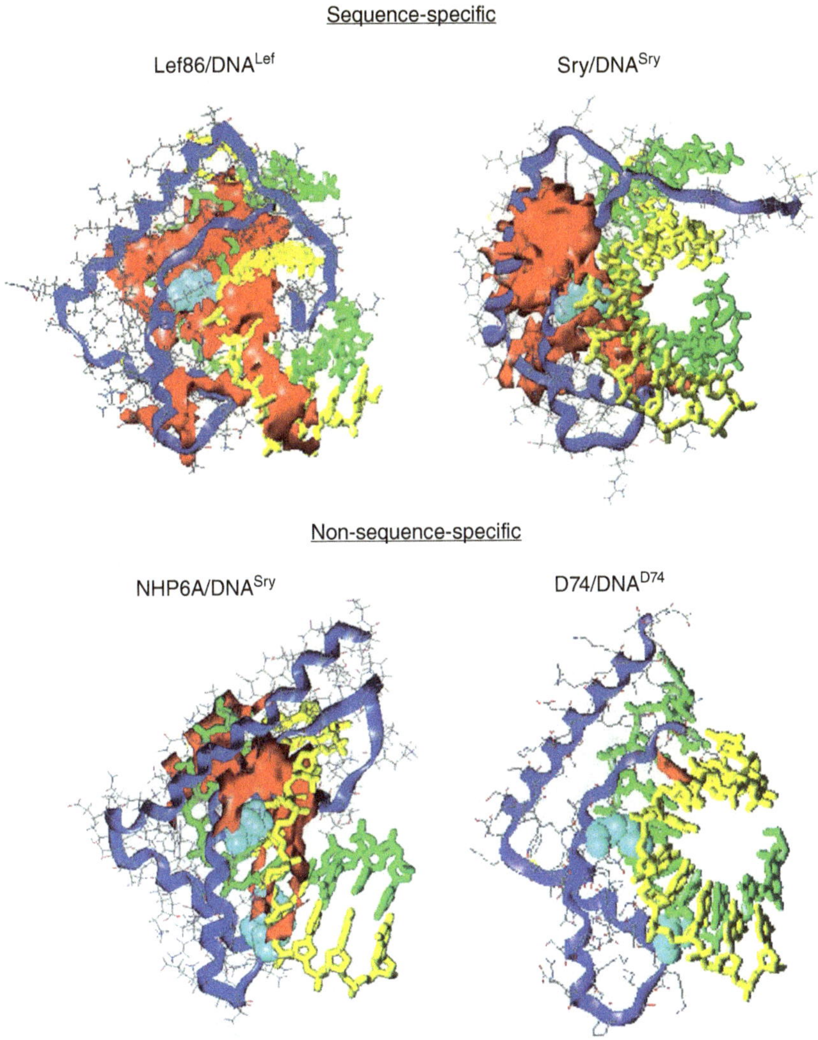

Figure 12.27. Packing densities at the interfaces of the SS complexes of Lef86 and Sry, and of the NSS complexes of NHP6A and D74. The red clusters are regions with packing density higher than 0.68. To help visualize the packing at the interface, only clusters that include groups of both protein and DNA are shown. Residues intercalating into the DNA are shown in pale blue. Although the Lef, Sry, and NHP complex structures were obtained by NMR and that of D74 by crystallography, the striking differences seen in the interfacial packing densities are too large to be explained solely by differences in the constraints applied during the structure determinations (Dragan et al., 2004b).

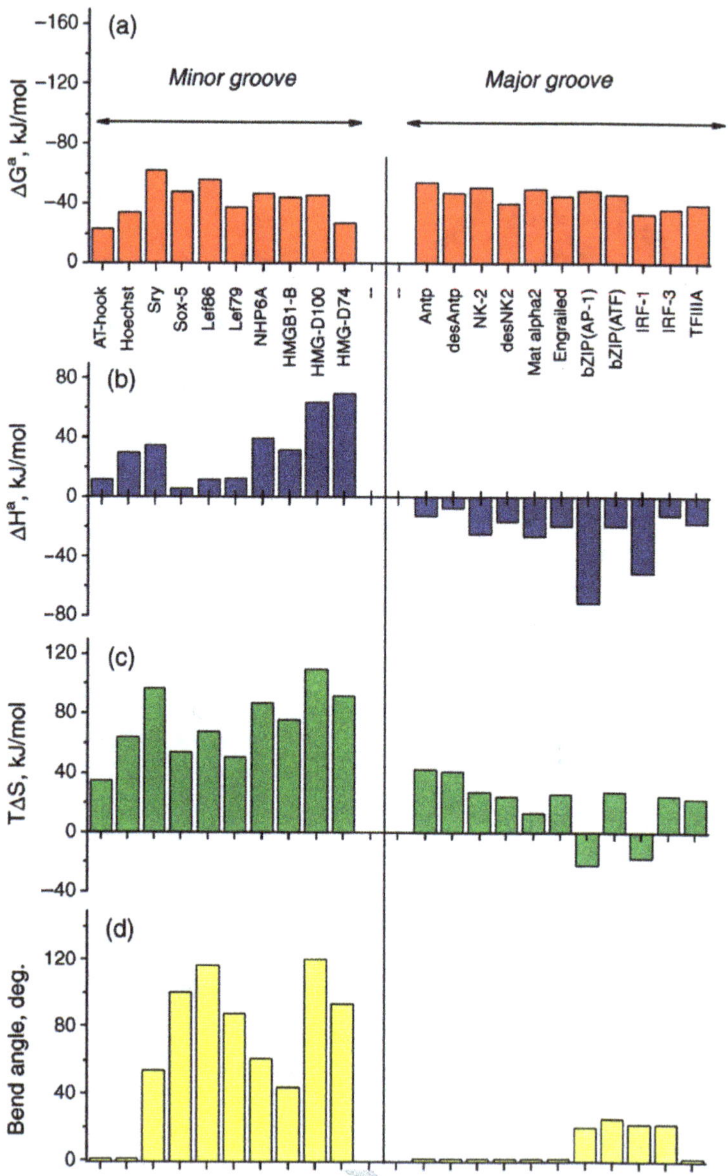

Figure 12.29. Thermodynamic parameters of association for various DBDs to the major and minor grooves of their cognate DNA: (a) Gibbs energies of association; (b) enthalpies of association; (c) entropy factors of association; (d) DNA bending angles (Privalov et al., 2007).

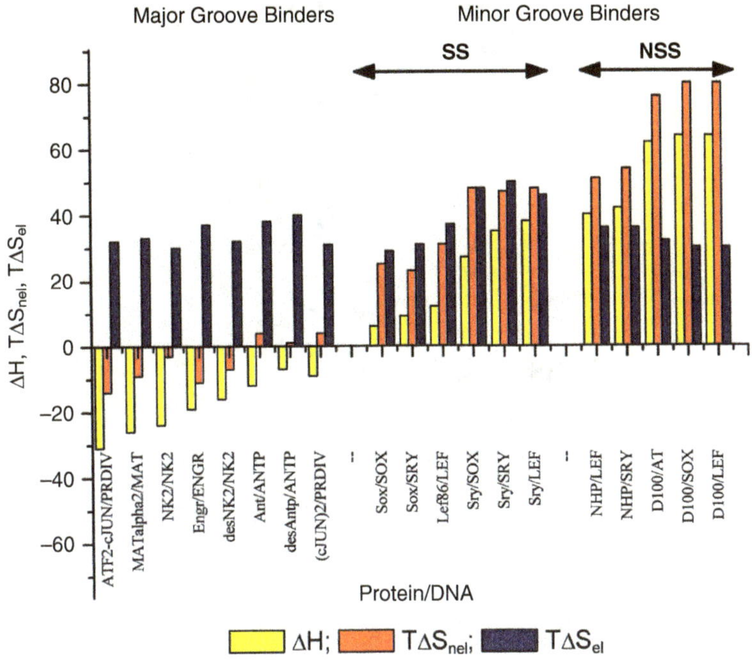

Figure 12.30. Enthalpies and entropy factors (electrostatic and nonelectrostatic) of binding proteins to the minor and major groove of their cognate and semicognate DNAs (Privalov et al., 2011).

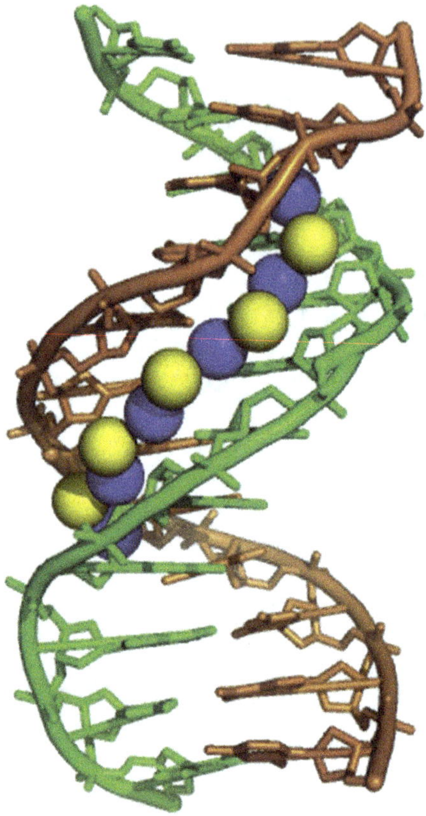

Figure 12.31. Display of primary (blue) and secondary (yellow) layers of the spine of water in the minor groove of the X-link dodecamer CGCAATTCGCG generated from the coordinates of NDB accession number bd0008 (Privalov et al., 2007).

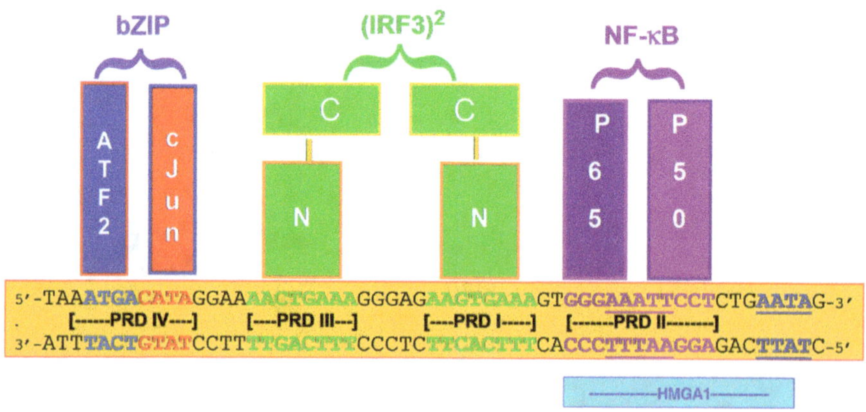

Figure 12.32. Schematic representation of the IFN-β enhanceosome with all its protein components assembled at the enhancer DNA.

10

MULTIDOMAIN PROTEINS

10.1. CRITERION OF COOPERATIVITY

10.1.1. Deviations from a Two-State Unfolding–Refolding

As shown in Chapter 7, upon increasing the denaturing action (e.g., the temperature, denaturant concentration, pH, or pressure) the small compact globular proteins unfold with sharp and simultaneous changes of all their characteristics. The sharpness of protein unfolding under extreme conditions is usually considered as indication of the extreme cooperativity of this process, that is, that it can be approximated as an all-or-none transition neglecting probabilities of all the intermediate states between the native and unfolded. This opens the prospect for thermodynamic analysis of the observed process based on equilibrium studies of protein unfolding and refolding by any means sensible for the state of the protein. In particular, the van't Hoff enthalpy and entropy of a process could be determined from the observed dependence of the optical characteristics (e.g., ellipticity) of a protein on temperature; from the dependence on the activity of some ions, the number of ions released or absorbed by a protein upon unfolding can be determined; from dependence on pressure, the volume change can be determined (see Section 2.2). However, in some

Microcalorimetry of Macromolecules: The Physical Basis of Biological Structures, First Edition.
Peter L. Privalov.

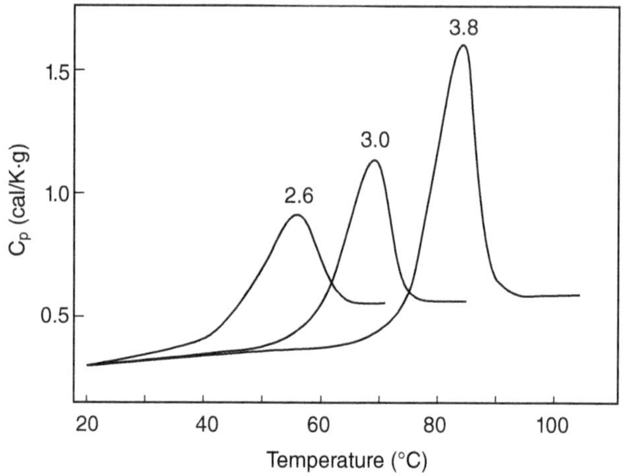

Figure 10.1. Partial specific heat capacity profile of papain in solutions with different pH (Tiktopulo and Privalov, 1978).

cases the sharpness of a process, and even the coincidence of the observed changes of various characteristics of the protein, might be misleading. The classical examples present unfolding of papain and pepsinogen, two rather compact globular proteins.

10.1.2. Papain

Papain is usually considered as a typical globular protein of modest size: its molecular weight is just 23,000 Da. Upon heating it denatures in a rather narrow temperature range with extensive heat absorption (Fig. 10.1). With the change of pH toward neutral the heat absorption peak moves to the higher temperature, showing an increase of protein stability. The observed heat absorption profile is very similar with that of barnase or ubiquitine (see Fig. 7.3): with the increase of protein stability the heat absorption peak becomes sharper, as one would expect for increasing with temperature the transition enthalpy. However, analysis of the heat capacity function of papain shows that the van't Hoff enthalpy of its unfolding, determined from the sharpness of transition, using Equation (3.24), is twice as small as the calorimetric enthalpy determined from the area of heat absorption peak (Table 10.1). This means unequivocally that the papain unfolding does not represent a two-state transition and what is observed as a single heat absorption peak in fact represents the overlap of two rather similar transitions (Tiktopulo and Privalov, 1978).

This is just what one could expect knowing papain structure: it is subdivided by a deep cleft on two lobes (Fig. 10.2). Thus, from calorimetric studies it follows that each of these lobes represents a more or less independent cooperative structural subunit of similar stability. Analysis of the course of the chain in each lobe suggests that they fold independently; that is, they are cooperative folding domains. These two lobes in this aggressive proteolytic enzyme work like a pair of scissors cutting the polypeptide chain entering into the cleft of papain.

TABLE 10.1. The van't Hoff and Calorimetric Enthalpies (in kcal/mol) of Papain and Pepsinogen Denaturation

Papain[a]					Pepsinogen[b]				
pH	T_d	ΔH^{vH}	ΔH^{cal}	$\Delta H^{cal}/\Delta H^{vH}$	pH	T_d	ΔH^{vH}	ΔH^{cal}	$\Delta H^{cal}/\Delta H^{vH}$
2.6	56.0	67	123	1.82	6.0	66.0	174	271	1.56
3.0	69.3	84	168	2.00	6.5	64.5	131	256	1.95
3.4	77.5	109	194	1.78	7.1	57.5	130	233	1.79
3.8	83.8	117	216	1.85	8.0	51.1	122	182	1.49

[a]Tiktopulo and Privalov (1978).
[b]Mateo and Privalov (1981).

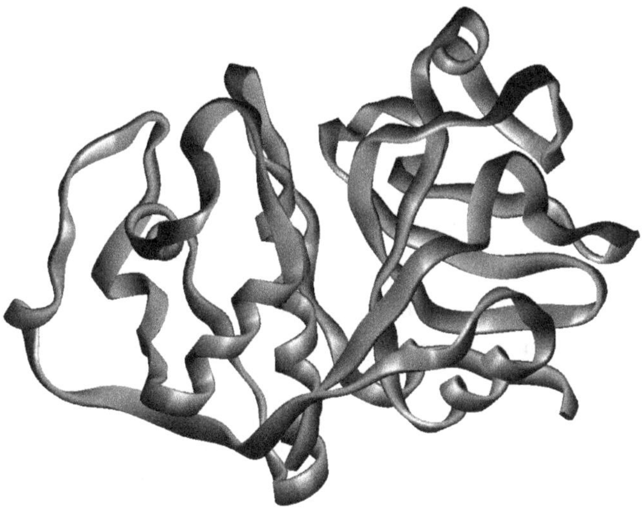

Figure 10.2. Backbone structure of papain showing deep cleft dividing this proteolytic protein on two folding domains.

10.1.3. Pepsinogen

A similar situation was found for pepsinogen, the protein with molecular mass of about 40,000 Da. The heat and urea denaturation of this protein are highly reversible processes proceeding with drastic changes of all its characteristics within a narrow range of denaturant concentration or temperature. The heat capacity profile of thermal denaturation of pepsinogen, as Figure 10.3 shows, does not differ visually from that of small compact globular proteins such as barnase or myoglobin. Therefore, it is obvious that the denaturation of pepsinogen was considered by many authors to be a two-state transition, and under this assumption were calculated the thermodynamic parameters of stabilization of its native structure based on its optically observed unfolding profiles (Ahmad and McPhie, 1978; McPhie, 1980). Analysis of the partial molar heat capacity function of this protein showed, however, that

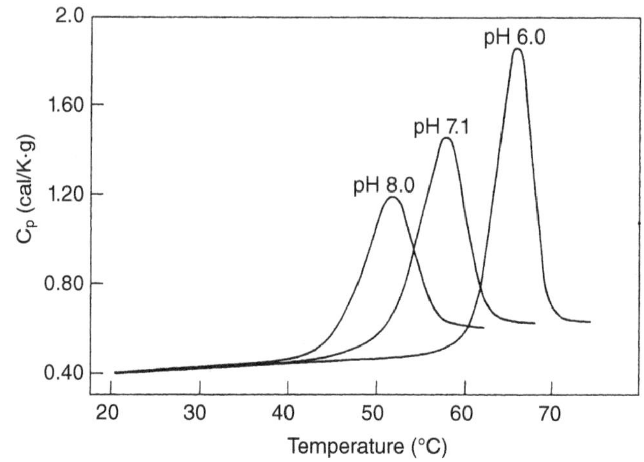

Figure 10.3. Specific partial heat capacity functions of pepsinogen at various pH values of solution (Privalov et al., 1981).

pepsinogen denaturation is not a two-state transition under any conditions (Mateo and Privalov, 1981; Privalov et al., 1981). Table 10.1 presents the calorimetric and effective van't Hoff enthalpies estimated for this protein at various pH values. In all cases, the ratio $\Delta H^{cal}/\Delta H^{vH}$ is far from 1.0 and approaches 2.0.

Deconvolution of the heat capacity function of pepsinogen into simple components corresponding to two-state transitions showed that denaturation of pepsinogen is approximated well by the two largely overlapping highly cooperative transitions slightly differing in transition temperature and enthalpies (Fig. 10.4a). On changing solvent conditions, these two peaks shift quite independently over the temperature scale, indicating that they correspond to the melting of structurally independent parts of the molecule.

Pepsinogen is a zymogene. On removing the 44-residue peptide from the N-terminal of its polypeptide chain, it is activated and becomes an aggressive proteolytic enzyme—pepsin. Activation of pepsinogen results in separation of the calorimetrically observed single heat absorption peak into two component peaks (Fig. 10.4b). One can expect that the heat absorption peak, which moved down upon removal of the N-terminal peptide, represents unfolding of the structure formed by the N-terminal part of polypeptide. Most interesting is that neither of the two observed heat absorption peaks of pepsin represents a two-state transition. Deconvolution analysis of the heat capacity function of pepsin shows that each of these peaks is perfectly approximated by a pair of quasi-independent two-state transitions. It follows that the structure of pepsin is subdivided into four more or less independent parts. These component transitions shift in temperature as the solution pH changes, and it was found that the enthalpies of these transitions significantly depend on temperature (Fig. 10.5), which means that they unfold with heat capacity increment; that is, their unfolding results in exposure of nonpolar groups to water. It

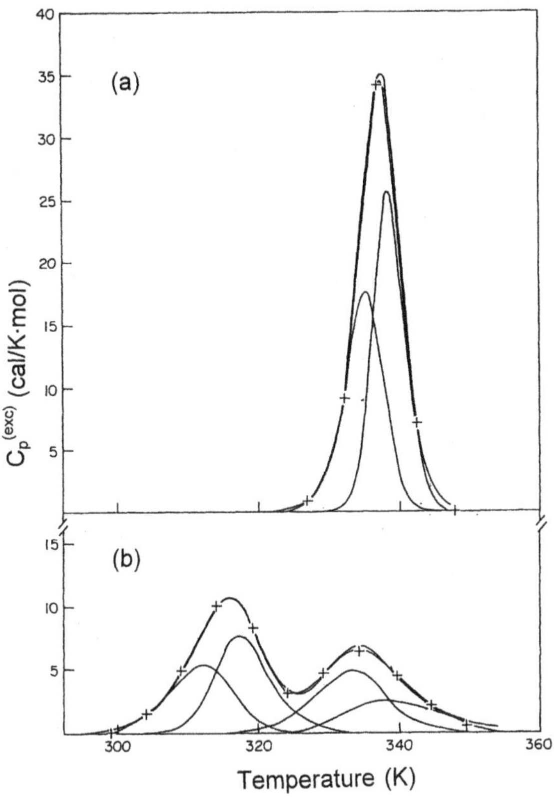

Figure 10.4. Deconvolution of the excess heat capacity function of pepsinogen (a) and pepsin (b). The experimental functions are given by the thick lines; the calculated components are shown by the light lines (Privalov et al., 1981).

follows that each of the revealed four submolecular structures in pepsin must have a well-developed hydrophobic core, which unfolds with the exposure of nonpolar groups. Thus, the structure of each of these subparts of pepsin is of a globular type, and each can unfold and refold independently from others. Such structures can be classified as **structural domains**.

One can expect that pepsinogen has the same domains as pepsin, but the pairs of domains in pepsinogen are less independent, being paired into two cooperative blocks. Perhaps they are connected by the removable N-terminal peptide filling the gap between the pairs of domains. It is interesting that the presence of such a specific substratelike inhibitor as pepstatin leads to the cooperation of its domains (Fig. 10.6). Thus, the pepsin is a more flexible molecule than pepsinogen and the presence of the substratelike inhibitor decreases its flexibility. This fact suggests that the inter-domain mobility of pepsin is important for its functioning as a proteolytic enzyme, that is, the device for cutting the polypeptides.

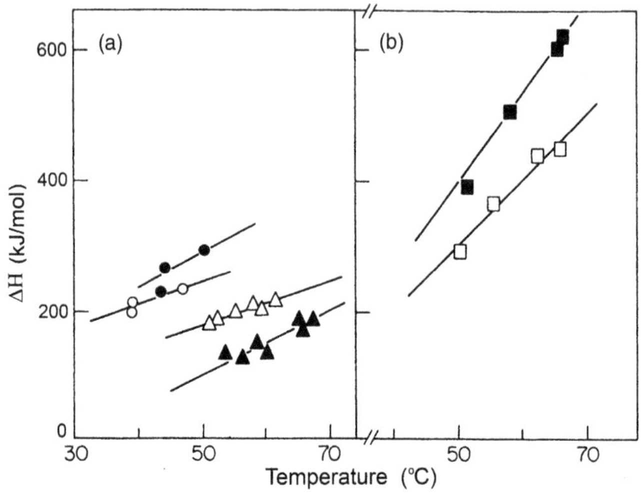

Figure 10.5. Temperature dependence of the transition enthalpies of the revealed domains in pepsin (a) and pepsinogen (b) in solutions differing in pH (Privalov et al., 1981).

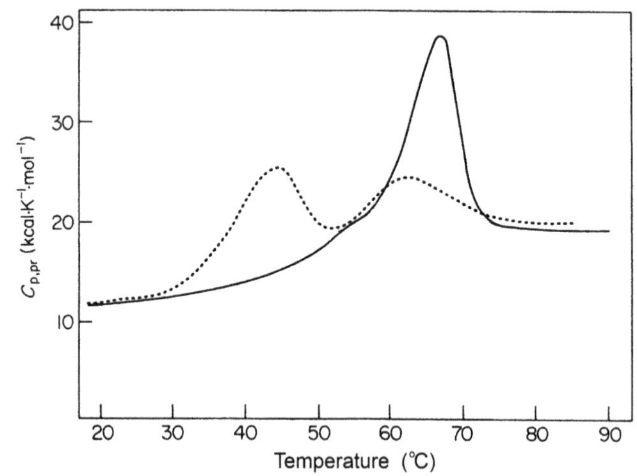

Figure 10.6. Heat capacity function of pepsin with pepstatin (solid line) and without pepstatin (dotted line) at pH 6.6, 100 mM NaCl (Privalov et al., 1981).

Analysis of the sequences of the polypeptide chain of acid proteases has revealed that the first and second halves of the chain have a similar pattern in each lobe and some homology in the sequence (Tang, 1979). Therefore each lobe of this protein appears as an "intramolecular dimer" built from two structurally equivalent parts. This finding is very intriguing, since it evokes an idea of gene duplication as a possible mechanism for the evolution of acid proteases (Fig. 10.7).

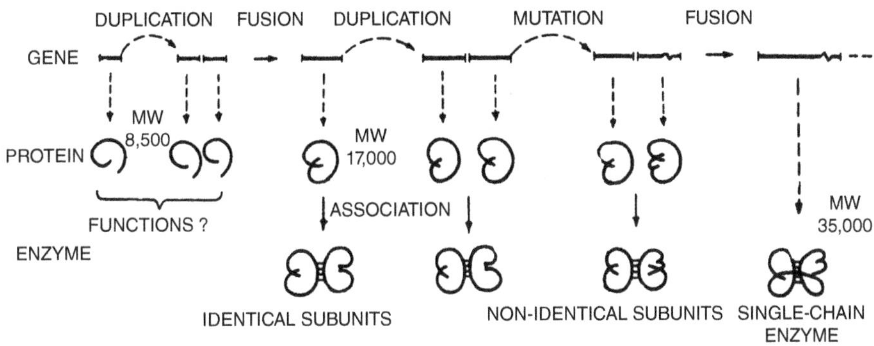

Figure 10.7. Schematic representation of genetic events in the evolution of carboxyl proteases according to Tang (1979). The ancestral gene was probably of one-fourth of the current size. After the first gene duplication and fusion, the primitive carboxyl protease may have contained two subunits. A second duplication and fusion produced a single-chain enzyme.

10.2. PROTEINS WITH INTERNAL HOMOLOGY

10.2.1. Evolution of Multidomain Proteins

There are many proteins that reveal sufficiently long homologous regions in their sequence. It is likely that they all appeared as a result of gene duplication and consequent fusion accompanied by a mutation. It is assumed that the history of their evolution is imprinted in the sequence of these proteins. The sequence, however, determines the folding of polypeptide into a unique compact structure, and, judging by the structural identity of not completely identical homologous proteins from various animals, a final three-dimensional structure is more conservative than the sequence. Therefore, one can expect that the history of protein evolution should be engraved in the structural organization; that is, the proteins with homologous repeats in their sequence should be multidomain, since each homologous region should fold more or less independently of the rest of the chain by the technology inherited from the ancestors. This is a very intriguing aspect of protein evolution and, as we see, calorimetry opens a new prospect for its investigation. This is why the thermodynamic study of unfolding of the multidomain proteins now attracts rapidly increasing attention.

10.2.2. Ovomucoid

This protein has three homologous regions (H) in the polypeptide chain (Beeley, 1972; Kato et al., 1976), which can be presented as $H^{I}(62)$–A(3)–$H^{II}(62)$–B(6)–$H^{III}(53)$ (Fig. 10.8). The number of residues in the corresponding regions of this polypeptide chain are indicated in parentheses. In a calorimetric study of temperature-induced unfolding of ovomucoid, it was found that the heat capacity profile of ovomucoid in the temperature range of its unfolding is complex and its shape

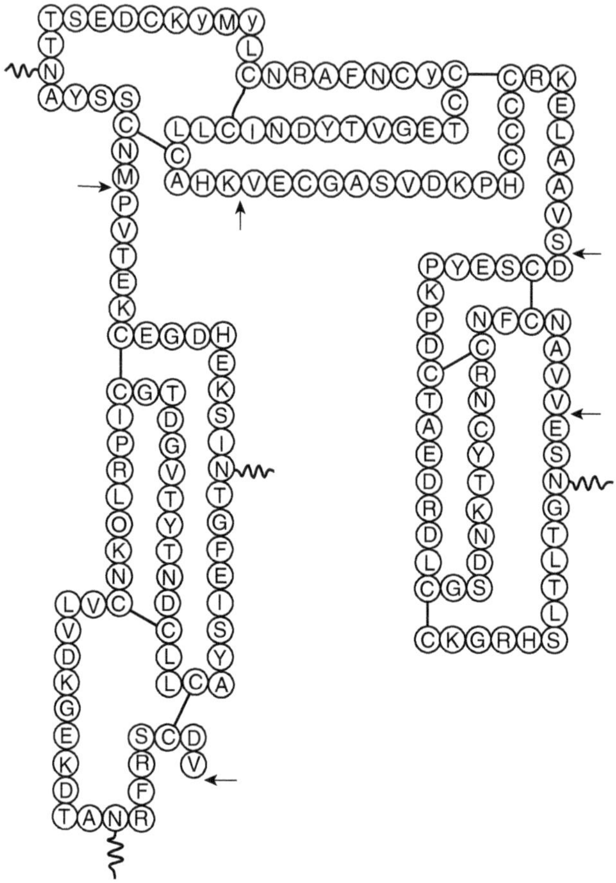

Figure 10.8. Primary structure of chicken ovomucoid according to Kato et al. (1978). The arrows show the position of introns in the ovomucoid gene as determined by Catterall et al. (1979).

changes significantly with changing solution pH (Fig. 10.9). Deconvolution of the excess heat capacity function shows that the observed complex heat capacity profile results from the overlap of three largely independent transitions (Fig. 10.10). Additional experiments on fragments corresponding to the homologous regions confirmed the conclusion on the independence of unfolding of the three homologous parts in ovomucoid and allowed identification of these parts of the protein, which have differing stabilities: the most unstable is the N-terminal part of this protein, while the C-terminal part is the most stable (Griko and Privalov, 1984). The heat capacity increase associated with the unfolding of ovomucoid led to the conclusion that each of the revealed cooperative parts has a well-developed hydrophobic core, that is, has a compact structure of the globular type. Thus, they can be considered as structural domains.

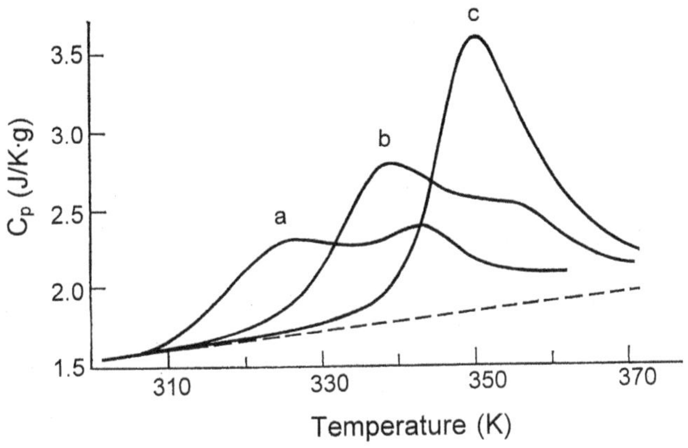

Figure 10.9. Partial specific heat capacity functions of ovomucoid in different solutions: (a) 50 mM glycine, pH 2.9; (b) sodium acetate, pH 3.6; (c) sodium phosphate, pH 6.4 (Griko and Privalov, 1984).

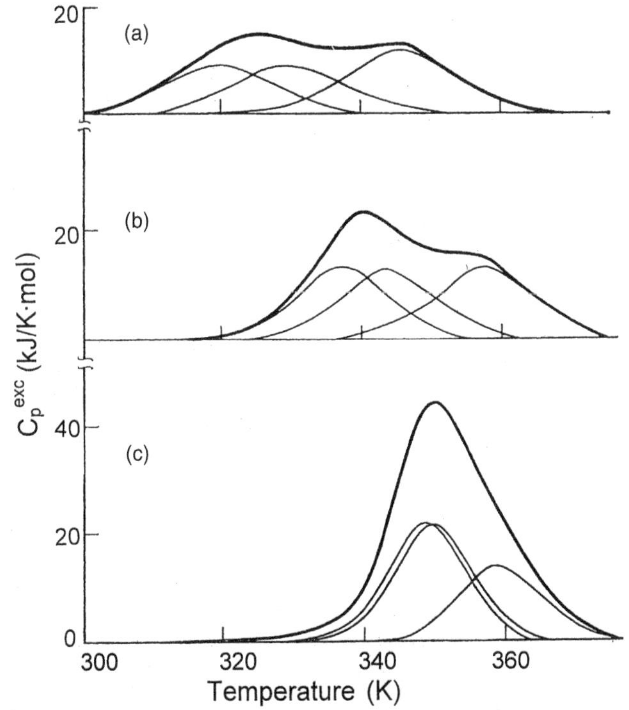

Figure 10.10. Deconvolution of the excess heat capacity function of ovomucoid in the solutions shown in Figure 10.9 (Griko and Privalov, 1984).

It is interesting that the gene coding ovomucoid includes introns, which separate the parts of this gene that code the cooperative domains of the protein (see the arrows in the scheme given in Fig. 10.8). This suggests that the protein-coding parts (exons) might represent the pra-gene coding the small pra-protein, the descendent of which figures as an independent domain in the ovomucoid.

10.2.3. Calcium-Binding Proteins

The most known representatives of this family of calcium-binding proteins are parvalbumin, troponin C, calmodulin (CaM), and the light chain of myosin. The primary structures of these four proteins include several homologous regions (H) and can be presented as follows:

Parvalbumin:

$$A_a(40)–H_a^I(30)–B_a(6)–H_a^{II}(30)$$

Troponin C:

$$A_b(15)–H_b^I(30)–B_b(6)–H_b^{II}(30)–C_b(10)–H_b^{III}(30)–D_b(6)–H_b^{IV}(30)$$

Calmodulin:

$$A_c(9)–H_c^I(30)–B_c(6)–H_c^{II}(30)–C_c(7)–H_c^{III}(30)–D_c(6)\text{-}H_c^{IV}(30)$$

Light chain:

$$A_c(8)–H_c^I(30)–B_c(6)–H_c^{II}(30)–C_c(9)–H_c^{III}(30)–D_c(6)–H_c^{IV}(30)$$

All these proteins have an initial segment, A, of varying length shown in brackets. The regions B, C, and D, which link the homologous regions, also vary in length, whereas the homologous regions, H, are all of equal length. Parvalbumin has two homologous regions, H_a^I and H_a^{II}, and it can be supposed that its gene appeared as a result of duplication of a gene of an ancestral protein. Three other proteins have four homologous regions, which could arise by further duplication of a double gene. This idea is confirmed by sequence analysis, which shows clear repeats between the first two and the last two homologous regions, with the regions H^I like H^{III} and H^{II} like H^{IV} (Collins et al., 1973; Weeds and McLachlan, 1974; Tufty and Kretsinger, 1975; Kretsinger, 1980). Crystallographic studies of parvalbumin structure show that the homologous region H forms two α-helices, E and F, and the loop joining these helices is coordinated by Ca^{2+} ion. Thus, this protein has two very similar calcium-binding domains. In the calmodulin and myosin light chain there are four homologous regions, H, which form four calcium-binding domains (Fig. 10.11).

Differential scanning calorimetric (DSC) studies of the thermal properties of parvalbumin showed that it unfolds upon heating with a single sharp peak of heat

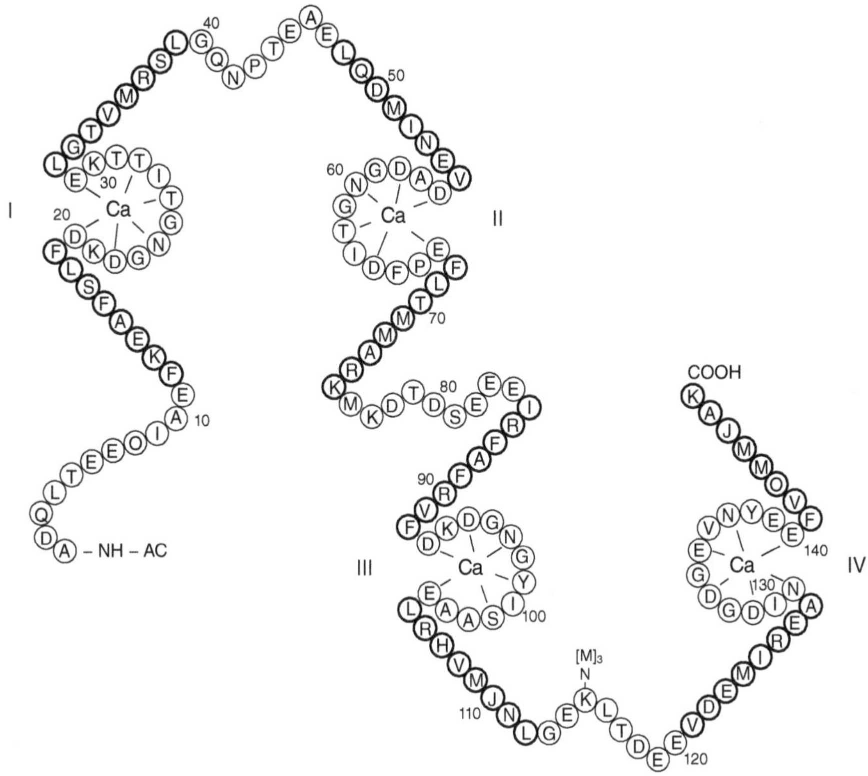

Figure 10.11. Chemical structure of calmodulin revealing four homologous regions. (Reproduced from Klee et al., 1980.)

absorption that is perfectly approximated by a two-state transition and largely depends on the presence of calcium: decrease of free calcium concentration in solution results in a drop in the stability of parvalbumin but does not affect the cooperativity of its unfolding (Fig. 10.12). Thus, both calcium-binding domains in this protein are consolidated into a single cooperative block (Filimonov et al., 1978).

Differential scanning calorimetric studies of troponin C and calmodulin showed that their unfolding is not a simple process and its complexity largely depends on the presence of calcium in solution (Tsalkova and Privalov, 1985). Therefore, calorimetric investigation of these proteins required their fragmentation by limited proteolysis (Fig. 10.13) and separate investigation of their fragments. Figures 10.14 and 10.15 present the excess heat capacity profiles of these proteins observed in the presence of various concentrations of free calcium. Deconvolution of these profiles on the component cooperative transitions show that in the presence of 2 mM EDTA, that is, practically in the absence of free divalent ions in solution, the troponin C unfolds in a single cooperative step, α (Fig. 10.14), but the calmodulin unfolds in

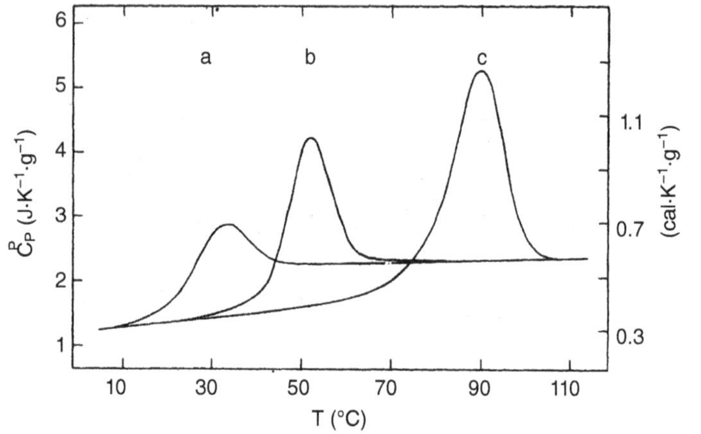

Figure 10.12. Temperature dependence of the partial specific heat capacity of parvalbumin: (a) in 10 mM sodium phosphate, 5 mM Na_2EDTA, pH 7.5; (b) in 10 mM sodium phosphate, 10 mM Na_2EDTA, 1 M NaCl, pH 7.5; (c) in 10 mM sodium cacodylate, 10^{-4} M $CaCl_2$, pH 7.0 (Filimonov et al., 1978).

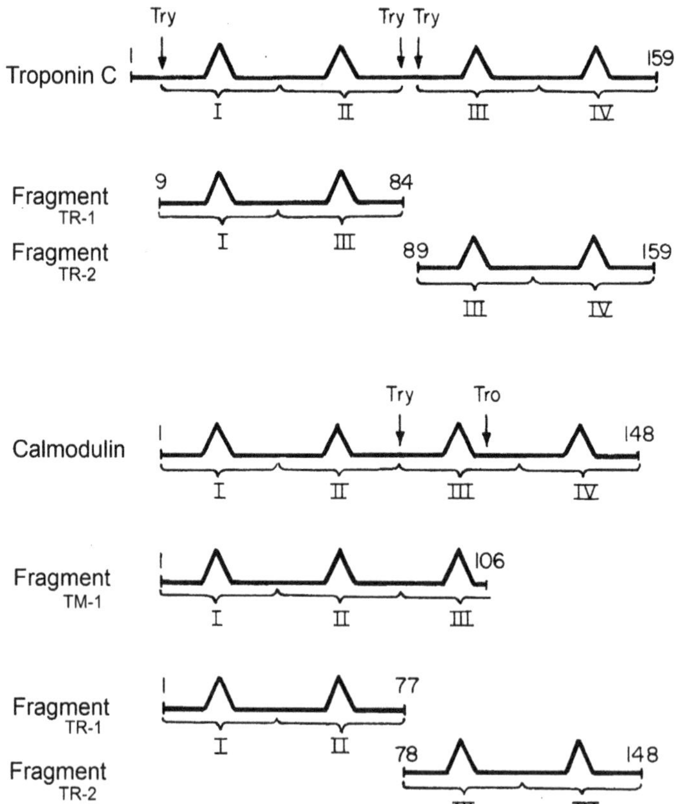

Figure 10.13. Diagram of troponin C and calmodulin fragments used in calorimetric experiments. Arrows indicate the points of trypsin (Try) and trombin (Thro) attack. Roman numbers denote homologous regions forming calcium-binding domains. Arabic numbers indicate the number of amino acids in the chain (Tsalkova and Privalov, 1985).

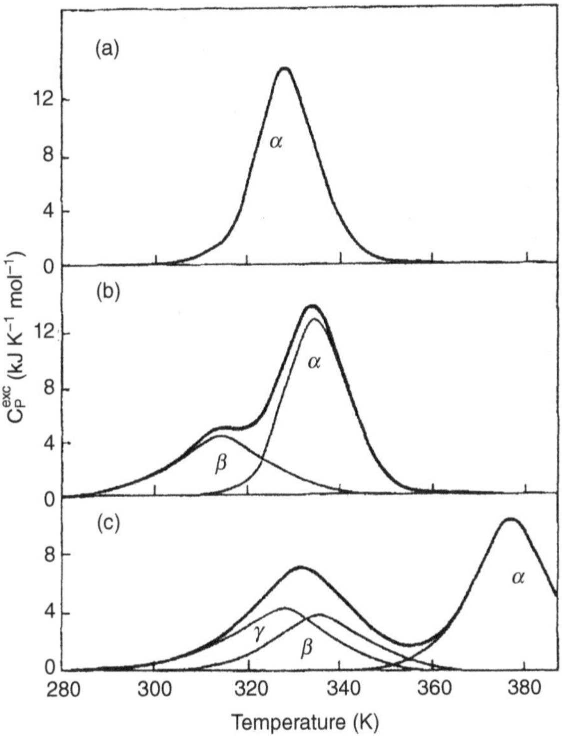

Figure 10.14. Temperature dependence of the excess molar heat capacity of troponin C at various concentrations of free Ca^{2+}: (a) 2 mM EDTA; (b) 10 mM EDTA + 7 mM $CaCl_2$; (c) 10 mM EDTA + 9.3 mM $CaCl_2$ (Tsalkova and Privalov, 1985).

two steps, α and β (Fig. 10.15). It is notable that the α steps in these two proteins are very similar. With the increase of free calcium ion concentration, the β transition in the troponin C profile also appears similar to the β transition in calmodulin. On further increase of free calcium concentration in both cases there appears the third transition, γ. The enthalpies of these three transitions, α, β, and γ, in both proteins are in the proportion 2 : 1 : 1.

Since both proteins consist of four calcium-binding domains, based on the preceding calorimetric information one can conclude that two of these domains are consolidated in a single cooperative block, as was found in the case of parvalbumin. To identify which of the four calcium-binding domains are coupled, the melting of the fragment of both proteins containing the first two calcium-binding domains have been studied calorimetrically. It was found that these fragments unfold in a single cooperative transition with enthalpy similar to that of the α-transitions in both proteins. Thus, first two calcium-binding domains in troponin C and calmodulin forming a single cooperative block, similar to that in parvalbumin. Calorimetric results, however, do not confirm the existence of a hydrophobic core that connects the four domains into a single system, as was suggested by Kretsinger and Barry (1975).

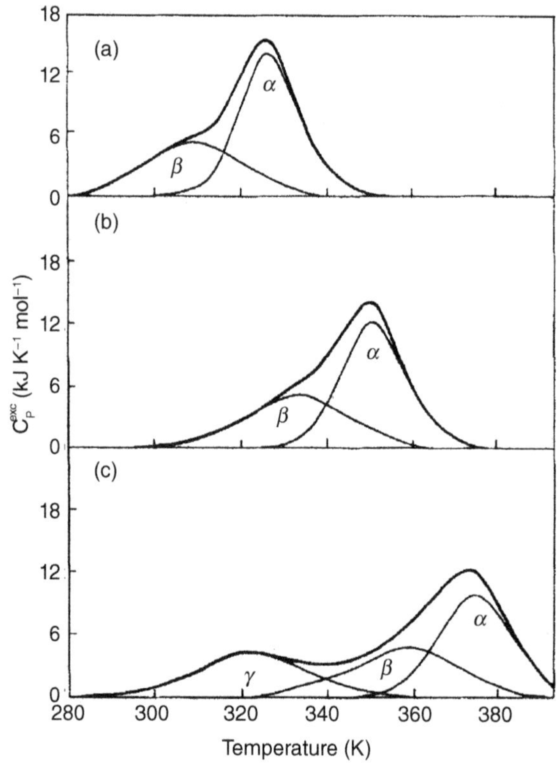

Figure 10.15. Temperature dependence of the excess molar heat capacity of calmodulin at various concentrations of free Ca^{2+}: (a) 2 mM EDTA; (b) 2 mM EDTA + 2 mM $CaCl_2$; (c) 10 mM EDTA + 9.25 mM $CaCl_2$ (Tsalkova and Privalov, 1985).

It is notable that the cooperative block consisting of the first two domains is specified with low-affinity calcium binding, which does not depend significantly on the environmental conditions (pH, ionic strength, concentration of divalent ions). The situation is quite different for the high-affinity calcium-binding domains III and IV. In troponin C, the removal of free divalent ions by dialysis in low-ionic-strength solution leads to such a great destabilization of domains II and IV that at room temperature they are already melted. As a result, at Ca^{2+} titration of troponin C, a large folding of the backbone is observed. The compact structure of domain IV is formed only in the presence of divalent ions. In calmodulin, domain IV is more stable than in troponin C and does not break down at room temperature in the absence of divalent ions, but domain III has a compact structure only in the presence of Ca^{2+}. Neither Mg^{2+} nor monovalent ions can stabilize its structure above room temperature.

It is remarkable that in the troponin C and calmodulin the same domain III is unstable and is stabilized only by Ca^{2+} ions. Thus, variation of the Ca^{2+} concentration within the range 10^{-8} to 10^{-7} M causes a dramatic change in the state of this domain:

at temperatures and ionic strengths close to physiological values, change in Ca^{2+} concentration induces reversible folding–unfolding of this compact structure. This change in conformation of this domain represents the essence of functioning of these regulatory proteins (see Section 11.3.2).

10.2.4. Plasminogen

The polypeptide chain of this protein, consisting of 790 amino acid residues, has five repeats in its sequence (Collen, 1980):

$$\mathrm{A(76)-H^{I}(96)-H^{II}(96)-H^{III}(96)-H^{IV}(96)-H^{V}(96)-B(230)}$$

Upon limited hydrolysis, the plasminogen loses its first 76 residues and is converted into Lys–plasminogen with the Lys residue at the N-terminus.

Figure 10.16 shows the original calorimetric recording of the heat effect at two consecutive heatings of the same plasminogen solution (Novokhatni et al., 1984). It appears that unfolding of plasminogen starts from the very beginning of heating and ends at 90°C. Upon repeated heating of the same sample the observed rather complicated heat capacity profile is perfectly reproduced, showing that the unfolding process is reversible and can be analyzed thermodynamically.

The excess molar heat capacities of the Lys–plasminogen and its various fragments, with a schematic indication of the fragments used, are presented in Figure 10.17. Deconvolution of these functions shows that Lys–plasminogen unfolds in seven transitions, five of which represent unfolding of the five kringle loops of this protein. Thus, each of the kringles forms a cooperative and largely independent domain. Two other transitions are associated with the C-terminal segment B(230) showing that this segment, which performs plasmolysis, consists of two cooperative domains.

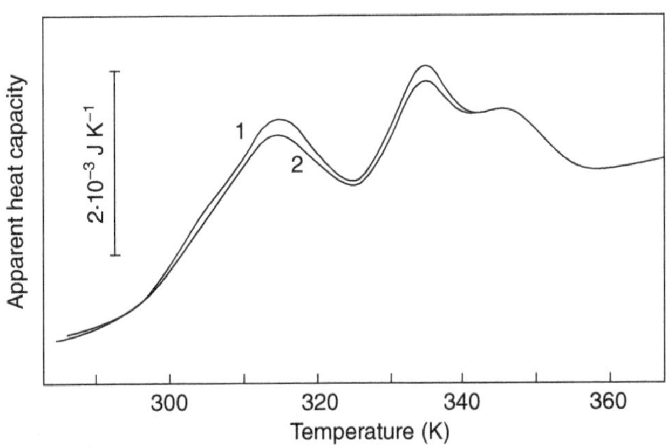

Figure 10.16. Original calorimetric recording of the heat effect at two consecutive headings of the Lys–plasminogen in 50 mM glycine, pH 3.4 (Novokhatni et al., 1984).

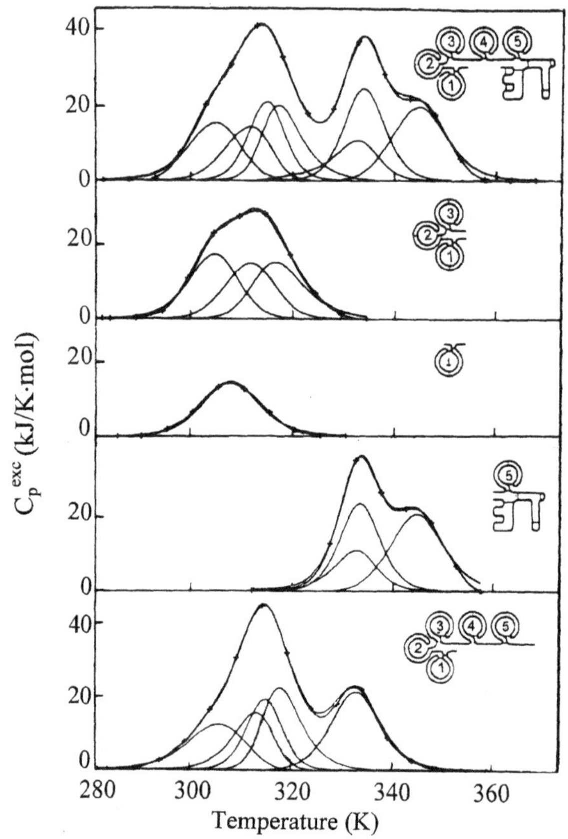

Figure 10.17. Deconvolution of the excess heat capacity function of (a) Lys–plasminogen, (b) fragment K 1–3 (residues 79–353), (c) fragment K 4 (residues 354–439), (d) miniplasminogen (residues 442–790), (e) heavy chain plasmin (residues 77–560) (Novokhatni et al., 1984).

10.2.5. Fibrinogen

Among the large and complex proteins, fibrinogen is one of the best studied calorimetrically. The molecular mass of this protein is 340,000 and it consists of two pairs of three different polypeptide chains (2α, 2β αvδ 3γ). All six of these chains are held together at the N-terminus by the knot of six disulfide cross-links (Fig. 10.18). On limited proteolysis it splits into several fragments. It is assumed that the readily hydrolyzed C-terminal portion of the α-chains does not have any compact structure, as for the D and E fragments, which have molecular weights of about 100,000 and 45,000, respectively, are quite stable against proteolysis and seemed to have a compact and ordered structure (Doolittle et al., 1978). This was confirmed by the DSC studies of fibrinogen and its fragments by Donovan and Mihalyi (1974), Medved' et al. (1980), and Privalov and Medved' (1982).

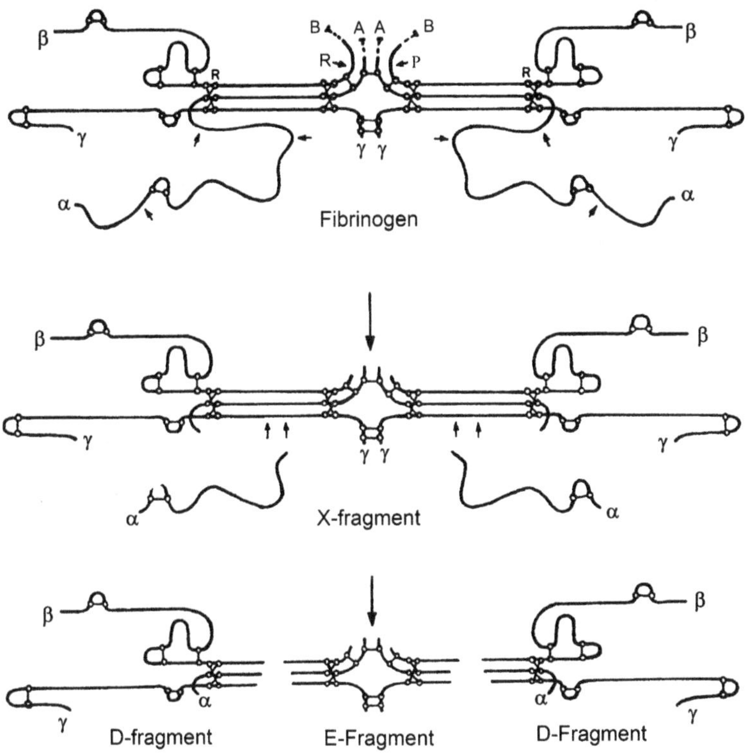

Figure 10.18. Chemical structure of fibrinogen (Privalov and Medved', 1982).

Figure 10.19 presents the heat capacity profiles of fibrinogen and its main proteolytic fragments in acidic solution (pH 3.5) in which fibrinogen and its fragments do not aggregate upon unfolding and their temperature-induced unfolding is highly reversible. The heating of intact fibrinogen proceeds with four heat absorption peaks of different magnitude. A comparison of the melting profiles of the intact molecule with those for the fragments shows that the first low-temperature peak [LT1(F)] and the second high-temperature peak [HT2(F)] are associated with the melting of some structures at the peripheral part of fibrinogen corresponding to the D fragment. The first high-temperature peak [HT1(F)] represents melting of the central part of the molecule corresponding to the E fragment. The second low-temperature peak [LT2(F)] might be connected only with melting of a structure formed by the C-terminal fragment of the α-chain, since it is observed only in the intact fibrinogen and is absent in its X fragment. Thus, one can conclude that the C-terminal part of the α-chain forms compact cooperative structures in the fibrinogen molecule and that the peripheral parts of this molecule consist of two structures with differing stabilities.

Analysis of the shape of the first high-temperature peak [HT1(F)] shows that it does not represent a two-state transition: the ratio of the calorimetric and van't Hoff

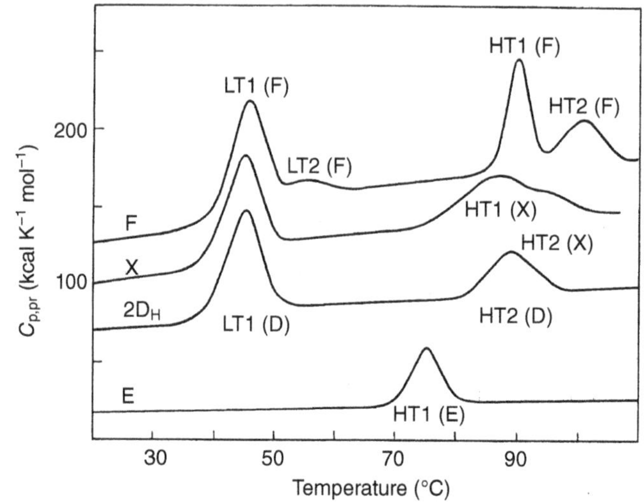

Figure 10.19. Heat capacity profiles of the fibrinogen (F) and its proteolytic fragments: X, D, and E in solution of pH 3.5, 50 mM glycine buffer (Privalov and Medved', 1982).

enthalpies for this peak in the intact fibrinogen and E fragment is 1.3, which is significantly higher than what one could expect for a single cooperative domain. It appears that unfolding of this part either proceeds through three rather stable intermediates or we have here unfolding of two strongly interacting domains. The latter appears more probable since this part indeed consists of two identical bunches of chains held together by a disulfide knot.

For the second high-temperature peak in the intact fibrinogen [HT2(F)] the ratio $\Delta H^{cal}/\Delta H^{vH}$ is 2.0, but for its D fragment [HT2(D)] it is 1.0. Thus, this thermostable structure represents a single cooperative unit in the D fragment, but in fibrinogen it is in two copies, as expected from the known chemical structure of fibrinogen.

The low-temperature process observed in fibrinogen and in the D fragment is more complicated. The $\Delta H^{cal}/\Delta H^{vH}$ ratio for the first low-temperature peak in the intact fibrinogen and its D fragment (LT1) was found to be 4.8 and 2.4, respectively. From these ratios one can conclude that there are at least three cooperative regions in the D fragment that melt simultaneously at low temperatures and one cooperative region that melts at high temperature. Proteolytic degradation of the D fragment revealed that it consists of four stable blocks with molecular mass 13,000, 36,000, 18,000, and 28,000. According to DSC studies performed at different pH values, the enthalpy of unfolding of the revealed fragments depends on temperature; that is, the fragments unfold upon heating and the heat capacity increases. This suggests that each of the revealed cooperative domains has a well-developed hydrophobic core. The overall scheme of fibrinogen construction based on calorimetric study of the intact molecule and its proteolytic fragments is presented in Figure 10.20.

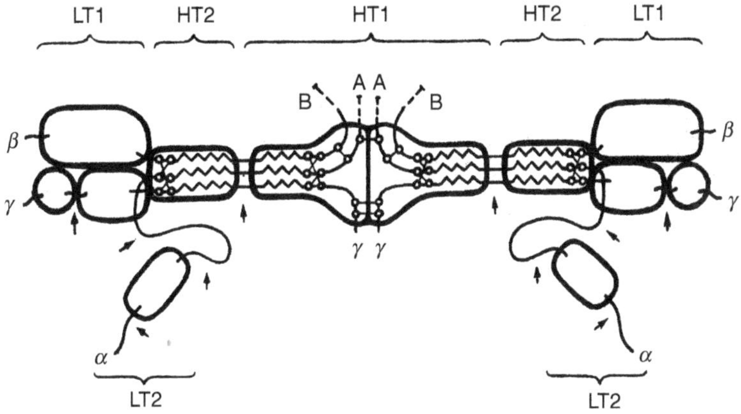

Figure 10.20. Diagram of the domain arrangement in the fibrinogen. Wavy lines are thermostable parts of the molecule with a high α-helical content (Privalov and Medved', 1982).

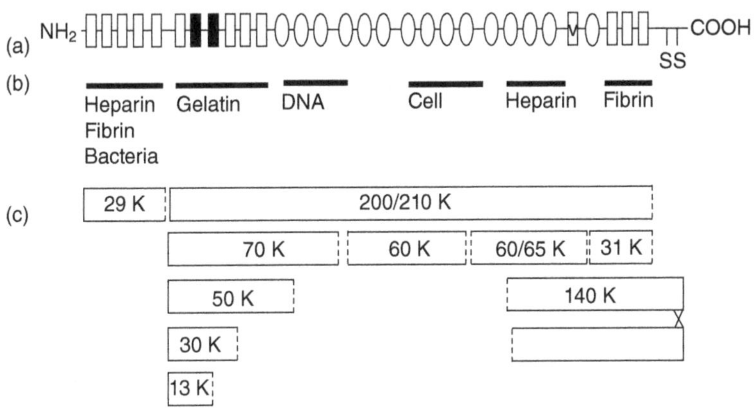

Figure 10.21. Fibronectin polypeptide chain: (a) three types of homologous repeats in the sequence of amino acid residues according to Skorstengaard et al., 1986; (b) binding functions, specific for various parts; (c) fragments obtained by limited proteolysis (Tatunashvili et al., 1990).

10.2.6. Fibronectin

Fibronectin, a 500,000 molecular weight glycoprotein of plasma and extracellular matrix, is a remarkable molecule with a variety of adhesive properties: it shows specific affinity for heparin, fibrin, collagen, DNA, and receptors on cell surfaces. This molecule consists of two identical polypeptide chains held by disulfide crosslinks near the C-terminus. In the sequence of these chains, three types of homologous repeat units, or modules, that are 45 and 90 amino acid residues long have been identified, as shown schematically in Figure 10.21. These properties suggest that the fibronectin polypeptide chain has evolved by manifold duplication and merging of

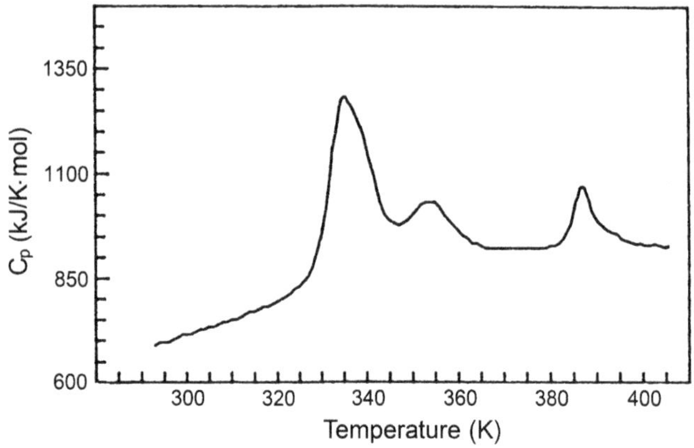

Figure 10.22. Apparent molar heat capacity of intact fibronectin in pH 8.0 solution (Tatunashvili et al., 1990).

several primordial genes. This idea is supported by the fact that proteases split fibronectin into definite fragments that show specific affinity for various ligands. Thus, it appears that the polypeptide chain of fibronectin forms several compact structures that are resistant to proteases. However, each of the isolated proteolytic fragments of fibronectin contains several modules and it is not clear whether each module, or combination of modules, forms stable and independent structural units that can be regarded as cooperative domains.

Differential scanning calorimetric study of the intact fibronectin showed too complicated a heat capacity profile to be quantitatively analyzed (Fig. 10.22). The excess heat capacity functions of the gelatin-binding regions (70k), which were split into three smaller fragments (50k, 30k, and 13k), are presented in Figure 10.23. Here the dashed lines show the cooperative components of the observed heat capacity profiles. The excess heat capacity profiles of the central part (60k) and the C-terminal part (65k) and the results of their deconvolution on the component transitions are shown in Figures 10.24 and 10.25. The information obtained by deconvolution analysis of the DSC-determined melting profiles of fibronectin fragments permitted identifcation the 12 cooperative domains in each of two identical polypeptide chains of this protein (Fig. 10.26). Some of these domains include two similar modules but some include three similar modules. The extreme C-terminal domains merge into a single cooperative highly stable block held by disulfide cross-links.

10.2.7. Discreteness in Protein Structure

It appears that the discreteness of the large protein structures results from their evolution: the multiplication, divergence, reshuffling, and fusion of primordial genes. It can be explained by the conservatism of living Nature, which in constructing new large molecules preferentially uses the well-tested standard blocks, the folding technology of which has been perfectly worked out during the long period

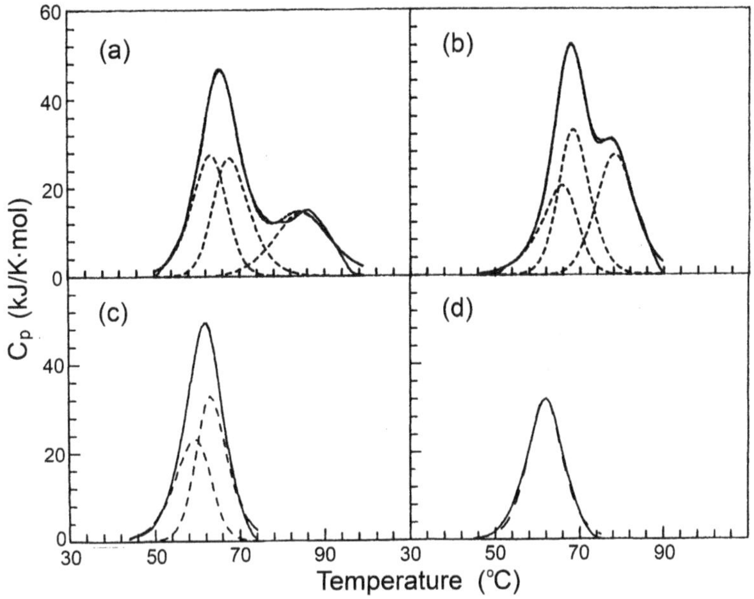

Figure 10.23. Deconvolution of the excess heat capacity functions of the fragments of gelatin binding region of fibronectin: (a) 70k fragment at pH 8.0, (b) 70k fragment at pH 4.0; (c) 50k fragment at pH 8.0; (d) 30k fragment at pH 4.0 (Tatunashvili et al., 1990).

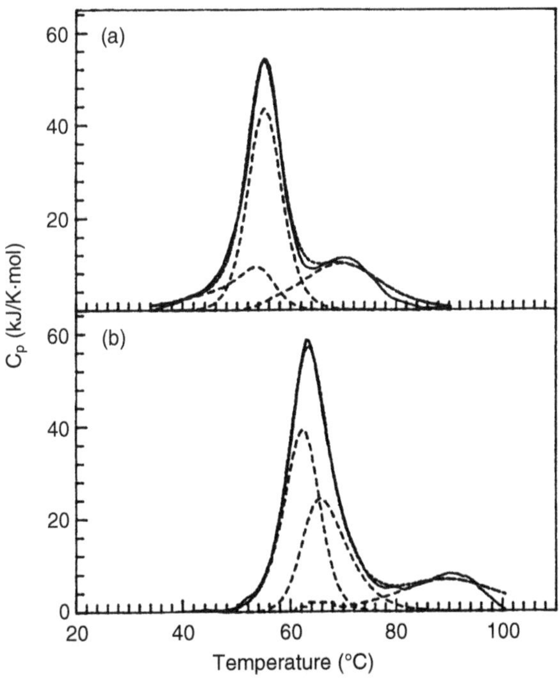

Figure 10.24. Deconvolution of the excess heat capacity function of the central 60k fragment of fibronectin at (a) pH 4.0 and (b) at pH 8.0 (Tatunashvili et al., 1990).

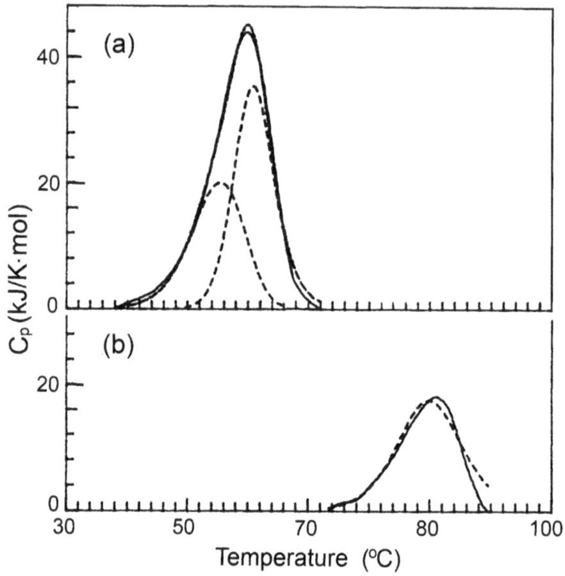

Figure 10.25. Deconvolution of the excess heat capacity function of the C-terminal of fibronectin at pH 8.0: (a) the 65k fragment; (b) the 31k fragment (Tatunashvili et al., 1990).

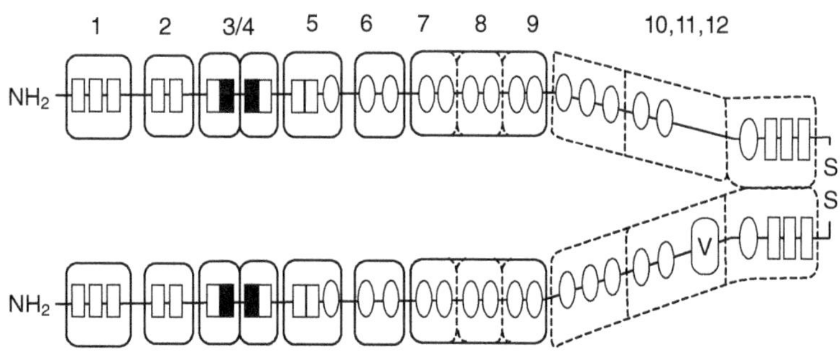

Figure 10.26. Calorimetrically identified cooperative domains in fibronectin (Tatunashvili et al., 1990).

of evolution. Since 20 different amino acids are used in nature, one can expect that the total possible number of polypeptides consisting of 100 amino acids with different sequences could be

$$(20)^{100} = 10^{130}$$

However, naturally occurring polypeptides use only a small portion of the possible sequences, perhaps not more than 10^6. Why?

This conservatism is dictated by the physics of such constructions. The physical requirement for the protein structure is evident: to fulfill its specific function protein should have a well-defined structure that is stable against fluctuations, which are severe at the molecular level; that is, it should be cooperative (see Section 9.1). The reserve of stability must considerably exceed the energy of thermal motion, $RT = 2.5$ kJ/mol, that is, it must be not less than 25 kJ/mol. This level of stability can be achieved only by integration of at least 50 residues in a single cooperative unit, that is, a domain.

The upper limit for the size of a cooperative unit, which is about 20,000, is set by difficulties of its design that increase rapidly with increasing size. These difficulties proceed not only from the necessity of proper adjustment of all elements of the structural unit, but also from the necessity of ensuring the cooperation of all these elements—without which the structure would be loose and would suffer from thermal fluctuations.

Since a stable compact structure can be formed only by polypeptides with certain unique sequences, it is clear that there are not too many possibilities for the creation of new sequences in constructing a new protein. Thus, the naturally occurring large proteins are constructed using the standard blocks, which have perfect folding technology, worked out during the billion years of evolution.

It appears that discreteness of structure is a general principle of organization of large molecular systems with aperiodic order, that is, biological macromolecules.

REFERENCES

Ahmad F. and McPhie P. (1978). *Biochemistry*, **17**, 241–246.

Beeley A.J. (1972). *Compr. Biochem.*, **26B**, 297–423.

Catterall J.F., Stein J.P., Lai E.C., Woo S.L., Dugaiczyk A., Mace M.L., Means A.K., and O'Malley B.W. (1979). *Nature*, **278**, 323–325.

Collen D. (1980). *Thromb. Homeostas.*, **43**, 77–89.

Collins J.H., Potter J.D., Horn M.J., Wilshire G., and Jackman N. (1973). *FEBS Lett.*, **36**, 268–272.

Donovan J.W. and Mihalyi E. (1974). *Proc. Natl. Acad. Sci. USA*, **71**, 4125–4128.

Doolittle R.F., Goldbaum D.M., and Doolittle L.R. (1978). *J. Mol. Biol.*, **120**, 311–325.

Filimonov V.V., Pfeil W., Tsalkova T.N., and Privalov P.L. (1978). *Biophys. Chem.*, **8**, 117–122.

Griko Yu.V. and Privalov P.L. (1984). *Biofizika (USSR)*, **29**, 194–198.

Kato I., Kohr W.J., and Laskowsky M., Jr. (1978). In: *Regulatory Proteolytic Enzymes and Their Inhibitors* (S. Magnusson, M. Otten, B. Foltman, K. Dano, and H. Neurath, Eds.), **47**, pp. 197–206. Pergamon, Oxford.

Kato I., Schrode J., Wilson K.A., and Laskowsky M. (1976). In: *Protides of the Biological Fluids* (H. Peeters, Ed.), **23**, pp. 235–243. Pergamon, Oxford.

Klee C.B., Crouch T.H., and Richman P.G. (1980). *Annu. Rev. Biochem.*, **49**, 489–515.

Kretsinger R.H. (1980). *CRC Crit. Rev. Biochem.*, **8**, 119–174.

Kretsinger R.H. and Barry C.D. (1975). *Biochim. Biophys. Acta*, **405**, 40–52.

McPhie P. (1980). *J. Biol. Chem.*, **255**, 4048–4052.

Mateo P.L. and Privalov P.L. (1981). *FEBS Lett.*, **123**, 189–192.

Medved' L.V., Tiktopulo E.I., Privalov P.L., and Varetskaya T.V. (1980). *Molek. Biol. (USSR)*, **14**, 835–842.

Novokhatni V.V., Kudinov S.A., and Privalov P.L. (1984). *J. Mol. Biol.*, **179**, 215–232.

Privalov P.L. and Medved' L.V. (1982). *J. Mol. Biol.*, **159**, 665–683.

Privalov P.L., Mateo P.L., Khechinashvili N.N., Stepanov V.M., and Revina L.P. (1981). *J. Mol. Biol.*, **152**, 445–464.

Skorstengaard K., Jensen M.S., Sahl P., Petersen T.E., and Magnusson S. (1986). *Eur. J. Biochem.*, **161**, 441–453.

Tang, J. (1979). *Mol. Cell. Biochem.*, **26**, 93–110.

Tatunashvili L.V., Filimonov V.V., Privalov P.L., Metsis M.L., Koteliansky V.E., Ingham K.C., and Medved' L.V. (1990). *J. Mol. Biol.*, **211**, 161–169.

Tiktopulo E.I. and Privalov P.L. (1978). *FEBS Lett.*, **91**, 57–58.

Tsalkova T.N. and Privalov P.L. (1985). *J. Mol. Biol.*, **181**, 533–544.

Tufty R.M. and Kretsinger R.H. (1975). *Science*, **187**, 167–169.

Weeds A.G. and McLachlan A.D. (1974). *Nature*, **252**, 646–649.

11

MACROMOLECULAR COMPLEXES

11.1. ENTROPY OF ASSOCIATION REACTIONS

11.1.1. Thermodynamics of Molecular Association

Association of biological macromolecules with their molecular partners, resulting in formation of specific macromolecular complexes, is the most basic biological function. This explains a rapid increase of interest in various aspects of this problem, both biological and physical. The biological aspect of this problem is very broad, since there are thousands of different proteins and nucleic acids forming a vast number of different complexes with different functions. The mechanisms of formation of these complexes differ in their details but they all have a common physical base.

It is more or less clear that association of several independent molecules into a single complex is driven by the same secondary interactions that are responsible for protein folding and proceed with similar dehydration effects on groups removed from contact with water. In some cases formation of a complex might proceed with conformational changes of the partners and these changes might have functional meaning. However, in contrast to the folding of monomeric proteins, the process of

Microcalorimetry of Macromolecules: The Physical Basis of Biological Structures, First Edition.
Peter L. Privalov.

complex formation results in a decrease in the number of the molecular kinetic units, that is, decrease in the degrees of freedom of the system—its translational entropy. This additional negative entropy effect of association opposes formation of the complex and actually determines the concentration dependence of its association–dissociation process. Thus this entropy presents the thermodynamic essence of the association process. Therefore, to understand various association reactions, they must be compared at standard concentration. In physical chemistry of low-molecular-weight compounds, a standard concentration of 1 M is conventionally assumed. The same standard is used in biophysical chemistry, although a 1 M solution in the case of macromolecules makes little sense. The magnitude of the standard translational entropy is still one of the most debatable subjects in the case of macromolecular solutions.

According to the view originally proposed by Gurney (1953) and later adopted by Kauzmann (1959), Tanford (1973), and many others, translational entropy is expressed by the cratic term, δS^{cratic}, which is just the entropy of mixing of the additional kinetic unit appearing upon dissociation of a complex with solvent and is assumed to be independent of the solution composition. For 1 M standard aqueous solution (containing 55 moles of water) $\delta S^{\text{cratic}} = R\ln(1/55) = -8.03$ cal/K·mol = −34.5 J/K·mol for formation of a dimer, and it is supposed to be independent of the molecular weight of the solute. According to Holtzer (1955), this cratic correction to the entropy of complex formation is groundless because even dilute solutions of macromolecules are not ideal (see also Gilson et al., 1997). The cratic entropy became a target of severe critics, also of statistical thermodynamics, as physically ungrounded. In the statistical mechanics of ideal gases each independent kinetic unit is specified by the translational–rotational enthalpy $H^0_{\text{tr+rot}}$ and entropy $S^0_{\text{tr+rot}}$. Although translational–rotational enthalpy depends only on temperature (Hill, 1986),

$$H^0_{\text{tr+rot}} = 6(RT/2) + P\,\Delta V = 3RT + RT = 4RT \tag{11.1}$$

translational–rotational entropy depends on concentration ρ_0 and some structural characteristics:

$$S^0_{\text{tr+rot}} = S^0_{\text{tr}} + S^0_{\text{rot}} = [2.5R - R\ln(\rho_0\Lambda^3] + [1.5R + R\ln\pi^{0.5}(8\pi^2 kT/h^2)^{3.2}\det(\mathbf{A})^{1/2}] \tag{11.2}$$

Here, $\Lambda = h/(2\pi MkT)^{0.5}$, where M is the molecular mass of the molecule, h is Plank's constant, k is Bolzmann's constant, and det(**A**) is the determinant of the inertial tensor. The first part of this equation, also called the Sacker–Tetrod equation, expresses the translational entropy, S^0_{tr}, which thus appears to depend on the mass of molecule (through Λ) and concentration ρ_0. The second part expresses the rotational entropy, S^0_{rt}. Unlike translational entropy, which is concentration dependent but is indifferent to the structure of the molecule, S^0_{rt} does not depend on concentration but depends on molecular structure through the term det(**A**).

Assuming that the translational entropy of macromolecules in aqueous solutions does not differ from that of small molecules in the gaseous phase and can be calculated by the simple Sacker–Tetrod equation based on the statistical mechanics of gases, Finkelstein and Janin (1989) obtained a value of 180–230 J/K·mol, depending on the molecular weight of the protein, for the translational entropy of a typical dimeric protein at 300 K and 1 M standard concentration. According to the same authors, the rotational entropy calculated from first principles is of the same order of magnitude. Therefore, the value of ($\delta S^{trans} + \delta S^{rot}$) amounts to 400 J/K·mol, with a positive sign for the dissociation of a dimer and a negative sign for its association.

Very similar values of the entropy effects of dimerization have been obtained by Tidor and Karplus (1994) using the statistical-thermodynamic approach suggested by Chandler and Pratt (1976). They calculated that dimerization of insulin results in a decrease of the translational entropy by 180 J/K·mol and decrease of rotational entropy by 200 J/K·mol, but they found that it should be accompanied by an increase of the vibrational entropy by 110 J/K·mol. Therefore, according to these authors the overall change of the external entropy (i.e., the entropy not associated with change in conformation or hydration) upon dimerization of insulin should amount to –270 J/K·mol. Values of entropy reduction in the range from 300 to 400 J/K·mol have been widely used by many authors in the thermodynamic analysis of the formation of protein–protein and protein–DNA complexes (see, e.g., Janin and Chothia, 1990; Janin, 1995; Searle et al., 1992).

Thus, the values of translational entropy derived by statistical mechanics differed from the cratic entropy by at least one order of magnitude. This was a matter of considerable concern. Amzel (1997) concluded that using the gas phase translational entropy given by the Sacker–Tetrod equation resulted in large overestimation of the translational entropy in aqueous solution, because most components of the movement of a ligand in the binding site of a protein and in solvent are similarly restricted and cannot be described by a harmonic oscillator. Other authors also came to some intermediate values for the translational–rotational entropy change upon association of molecules into a complex. Horton and Lewis (1992) calculated the association energy of a quaternary complex starting from atomic coordinates and concluded that the entropy cost of fixing one subunit relative to the other is –80 J/K·mol. Brady and Scharp (1997) analyzed the crystal packing and solvation of cyclic dipeptides and found that the association entropy penalty should be –65 J/K·mol. Murphy et al. (1994) assumed that a loss of 8 cal/K·mol = 33.3 J/K·mol in translational entropy for the formation of a dimer is the most appropriate for energy parameterization calculations.

11.1.2. Experimental Verification of the Translational Entropy

There have been many attempts to verify experimentally the theoretically predicted translational–rotational entropy values. However, experimental determination of the translational–rotational entropy is not simple because it is only a part of the overall entropy of an association reaction, which includes the entropy of dehydration of the groups removed from water upon complex formation and the entropy of conformational changes in both partners upon association. In the case of macromolecular binding reactions, these two effects might greatly exceed the translational–rotational entropy effect.

Studies of reactions such as the formation of acetic anhydride versus the formation of succinic anhydride led to a translational–rotational entropy value of about 200 J/K·mol (Page and Jenks, 1971; Jenks, 1975). However, it was unclear whether the results obtained on small compounds in a reaction involving the rearrangement of covalent bonds could be extrapolated to the case of formation of macromolecular complexes, which usually proceed without changing chemical structure. It is therefore of paramount importance to determine experimentally the translational entropy for macromolecular complexes.

The appropriate procedure for measuring the translational entropy is to compare the entropy change of an intermolecular reaction with that of an intramolecular reaction in which the two reactants are linked by a covalent bond. Namely, the translational entropy on formation of a macromolecular complex can be determined by comparing the entropy of unfolding–dissociation of the complex with the entropy of its unfolding without dissociation, that is, unfolding of the same complex having covalently linked subunits:

$$
\begin{array}{ccc}
 & \Delta S_{\text{unfolding}}^{\text{cross-linked dimer}} & \\
\text{[Folded cross-linked dimer]} & \rightarrow & \text{[Two unfolded cross-linked chains]} \\
\uparrow \Delta S_{\text{cross-linking}}^{\text{folded dimer}} & & \uparrow \Delta S_{\text{cross-linking}}^{\text{2chains}} \\
\text{[Folded dimer]} & \rightarrow & \text{[Two unfolded chains]} \\
 & \Delta S_{\text{unfold/dissociate}}^{\text{dimerr}} &
\end{array}
$$

Most of the specific macromolecular associates, such as protein–protein and protein–DNA complexes, dissociate upon unfolding of the subunits, if these subunits are not covalently linked. If their unfolding is induced by temperature increase (heat denaturation), the entropy of this process can be measured directly by heat capacity calorimetry (see Sections 2.1.7 and 3.2.7).

The idea of such an experiment is simple, but there are two practical problems in its realization. First, the heat capacity calorimetric instrument should be precise enough to reliably register the small differences between the large entropies of unfolding of the two species, when covalently bound and when not. Second, the species studied should differ by only a single covalent cross-link, which holds the subunits together without any deformation, and the temperature-induced unfolding of these species should be highly reversible so as to be able to treat it thermodynamically. The first requirement has been met by the appearance of sensitive scanning microcalorimeters having a highly stable baseline (the Nano-DSC), which is especially important for the precise determination of the excess heat of reaction that proceeds over a broad temperature range. More difficult is the problem of finding an appropriate object for such experiments: a dimeric macromolecule that unfolds and dissociates reversibly upon heating and can be cross-linked without noticeable stress. One such object appeared to be the dimeric globular protein *Streptomyces* subtilisin inhibitor (SSI); another was the synthetic α-helical coiled-coil.

11.2. CALORIMETRY OF ASSOCIATION ENTROPY

11.2.1. SSI Dimer Dissociation

In the native state SSI exists as a homodimer and dissociates to monomers upon heat denaturation (Takahashi and Sturtevant, 1981; Tamura et al., 1991a,b). With replacement of Asp83 by Cys at its flat surface forming an interface, this mutated dimer (D83C) (Fig. 11.1) might be cross-linked with a disulfide bond (Tamura et al., 1994). Detailed optical (Fig. 11.2) and nuclear magnetic resonane (NMR; Fig. 11.3) studies of the wild-type protein WT SSI and its mutant D83C in the folded and unfolded states showed that disulfide cross-link formation does not induce any noticeable changes in its conformation (Tamura and Privalov, 1997). Therefore the WT SSI and its D83C mutant appeared to be perfect objects for calorimetric study of the thermodynamic effect of cross-linking the subunits of a complex for evaluation of the translational entropy.

Figure 11.4 shows the partial molar heat capacities of WT SSI and its D83C mutant at various concentrations of proteins in pH 6.0 and pH 3.0 solutions. One can see that, although variation of concentration does not shift the melting profile of the D83C mutant, it has a noticeable effect on the melting profile of the WT SSI, as one can expect for the bimolecular reaction: with decrease of concentration of protein, the heat absorption peak shifts to lower temperatures and slightly decreases

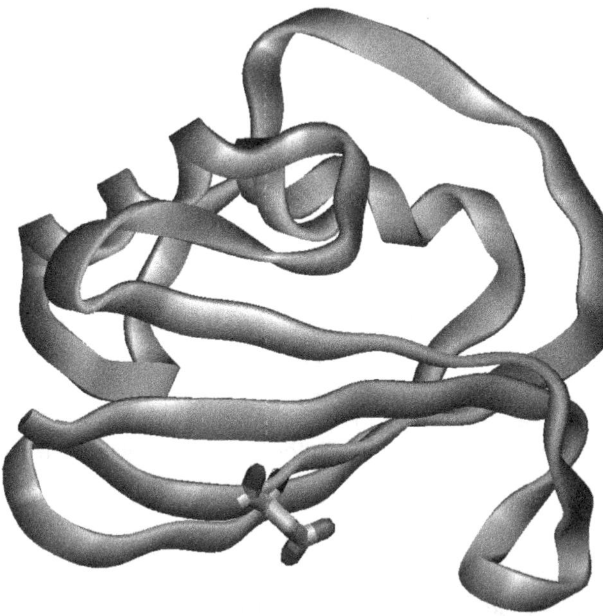

Figure 11.1. One of two subunits of the *Streptomyces* subtilisin inhibitor with the Asp83 residue, shown by the stick, that is replaced in the mutant form by Cys for cross-linking the dimer.

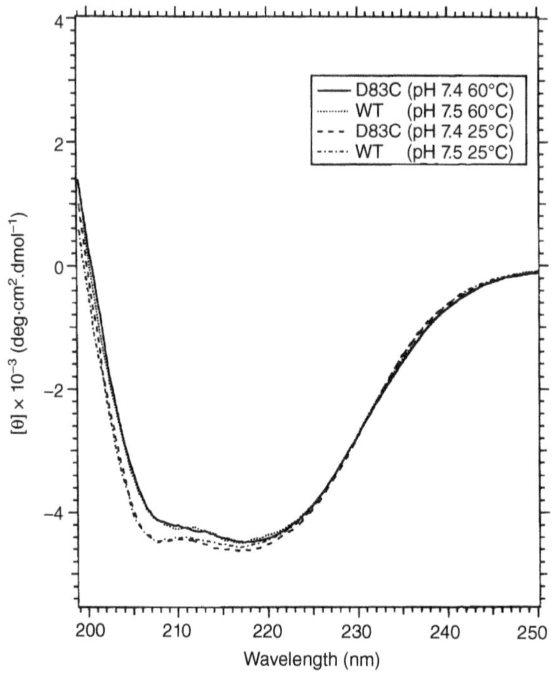

Figure 11.2. CD spectra in the far UV of WT SSI and the D83C mutant under conditions where both are in the folded (native) state: at 25°C and 60°C in solutions with neutral pH (Tamura and Privalov, 1997).

in size. Analysis of the excess heat absorption profile showed that in the case of D83C mutant it is perfectly described by a two-state monomolecular transition, whereas in the case of WT SSI it is perfectly described by a bimolecular two-state transition (Fig. 11.5).

To determine the heat effect of unfolding, the heat capacity of the native state was extrapolated linearly to higher temperatures and the heat capacity of the unfolded state was calculated by summing the heat capacities of the amino acid residues constituting the polypeptide chains of these two proteins. Both these functions are almost identical for the WT SSI and its D83C mutant (Fig. 11.6).

For the monomolecular two-state transition, unfolding entropy can be determined by the following modification of the general Equation (2.27):

$$\Delta S(T_t) = \Delta H(T_t)/T_t \tag{11.3}$$

The entropy of dissociation of a homodimer depends on the concentration, and at concentration N^0 the standard entropy is given by Equation (3.29):

$$\Delta S^0(T_t) = \Delta H(T_t)/T_t + R\ln\{2[\mathrm{N}]\} \tag{11.4}$$

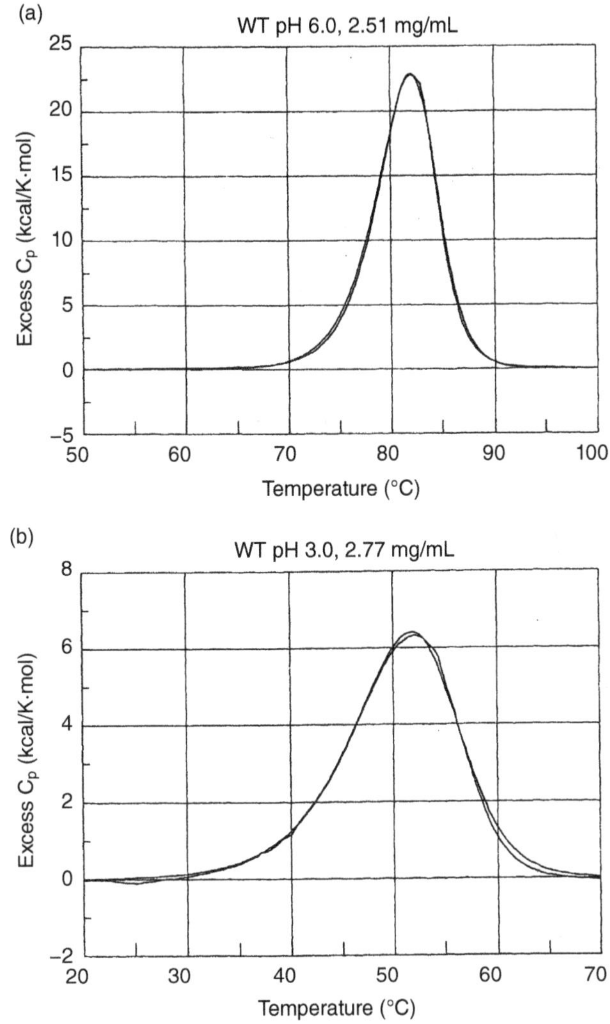

Figure 11.5. Comparison of the calorimetrically determined excess heat capacity function of WT SSI calculated under the assumption that the process represents a bimolecular two-state transition with $T_t = 81.3°C$ and $\Delta H_t = 182.5$ kJ/mol (Tamura and Privalov, 1997).

of the non-cross-linked and cross-linked dimers corresponds to the entropy of their cross-linking.

It appears that the entropies of unfolding of the D83C mutant at standard temperature of 80°C (in the case of pH 6.0 experiment) or 60°C (in the case of pH 3.0 experiment) do not depend on concentration, which is as it should be for a monomolecular reaction. For the WT SSI, unfolding of which proceeds with the dissociation, the entropy of unfolding extrapolated to standard temperature shows dependence on concentration but the standard entropy of unfolding at 1 M standard concentration

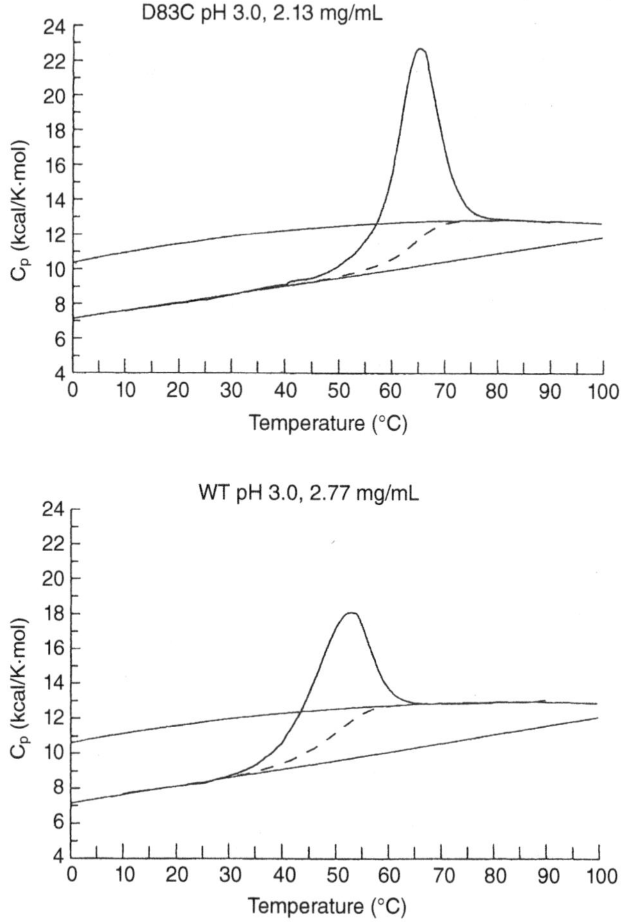

Figure 11.6. Temperature dependence of the partial molar heat capacity of WT SSI and D83C mutant at pH 3.0 (solid line) and the heat capacity functions for the native and unfolded states of these proteins (light lines) (Tamura and Privalov, 1997).

and the chosen standard temperature shows no dependence on concentration. It was most surprising that its value appeared to be very close to the entropy of unfolding of the D83C mutant at that temperature, which can be regarded as a standard since it does not depend on concentration. The difference between these two standard entropies, that of the D83C mutant and that of WT SSI, in the experiments at pH 6.0 and pH 3.0 are 4 cal/K·mol and 6 cal/K·mol, respectively, that is, about 5 cal/K·mol = 20 J/K·mol. Analysis of the possible errors shows that uncertainty in determination of this entropy difference does not exceed ±4 cal/K·mol = ±12 J/K·mol (Tamura and Privalov, 1997).

TABLE 11.1. Calorimetric Data for the Wild-Type SSI and S–S Cross-Linked Mutant D83C[a]

Protein	$N \times 10^{-6}$	T_t (°C)	ΔH_t (kcal/mol)	ΔH(80°C) (kcal/mol)	ΔS(80°C) (cal/K·mol)	ΔS^0(80°C) (cal/K·mol)
WT	445	83.6	185.9	178.8	501	487
pH 6.0	223	82.8	184.8	179.2	503	488
	111	82.1	183.5	179.3	504	487
	55.7	81.3	181.0	178.4	505	486
	27.8	80.4	179.2	178.8	506	486
	Averaged			**178.8**		**487**
D83C	98	94.2	198	179.1	482	
pH 6.0	113	94.0	199	180.1	485	
	11	94.2	–			
	Averaged			**179.6**	**483**	**483**
				ΔH(60°C)	ΔS(60°C)	ΔS^0(60°C)
WT	455	59.1	118.4	120.7	342	328
pH 3.0	242	54.9	106.9	120.7	347	332
	120	52.1	99.3	121.4	348	331
	60	49.8	92.1	120.7	348	330
	30	48.1	87.1	120.4	349	329
	15	46.3	–			
	Averaged			120.9		**330**
D83C	92.4	65.1	131.0	117.8	324	324
pH 3.0	11.6	65.1	–			
	Averaged					**324**

[a]Tamura and Privalov (1997).

11.2.2. Dissociation of the Coiled-Coil

This experiment used specially designed peptides with repetitive heptad units capable of forming a homodimeric α-helical coiled-coil, the leucine zipper (Yu et al., 1999).

Ac–Y–KSKSLES–KVKSLES–KAKSLES–KVKSLES–KVKSLES–AM

Ac–Y–KSKSLES–KVKSLES–KAKSLES–KVKSLES–KVKSLES–AM

A cross-link was then introduced replacing the serine residue in the first heptad by the cysteine, thus forming a disulfide bond at the hydrophobic interface:

Ac–Y–KCKSLES–KVKSLES–KAKSLES–KVKSLES–KVKSLES–AM

|

Ac–Y–KCKSLES–KVKSLES–KAKSLES–KVKSLES–KVKSLES–AM

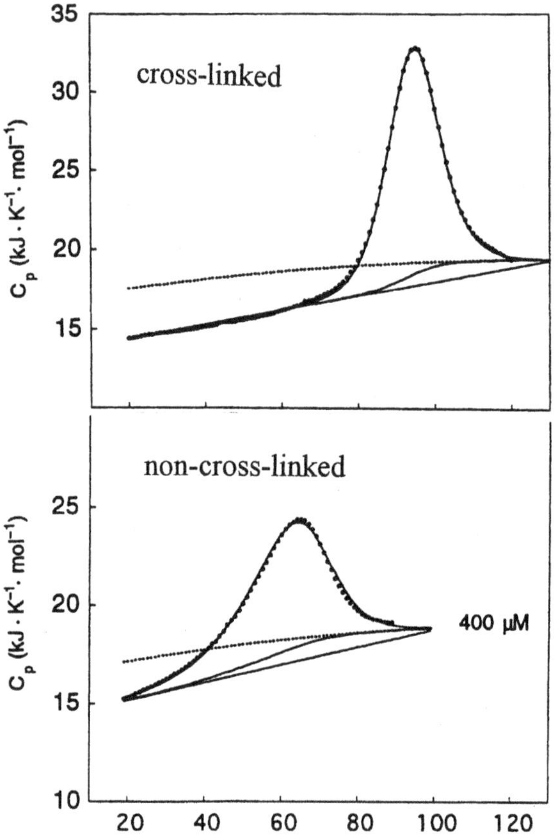

Figure 11.7. Partial molar heat capacity of the cross-linked and non-cross-linked peptides. Dark dots are the experimental data, and the solid line through the dark dots is the fitting of the two-state transition model. The dashed line is the fitting of the heat capacity of the unfolded state, and the solid line is the fitting of the heat capacity of the native state. (Yu et al., 1999).

Circular dichroism (CD) spectroscopy and ultracentrifugation showed that both species are forming the α-coiled-coil and the non-cross-linked one unfolds upon heating and strands dissociate. Upon heating these two α-helical coiled-coils unfold, but while the non-cross-linked one dissociates into two separated strands, the cross-linked one does not dissociate. Calorimetric studies showed that unfolding of both species proceeds with extensive heat absorption but, while the unfolding temperature of the cross-linked one does not depend on concentration, that of the non-cross-linked coiled-coil does depend on concentration (Fig. 11.7).

The main complication in analyzing these results is that the helicity of the non-cross-linked coiled-coil is 17% lower than that of the cross-linked coiled-coil, presumably due to end fraying at the N-terminal of the non-cross-linked coiled-coil.

TABLE 11.2. Thermodynamic Data of Unfolding of the Cross-Linked and Non-Cross-Linked Coiled-Coils[a]

Peptide	$N \times 10^{-6}$	T_t (°C)	ΔH_t (kJ/mol)	ΔH(70°C) (kJ/mol)	ΔS^0(70°C) (J/K·mol)
Cross-linked	350	94.4	250	174	470
	223188	94.3	249	170	460
	66	94.4	–		
	Averaged				**465**
Non-cross-linked	877	69.1	195	195	522
	466	66.0	188	194	522
	400	64.1	186	196	525
	217	61.6	–	–	–
	103	57.7	–	–	–
	55	54.0	–	–	
	Averaged				**522**

[a]Yu et al. (1999).

Therefore, comparison of the enthalpy and entropy of unfolding of the cross-linked and non-cross-linked coiled-coils required correction for this difference in the helicity. The corrected calorimetric data for these two coiled-coils are listed in Table 11.2. It shows that the difference in the entropy of unfolding is about (57 ± 36) J/K·mol = (13 ± 8) cal/K·mol. The experimental error in this calorimetric study of coiled-coil unfolding is significantly larger than in studying unfolding of SSI dimer. However, within the experimental errors, the entropy effects of cross-linking in these two experiments are comparable.

> *The averaged value of the cross-linking entropy that follows from these experiments is (30 ± 20) J/K·mol = (8 ± 5) cal/K·mol.*

11.2.3. Entropy Cost of Association

The question now is how should one interpret the estimated entropy effect of cross-linking the subunits in a dimer? It is clear that unfolded cross-linked dimer does not dissociate and its translational freedom is the same as that of the monomer. But its rotational freedom is also not twice that of the monomer. Therefore, contributions to the entropy effect of cross-linking should come from not only restrictions on the translational freedom but also, at least to some extent, restrictions on the rotational freedom; that is, the measured entropy of cross-linking includes both translational and rotational effects (Tamura and Privalov, 1997; P. Privalov and Tamura, 1999; Yu et al., 2001). Nevertheless:

> *The value of the cross-linking entropy measured in these experiments is one order magnitude smaller than the value predicted by statistical mechanics consideration of the loss of translational entropy (−200 J/K·mol) and more so for the*

(translational + rotational) entropy (−400 J/Kmol) on the formation of protein dimer at the standard 1 M concentration (Finkelstein and Janin, 1989; Tidor and Karplus, 1994).

The value of the calorimetrically measured cross-linking entropy is close in magnitude to the value suggested by the classical mixing theory for the loss in translational entropy on formation of a dimer, −8 cal/K·mol = −33 J/K·mol.

Thus, the calorimetric experiments showed that statistical mechanics, which has been successfully developed for low-molecular-weight gases, is not applicable to macromolecules, especially their aqueous solutions. There are many reasons for that. These might be the restrictions on the motion in the solvent, the inharmonic nature of these motions (Amzel, 1997), the change in the vibrational modes upon association (Tidor and Karplus, 1993), among others. It would be surprising indeed if the macromolecular systems, more so in the liquid environment, could be described by the rather simplistic theory of gases. It is encouraging, however, that the theoretically calculated entropy loss upon protein association decreases in magnitude with the elaboration of statistical-mechanical analysis of the process (Brady and Scharp, 1997). Therefore one would expect that at some time the calculated values will come close to the experimentally determined ones. A critical review of the statistical-thermodynamic basis for computations of binding was given by Gilson et al. (1997) and Yu et al. (2001).

11.3. THERMODYNAMICS OF MOLECULAR RECOGNITION

11.3.1. Calorimetry of Protein Complex Formation

One of the most remarkable features of specific macromolecular complexes is a perfect recognition of their partners. One can expect that this high selectivity of the partners is based on the complementarity of the interacting surfaces of the partners. However, it is known that in many cases association of the molecular partners results in changes of their structures and it looks as though in many cases just these changes are playing essential functional roles. Such examples led to a surprising conclusion:

What is recognized is not the initial structures of the reactants but their final structures in the complex.

Traditionally investigation of the formation of protein complexes was concentrated on equilibrium studies of the binding–dissociation constants at different conditions and, deduced from that, the Gibbs energies of formation of the complexes. The main reason for concentrating on studying the equilibrium constants was experimental: the specificity of recognition of partners by the biological macromolecules, particularly proteins, is usually very high. Correspondingly the association constant between the molecular partners in biology is extremely high, while the dissociation constant is extremely low, in the picomolar concentration range. Therefore, quantitative studies of their binding reaction require the use of highly sensitive methods for measuring association in highly dilute solutions, which are mostly optical. However, the binding constant and the Gibbs energy of binding do not reveal the forces involved in complex formation. For that we need to know the binding enthalpy, entropy, and the heat capacity increment of binding, which could only be determined calorimetrically. As discussed in Section 3.1.4, in contrast to determination of the binding constant, determination of the binding enthalpy does not require the use of highly dilute solutions. The sensitivity of new Nano-ITC instruments is quite sufficient to measure the enthalpy of binding of the macromolecular partners. This is considered next for the example of calmodulin interaction with its molecular target; Chapter 12 is dedicated to the thermodynamics of protein–DNA complexes.

11.3.2. Target Peptide Recognition by Calmodulin

Calmodulin (CaM) is an important regulatory protein that is involved in a wide variety of Ca^{2+}-modulated cellular processes. When activated by binding Ca^{2+}, calmodulin is capable of interacting with a large number of intracellular proteins that, in turn, become activated and regulate various processes such as muscle contraction and cytoskeletal activity (Klee and Vanaman, 1982; Weinstein and Mehler, 1994). The ability to recognize and distinguish between a large number of potential substrates, while avoiding unwanted binding events, is crucial to the proper functioning of calmodulin. Thus it is an appropriate model system for the molecular recognition study.

As discussed in Section 10.2.3, calmodulin consists of two Ca^{2+}-binding blocks (Fig. 10.11). Each block contains two Ca^{2+}-binding domains, which are largely unoccupied at Ca^{2+} concentrations that are typical for a resting cell (100 nM) but are largely occupied at higher concentrations that can be transiently induced by external stimuli (Alberts et al., 1983).

Calmodulin substrates are short amphiphilic helices in kinases (Meador et al., 1992). In particular, the peptide from skeletal muscle light chain kinase (smMLCKp),

ARRKWQKTGHAVRAIGRKSS

binds to Ca^{2+} – CaM with picomolar affinity. According to the optical studies in 100 mM NaCl and 2 mM $CaCl_2$ solution at 30°C, the Gibbs energy of smMLCKp binding to calmodulin is −51 kJ/mol (Blumenthal and Krebs, 1988). When free, this peptide is in a random-coiled conformation but folds into an α-helix upon association with calmodulin (Cachia et al., 1986). Binding also involves a significant conformational change of calmodulin (Meador et al., 1992): in the free state, the two calcium-binding blocks are separated by the central helix; in the bound state, these two calcium-binding blocks come together to engulf smMLCKp and form a single compact structure (Fig. 11.8). The segment of polypeptide chain between the two blocks forms a largely nonpolar central space in which the peptide can reside. Upon binding Ca^{2+}, calmodulin undergoes a large conformational change in the

(a)

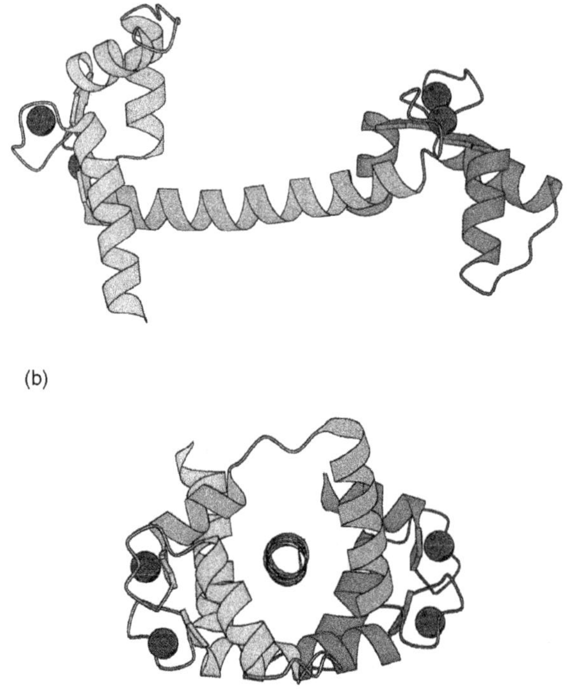

(b)

Figure 11.8. Molscript diagrams of (A) calmodulin and (B) the calmodulin–smMLCKp complex. The N-terminal domain of calmodulin is shown in light gray, the central helix in medium gray, and the C-terminal domain in dark gray (Wintrode and Privalov, 1997).

calcium-binding domains (Finn et al., 1995). Changes in the helix packing in the Ca^{2+}-binding domains result in a significant amount of nonpolar surface area in each domain being exposed to solvent (Kuboniwa et al., 1995; Zhang et al., 1995). Extensive van der Waals contacts are formed between the two molecules. Side chains of nonpolar residues in the peptide occupy the pockets of each Ca^{2+}-binding domain. The fact that Ca^{2+} binding to calmodulin induces the exposure of nonpolar surface area, and that this surface area is buried upon binding smMLCKp, leads to the conclusion that the hydrophobic force is playing the main role in that.

The structural studies provide us with an atomic-resolution view of extensive contacts that are formed between calmodulin and its target peptides. However, understanding of the high affinity that is observed between calmodulin and its targets requires knowledge of its energetics, that is, not only of the binding constant, which was determined from the equilibrium studies, but also the enthalpy and entropy of binding, which are needed to specify the forces involved in binding of the target peptide by calmodulin.

From the existing structural information the changes in the apolar and polar water-accessible surface areas (ΔWASA) upon formation of the Ca^{2+}–CaM–smMLCKp

TABLE 11.3. Thermodynamic Parameters of Dehydration upon smMLCKp Association with Calmodulin at 25°C[a]

Type of Surface	ΔWASA (Å²/mol)	ΔH (kJ/mol)	ΔS (kJ/K·mol)	$T\Delta S$ (kJ/mol)	ΔG (kJ/mol)	ΔC_p (J/K·mol)
Polar	−1443	974	0.734	219	772	1265
Aromatic	−235	35	0.074	22	12	−3825
Aliphatic	−1787	218	1.033	308	−89	−364
Backbone	−252	429	0.248	74	338	457
Total	−3717	1656	2.091	623	1033	−2467

[a]Wintrode and Privalov (1997).

complex were estimated (Table 11.3). Using this structural information and known specific values of the enthalpy, entropy, and heat capacity effects of dehydration of the considered surfaces, these dehydration characteristics upon formation of the Ca^{2+}–CaM–smMLCKp complex were estimated. It appeared that the expected enthalpy of dehydration, ΔH^{hyd}, upon formation of this complex amounts to 1650 kJ/mol. The overall entropy of removal from solvent of polar aromatic and aliphatic groups proceeds with a positive entropy, which amounts to $\Delta S^{hyd} = 2.1$ kJ/K·mol. However, for polar and aromatic groups there is a large unfavorable enthalpic term that easily cancels the favorable entropy. Ultimately, only aliphatic groups make a favorable contribution to the hydration term of the Gibbs energy of binding ($\Delta G^{hyd} = -89$ kJ/mol for aliphatic groups).

Figure 11.9 shows a typical ITC experiment of binding the target smMLCK peptide by calmodulin in pH 7.5 solution with 5 mM imidazole, 100 mM NaCl, and 2 mM $CaCl_2$ at 30°C (Wintrode and Privalov, 1997). This experiment gives the enthalpy of binding as $\Delta H = -68$ kJ/mol. Since at the indicated conditions the Gibbs energy of smMLCKp binding to calmodulin is −51 kJ/mol, that gives for the entropy factor of binding the value $T\Delta S^{exp} = -17$ kJ/mol, and for the binding entropy at this temperature the value $\Delta S^{exp} = -57$ J/K·mol.

Figure 11.10 shows the enthalpies of binding plotted versus the temperature of binding at two different concentrations of $CaCl_2$. According to DSC studies of free calmodulin (see Fig. 10.15 in Section 10.2.3), upon heating calmodulin starts to unfold just above 25°C. Thus, titration at 25°C, 30°C, and 35°C will measure the enthalpy of smMLCKp binding to different conformational states of calmodulin, in which the central helix is progressively more disordered. According to Kirchhoff's relation, the slope of a plot of ΔH versus temperature gives the heat capacity effect of the considered reaction (see Section 2.1.6). This, however, holds if we measure ΔH for the same process at different temperatures. For Ca^{2+}–CaM–smMLCKp binding this will clearly not be the case when we move the temperature above 25°C. Accordingly, one has to exclude the points above 25°C from the linear regression of data estimating the heat capacity increment of binding. Without these points the linear regression gives $\Delta C_p = -2.7$ kJ/K·mol for the heat capacity increment of Ca^{2+}–CaM–smMLCKp binding. This value is close to that estimated from the

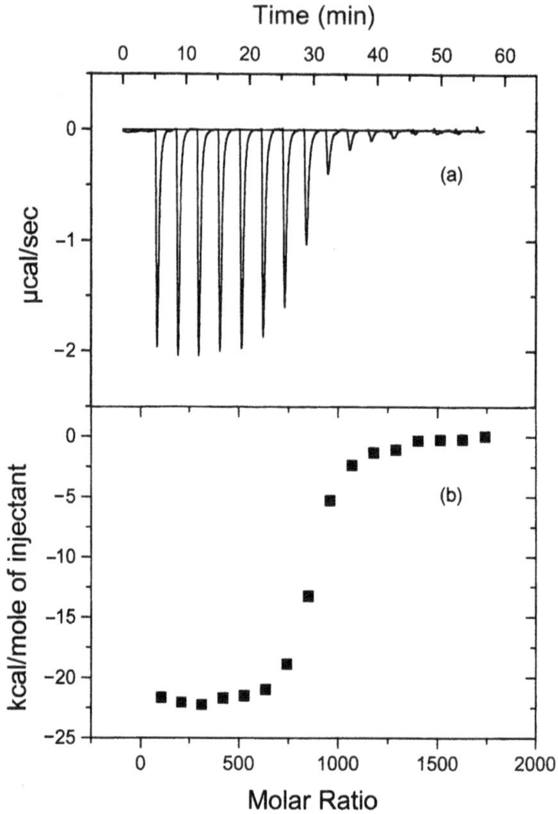

Figure 11.9. (a) Calorimetric titration of calmodulin with smMLCKp at 30°C, pH 7.0, 5 mM imidazole, 100 mM NaCl, 2 mM $CaCl_2$. (b) Binding isotherm obtained from the experiment shown in (a) (Wintrode and Privalov, 1997).

expected dehydration of the heat capacity effect of binding, −2.45 kJ/K·mol (Table 11.3).

According to the ITC data the enthalpy of Ca^{2+}–CaM–smMLCKp complex formation at 25°C amounts to $\Delta H^a = -68$ kJ/mol. This, however, includes the dehydration effect. To obtain the net enthalpy of binding without dehydration effect, we have to exclude from the calorimetrically measured binding enthalpy the expected enthalpy of dehydration, which amounts to −1600 kJ/mol (Table 11.3). Thus, the net enthalpy of binding without hydration effect comes to be 1700 kJ/mol.

Similarly, in considering the entropy effects we have to take into account that the experimentally measured entropy represents the sum of three components:

$$\Delta S^{\text{exp}} = \Delta S^{\text{conf}} + \Delta S^{\text{hyd}} + \Delta S^{\text{trans+rot}}$$

where ΔS^{conf} is the entropy change due to the loss of conformational freedom upon binding, ΔS^{hyd} is the entropy change due to changes in the hydration state of protein

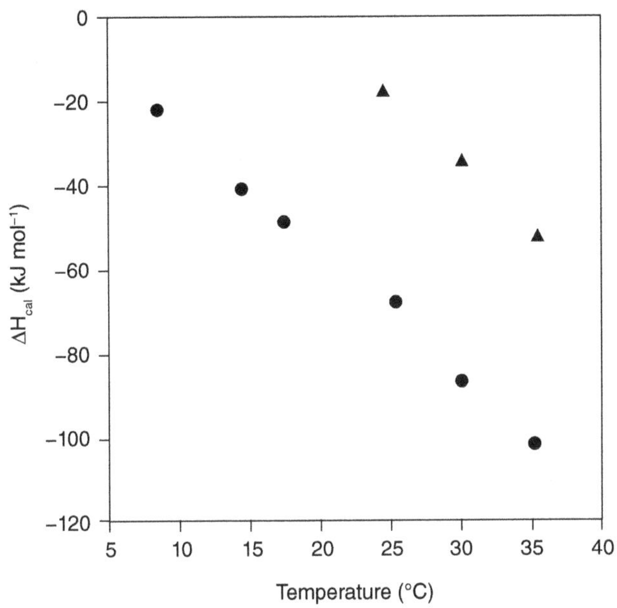

Figure 11.10. Plot of ΔH versus temperature for CaM – smMLCKp binding at pH 7.0 in 5 mM cacodylate 100 mM NaCl, 2 mM $CaCl_2$ (circles) and 5.mM $CaCl_2$ (triangles). The slope in both cases suggests that the heat capacity effect of binding is about −2.7 kJ/K·mol (Wintrode and Privalov, 1997).

groups, and $\Delta S^{trans+rot}$ is the entropy change associated with the loss of the translational and rotational degrees of freedom upon complex formation. As found previously, for the complex formation $\Delta S^{exp} = -57$ J/K·mol, $-\Delta S^{hyd} = 2.10$ kJ/K·mol, and the entropy of reduction of translational–rotational freedom $\Delta S^{trans+rot}$ at the complex formation is about −30 J/K·mol (see Section 11.2.2). It appears thus that the conformational entropy amounts to −2 kJ/K·mol at 25°C. This large negative conformational entropy results from folding of the peptide into a helix upon binding to calmodulin and also the binding-induced conformational changes of calmodulin itself. In particular, if the central helix of calmodulin is already partially disordered, binding of the peptide might lead to its ordering. In any case, it appears that the overall binding entropy, which is calorimetrically measured, is negative, that is, is thermodynamically unfavorable for binding. Thus, binding is provided by the overall enthalpy of binding, which is negative ($\Delta H^{exp} = -68$ kJ/mol). It is notable that it is negative notwithstanding that one of its components, the dehydration enthalpy, is positive ($\Delta H^{deh} = 1600$ kJ/mol).

It appears, therefore, that the real driving force of complex formation is not the negative entropy (which is assumed to be the driving force in the hydrophobic effect) but the large positive net enthalpy of binding resulting from tight van der Waals contacts between the helical peptide and calmodulin (Wintrode and Privalov, 1997).

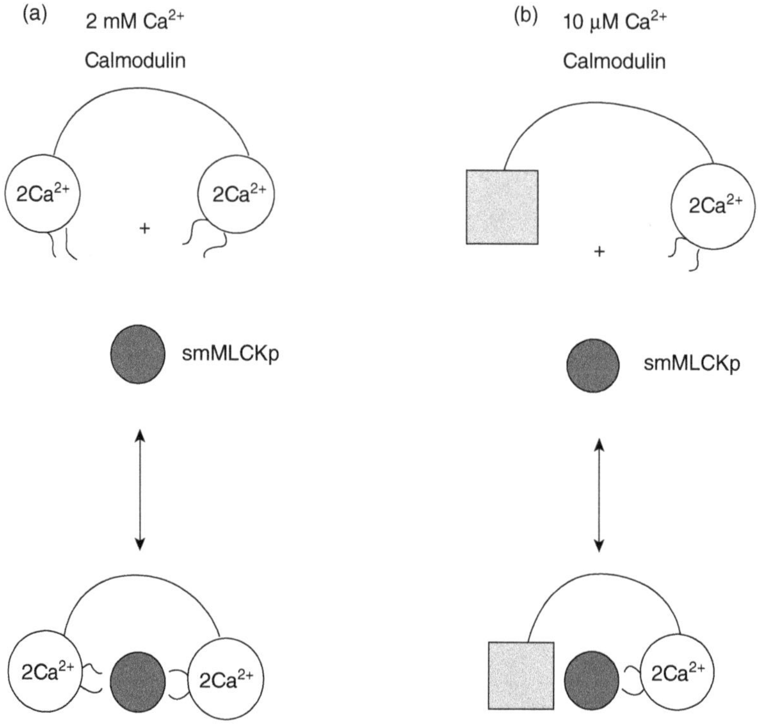

Figure 11.11. Schematic representation of calmodulin smMLCKp binding at (a) 2 mM Ca^{2+} and (b) 10 μM Ca^{2+}. The circles represent calcium-binding domains of calmodulin in the Ca^{2+}-bound conformation with hydrophobic side chains (represented by wavy lines) exposed. Squares represent these domains in the apo conformation with hydrophobic side-chains buried (Wintrode and Privalov, 1997).

The effect of the presence of Ca^{2+} on binding of a target peptide by calmodulin is of particular interest. As Figure 11.10 shows, the enthalpy of formation of the Ca^{2+}–CaM–smMLCKp complex depends dramatically on the concentration of Ca^{2+} in solution. It is known from structural studies of calmodulin that Ca^{2+} binding induces the exposure of apolar surface area to solvent (Finn et al., 1995; Kuboniva et al., 1995; Zhang et al., 1995) and that this surface area makes extensive contacts with smMLCKp peptide (Meador et al., 1992). It appears that Ca^{2+} modulates calmodulin binding activity by changing the exposed apolar surface area that is available to form contacts with its substrates (O'Neil and DeGrado, 1990). In the presence of 10 μM Ca^{2+}, two low-affinity Ca^{2+}-binding sites of calmodulin should not be saturated and the total amount of exposed hydrophobic surface area should be less than at 2 mM Ca^{2+}. Binding of the smMLCKp polypeptide by calmodulin at these two concentrations of calcium in solution is shown schematically in Figure 11.11: at low concentration of calcium the interaction between the peptide and

N-terminal binding sites of calmodulin is lost, and this results in the calorimetrically observed drastic changes in the enthalpy of binding.

11.3.3. Thermodynamic Analysis of Macromolecular Complexes

The fact that the thermal properties of protein complexes with other proteins or nucleic acids change gradually on heating and that this occurs in the temperature range over which the formation of the complexes is usually studied by ITC complicates considerably the analysis of the energetics of formation of these complexes. The situation becomes especially complicated if the binding is not strong and the heating results in partial dissociation of the complex, as shown schematically for protein–DNA complex in Figure 11.12. The equilibrium in that case is expressed by the following set of equations (G. Privalov and P. Privalov, 2000):

$$[\mathrm{PD}] \leftarrow K_A \rightarrow [\mathrm{P}] + [\mathrm{D}]$$

$$[\mathrm{P}] \leftarrow K_P \rightarrow [\mathrm{P}']$$

$$[\mathrm{D}] \leftarrow K_D \rightarrow [\mathrm{D}'] + [\mathrm{D}'']$$

where [PD] is the concentration of the protein–DNA complex, [P] is the concentration of the free folded protein, and [D] is the concentration of free DNA duplex. The concentration of free unfolded protein is [P′] and concentration of free DNA strands is [D′] and [D″], which are usually equal. The equilibrium constant can be expressed by the following set of equations:

$$K_A = \frac{[\mathrm{PD}]}{[\mathrm{P}][\mathrm{D}]}$$

$$K_P = \frac{[\mathrm{P}']}{[\mathrm{P}]}$$

$$K_D = \frac{[\mathrm{D}'][\mathrm{D}'']}{[\mathrm{D}]}$$

$$[\mathrm{P}] + [\mathrm{P}'] + [\mathrm{PD}] = \mathrm{P}_{\mathrm{total}}$$

$$[\mathrm{D}] + [\mathrm{D}'] + [\mathrm{PD}] = \mathrm{D}_{\mathrm{total}}$$

By solving this set of equations the free two-stranded DNA concentration can be determined from

$$[\mathrm{D}] + ([\mathrm{D}]K_D)^{1/2} + K_A[\mathrm{D}]\frac{\mathrm{P}_{\mathrm{total}}}{1 + K_P + K_A[\mathrm{D}]} - \mathrm{P}_{\mathrm{total}} = 0 \qquad (11.5)$$

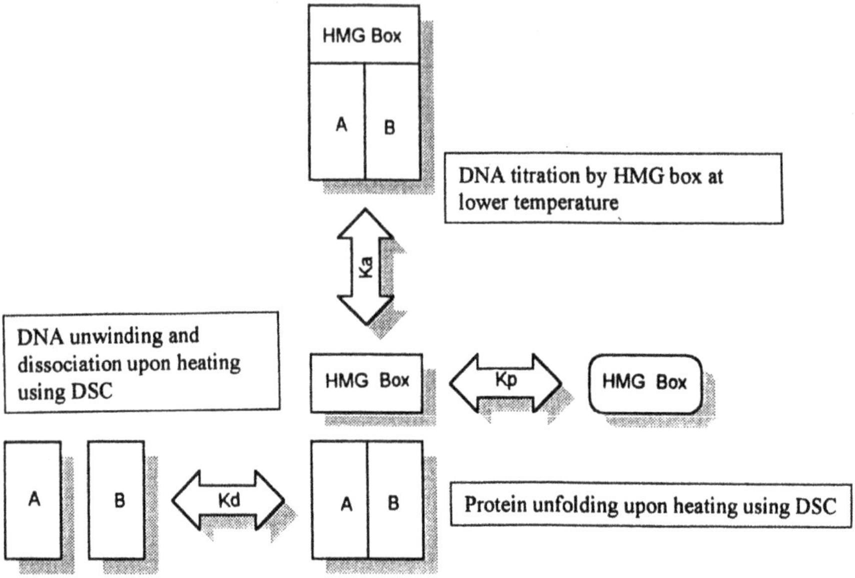

Figure 11.12. Thermodynamic pathways of heat-induced dissociation–unfolding of protein–DNA complex (G. Privalov and P. Privalov, 2000).

The following equations give the concentration of the rest of the components:

$$[\mathrm{P}] = \frac{\mathrm{P}_{\mathrm{total}}}{1 + K_P K_A [\mathrm{D}]} \tag{11.6}$$

$$[\mathrm{PD}] = K_A [\mathrm{P}][\mathrm{D}] \tag{11.7}$$

Binding parameters $K_A(T)$ and $\Delta H_A(T)$ can be evaluated over a wide temperature range using values determined at a given temperature T_0:

$$K_A(T) = K_A(T_0)\exp\left\{-\frac{\Delta H_A}{R}\left(\frac{1}{T} - \frac{1}{T_0}\right) + \frac{\Delta C_{pA}}{R}\left[\ln\left(\frac{T}{T_0}\right) + \frac{T_0}{T} - 1\right]\right\} \tag{11.8}$$

For K_P and K_D we have

$$K = \exp\left[-\frac{\Delta \mathrm{H(T)} - \mathrm{T}\Delta \mathrm{S(T)}}{R\mathrm{T}}\right] \tag{11.9}$$

The apparent heat capacity is determined by the concentration of intermediates:

$$C_p(T) = \Delta H_a \frac{\partial F_c(T)}{\partial T} + \Delta H_P \frac{\partial[1 - F_P(T)]}{\partial T} + \Delta H_D \frac{\partial[1 - F_D(T)]}{\partial T} + C_p^b(T) \tag{11.10}$$

where ΔH_a is the enthalpy of the DNA–protein association, ΔH_p is the enthalpy of protein unfolding, ΔH_{DNA} is the enthalpy of DNA unfolding, and C_p^b is the baseline heat capacity:

$$C_p^b(T) = C_p(T)^{\mathrm{PN}} \times F(T)^{\mathrm{PN}} + C_p(T)^{\mathrm{DN}} \times F(T)^{\mathrm{DN}} + C_p(T)^{\mathrm{PU}} \times F(T)^{\mathrm{PU}} + C_p(T)^{\mathrm{DU}} \times F(T)^{\mathrm{DU}} + C_p(T)C \times F(T)^{\mathrm{C}} \tag{11.11}$$

where $C_p(T)^{\mathrm{PN}}$ is the heat capacity of protein in the native state, $F(T)^{\mathrm{PN}}$ is the concentration of free protein in the native state, $F(T)^{\mathrm{C}}$ is the concentration of protein–DNA complex, $C_p(T)^{\mathrm{DN}}$ is the concentration of the DNA duplex, and $C_p(T)^{\mathrm{DU}}$ and $F(T)^{\mathrm{DU}}$ are the heat capacity and concentration of single-stranded DNA. By modeling the observed heat capacity function of the complex and its free components, all thermodynamic characteristics of the systems can be evaluated. Examples are considered in the next chapter.

REFERENCES

Alberts M., Caves L.S.D., Guimard L., Habbard R.E., Calas B., Gressy G., and Haieck J. (1983). *J. Mol. Biol.*, **244**, 554–571.

Amzel L.M. (1997). *Proteins: Struct. Funct. Genet.*, **29**, 1–6.

Blumenthal D.K. and Krebs E.G. (1988). In: *Calmodulin* (P. Cohen and C.B. Klee, Eds.), pp. 35–53. Elsevier, Amsterdam.

Brady G.P. and Scharp K.A. (1997). *Biophys. J.*, **72**, 913–927.

Cachia P.J., Van Eyk J., Ingraham R.H., McCubbin W.D., Kay C.M., and Hodges R.S. (1986). *Biochemistry*, **25**, 3553–3562.

Chandler D. and Pratt L.R. (1976). *J. Chem. Phys.*, **65**, 2925–2940.

Finkelstein A.V. and Janin J. (1989). *Protein Engineering*, **3**, 1–3.

Finn B.E., Drakenberg T., Waltho J.P., Thulin E., Evanas J., and Forsen S. (1995). *Nature Struct. Biol.*, **2**, 777–783.

Gilson M.K., Given J.A., Bush B.L., and McCommon J.A. (1997). *Biophys. J.*, **72**, 1047–1069.

Gurney R.W. (1953). *Ionic Processes in Solution.* McGraw-Hill, New York.

Hill T.H. (1986). *An Introduction to Statistical Thermodynamics.* Dover, New York.

Holtzer A. (1955). *Biopolymers*, **35**, 595–602.

Horton N. and Lewis M. (1992). *Protein Sci.*, **1**, 169–181.

Janin J. (1995). *Proteins: Struct. Funct. Genet.*, **21**, 30–39.

Janin J. and Chothia C. (1990). *J. Biol. Chem.*, **265**, 16027–16030.

Jenks W.P. (1975). *Adv. Enzymol.*, **43**, 219–410.

Kauzmann W. (1959). *Adv. Protein Chem.*, **14**, 1–63.

Klee C.B. and Vanaman T.C. (1982). *Adv. Protein Chem.*, **35**, 213–321.

Kuboniwa H., Tjandra N., Grzesiek S., Ren H., Klee C.B., and Bax A. (1995). *Nature Struct. Biol.*, **2**, 768–776.

Meador W.E., Means A.R., and Quiocho F.A. (1992). *Science*, **257**, 1251–1257.

Murphy K.P., Xie D., Thompson K.S., Amzel L.M., and Freire E. (1994). *Proteins: Struct. Funct. Genet.*, **18**, 63–67.

O'Neil K.T. and DeGrado W.F. (1990). *Trends Biochem. Sci.*, **15**, 59–64.

Page M.I. and Jenks, W.P. (1971). *Proc. Natl. Acad. Sci. USA*, **68**, 1678–1683.

Privalov G.P. and Privalov P.L. (2000). *Methods Enzymol.*, **323**, 31–62.

Privalov P.L. and Tamura A. (1999). *Protein Engineering*, **12**, 178–187.

Searle M.S., Williams D.H., and Gerhard U. (1992). *J. Am. Chem. Soc.*, **114**, 10697–10704.

Takahashi K. and Sturtevant J.M. (1981). *Biochemistry*, **20**, 6185–6190.

Tamura A. and Privalov P.L. (1997). *J. Mol. Biol.*, **273**, 1048–1060.

Tamura A., Kimura K., and Akasaka K. (1991b). *Biochemistry*, **30**, 11313–11320.

Tamura A., Kimura K., Takahara H., and Akasaka K. (1991a). *Biochemistry*, **30**, 11307–11313.

Tamura A., Kojima S., Miura K., and Sturtevant J.M. (1994). *Biochemistry*, **33**, 14512–14520.

Tanford C. (1973). *The Hydrophobic Effect: Formation of Micelles and Biological Membranes*. Wiley, New York.

Tidor B. and Karplus M. (1993). *Proteins: Struct. Funct. Genet.*, **15**, 71–90.

Tidor B. and Karplus M. (1994). *J. Mol. Biol.*, **238**, 405–414.

Weinstein H. and Mehler E.L. (1994). *Annu. Rev. Physiol.*, **56**, 213–236.

Wintrode P.L. and Privalov P.L. (1997). *J. Mol. Biol.*, **266**, 1050–1062.

Yu Y.B., Privalov P.L., and Hodges R.S. (2001). *Biophys. J.*, **81**, 1632–1642.

Yu Y.B., Lavigne P., Kay C.M., Hodges R.S., and Privalov P.L. (1999). *J. Phys Chem.*, **103**, 2270–2278.

Zhang M., Tanaka T., and Ikura M. (1995). *Nature Struct. Biol.*, **2**, 758–767.

12

PROTEIN–DNA INTERACTION

12.1. PROBLEMS

12.1.1. Two Approaches

Interest in studying the energetics of protein association with DNA rose with the understanding that DNA is the bearer of all genetic information and that reading of this information is controlled by special proteins, the transcription factors, which recognize definite DNA sequences. This assumes formation of specific protein–DNA complexes. One can expect that interaction between the highly charged DNA and the DNA-binding proteins should be largely electrostatic. However, the highly charged DNA is surrounded with counterions, which have to be removed by the protein when binds to the DNA. Mixing of these ions with the ions in bulk solution produces positive entropy. According to the **counterion condensation theory** (CC theory), this entropic effect of mixing might be the main force responsible for the formation of protein–DNA complexes (Manning, 1969, 1978; Record et al., 1978, 1998).

At relatively low concentrations of salt in aqueous solution, when the activity of water is little affected by the presence of salt, this entropy effect is simply proportional to the number of released counterions, mostly from DNA, since the low

Microcalorimetry of Macromolecules: The Physical Basis of Biological Structures, First Edition.
Peter L. Privalov.

density of charges on protein surfaces does not result in a tight coat of counterions. Correspondingly, the logarithm of the protein–DNA association constant, K^a, is presented in just two terms:

$$\log(K^a) = \log(K^a_{\text{nel}}) - (Z \cdot \psi + \beta) \cdot \log[\text{NaCl}] \qquad (12.1)$$

The first term on the right-hand side results from the salt-independent interactions between DNA and protein; the second, the salt-dependent term, results from the release of counterions. Here Z is the number of DNA phosphates that interact with the protein, ψ is the number of cations per phosphate released upon protein binding, and β is the number of anions released from the protein upon DNA binding. Thus the slope of this function, that is, $\partial \log(K^a)/\partial \log[\text{NaCl}] = (Z \cdot \psi + \beta) = N$ actually gives the total number N of counterions released upon protein binding to DNA.

When the salt concentration approaches 1 M, that is, $\log[\text{NaCl}] = 0$, the second term in Equation (12.1) vanishes and $\log(K^a)$ approaches $\log(K^a_{\text{nel}})$. This **salt-independent component** of the Gibbs free energy of association is usually considered as **nonelectrostatic**: $\Delta G^{\text{nel}} = -2.3 \cdot RT\log(K^a)$. Correspondingly, the **salt-dependent component** of the Gibbs energy of binding is usually regarded as **electrostatic**, ΔG^{el}. Therefore, this approach to protein–DNA interaction permits splitting the observed Gibbs energy of association into two components—salt independent (nonelectrostatic) and salt dependent (electrostatic): $\Delta G^a = \Delta G^{\text{nel}} + \Delta G^{\text{el}}$.

According to the CC concept, the salt-dependent component of the Gibbs energy of binding is completely entropic. It follows then that the binding enthalpy should not depend on the salt concentration; that is, it should be nonelectrostatic.

This approach, which assumes that the counterions are tightly bound to the DNA phosphates, was criticized as too simplified. A much more sophisticated approach was suggested, based on application of the nonlinear Poisson–Boltzmann (PB) equation to the protein–DNA complex in aqueous media (Misra et al., 1994a,b; Misra and Honig, 1995; Sharp and Honig, 1990; Hecht and Honig, 1995). According to the PB theory developed for protein–DNA interaction, the salt effect is described in terms of three free energy contributions: the Gibbs free energy of electrostatic ion–molecule interaction, of electrostatic ion–ion interaction, and of entropic ion organization. It is assumed that the first two Gibbs energy terms have both enthalpic and entropic components, while the third term is purely cratic entropy of dissolution of released ions. The main practical problem of this approach is that it does not suggest a clear procedure for the experimental evaluation of each of these terms, which is important for its verification. On the other hand, the complexity of the system being considered, which includes not only protein and DNA but also the water between them, induces a certain doubt in the efficiency of using such a sophisticated theory for its analysis. Indeed, how can one apply the PB equation to such a heterogeneous system as a protein–DNA interface, where water is no longer a liquid solvent and the dielectric constant does make any sense?

On the other hand, the CC approach to protein–DNA interactions, which proposes a split of the Gibbs free energy of association into salt-dependent (electrostatic) and salt-independent (nonelectrostatic) components, in itself poses questions in order to decide if it effectively describes the energetic basis of forming such complexes:

(a) Is the salt dependence of the binding constant indeed linear on the logarithmic scale and does the slope of this functional dependence really represent the number of released counterions?
(b) What is the meaning of $\log(K^a)$ at 1 M salt concentration, where the second term in Equation (12.1) drops to zero?
(c) Is the salt-dependent electrostatic component of the Gibbs energy of association indeed purely entropic?
(d) Is the binding enthalpy indeed salt independent?
(e) What does the nonelectrostatic component of binding energy represent?

As we will see from the discussion in this chapter, calorimetry has played an essential role in answering these questions and thus in elucidating the physics of protein–DNA interaction.

12.1.2. Protein Binding to the DNA Grooves

The important feature of protein–DNA interactions is that the DNA-binding proteins, which recognize their cognate sites with highest accuracy, are interacting tightly with one of the two grooves of the DNA, the major or minor (Fig. 4.17). Correspondingly, the DNA-binding proteins are divided into two groups: major-groove binders and minor-groove binders. The main difference between these two grooves of DNA is in their width: the major groove is significantly wider than the minor one (see Section 4.4). The major groove is wide enough to accommodate without significant distortions some structural elements of proteins, particularly the α-helix. Binding to the minor groove induces deformation not so much of the protein as of the DNA: in some cases it results in DNA bending by about 120°, in contrast to protein binding to the major groove, which usually occurs without noticeable deformation of the DNA. One could expect therefore that a strong bending of DNA upon protein binding to the minor groove should require considerable work. Classical studies characterized DNA as an elastic rod with a persistence length L_p on the order of 50–60 nm (~150 bp) (Wiggins et al., 2005; Baumann et al., 2000; Cloutier and Widom, 2004). The wormlike chain model (Landau and Lifshitz, 1970) predicts a free energy expenditure of ~70 kJ/mol in bending a 10-bp duplex through 50 degrees, that is, ~1.5 kJ per degree of bend. The question is then: why is binding of proteins to the minor groove used in nature? Is it because bending of DNA is functionally required? Then, how is the work of DNA-bending covered?

A remarkable specificity of the minor-groove binder proteins is their preference for the AT-rich sequences in DNA. Bearing in mind the absence of the 2-amino group in adenosine (relative to guanosine), one might argue that this groove is more apolar than the major groove (Fig. 4.16). It is tempting to assume that the hydrophobic interactions play a significant role in binding to the minor groove rather than to the major one. Therefore, comparison of the energetics of protein binding to the major and minor grooves of DNA presents a certain interest.

First we consider proteins binding to the major groove of DNA, which does not induce significant distortions in the structure of both interacting partners, such as occur upon formation of the DNA complex with homeodomains (Section 12.2.1), or induce considerable changes in the protein but not DNA, as happens upon formation of DNA complexes with the bZIP proteins (Section 12.2.2). After that we consider proteins binding to the minor groove: first, AT-hooks, which enter easily into the minor groove of DNA without noticeably distorting it (Section 12.3.1); then examples of the transcription factors of the high-mobility group (HMG) family, which induce significant bending of DNA (Section 12.3.2).

12.2. BINDING TO THE MAJOR GROOVE OF DNA

12.2.1. Homeodomains

Eukaryotic homeotic transcription factors and bacterial repressors contain DNA-binding domains (DBDs) of the helix-turn-helix type that exhibit sequence-specific DNA binding by insertion of their "recognition" α-helices into the major groove of DNA and a short extended N- or C-terminal arm into the adjacent minor groove (Fig. 12.1). Structural studies showed that binding does not result in noticeable deformation of the DNA, though some changes in the structure of the homeodomains were noticed (Fraenkel et al., 1998; Gruschus et al., 1997; Kissinger et al., 1990; Tsao et al., 1994; Wolberger et al., 1991). Several calorimetric studies have been carried out to determine the energetics of association of the homeodomains and bacterial repressor DBDs with DNA duplexes containing their recognition sequences (Carra and Privalov, 1997; Gonzales et al., 2001; Ladbury et al., 1994; Merabet and Ackers, 1995; Takeda et al., 1992). These have shown a broad range of enthalpies and entropies of binding that do not obviously correlate with their association constants, which in fact do not vary greatly. To solve the problem, four different examples of homeodomains and their interaction with the target DNA duplexes were studied by calorimetric and optical methods under identical conditions: *Drosophila* Engrailed, *Saccharomyces cerevisiae* MATα2, *Drosophila* Antennapedia, and *Drosophila* NK2, the last two of which were also investigated in truncated forms lacking their N-terminal arms.

Figure 12.2 shows the partial specific heat capacity functions of NK-2 and Antennapedia homeodomains and their truncated forms in solutions of 100 mM NaCl, 10 mM sodium acetate, pH 5.0, determined by differential scanning calorimetry (DSC). These profiles are perfectly reproducible in subsequent heating–cooling experiments if the solution is not heated above 70°C. One can therefore conclude that the temperature-induced unfolding of these proteins is highly reversible. This figure also shows the heat capacity functions of the fully unfolded homeodomain,

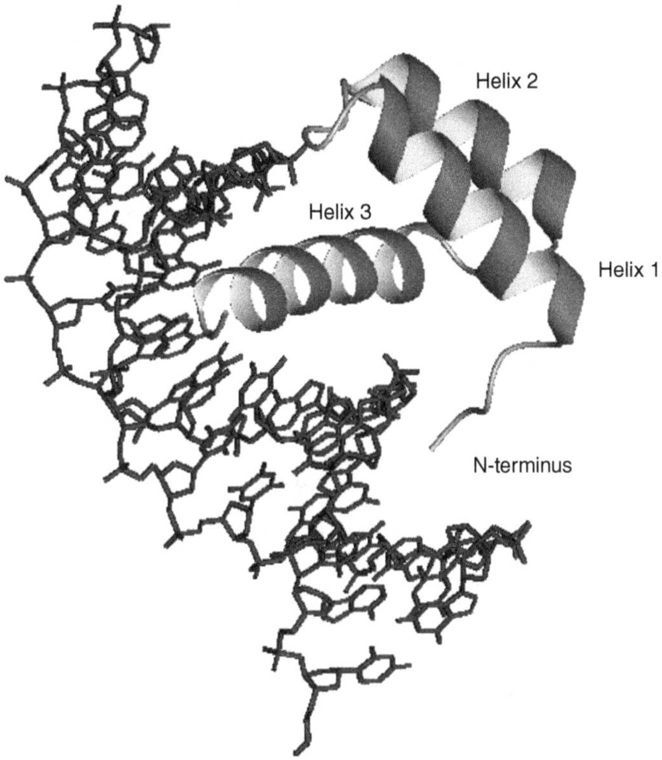

Figure 12.1. Association of the Antennapedia homeodomain with its target DNA. The DNA-recognition helix sits in the major groove of DNA without its essential distortion, while the N-terminal tail enters into the minor groove (PDB 9ANT). (See insert for color representation of the figure.)

C_p^{U}, calculated by summing the known heat capacities of individual amino acids (see Section 7.2.4), which appeared to be very similar for all homeodomains studied. The heat capacities of the native homeodomains increase gradually, with slopes that do not differ much from one expected for a fully folded protein, C_p^{F}, with a compact rigid structure, such as bovine pancreatic trypsin inhibitor (BPTI) (see Section 8.2.1); but they are somewhat higher in absolute magnitude, as one would expect, since the structure of homeodomains is not 100% compact but contains an unfolded terminal segment. The $C_p(T)$ functions at low temperatures show that the extent of this unfolded part does not significantly increase with heating below the point of temperature-induced cooperative unfolding of the compact domain, which proceeds with an extensive heat absorption and significant heat capacity rise to the level expected for fully unfolded polypeptides. Analysis of the excess heat absorption shows that this process is well approximated by a two-state transition. It is notable that these homeodomains differ significantly in stability: among them the NK-2 is the most unstable and starts to unfold upon heating from 20°C, and its transition

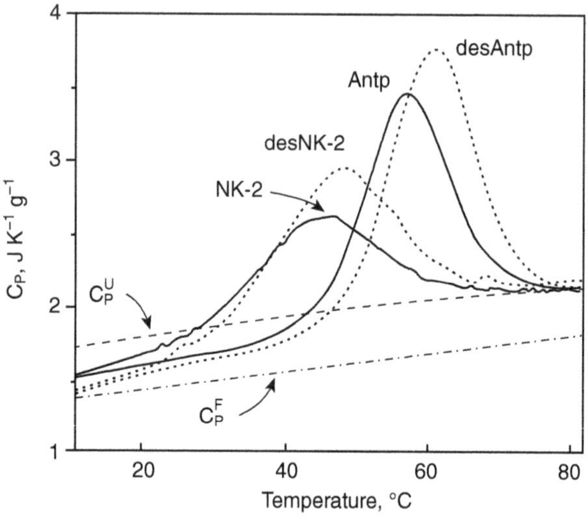

Figure 12.2. Partial specific heat capacity functions of the NK-2, Antp (solid lines) and their truncated forms desNK-2 and desAntp (dashed lines) in 100 mM NaCl, 10 mM sodium acetate buffer, pH 5.0. The dashed-and-dotted lines show the heat capacity function expected for the fully folded proteins, C_p^F, and fully unfolded proteins, C_p^U (Dragan et al., 2006).

temperature is 42.1°C. The most stable is desAntp, the transition temperature of which is 59°C. It is striking that the transition temperatures of the truncated forms of Antp and NK-2, lacking their N-terminal arms, are about 5 K higher than the full-length parent homeodomains. Thus, stabilities of the truncated forms, expressed in the Gibbs energies, are higher by 2 kJ/mol. It appears that:

The N-terminal arm is destabilizing the globular part of these proteins.

Figure 12.3a shows the enthalpies of binding the homeodomains to their cognate DNAs measured by isothermal titration calorimetery (ITC) at different temperatures (Dragan et al., 2006). The binding enthalpies are largely negative over the temperature range considered, and their magnitude increases with increasing temperature. The values of these enthalpies vary considerably for the different complexes, lying between −7 and −50 kJ/mol at 20°C, but their temperature dependence is similar. This is because the DNA binding of MATα2 and NK-2 is accompanied by an increase in the length of their recognition helices by about seven residues. Taking the enthalpy of folding of an α-helix to be about 3 kJ per mole of residue (see Section 5.1.2), refolding of these proteins upon binding to DNA results in a release of ~21 kJ/mol; therefore a positive correction of this magnitude has been applied to these two homeodomains. A small additional correction for refolding must also be introduced for NK-2, which starts to unfold at temperatures above 20°C but refolds upon binding to DNA (see Fig. 12.2). Introducing all these corrections (see Section 3.2.8 for the correction method), one finds that the enthalpies of association of the completely

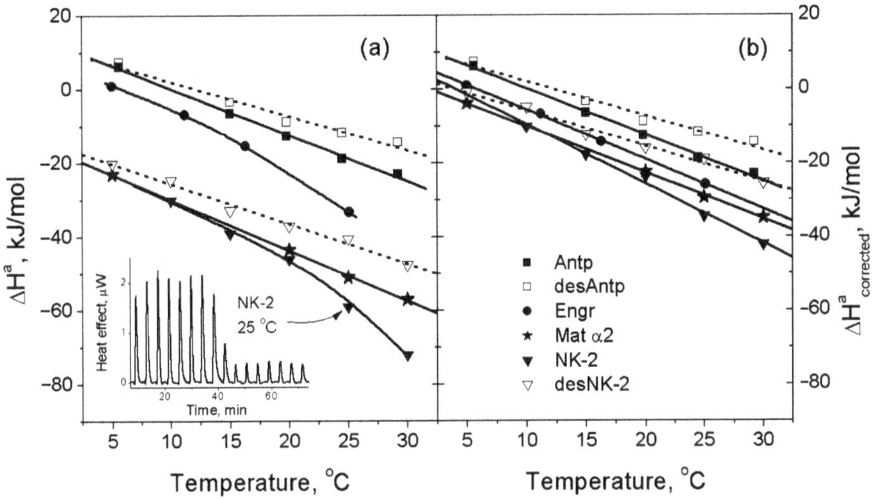

Figure 12.3. (a) Enthalpies of association of the homeodomains (solid lines) and their truncated forms (dashed lines) with their cognate DNAs measured by ITC at different temperatures; inset, traces of the calorimetric titration of 16 μM DNA by NK-2 at 25°C. (b) Enthalpies of association of the homeodomains after correcting NK-2 and MATα2 for the heats of refolding of seven residues in their recognition helices and in the case of NK-2 for some further protein refolding that occurs upon binding. All the data shown were obtained in 100 mM NaCl, 10 mM sodium acetate, pH 5.0 (Dragan et al., 2006).

folded homeodomains with DNA are rather similar and have similar temperature dependence, that is, the heat capacity effects of their binding to DNA (Fig. 12.3b).

If the binding enthalpies for the Antp and NK-2 DBDs are compared with those of their truncated forms, it is seen that they differ significantly. For Antp, removal of the six N-terminal residues (RKRGRQ) reduces the binding enthalpy at 20°C from −12 to −7 kJ/mol. In the case of NK-2, removal of its longer N-terminal arm, of which only the seven residues KKRKRRV make contact with DNA, reduces the binding enthalpy at this temperature from −24 to −16 kJ/mol. It appears that the N-terminal arms, which enter into the minor groove of DNA, contribute a significant negative enthalpy to binding. It is notable that the N-terminal arms also change the slopes of the enthalpy functions: the temperature dependence of the enthalpy is less for desAntp and desNK-2 than for the full-length homeodomains. Thus, the heat capacity effect (ΔC_p) of binding the arms to the minor groove is negative (Crane-Robinson et al., 2006).

The DNA affinities of all the homeodomains are quite high: the K_d values being in the nanomolar range. Therefore, they should be determined at nanomolar concentrations of the reactants. So the binding constants of homeodomains with their cognate DNAs cannot be determined by ITC experiments, which require the use of significantly higher concentrations of the reactants. The binding isotherms of the homeodomains to their target DNAs were therefore determined by fluorescence

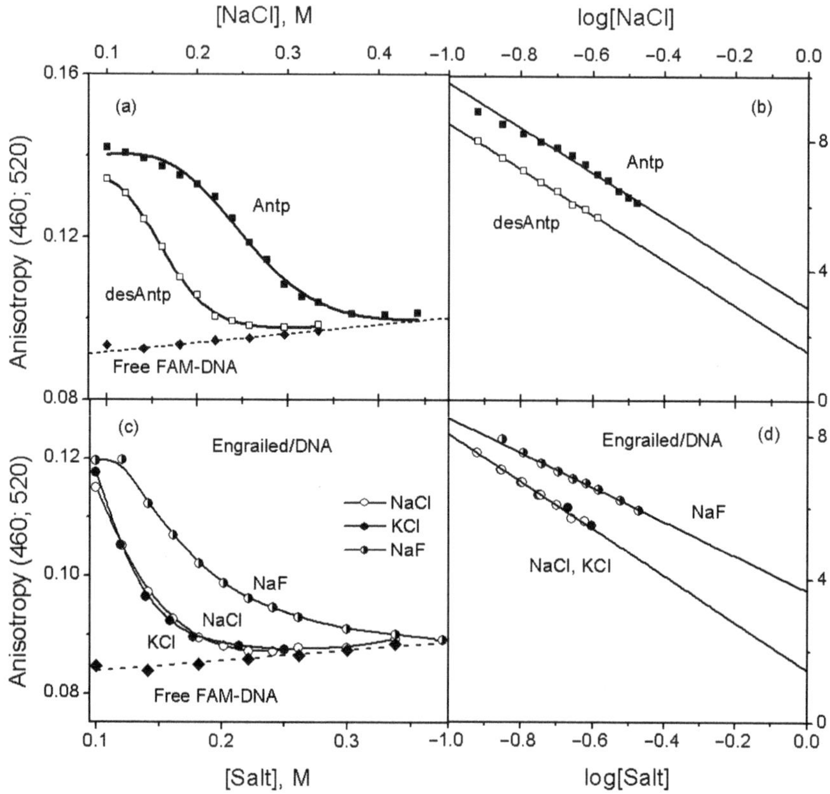

Figure 12.4. (a) The change of anisotropy of the Antp and desAntp complexes at 20°C in 10 mM sodium acetate, pH 5.0, upon increase of NaCl concentration. The initial concentration of the complex was 50 nM. The dotted line represents the anisotropy of the free FAM-labeled DNA. (b) Dependence of log(K^a) on log[NaCl] for the binding of the full and truncated Antp with their cognate DNA at 20°C. (c) The change of anisotropy of the Engrailed complex at 20°C in 20 mM sodium acetate, pH 5.0, upon increase of NaCl, KCl, and NaF concentrations. The dotted line represents the anisotropy of the free FAM-labeled DNA. (d) The dependence of log(K^a) of the Engrailed homeodomain on the logarithm of the NaCl, KCl, and NaF salt concentrations, showing the importance of the anion (Dragan et al., 2006).

anisotropy titration, using DNA duplexes 5′-labeled with FAM fluorophore. Figure 12.4a shows the salt titrations of the Antp and desAntp complexes, from which it is seen that the desAntp complex starts to dissociate at significantly lower salt concentration than the full Antp. The association constants calculated from these experiments are presented on a logarithmic scale in Figure 12.4b, showing that the dependence of log(K^a) on log[NaCl] is linear. It is particularly interesting that removal of the N-terminal arm does not change the slope of the log(K^a) dependence on log[NaCl] but leads to a parallel shift of this function. Thus, at log[NaCl] = 0, that is, at 1 M NaCl solution, the value of log(K^a)$_0$ is smaller. A similar situation was

TABLE 12.1. Binding Characteristics of Homeodomains Binding to their Cognate DNAs[a,b]

Protein	ΔH^a	ΔC_p	log(*K*)	$\frac{(\log(K))}{(\partial \log[\mathrm{NaCl}])}$	log(*K*)$_0$
Antp	**−12.0**	**−1.26**	**9.7**	−6.9	2.8
desAntp	**−7.0**	**−0.93**	**8.5**	**−7.0**	1.5
NK-2	−24.0	−1.60	9.1	−5.5	3.6
desNK-2	−16.0	−0.97	7.2	−5.6	2.9
MATα2	−26.0	−1.25	8.9	−5.8	3.1
Engrailed	−19.0	−1.35	8.1	−6.6	1.5
σ	±1.0	±0.10	±0.2	±0.3	±0.2

[a]At 20°C in 100 mM NaCl, 20 mM sodium acetate buffer, pH 5.0; ΔG and ΔH in kJ/mol; ΔC_p in kJ/K·mol.
[b]ΔH values obtained from ITC measurements; ΔC_p values obtained as the slopes of the $\Delta H/T$ functions after correction for protein refolding; values of association constant obtained from binding curves determined by fluorescence anisotropy titrations at different NaCl concentrations; $(K)_0$ is the value of the association constant extrapolated to 1 M NaCl concentration.

found for truncation of the N-terminal arm of NK-2. Figure 12.4c shows, for example, that the effect of NaF on the dissociation of the Engrailed homeodomain complexes is much lower than that of either NaCl or KCl. Thus, the dependence of $\log(K^a)$ on the logarithm of the NaF concentration is less steep than for KCl or NaCl (Fig. 12.4d). It appears that the association constants do not depend on the monovalent cations used, but do depend on the anion, namely Cl^-. One can conclude from this that the Cl^- ion is specifically associated with the Engrailed homeodomain and is released upon its association with DNA (see also Lohman et al., 1996; Arosio et al., 2004).

The logarithm of association constants in 100 mM NaCl, the slopes of their functional dependence on the logarithm of the salt concentration, that is, $\partial \log(K^a)/\partial \log[\mathrm{NaCl}]$, and the extrapolated values of the logarithm of the association constant at $\log[\mathrm{NaCl}] = 0$, that is, $\log(K^a)_0$, for all these homeodomains and their truncated forms are all listed in Table 12.1. From this table is seen that the association enthalpies of all the homeodomains are negative and the entropy factors, $-T\Delta S$, are also negative, that is, the binding entropies are positive. Thus the enthalpy and entropy factors both favor binding:

The formation of DNA–homeodomain complexes is enthalpy and entropy driven, but the entropy factor dominates.

From the large positive overall entropy increase on binding it is tempting to conclude that hydrophobic interactions play a significant role in binding the homeodomain to its target DNA; that is, their association results in dehydration of contacting apolar groups with a consequent increase in the water entropy. This conclusion seems to be confirmed by the negative heat capacity effect of binding the homeodomains, a feature specific to the dehydration of apolar groups (see Section 2.4.1). In fact, as follows from Table 12.2:

TABLE 12.2. Thermodynamic Parameters of Association of the Homeodomains with Their Cognate DNAs[a,b]

	Total			Electrostatic	Nonelectrostatic		
Protein	ΔG	ΔH	$-T\Delta S$	ΔG_{el}	ΔG_{nel}	ΔH_{nel}	$-T\Delta S_{nel}$
Antp	−54.3	−12.0	−42.3	−38.6	−15.7	−12.0	−3.7
desAntp	−47.6	−7.0	−40.6	−39.2	−8.4	−7.0	−1.4
NK-2	−51.3	−24.0	−27.3	−30.3	−21.0	−24.0	3.0
desNK-2	−40.6	−16.0	−24.6	−31.3	−9.3	−16.0	6.7
MAT 2	−49.8	−26.0	−23.8	−32.5	−17.3	−26.0	8.7
Engrailed	−45.1	−19.0	−26.1	−36.9	−8.2	−19.0	10.8
σ	±1.0	±1.0	±2.0	±2.0	±2.0	±1.0	±3.0

[a]At 20°C in 100 mM NaCl, 20 mM sodium acetate, pH 5.0; ΔG, ΔH, and $T\Delta S$ in kJ/mol.
[b]ΔH values obtained from ITC measurements; ΔG values calculated from the K^a values obtained from fluorescence anisotropy titrations; $T\Delta S$ values obtained by calculation from the equation $T\Delta S = \Delta H - \Delta G$. ΔG_{nel} is calculated from $(K^a)_0$ since this represents the binding constant in 1 M NaCl, where electrostatic contributions have become zero. Subtraction of ΔG_{nel} from ΔG gives ΔG_{el}, which equals $-T\Delta S_{el}$ since there is no enthalpic contribution to the electrostatic interactions. Since the measured enthalpy is all nonelectrostatic, combining ΔH_{nel} with ΔG_{nel} yields the nonelectrostatic entropy factor $T\Delta S_{nel}$.

The positive entropy of binding is provided largely by the release of counterions, whereas the nonelectrostatic component of the binding entropy, which includes the effect of dehydration of the groups at the interface, is close to zero and is even negative in the case of the NK-2, MATα2, and Engrailed homeodomains.

Since the enthalpy of hydrophobic interactions (i.e., the net enthalpy of formation of van der Waals contacts between apolar groups and the dehydration of these groups) is known to be close to zero at 20°C (see Section 8.3.1 and Fig. 8.12), the observed negative enthalpies of binding at that temperature must result mainly from hydrogen bonding between the protein and DNA. The enthalpy of hydrogen bond formation at 20°C is about −3 kJ/mol (see Sections 5.1.2 and 8.3.2). Therefore, one can expect formation of three to nine hydrogen bonds upon association of the homeodomains with DNA. Indeed, the crystal structure shows that in the case of the Antp/DNA complex three residues make direct hydrogen bonds to the DNA: Arg5 in the minor groove; Ile47 and Asn51 in the major groove, the last being bifurcated (Fraenkel and Pabo, 1998). If that is taken to represent four hydrogen bonds, the binding enthalpy at 20°C should be about −12 kJ/mol, which corresponds well with the observed binding enthalpy value (Table 12.1).

The electrostatic component of the total free energy of homeodomain binding to DNA varies between −30 kJ/mol and −39 kJ/mol, representing 60–80% of the total binding energy.

Thus the electrostatic component dominates over the nonelectrostatic component.

Nevertheless, the nonelectrostatic component is of special interest, as will be discussed later (see Section 12.4.2), since just this component of the binding energy determines the specificity of DNA recognition.

12.2.2. Binding of the GCN4 bZIP to DNA

The discovery of a novel scissors-grip motif of DNA recognition of eukaryotic transcription factors (Landschulz et al., 1988) induced a wave of research on their structure and interaction with DNA. The suggested dimeric bZIP model (Vinson et al., 1989; Weis, 1990) was almost immediately confirmed by a number of crystallographic studies of the complexes of various representatives of this large protein family associated with their target DNAs (Ellenberger et al., 1992; Ellenberger, 1994; König and Richmond, 1993; Keller et al., 1995; Glover and Harrison, 1995; Schumacher et al., 2000; Fujii et al., 1999; Miller et al., 2003). These structural studies showed that all bZIP–DNA complexes are rather similar, consisting of the dimerization domain (leucine zipper) and the C-terminal basic segments responsible for DNA recognition and binding (Fig. 12.5). The positively charged basic segments are largely unfolded in the absence of the target DNA, but, being neutralized by the negatively charged groups of the DNA, they fold and enter into the major groove of DNA. The association constant of the bZIP, particularly of the GCN4 transcription factor, with different DNA binding sites was estimated to vary between 2×10^7 and 7×10^8 M^{-1} at 25°C (Weiss et al., 1990; König and Richmond, 1993; Berger et al., 1996; Cranz et al., 2004).

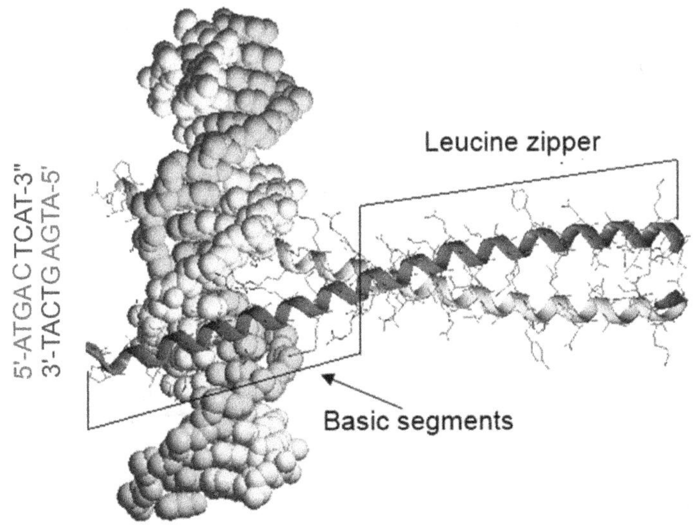

Figure 12.5. The model of bZIP-GCN4 complex with the pseudopalindromic AP-1 DNA showing the DNA-recognizing basic segments entering into the major grooves of DNA. (See insert for color representation of the figure.)

Determination of the binding constant of the bZIP to target DNA is not simple because it is too high for the dimeric bZIPs, whereas the dimerization constant of the strands forming bZIP is rather low. If the association constant of the bZIP to the target DNA is on the order of 10^8 M^{-1}, to get a well-resolved binding isotherm the titration experiment should be carried out at concentrations of DNA and titrant on the order of 10^{-8} M. However, the dimeric titrant, that is, the bZIP, even at a hundred times higher concentration (i.e., at about 10^{-6} M) might be partly dissociated. In that case the observed association process of the bZIP with DNA would not be a simple reaction but would proceed through a dimeric and monomeric pathway (Patel et al., 1994; Kohler and Schepartz, 2001; Cranz et al., 2004). The detailed thermodynamic characterization of this complicated process requires not only determination of the proportion of the monomeric and dimeric proteins, which changes due to the course of association, but also to account for all concomitant effects, for example, the loss in translational freedom by the separated strands upon association, refolding, and so on (see Section 11.3.3). This complication can be avoided, however, by the S–S cross-linking bZIP at the N-terminus: this will not change the ability of the bZIP to bind to DNA, but will prevent its dissociation–unfolding at the low concentrations used in the titration experiment.

Figure 12.6a shows the results of fluorescence anisotropy titration of AP-1 and ATF/CREB sites by the S–S cross-linked GCN4 $bZIP^{SS}$ at 20°C. The DNA concentration in both cases was 10 nM. One can see the clear difference between the two isotherms, suggesting two different dissociation constants: K_d for the bZIP complex with the AP-1 site was found to be 1.5 nM; for the complex of the $bZIP^{SS}$ with ATF/CREB the K_d is 6.4 nM. The logarithmic plot presented in Figure 12.6b shows that:

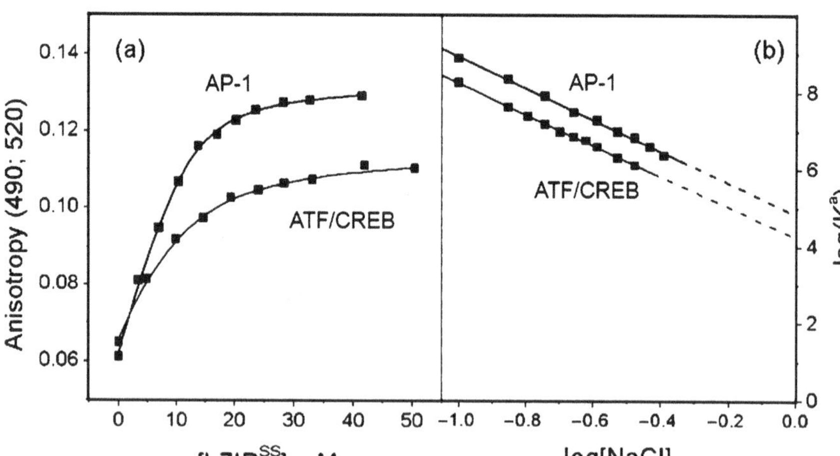

Figure 12.6. (a) Binding isotherms of GCN4-bZIPSS with the 15-bp FAM-labeled AP-1 and ATF/CREB sites measured by fluorescence anisotropy at 10 nM concentration of DNA. (b) Dependence of the logarithm of the association constant of bZIPSS with the AP-1 and ATF/CREB sites plotted against the logarithm of the NaCl concentration (Dragan et al., 2004a).

On the logarithmic scale the association constant is a perfectly linear function of salt concentration.

Extrapolating $\log(K^a)$ to $\log[\text{NaCl}] = 0$, that is, to 1 M salt concentration, where the electrostatic component drops to zero, one can determine the nonelectrostatic component of binding energy at this standard condition. By excluding it from the binding energy at 0.1 M NaCl, one gets the electrostatic component of binding energy at this salt concentration.

The enthalpy of association of bZIP with target DNA duplexes was studied by ITC at several fixed temperatures (Dragan et al., 2004a). The sensitivity of the ITC instrument does not permit experiments at a concentration low enough to resolve the binding isotherm in detail, but the heat effects prior to saturation reliably provide the enthalpy of association. It appeared that the enthalpies of association of bZIP with AP-1 and ATF/CREB sites are very different: in both cases they are negative and increase in magnitude with increasing temperature, with slopes of about −2.5 kJ/K·mol, but differ in value by about 50 kJ/mol (Fig. 12.7a). This surprising difference in the enthalpies of binding GCN4 bZIP to these two DNA binding sites was observed also by Berger et al. (1996).

The dependence of association enthalpy on temperature is usually interpreted as the heat capacity of binding in accordance with Kirchhoff's relation, $\Delta C_p^a = d\Delta H^a/dT$ (see Section 2.1.6). The heat capacity effect of binding is an important parameter since it indicates the extent of dehydration of the groups at the interface that is formed. However, Kirchhoff's relation is valid only if the components of reaction

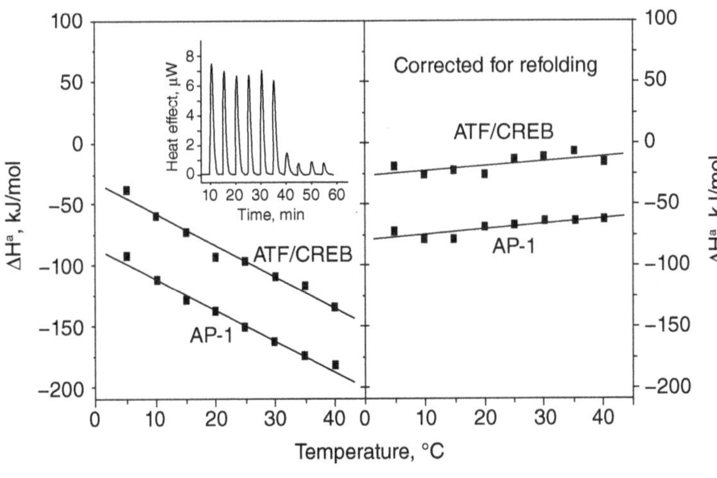

Figure 12.7. (left) The enthalpy of association of the bZIP with the AP-1 and ATF/CREB sites measured by ITC at different temperatures. Inset: the ITC-observed heat effect of titration of the 10 nM AP-1DNA by the 300 nM solution of bZIP injected in 5 μL portions. (right) The net enthalpies of association of the bZIP with the AP-1 and ATF/CREB binding sites, corrected for the refolding of the basic segments (Dragan et al., 2004a).

do not change in the temperature range being considered. It is known, however, that the basic segments of the bZIP unfold in that temperature range and they refold upon binding to the DNA. Thus, the calorimetrically measured enthalpy of the bZIP association with DNA includes the heat of refolding of the basic segments and, therefore, the observed temperature dependence of this enthalpy does not represent the heat capacity effect of binding. If we want to determine the net enthalpy of association of the bZIP with completely folded basic segments to the DNA, we have to exclude this temperature-dependent heat of refolding of the basic segments from the calorimetrically measured apparent enthalpy of association. As is evident, to understand the energetics of the bZIP interaction with DNA we need just this net enthalpy of association of the fully folded bZIP with the target DNA. If one takes into account that the mid-transition temperature of the basic segments is at room temperature, that is, in the considered temperature range, the Gibbs energy of their unfolding–refolding is close to zero and does not contribute much to the Gibbs energy of the bZIP binding to DNA. Thus, we only need to correct the enthalpy of association for refolding, but not the Gibbs energy of binding, and use this corrected enthalpy to determine the entropy of binding, which then would be automatically corrected for refolding.

The DSC studies of unfolding–refolding of the DNA-binding basic segment of the GCN4 bZIP showed that it is highly unstable when isolated from the leucine zipper part of the GCN4, but when attached to the leucine zipper it melts with considerable heat effect in the broad temperature range, from −10°C to 70°C, with a peak at room temperature (Fig. 12.8). Using the heat capacity function of the basic segment attached to the leucine zipper, one can estimate for each temperature the

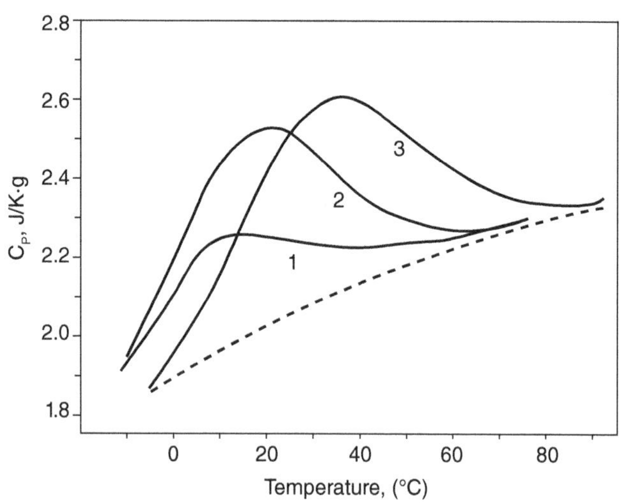

Figure 12.8. DSC study of melting of the basic segments of GCN4 bZIP: (1) the DNA-binding basic segment isolated from the GCN4 bZIP; (2) the basic segment attached to the GCN4 bZIP; (3) the isolated basic segment with covalently closed terminal loops (Dragan et al., 2004a).

heat of refolding of the basic segment and use it for correcting the heat effect of binding the bZIP to the DNA (see Section 3.2.8 for the correction method). The net enthalpies of association of the bZIP with the AP-1 and ATF/CREB binding sites corrected for refolding, $\Delta\hat{H}^a$, are shown in Figure 12.7b. It appears that the correction for the heat of refolding the basic segment has changed drastically the magnitudes of the enthalpies of association and their dependence on temperature: they have become smaller in magnitude (although still negative) and their dependence on temperature has changed sign and become small. It appears thus that the heat capacity effect of binding the fully folded bZIP to the target DNA is positive. The positive heat capacity effect means that in the binding process the dehydration of the polar and charged groups dominates over the dehydration of the apolar groups (Privalov and Makhatadze, 1992). This is just what one would expect when the positively charged helical segments interact with the negatively charged DNA.

All binding characteristics of GCN4 bZIP determined for standard temperature 25°C are listed in Table 12.3, which also includes the electrostatic and nonelectrostatic components of the Gibbs energy of binding. It appears that the Gibbs energies of binding of GCN4 bZIP to the AP-1 and ATF/CREB sites are rather similar and are also similar in their electrostatic and nonelectrostatic components, each amounting to 50% of the total binding energy.

Most surprising, however, is the very large difference in the enthalpies of binding and, correspondingly, in the nonelectrostatic entropies of binding.

TABLE 12.3. Thermodynamic Parameters of Association of the GCN4 bZIP^{SS} with AP-1 and ATF/CREB Sites[a,b]

	T	ΔH^a	ΔG^a	ΔG^a_{nel}	ΔG^a_{el}	ΔS^a	ΔS^a_{nel}	ΔS^a_{el}
AP-1	5	−77.3	−44.7	−22.4	−22.3	−117	−197	80
	10	−75.1	−46.6	−23.3	−23.3	−101	−183	82
	15	−73.0	−48.2	−25.9	−22.3	−86	−166	80
	20	−70.8	−49.2	−26.9	−22.3	−74	−150	76
	25	−68.6	−50.0	−27.7	−22.3	−62	−137	75
	30	−66.5	−50.3	−28.0	−22.3	−53	−127	74
	35	−64.3	−50.2	−27.9	−22.3	−46	−118	72
	40	−62.1	−49.7	−27.40	−22.3	−40	−111	71
ATF/CREB	5	−24.7	−42.6	−19.5	−23.1	65	−19	84
	10	−22.7	−44.1	−21.0	−23.1	76	−6	82
	15	−20.7	−45.5	−22.4	−23.1	86	6	80
	20	−18.7	−46.5	−23.4	−23.1	95	16	79
	25	−16.7	−47.9	−24.8	−23.1	105	27	78
	30	−14.7	−49.0	−25.9	−23.1	113	37	76
	35	−12.7	−50.0	−26.9	−23.1	121	45	76
	40	−10.7	−51.0	−27.9	−23.1	129	53	76

[a]Dragan et al. (2004a).

[b]In 100 mM NaCl, 30 mM sodium phosphate, pH 7.4; T in °C; ΔH and ΔG in kJ/mol; ΔS in J/K·mol.

This difference in the energetics of these two complexes is surprising because, according to the crystal structures of GCN4-bZIP bound to AP-1 (Elenberger et al., 1992) and ATF/CREB (König and Richmond, 1993; Keller et al., 1995), the protein–DNA contacts are virtually identical in these two complexes except for the interaction of the invariant Arg243 with the central base pairs. When bound to the inherently asymmetric AP-1 site, Arg243 from the left monomer forms two hydrogen bonds to the guanine at the central base pair; Arg243 from the right monomer contacts phosphates on the central base pair and the adjacent residue on the opposite strand. In contrast, GCN4-bZIP binds symmetrically to the ATF/CREB site with both Arg243 residues interacting with the equivalent guanidine residues at the two central base pairs (Kim and Struhl, 1995). This difference in interactions certainly cannot explain the large difference in the enthalpies of formation of these two complexes, not saying anything about the entropy difference:

Although the entropy of binding to the AP-1 site is negative, binding to the ATF/CREB site is positive.

The observed −50 kJ/mol difference in the enthalpy assumes quite a significant difference in the number of enthalpic interactions stabilizing these complexes: if the enthalpy of hydrogen bonding is about 3 kJ/mol (see Section 5.1.2), the observed difference in enthalpy can be explained by formation of 15 additional hydrogen bonds in the AP-1 complex. The observed difference in the entropy factors of these two complexes is equivalent to ordering 15 residues if the entropy cost of folding of one residue is −4 kJ/mol (see Section 4.2.2). In view of the known structures of these complexes, this is absolutely improbable.

The only explanation suggested for the observed large differences in the enthalpy and entropy of formation of these two complexes of GCN4 bZIP with the AP-1 and ATF/CREB sites of DNA is that they result in incorporation of different numbers of water molecules in these complexes (Dragan et al., 2004a).

A strong argument that the difference in the enthalpy and entropy of binding bZIP to AP-1 and ATF/CREB sites comes from the incorporated water is that these differences almost completely balance each other and the Gibbs energies of formation of these two complexes are almost similar. Such perfect enthalpy–entropy compensation is a privilege of the water ordering–disordering process; for example, the Gibbs energy of melting of ice is zero, whereas the enthalpy and entropy are very large.

The water molecules incorporated in the structure of these two complexes have been revealed by crystallography. In the complex of bZIP with ATF/CREB, 14 water molecules were found bridging the charged groups of the DNA and protein (Keller et al., 1995). In the complex of bZIP with the AP-1 site the number of incorporated water molecules is unclear because the structure was obtained at lower resolution (2.56 Å; Elenberger et al., 1992). However, judging by the negative nonelectrostatic entropy value, the number of incorporated water molecules in that case should be much larger than in the ATF/CREB complex. If we assume that the entropy loss upon immobilization of a water molecule is similar to that for water freezing, that is, is on the order of 22 J/K·mol, then we should expect that the AP-1 complex has eight more incorporated water molecules than the ATF/CREB complex. Consideration of the unusually large negative enthalpy of association of AP-1 with bZIP gives us a similar number: assuming that the enthalpy of water incorporation is similar to that of water freezing, that is, about 6 kJ/mol, one can find that eight additionally incorporated water molecules can explain the observed 50 kJ/mol difference in the enthalpy of formation of AP-1 and sites. If so, and if the ATF/CREB complex has 14 incorporated water molecules, one can expect that in the AP-1 complex there should be 22 incorporated water molecules. It certainly would be interesting to know where the AP-1/bZIP complex accommodates 22 water molecules. The location of all these incorporated water molecules is important for understanding the structure and energetics of the bZIP complexes and is a challenging problem for high-resolution crystallography.

Another important piece of information, which one can get by analyzing data in Table 12.3, is that:

> *Although the nonelectrostatic component of the Gibbs energy of binding increases with increasing temperature, the electrostatic component does not show any dependence on temperature. This is just what one would expect if* ***the electrostatic component is purely entropic and its enthalpy is close to zero.***

It is remarkable that the electrostatic components of entropy are identical for both of the complexes considered here—as they should be if they result from mixing of the released counterions with the ions in the bulk solution as suggested by CC theory.

12.2.3. Heterodimeric bZIP Interactions with the Asymmetric DNA Site

The heterodimeric bZIP is formed particularly by the ATF2 and cJun transcription factors, which interact with the asymmetric binding site PRDIV (Dragan et al., 2008; Carrillo et al., 2010). Formation of this complex has a certain interest because the ATF2 and cJun transcription factors are the components of the interferon beta (IFN-β) enhanceosome, which is formed on the enhancer DNA and controls synthesis of interferon upon virus infection (see Section 12.5.1). The ATF2 and cJun are able to form two homo and one hetero bZIPs: $(ATF2)_2$, $(cJun)_2$, and (ATF2/cJun), while their binding site on the enhancer DNA, PRDIV, is asymmetric:

$$5'\text{-ATGACATAGGC-}3'$$
$$3'\text{-TACTGTATCCG-}5'$$

The asymmetry in the PRDIV sequence is critical for the proper orientation of the ATF2/cJun heterodimer on the enhancer DNA.

The results of isothermal calorimetric titration of binding of the considered bZIPs to the 15-bp DNA duplex containing the PRDIV site are presented in Figure 12.9. In all three cases the ITC-measured enthalpies of binding are negative and depend

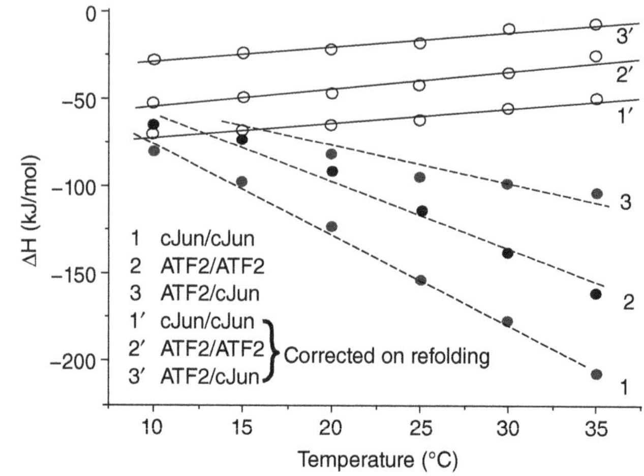

Figure 12.9. Binding enthalpy of the homo and hetero bZIPs formed by the ATF2 and cJun transcription factors.

TABLE 12.4. Thermodynamic Parameters of Association of the Homo and Hetero bZIPs Formed by the ATF2, cJun Transcription Factors to the PRDIV Binding Site of the IFN-β Enhancer DNA[a]

	$-\Delta G^a$			$-\Delta H^a$			$-T\Delta S^a$		
T	$(cJun)_2$	$(ATF2)_2$	ATF2/cJun	$(cJun)_2$	$(ATF2)_2$	ATF2/cJun	$(cJun)_2$	$(ATF2)_2$	ATF2/cJun
10	42.3	44.5	48.0	28.0	71.0	52.9	14.3	−26.5	−4.9
20	43.9	44.7	49.0	22.0	65.0	47.0	21.5	19.9	2.0
30	46.2	45.3	50.0	9.8	55.0	35.0	36.3	−9.7	15.0

[a]Carrillo et al. (2010).
T in °C; ΔG^a, ΔH^a, and $T\Delta S^a$ in kJ/mol.

strongly on temperature, and this dependence is different for the three dimers considered: it is much steeper for the $(ATF2)_2$ and less steep for the $(cJun)_2$ homodimers.

Using the DSC-measured enthalpies of unfolding the basic segments of these bZIPs the apparent heats of binding can be corrected for refolding of the basic segments (see Section 3.2.8 for the correction method). Correction for refolding of the basic segments drastically changes the binding enthalpy functions: they decrease in magnitude and their slopes change sign and become almost identical. The corrected binding characteristics of these three bZIPs to the PTDIV site are listed in Table 12.4. It is interesting that the Gibbs energies of binding in these three cases do not differ much, being only 10% larger for binding the heterodimer to the asymmetric binding site. However, these three cases differ significantly in their binding enthalpies and, correspondingly, in binding entropies: the difference in enthalpy amounts 40 kJ/mol, that is, 80% of the maximal enthalpy value! The existing structural information does not provide an explanation for this large difference in the enthalpy and entropy of binding these tree bZIPs to the same DNA.

However, the fact that these differences in the enthalpy and entropy efficiently compensate each other in the Gibbs energies of binding suggests that they result from the hydration effects, bearing in mind that the hydration enthalpy and entropy efficiently balance each other at temperatures not far from that of melting ice.

Of particular interest is the orientation of the heterodimer ATF2/cJun bound to the asymmetric binding site PRDIV with two nonidentical half-sites. Figure 12.10 presents the results of fluorescence anisotropy titration with the three bZIPs and the FAM-labeled 15-bp DNA duplex containing the PRDIV site:

FAM–5′–CGATGACATAGGCTC–3′
3′–GCTACTGTATCCGAG–5′

It is notable that the asymptotic value of the anisotropy for the ATF2/cJun heterodimer is very similar to that of the ATF2/ATF2 homodimer. If these differences were due to differences in the molecular weight, the heterodimer would be intermediate; however, since this is not the case, it follows that the increased anisotropy of the $(cJun)_2$ homodimer must result from some direct preferential effect of the cJun protein on the FAM moiety. Since the heterodimer does not show this effect, it follows that the strand closest to the FAM is not cJun—it is the ATF2 strand in the

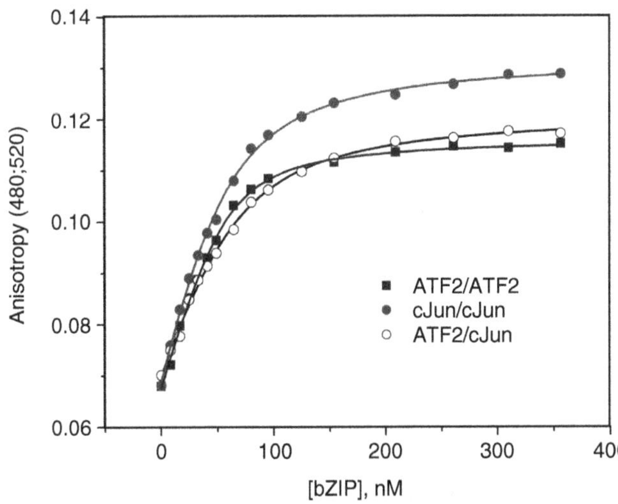

Figure 12.10. Binding isotherms of the S–S cross-linked bZIPs, the ATF2/cJun, $(ATF2)_2$, and $(cJun)_2$, to 15-bp FAM-5″-DNA duplexes containing the PRDIV site determined by fluorescence anisotropy titration at 25°C in 100 mM NaCl, 10 mM sodium phosphate, pH 7.4 solution (Dragan et al., 2008).

heterodimer that is closer to the FAM; that is, it is the ATF2 strand in the heterodimer that is binding to the upstream half-site (Fig. 12.11). The ratios of the asymptotic values of the anisotropies show that the probability of this orientation of the ATF2/cJun occurring at PRDIV amounts to 75%. One may therefore conclude that the Gibbs binding energies of the two orientations differ by about 3 kJ/mol. The average dissociation constant of the ATF2/cJun heterodimer is (3 ± 2) nM, that is, the binding Gibbs energy is (49 ± 0.5) kJ/mol at 25°C. The dissociation constants of the $(ATF2)_2$ and $(cJun)_2$ homodimers are (12 ± 3) and (17 ± 5) kJ/mol, respectively. It is remarkable that the binding energy of the heterodimer is significantly greater than that of the homodimers. This is just what one would expect for the binding of the heterodimer to its cognate asymmetric site. This explains why this site is asymmetric: there are several dozens of genes in the human genome encoding bZIPs; however, for functional reasons only one of the many possible combinations is used for the IFN-β enhanceosome—the ATF2/cJun pair, which is selected by its asymmetric binding site at the enhancer DNA. As follows from the preceding discussion, the mechanism of selecting this particular heterogeneous combination and its proper positioning on the enhancer is straightforward: it is provided by a larger binding Gibbs energy.

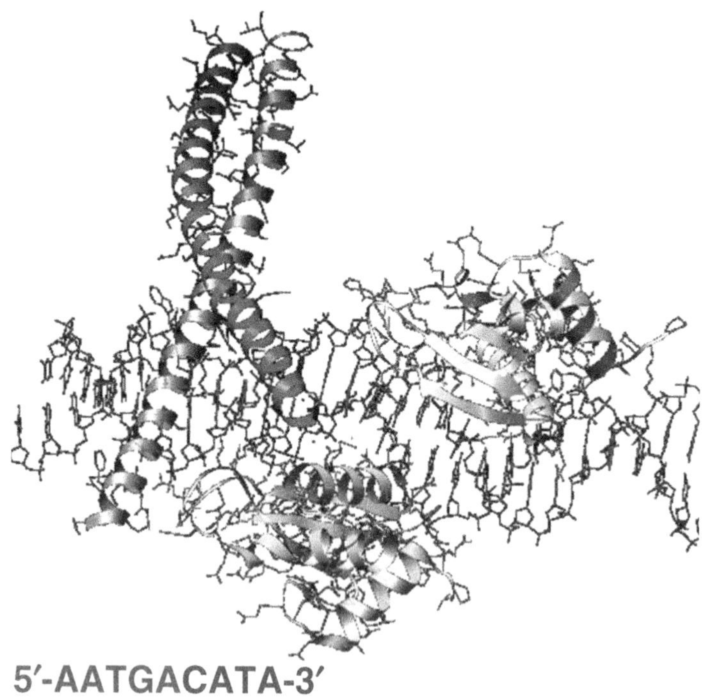

Figure 12.11. The bZIP ATF2/cJun positioned at the PRDIV asymmetric site. The N-terminal domains of IRF-3 bound to the PRDIII and PRDI sites are shown in gray. (See insert for color representation of the figure.)

12.2.4. IRF Transcription Factors

A prompt and regulated cellular reaction (the innate immune response) is central to host defense against extracellular pathogens. It is coordinated by a genetic network controlled by various transcription factors, in particular the **interferon regulatory factors** (IRFs). Among the many members of this family the best studied biochemically and structurally is IRF-3. Normally in a latent state, upon virus infection IRF-3 is activated by the IKKε and TBK1 kinases that phosphorylate specific Ser and Thr residues. Phosphorylated IRF-3 is translocated to the nucleus and binds as a dimer to its target DNA sequences containing two consensus binding sites, PRDIII and PRDI, in the enhancer of the IFN-β gene. Together with two other transcription factors, the ATF2/cJUN bZIP dimer and nuclear factor κB (NF-κB), bound to the PRDIV and PRDII sites, respectively, they form the **IFN-β enhanceosome**.

According to structural studies (Panne et al., 2004; Qin et al., 2005; Takahashi et al., 2003; Escalante et al., 1998; Fujii et al., 1999; Lin et al., 2000), IRF-3 has two compact domains: the N-terminal DNA-binding domain (NTD) and the C-terminal dimerization domain (CTD) connected by a 75-residue linker, assumed to be unfolded. The NTD binds to the major groove of DNA at the AANNGAAA sequence and bends it by about 20° toward the protein (Fig. 12.12).

The NTD of IRFs is positively charged and the CTD is negatively charged. Phosphorylation of Ser/Thr residues substantially increases the negative charge of the CTD. This should increase repulsive electrostatic interactions between these domains of IRFs and also between the IRFs and the negatively charged DNA. Nevertheless, phosphorylation results in some rearrangements in the CTD and binding to the similar CTD. The dimeric ITF-3 then binds to the tandem PTDIII and PRDI sites located on the opposite faces of the IFN-β enhancer DNA (Fig. 12.13). It was shown that phosphorylation of the IRF-3 can be mimicked by replacing certain Ser and Thr residues by Asp(D): S396D, S398D, S402D, T404D, and S405D. The 5D phosphomimetic mutant of IRF-3, which for simplicity is denoted here as IRF-3*, dimerizes and binds to the tandem binding sites PRDI and PRDIII as a dimer (Lin et al., 1998; Servant et al., 2003; Hargreaves et al., 2005; Dragan et al., 2007).

The ITC studies of interaction of the single isolated N-terminal domain and the full-length 5D phosphomimetic mutant IRF-3* showed that the enthalpy ΔH^a and Gibbs energy ΔG^a of their association to the 12-bp DNA duplex containing only one PRDI site,

5′–GAG**AAGTGAAA**GT–3′
3′–CTC**TTCACTTT**CA–5′

are very similar (Hargreaves et al., 2005; Dragan et al., 2007, 2008). This suggests that:

Phosphorylation of the Ser/Thr residues in the C-terminal domain does not affect the DNA-binding ability of the N-terminal domain.

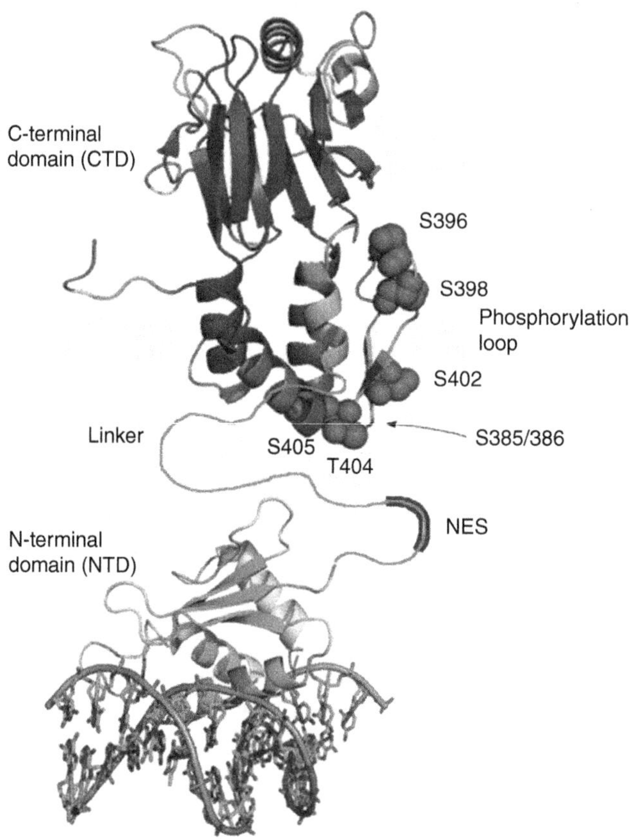

Figure 12.12. Structures of the DNA-binding N-terminal domain and C-terminal dimerization domains of IRF-3 connected by the hypothetical 75-residue linker. The Ser/Thr residues corresponding to the *in vivo* specific sites of phosphorylation are shown in creatinine phosphokinase (CPK) (Dragan et al., 2008). (See insert for color representation of the figure.)

On the other hand, studies of binding the IRF-3* to the 26-bp DNA containing PRDIII and PRDI tandem binding sites,

5′–GG**AACTGAAA**GGGAG**AAGTGAAA**GT–3′
3′–CC**TTGACTTT**CCCTC**TTCACTTT**CA–5′

showed that the association constant has increased manyfold but the binding enthalpy calculated per mole of NTD has not changed (Table 12.5). It appeared thus that:

Increase of the dimer binding results solely from its smaller translational entropy at binding to the tandem binding site.

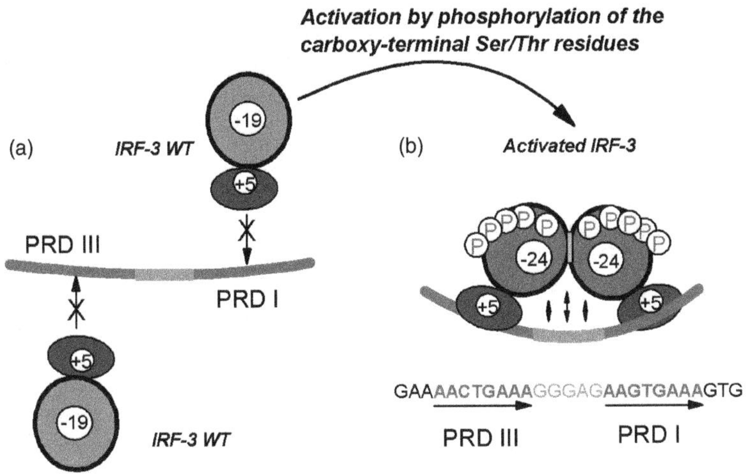

Figure 12.13. Cartoons showing: (a) the independent binding of two monomeric IRF-3 to the PRDI and PRDIII binding sites located on the opposite faces of the IFN-β enhancer DNA; (b) phosphorylation of the IRF-3 resulting in its dimerization and binding as a dimer to the PRDI and PRDIII sites which should result in bending of the shown in green spacer. (See insert for color representation of the figure.)

TABLE 12.5. Binding Characteristics of IRF-3 N-Terminal Domain (NTD), Full-Length IRF-3 WT and 5D Mutant with 26-bp DNA Containing the PTDIII- and PRD- Binding Sites, and with the 13-bp DNA Containing the Single PRDI Site[a]

	26-bp DNA Containing the PRDIII- & PRDI-Binding Sites			13-bp DNA Containing the PRG1-Binding Site		
Protein	K_d	$-\Delta G^a$	$-\Delta H^a$	K_d	$-\Delta G^a$	$-\Delta H^a$
NTD	280	36.7	11.5	280	36.7	11.5
IRF-3 WT	1226	33.1	23.0	800	34.1	23.0
IRF-3 5D	5.8	46.1	23.0	1200	33.1	23.0

[a]Dragan et al. (2007).
K_d in nM; ΔG^a and ΔH^a in kJ/mol.

It is remarkable, however, that the Gibbs energy of binding of the dimeric 5D IRF-3* phosphomimetic mutant to the 12-bp DNA (containing only the PRDI site and thus being able to accommodate only one NTD) is –(33.1 ± 0.3) kJ/mol and binding to the 26-bp DNA (containing two IRF-3 binding sites, PRDI and PRDIII) is only –(46.1 ± 0.3) kJ/mol, instead of the expected twice-larger value, –66 kJ/mol. A possible reason for the Gibbs energy of dimeric IRF-3* binding being less than twice that of the monomeric could be the work needed to distort the DNA at binding the dimeric ITF-3* to the tandem binding sites located on the opposite faces of the enhancer DNA (Fig. 12.13).

12.2.5. Binding of NF-κB to the PRDII Site

The NF-κB, another component of the IFN-β enhanceosome, consists of p50 and p65 subunits, the various combinations of which result in formation of two homodimers (p50/p50 and p65/p65) and one heterodimer (p50/p65). They bind to the major groove of the asymmetric PRDII site, —AAATTCC— of the IFN-β enhancer DNA (Fig. 12.14).

The results of the fluorescence anisotropy titrations of the FAM-labeled 12-bp DNA duplex containing the PRDII site,

FAM–5′–GGGAAATTCCGC–3′
3′–CCCTTTAAGGCG–5′

are presented in Figure 12.15. These isotherms show that, while the asymptotic values of the anisotropy of the p50/p65 heterodimer and p65/p65 homodimer are comparable, it is significantly larger for the p50/p50 homodimer. As with bZIP dimers, this information allows us to conclude which component is closer to the FAM-labeled 5′ end of the 12-bp duplex containing the PRDII site.

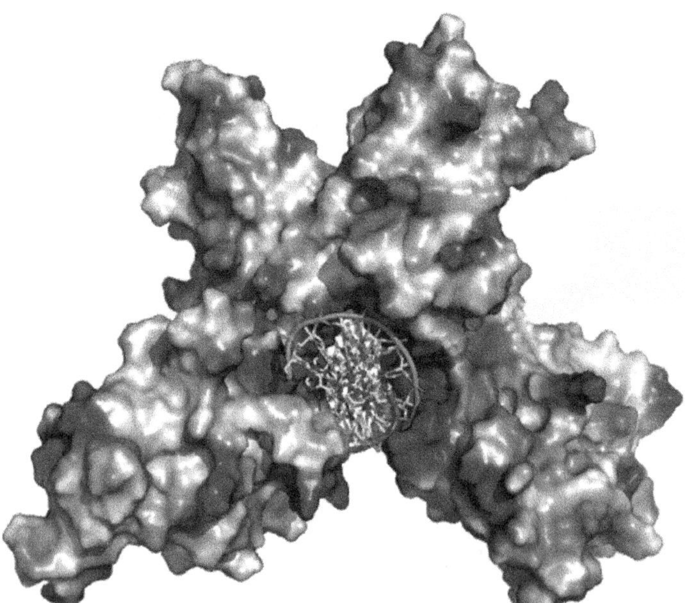

Figure 12.14. Charge distribution in the complex of NF-κB (p50/p65) with DNA containing the PRDII site. Positive charges are shown in blue, and negative charges are shown in red. It is seen that the positively charged surface of NF-κB is in contact with the negatively charged DNA. The model was constructed using the program PyMol and coordinates from the Protein Data Bank file 1VKx (Dragan et al., 2008). (See insert for color representation of the figure.)

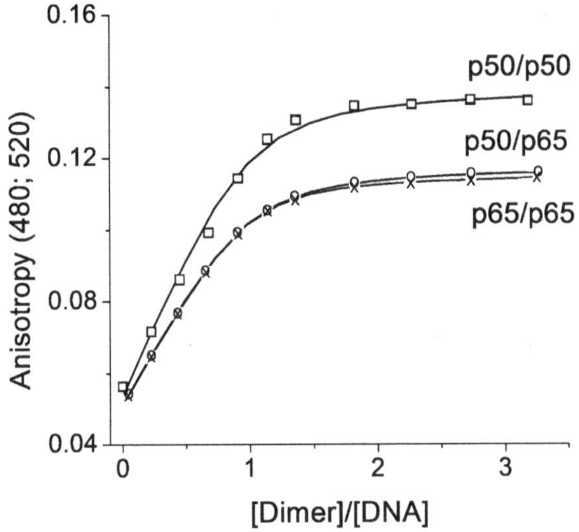

Figure 12.15. Fluorescence anisotropy titration of the 12-bp DNA containing the PRDII site by the (p50/p650) and (p65/p65) homodimers and by the (p50/p65) heterodimer at 25°C in 100 mM NaCl, 10 mM sodium phosphate (pH 7.4), and 1 mM DTT (Dragan et al., 2008).

From the asymptotic values of the anisotropy, one can conclude that the fraction of p65/p50 bound with p65 close to FAM amounts to $F_1 = 0.92$ and thus for binding in the opposite direction $F_2 = 0.08$. The ratio of these fractions determines the Gibbs energy difference of binding p50/p65 to PRDII in two orientations: $\delta\Delta G$(p50/p65) $= -RT\ln(F_1/F_2) = -6.1$ kJ/mol at 25°C. The apparent Gibbs energies of binding (ΔG^a) of (p50/p50), (p65/p65), and (p50/p65) to this site, derived from the observed isotherms, are –(46.5 ± 0.2), –(45.3 ± 0.2), and –(46.6 ± 0.3) kJ/mol, respectively. The Gibbs energy of binding the (p50/p65) heterodimer in a definite orientation can then be determined from the apparent Gibbs energies of binding, bearing in mind that

$$\Delta G^a(\text{p50/p65}) = 0.92\ \Delta G_1 + 0.08[\Delta G_1 - \delta\Delta G(\text{p50/p65})] \tag{12.2}$$

It follows that the Gibbs energy of binding (p50/p650) with p65 at the upstream half-site of PRDII is –47.1 kJ/mol, while it is –41.0 kJ/mol in the opposite orientation. Correspondingly, the dissociation constant for the two orientations of this heterodimer are 6.3 and 71, that is, they differ by one order of magnitude.

It should be noted that this conclusion, that the p50/p65 heterodimer binds to the asymmetric PRDII site with p65 preferentially at the upstream half-site, is opposite to that found in the crystal structure of this complex, according to which it is p50 at the upstream half of PRDII (Berkowitz et al., 2002; Escalante et al., 2002). This raises the question as to which of these two orientations is realized in the functioning enhanceosome, that is, in the living cell.

The orientation of the p50/p65 heterodimer, which follows from fluorescence anisotropy experiments, is realized in solution and certainly should be closer to orientation realized in the living cell, while in the case of the crystal structures, especially of the complexes, we always face a dilemma if it is not an artifact of crystallization. As shown previously, both orientations of this heterodimer are possible, but the population of dimers with p50 at the upstream half of the PRDII site is one order of magnitude lower than the opposite. However, the growing crystal might select complexes that are more suitable for the neighbors in the crystal lattice.

12.3. BINDING TO THE MINOR GROOVE OF DNA

12.3.1. AT-Hooks

Surprisingly, the specific binding of protein to the minor groove usually takes place at the AT-rich sequences. One of the most impressive representatives of such binders is the so-called **AT-hook**. This is a small DNA-binding protein motif that binds selectively to the AT-rich DNA sequences. This DNA-binding motif was first discovered in the high-mobility group nonhistone chromosomal protein, HMGA1, but then was found in a wide range of other proteins, known to be chromosomal (Aravind and Landsman, 1998; Reeves and Nissen, 1990; Johnson et al., 1988). The HMGA1, a protein of modest size consisting of 107 amino acid residues, is unfolded when free in solution. Surprisingly, however, it regulates the expression of a large number of genes.

HMGA1 has three copies of the AT-hook motif, separated by short spacers:

1–[23-ATH1-31]–[55-ATH2-70]–[81-ATH3-89]–107

The sequences of these AT-hooks are rather specific: the core region of each AT-hook contains a central arginine–glycine–arginine (RGR) tripeptide sequence, which usually has proline residues at both its N- and C-ends, and the resultant PRGRP pentapeptide tends to be flanked at both termini by pairs of positively charged residues (K/R):

ATH1: –KR**P<u>RGR</u>P**K–
ATH2: –PKRP**<u>RGR</u>**PKGSKNKGA–
ATH3: –RKP**<u>RGR</u>**PRK–

The interactions of AT-hooks with target DNA duplexes is of considerable interest since the AT-hook penetrates deep into the minor groove without noticeable deformation of the DNA duplex (Fig. 12.16). The AT-hooks are thus particularly suitable for obtaining a better understanding of the forces responsible for the formation of specific DNA–protein complexes in the minor groove. It is also interesting to determine whether there are some special features of the minor groove, particularly at AT-rich regions, that makes it an attractive harbor for the extended families of minor-groove DNA-binding proteins.

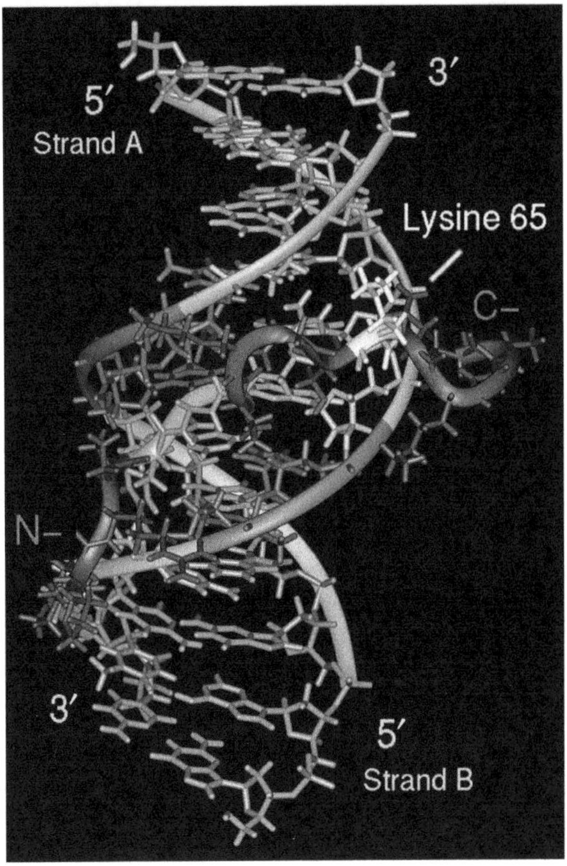

Figure 12.16. Ribbon drawing of the DBD2/DNA complex. The five AT base pairs, which make up the binding site, are shown in yellow (A) and green (T); the rest of the DNA duplex is shown in orange. The central RGR motif, which penetrates deep into the minor groove of DNA, is shown in red; the other amino acids in the peptide are shown in blue (Dragan, 2003b). (See insert for color representation of the figure.)

The target DNA duplex corresponding to the binding site PRDII of the ATH2 and ATH3 on the IFN-β enhancer is 5′–GGGAAATTCCGC–3′. The following peptides mimicking AT-hooks were used in this study:

DBD2: *N*-acetyl–PKRP**<u>RGR</u>**PKGSKNKGA–NH_2
Core DBD2: *N*-acetyl–PKRP**<u>RGR</u>**PKG–NH_2
DBD3: *N*-acetyl–RKP**<u>RGR</u>**PRK–NH_2

Figure 12.17 presents the logarithmic salt dependence of the association constant of ATH2, ATH3, and CorATH2 with their target DNA duplex corresponding to the PRDII-binding site of these hooks on the IFN-β enhancer

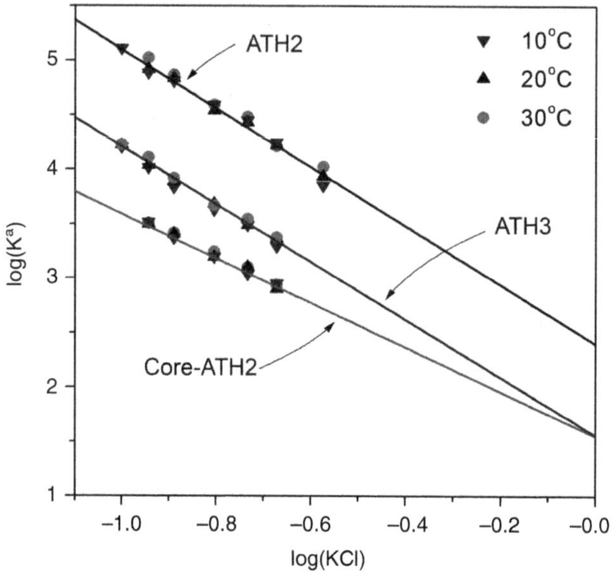

Figure 12.17. The dependence of the logarithm of the association constant of the peptides with DNA on the logarithm of the KCl concentration, determined by fluorescence anisotropy titration at different temperatures and salt concentration (Dragan et al., 2003b).

(5′–GGGAAATTCCGC–3′). These three hooks show perfectly linear dependence between the logarithm of the binding constant and the logarithm of salt concentration as suggested by CC theory. The other notable fact is that the logarithmic dependence of association constant and salt concentration does not depend on temperature. From this fact immediately follows that:

> *The electrostatic part of the binding energy does not include an enthalpic component, which also means that the electrostatic component of binding energy is fully entropic and the binding enthalpies are totally nonelectrostatic, as is expected from CC theory.*

The results of calorimetric isothermal titration of the ATH peptides into target DNA are shown in Figure 12.18. It is most remarkable that these enthalpies are positive. However, the positive enthalpy opposes binding. Therefore, **binding of these AT-hooks to their target DNAs is completely entropy driven.** The other notable fact is that the positive binding enthalpy decreases with increasing temperature; that is:

> *The heat capacity effect of binding is negative. Thus, binding results in significant decrease of the heat capacity of these systems.*

According to the CC concept, the slope of the logarithmic salt dependence of the association constant represents the number of tightly bound counterions (cations) released from the DNA upon protein binding, and this is assumed to be proportional

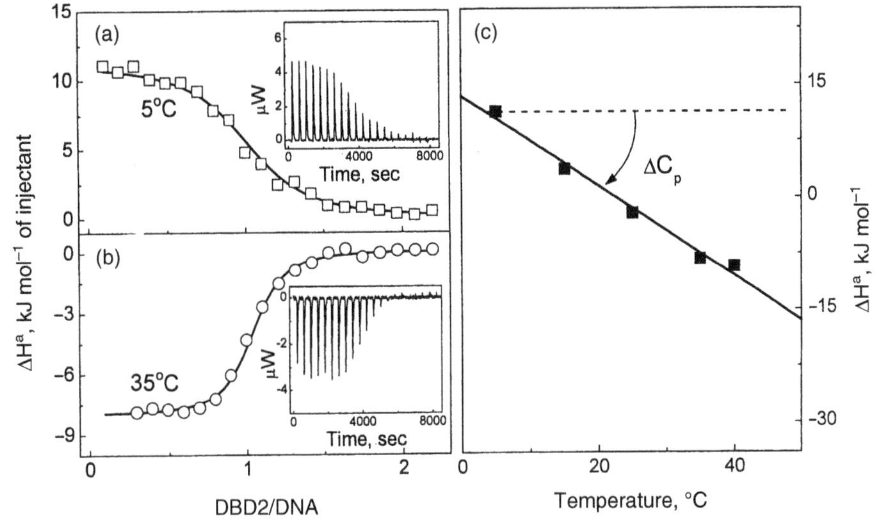

Figure 12.18. Calorimetric titration of a 16-bp DNA with the AT-hook peptide DBD2 at (a) 5°C and (b) 35°C. The volume of each injection is 10 μL. (c) Association enthalpy measured at different fixed temperatures: the slope of this function yields the heat capacity effect of association (Dragan et al., 2003b).

to the number of ionic contacts made (see Section 12.1). As seen in Figure 12.17, the slopes of ATH2 and ATH3 are almost identical, showing that they both make four ionic contacts with phosphates, while the slope of Core-ATH2 is smaller, corresponding to three contacts, as one would expect since in this case the C-terminal lysine is replaced by the glycine. These numbers of contacts accord with the structural studies (Huth et al., 1997). It follows that the extension in ATH2 makes one additional contact above the 3 of CoreATH2 using one of the lysine residues in the extension.

Extrapolation of the logarithmic salt dependence of the binding constants of ATH3, ATH2, and CoreATH2 to 1 M salt shows that for ATH3 and Core-ATH2 these functions focus at the same point (as expected from their closely related sequences) corresponding to their nonelectrostatic Gibbs energies of −8.5 kJ/mol. However, for ATH2 the extrapolation gives a value of −12 kJ/mol, showing that its seven-residue extension contributes a further −3.5 kJ/mol to the nonelectrostatic component of the binding free energy. When the salt concentration approaches 1 M, the electrostatic term in Equation (12.1) vanishes and $-RT\ln(K^a)$ approaches the nonelectrostatic part of the Gibbs energy of complex formation, $-RT\ln(K^a_{\text{nel}})$. This permits the observed Gibbs energies of binding to be split into two components: the nonelectrostatic and the electrostatic. The results of such an analysis for the standard solvent conditions (100 mM KCl, 10 mM phosphate, pH 6.0, 1 mM EDTA) are presented in Table 12.6. These results are in perfect correspondence with that expected from the theory: two extrapolations for the Core-DBD2 and DBD 3 cross each other exactly at $\log[\text{KCl}] = 0$, that is, at 1 M salt, where $\log K^a$ reaches the value (1.55 ± 0.03). Excluding this value from the total Gibbs energy of binding, which is determined

TABLE 12.6. Association of the AT-Hook Peptides with 12-bp DNA[a,b]

Peptide	T	log K^a 0.1 M KCl	log K^a 1.0 M KCl	$\frac{\partial \log K}{\partial \log[\text{KCl}]}$	$-\Delta G^a$ Total	$-\Delta G^a$ Electr	$-\Delta G^a$ Nel	ΔH^a Nel	ΔS^a Nel	ΔS^a Electr
Cor-DBD2	10	3.56	1.56	2.03	19.4	10.9	8.5	11.8	71.7	38.5
	20	3.63	1.54	2.06	20.5	11.8	8.7	8.8	59.7.4	39.6
	30	3.57	1.56	2.07	20.8	11.7	9.1	5.8	57.7	38.6
DBD3	10	4.12	1.50	2.68	22.4	14.2	8.2	15.0	82.0	50.5
	20	4.15	1.58	2.60	23.4	15.0	8.9	11.6	68.2	51.2
	30	4.14	1.57	2.62	24.1	15.0	9.1	8.2	67.7	48.8

[a]Dragan et al. (2003b).
[b]In 0.1 M and 1.0 M KCl; T in °C; ΔG^a, ΔH^a in kJ/mol^{-1}; ΔS^a in J/K·mol.

from $\log K^a$ at 0.1 M KCl (column 6), one finds the electrostatic component of binding energy at 0.1 M KCl (column 7).

Since the electrostatic component of binding energy is fully entropic, the entropy of this component can be determined just by dividing the electrostatic Gibbs energy of binding by the absolute temperature; the entropy of the nonelectrostatic component can be determined by division of the difference between the ITC-measured enthalpy of association and nonelectrostatic Gibbs energy of association by the absolute temperature: $\Delta S^{nel} = (\Delta H - \Delta G^{el})/T$ (Table 12.6, columns 10 and 11). It appears that both entropies are positive but, while the electrostatic one does not depend on temperature, the nonelectrostatic entropy of binding decreases with increasing temperature. In the case of electrostatic entropy this is just what is expected if it represents the cratic entropy of mixing of the counterions released on binding with the ions in the bulk solution: it should be positive and independent of temperature. However, why is the nonelectrostatic entropy of binding positive and why does it decrease in magnitude with increasing temperature? Since binding of peptides to DNA results in reduction of the conformational and translational–rotational freedom of the components, one would expect that upon fixing the conformation of the unfolded peptide and tightening the DNA duplex the entropy change should be negative but not positive.

The other surprising fact is that the nonelectrostatic entropy significantly exceeds in magnitude the electrostatic one. Since the positive entropy of binding is the only force of binding, it appears that binding is provided mostly by the nonelectrostatic component of the binding entropy. These two paradoxes of the binding of AT-hooks to the AT-rich minor groove of DNA are discussed in Section 12.4.3.

12.3.2. HMG Boxes

The HMG box represents an example of a versatile eukaryotic DNA-binding domain (DBD) of essentially conserved L-shaped fold (Love et al., 1995; Werner et al., 1995), which binds to the minor groove of the AT-rich DNA sequences and bends DNA considerably, while its C-terminal basic tail enters into the major groove of DNA (Fig. 12.19). Two distinct categories exist: sequence-specific (SS) DBDs, which

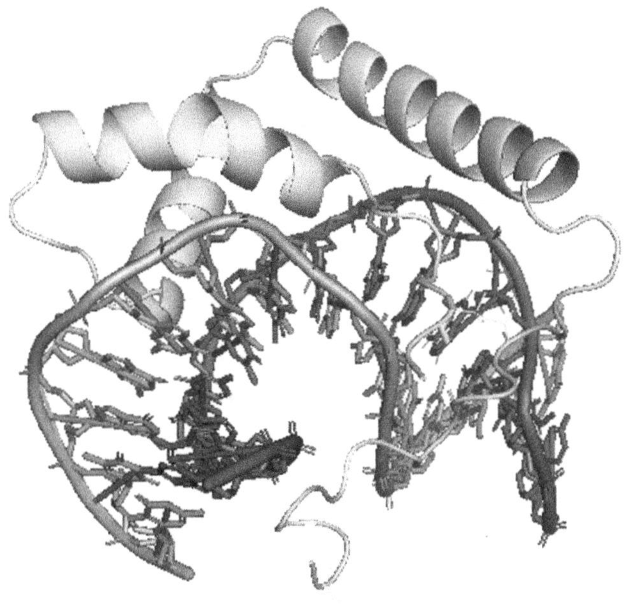

Figure 12.19. HMG DBD sits in the minor groove of DNA and bends it considerably, while its C-terminal basic tail enters into the major groove of DNA. (See insert for color representation of the figure.)

exhibit strong DNA sequence specificity, and non-sequence-specific (NSS) DBDs (Grosschedl et al., 1994; Bustin and Reeves, 1996; Giese et al., 1991, 1992; Ferrari, 1992).

These two categories of HMG box differ also in their thermodynamic properties, first of all in their stability: proteins of both categories are rather unstable and upon heating start to unfold from the very beginning of heating. It is remarkable, however, that the SS DBDs unfold in three cooperative steps, while the NSS DBDs unfold in two steps (Fig. 12.20). Thus, at physiological temperature these DBDs are partly unfolded, but the SS DBDs, in which unfolding is more extensive at low temperatures, are more unfolded at the physiological temperature. These partly unfolded DBDs refold upon binding to the cognate DNA, and their complex with DNA is rather stable and melts as a single cooperative unit above physiological temperature (Fig. 12.21).

In the case of HMG DBDs, obtaining comprehensive information on the thermodynamics of binding is not a straightforward matter due to their low stability. Clearly, the ITC-measured heat of HMG DBD association with DNA includes the heat of protein refolding upon binding to DNA, which increases with the temperature of the binding experiment. As a result, the ITC-measured heat of HMG–DNA binding largely depends on temperature (Fig. 12.22a). This heat of refolding should be taken into account in the calorimetric determination of the enthalpy of association since the main interest is the enthalpy of DNA association with the fully folded HMG

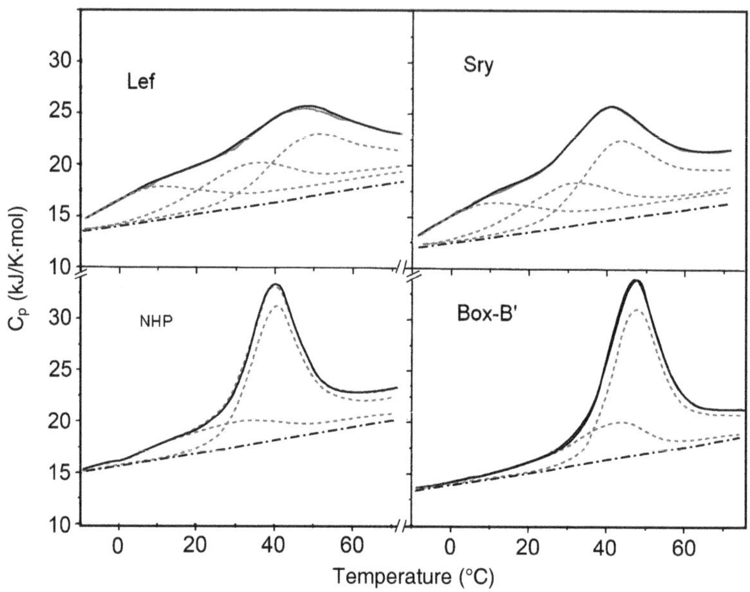

Figure 12.20. The partial molar heat capacity functions of two SS boxes (Lef and Sry) and two NSS boxes (NHP and Box-B′) under standard solvent conditions (10 mM potassium phosphate, pH 6.0, 100 mM KCl). The black broken line shows the baseline representing the expected partial molar heat capacity of folded protein; the gray broken line shows the results of deconvolution analysis of the heat capacity profile (Dragan et al., 2004b).

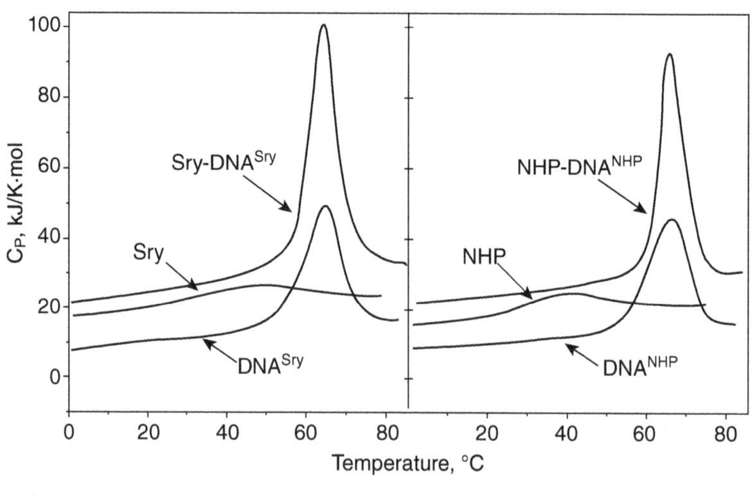

Figure 12.21. The partial molar heat capacity functions of free Sry and NHP (in green), their free cognate DNA duplexes (in red), and their complexes (in black). The dot-dashed lines show the sums of the heat capacities of the free proteins and the free DNAs. The black broken line shows heat capacity function expected for the fully folded protein. All results obtained in 10 mM potassium phosphate pH 6.0, 100 mM KCl (Dragan et al., 2004b).

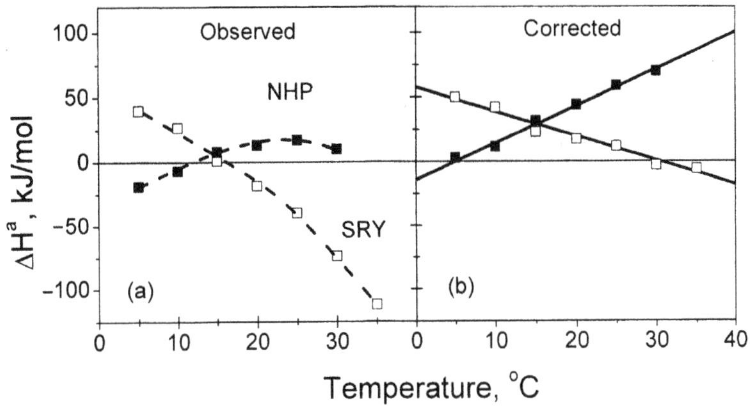

Figure 12.22. (a) ITC-measured binding enthalpies of the sequence-specific SRY and non-sequence-specific NHP with their optimal DNAs, plotted as functions of temperature; (b) binding enthalpy functions corrected for the heat of protein refolding upon binding to DNA.

DBD. This can be done using the DSC-measured heat of protein unfolding up to the temperature at which the ITC experiment is performed, and this heat effect should be added to the ITC-measured enthalpy of association (see Section 3.2.8). Correction for refolding not only changes the values of the binding enthalpies but also their temperature dependence (Fig. 12.22b).

The association constants of the HMG DBDs with their target DNAs are rather high; therefore they can be measured efficiently only by optical methods, in particular by fluorescence anisotropy or FRET titration using DNA duplexes labeled at both ends by the FAM and TAMRA fluorophores. Since FRET is sensitive to the distance between the fluorophores placed at two ends of the DNA duplex, it can give not only the binding constant but also information on the protein-induced DNA deformation (Dragan et al., 2004a; Dragan and Privalov, 2008).

As Figure 12.23a shows, the logarithmic dependence of the binding constant on salt concentration is perfectly linear, and binding to different DNA sequences does not alter the slope of the $\log(K^a)/\log[\mathrm{KCl}]$ plot but gives a parallel shift and therefore results in different extrapolated values at $\log[\mathrm{KCl}] = 0$, that is, different $\log(K^a)_0$. For the binding of Lef86 and its truncated form, Lef79, to their cognate $\mathrm{DNA^{Lef}}$ (filled squares, Fig. 12.23b), both show linear dependence but with **different slopes that extrapolate to the same point** at $\log[\mathrm{KCl}] = 0$. This extrapolation shows that both Lef boxes, full and truncated, exhibit the same nonelectrostatic interactions with their cognate DNA, thereby demonstrating that the eight C-terminal residues of Lef86 have no nonelectrostatic interactions with the DNA. However, the association of Lef86 and Lef79 with the noncognate $\mathrm{DNA^{Sry}}$ (open squares, Fig. 12.23b) is significantly weaker and, correspondingly, their $\log(K^a)$ functions are shifted downward to give lines parallel to those found for the association with the cognate DNA; these likewise converge at $\log[\mathrm{KCl}] = 0$.

It is remarkable that although the association constant is highly sensitive to salt concentration the enthalpy of association is quite invariant and depends only on

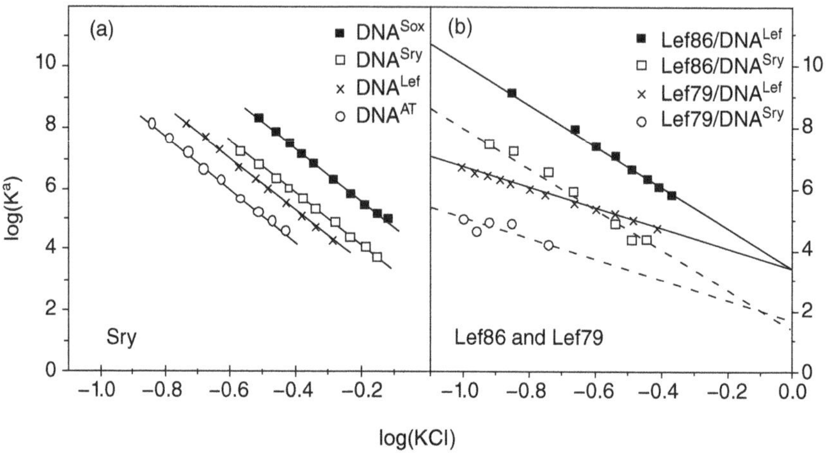

Figure 12.23. (a) Dependence of the logarithms of the association constants of Sry with four different DNAs on the KCl concentration, at 20°C in 10 mM potassium phosphate, pH 6.0. (b) Dependence of the logarithms of the association constants of Lef86 and Lef79 with their cognate and noncognate DNAs on the KCl concentration, obtained from changes in the intrinsic tryptophan fluorescence of the proteins, at 20°C in 10 mM K/phosphate pH, 6.0 (Dragan et al., 2004b).

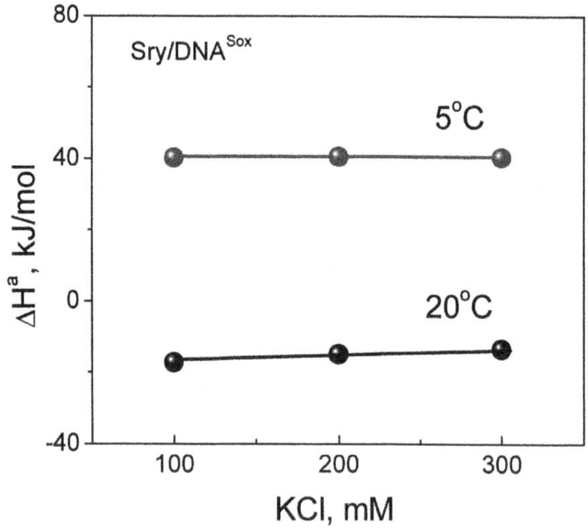

Figure 12.24. Enthalpies of binding of the SS HMG box from Sry to its optimal target sequence, DNA SRY, measured by ITC in 100, 200, and 300 mM KCl at temperatures 5°C and 20°C (Privalov et al., 2011).

temperature. For example, for an increase of KCl concentration from 100 mM to 300 mM the association constant of Sry box to its cognate DNA drops by four orders of magnitude (Fig. 12.23a); however, the enthalpy of association, measured by ITC and corrected for protein refolding, was found to be essentially the same—about −20 kJ/mol at 20°C and +40 kJ/mol at 5°C (Fig. 12.24), as predicted by the CC theory.

Table 12.7 lists all characteristics of the calorimetrically studied HMG–DNA complexes: the induced bend angle; enthalpy (corrected for refolding); Gibbs energy of binding at 0.1 M KCl, with its nonelectrostatic component determined by extrapolation of log(K^a) to log[KCl] = 0, that is, to 1 M KCl; and the electrostatic component of the Gibbs energy of binding obtained upon exclusion of the nonelectrostatic component from the total Gibbs energy of binding.

12.4. COMPARATIVE ANALYSIS OF PROTEIN–DNA COMPLEXES

12.4.1. Sequence-Specific versus Non-Sequence-Specific HMGs

Thermodynamic characteristics of binding the sequence-specific (SS) and non-sequence-specific (NSS) HMG DBDs to their optimal and quasi-optimal DNAs are illustrated by the diagram in Figure 12.25. The numbers above the energy bars give the DNA bend angles induced by binding the indicated DBDs. The electrostatic components of the Gibbs energy of binding are given in blue and the nonelectrostatic in yellow. In all cases, both SS and NS:

The electrostatic component dominates the nonelectrostatic, reflecting the high positive charge of HMG boxes.

Furthermore, the electrostatic component is independent of the actual DNA sequence bound, meaning that interactions between basic side chains and phosphate groups do not depend on sequence (as expected). Thus:

The specificity of binding is determined by the nonelectrostatic component.

Analysis of this component is therefore of particular interest since it is directly associated with the formation of new enthalpic interactions (van der Waals and hydrogen bonding) and dehydration of the interface that is formed.

Figure 12.26 presents the dependence of the nonelectrostatic component of the Gibbs binding energy and its enthalpic and entropic constituents on the protein-induced bend angle of the DNA duplexes (Privalov et al., 2009). The numbers in the boxes indicate the cases listed in Table 12.7. For both SS and NSS HMG boxes the enthalpies ΔH (which are entirely nonelectrostatic) are positive; the nonelectrostatic entropy factors, $-T\Delta S_{\text{nel}}$, are negative and it is notable that the spread of their values is larger than the spread of the ΔG_{nel} values. This shows that there are some components in the enthalpy and the entropy factors which efficiently compensate each other at the considered temperature, 20°C. This can only be the enthalpy and entropy of DNA dehydration: the fact that they efficiently compensate each other, that is, the Gibbs energy of dehydration is small, indicates that the dehydration process that occurs on forming these complexes closely resembles a phase transition that, in general, is specified by a zero Gibbs energy change at the transition temperature.

A striking feature of the bend angle dependence of ΔH and $T\Delta S_{\text{nel}}$ is that for the SS DBDs these two parameters decrease in magnitude as the bend increases, while for the NSS boxes they increase. For the SS DBDs, the decrease of these parameters

TABLE 12.7. Binding and Bending Characteristics of DNA–Protein Complexes[a,b]

Protein			DNA Binding Site	Case No.	Bend Angle	ΔG^{tot}	ΔG^{nel}	ΔH	$T\Delta S^{nel}$
Minor groove binders	SS HMG BOX DBD	Sry	SOX	1	75	27	48	27	48
			SRY	2	54	35	47	35	47
			LEF	3	59	38	48	38	48
		Sox	SOX	4	101	6	25	6	25
			SRY	5	40	9	23	9	23
			LEF	6	36	17	31	17	31
		Lef86	LEF	7	117	12	31	12	31
			SOX	8	50	20	32	20	32
			SRY	9	42	43	51	43	51
		Lef79	LEF	10	88	13	32	13	32
			SRY	11	–	7	20	7	20
	NSS HMG BOX DBD	NHP	LEF	12	62	40	51	40	51
			SRY	13	60	42	54	42	54
		Box-B′	LEF	14	39	32	37	32	37
			SOX	15	44	38	43	38	43
		D100	LEF	16	121	64	80	64	80
			AT	17	110	62	76	62	76
			SOX	18	106	64	80	64	80
		D74	LEF	19	94	65	78	65	78
			AT	20	92	70	85	70	85
			SOX	21	90	64	76	64	76
Major groove binders	Homeodomains	Antp	ANTP	22	0	−54	−16	−12	+4
		desAntp	ANTP	23	0	−48	−8	−7	+1
		NK2	NK2	24	0	−51	−21	−24	−3
		desNK2	NK2	25	0	−41	−9	−16	−7
		MATα2	MAT	26	0	−50	−17	−26	−9
		Engrailed	ENGR	27	0	−45	−8	−19	−11
	bZIPs	$(GCN4)_2$	ATF/CREB	28	13	−46	−23	−19	+5
		$(GCN4)_2$	AP-1	29	9	−49	−27	−71	−44
		$(ATF2)_2$	PRDIV	30	–	−45	−14	−66	−52
		$(cJun)_2$	PRDIV	31	–	−44	−13	−9	+4
		ATF2/cJun	PRDIV	32	0	−49	−17	−31	−14
	IRF	IRF-1	PRD1	33	25	−34	−26	−51	−25
		IRF-3	PRD1	34	20	−37	−19	−12	+7
		TFIIIA	5S	35	20	−45	−30	−20	+10

[a]At 20°C, pH 6.0, 100 mM NaCl. All data are corrected for protein refolding upon binding. Bend angle in degrees; ΔG, ΔH, and $T\Delta S$ in kJ/mol.

[b]References: cases 1–21, Dragan et al. (2004a,b); cases 22–27, Dragan et al. (2006); cases 28–32, Dragan et al. (2008); cases 33–34, Hargreaves et al. (2005); case 35, Liggins and Privalov (2000).

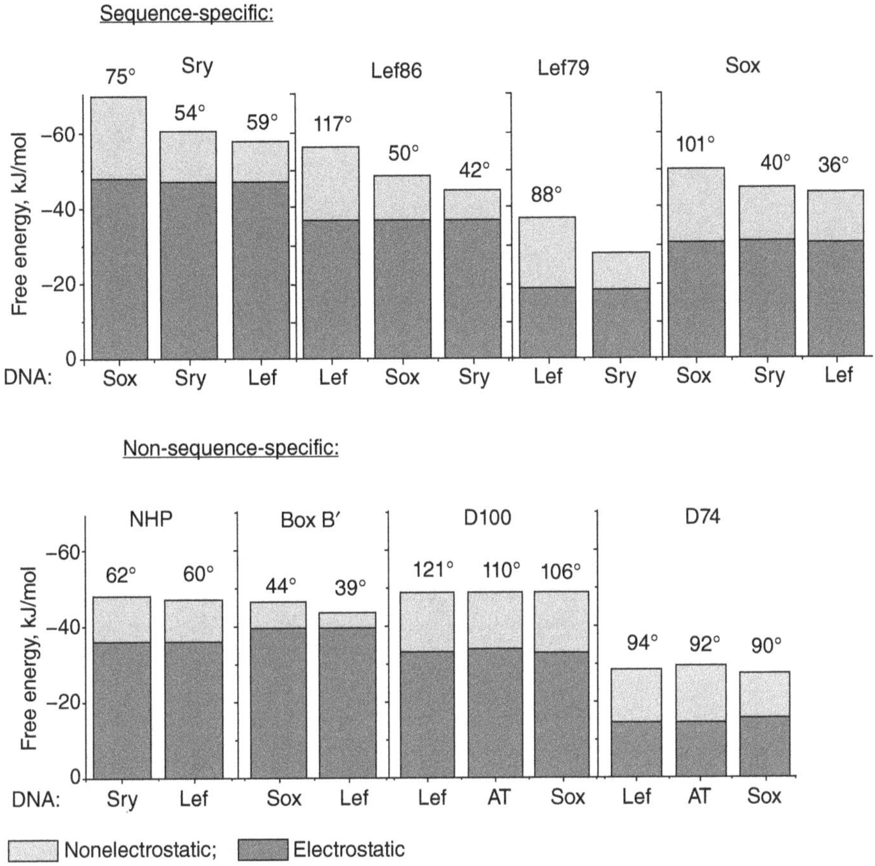

Figure 12.25. Electrostatic (blue) and nonelectrostatic (yellow) components of the total Gibbs free energy of binding the DBDs with various DNAs. The numbers above the bars indicate the induced DNA bend angles measured in the standard buffer (Dragan et al., 2004b). (See insert for color representation of the figure.)

shows that, in addition to the positive enthalpy and entropy of dehydration, there are increasingly negative contributions to the enthalpy and entropy associated with the formation of more contacts at the interface as the DNA sequence becomes more appropriate and the bend angle increases. Indeed, the structures of the complexes formed by the representatives of these two groups differ significantly in their packing densities (Fig. 12.27):

The complexes of the SS DBDs have more densely packed interfaces.

Correspondingly, they make a greater number of van der Waals contacts and hydrogen bonds, which provide negative enthalpy counterbalancing the effects of dehydration, thereby decreasing the overall positive enthalpy of binding.

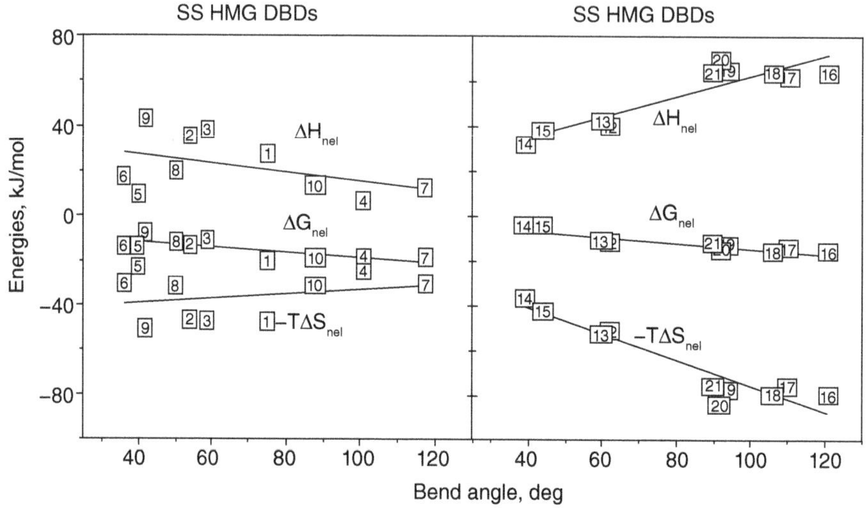

Figure 12.26. Relationship of bend angle and binding energetics. The correlation between the bend angles of various DNA sequences induced by binding SS and NSS HMG DBDs and the nonelectrostatic characteristics of complex formation: Gibbs energy; enthalpy; and entropy factor. The numbers in the boxes indicate the cases listed in Table 12.7 (Privalov et al., 2009).

> *The binding enthalpies of NSS DBDs are thus more positive due to poorer interfacial complementarity, and therefore have a lower contribution from van der Waals and other contacts that are present in the SS complexes.*
>
> *The large positive binding entropies, when electrostatic effects are excluded, can derive only from the release of water on complex formation, that is, dehydration of the groups at the interface, and this must be the dominant contribution to the binding entropy.*

A reason for the larger positive entropy upon forming NSS complexes could be a smaller loss of conformational freedom as a result of poor complementarity at the interface, that is, a reduced negative conformational entropy contribution, as one expects from the less tightly packed interface. Another reason might derive from the differences in intercalation of these two groups of DBDs: in the SS DBDs there is only one large intercalating wedge (Met13 in Lef, and Sox and Ile13 in Sry). However, the complexes of D74 and NHP have an additional intercalating wedge: Val32 and Thr33 for D74, and Phe48 for NHP (Murphy et al., 1999; Churchill et al., 1999; Klass et al., 2003; Masse et al., 2002). Dehydration of large protruding hydrophobic wedges will yield substantial positive entropy.

It should be noted that the positive entropy of dehydration of hydrophobic wedges can hardly be regarded as a driving force forming this complex. This is because insertion of the amino acid side chain between the DNA bases is a thermodynamically unfavorable process: the isolated amino acid side chains do not readily

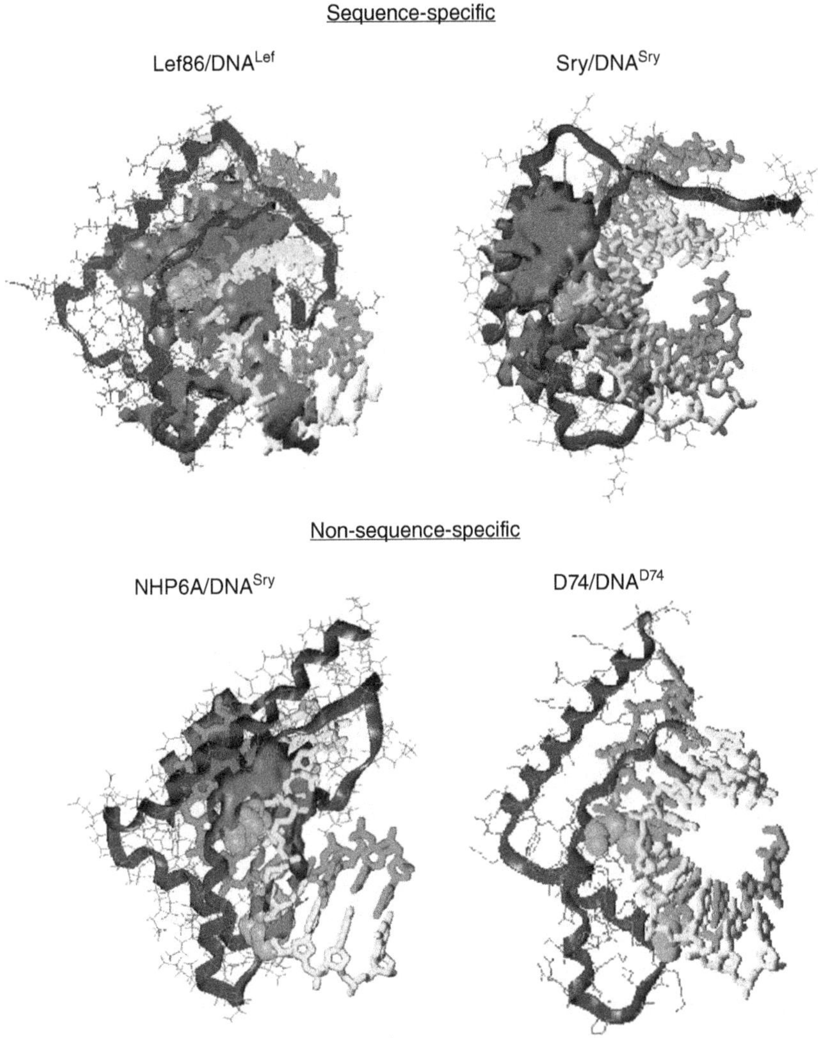

Figure 12.27. Packing densities at the interfaces of the SS complexes of Lef86 and Sry, and of the NSS complexes of NHP6A and D74. The red clusters are regions with packing density higher than 0.68. To help visualize the packing at the interface, only clusters that include groups of both protein and DNA are shown. Residues intercalating into the DNA are shown in pale blue. Although the Lef, Sry, and NHP complex structures were obtained by NMR and that of D74 by crystallography, the striking differences seen in the interfacial packing densities are too large to be explained solely by differences in the constraints applied during the structure determinations (Dragan et al., 2004b). (See insert for color representation of the figure.)

intercalate DNA, in contrast to the typical intercalators, which are planar, polycyclic aromatic cations. Thus:

> *The side chain must be forced between the bases by the cumulative action of various protein–DNA interactions (Dragan et al., 2003a).*

These interactions can be electrostatic or nonelectrostatic, depending on the interface formed. Since the complexes of SS DBDs with their cognate DNAs are specified by a tighter interface, the intercalation proceeds with the formation of more extensive enthalpic contacts. In the case of NSS complexes, however, the lower flexibility of these DBDs does not permit formation of extensive enthalpic contacts and insertion of the intercalators into DNA is then driven mostly by long-range electrostatic interactions.

> *The indifference of electrostatic (i.e., entropic) interactions to the DNA sequence explains the low sequence specificity of NSS DBDs for DNA binding and bending.*

In contrast to the binding enthalpy and the binding entropy factor, the Gibbs energy of binding does not depend much on the bend angle (Fig. 12.26). In the case of Lef86, for example, an increase of bend angle by 75°, induced by using a more appropriate DNA sequence, is associated with a nonelectrostatic Gibbs energy change of only on about −10 kJ/mol. This, however, is not the increasing work of DNA bending, which would result in positive changes to the Gibbs energy, but presents the balance of the work on binding and bending:

> *The negative sign of this balance shows clearly that the binding contribution dominates the bending.*

Since the Gibbs energy of binding without bending (i.e., the value of ΔG_{nel} extrapolated to zero bend angle) is of the order −10 kJ/mol, the work of bending must be much smaller than this. Moreover, this bending work must be substantially smaller than that for bending the naked DNA, for which the wormlike chain model (Landau and Lifshitz, 1970) predicts the Gibbs free energy expenditure of about 70 kJ/mol for bending a 10-bp DNA duplex through 50°, that is, about 1.5 kJ per degree of bend.

> *Thus the bending of DNA by HMG DBDs requires a great deal less free energy than predicted from the properties of free DNA.*

The next section substantiates this surprising conclusion.

12.4.2. Salt-Dependent versus Salt-Independent Components of Binding

Since salt addition leads to redistribution of the ionic interactions, it is sensible to consider the salt-dependent part of the binding energy as its electrostatic component.

The straightforward interpretation of Equation (12.1) is that all electrostatic interactions drop to zero at 1 M salt concentration and so the value of K^a at this point represents the association constant resulting from only nonelectrostatic forces—leading to the corresponding nonelectrostatic Gibbs free energy (ΔG^{nel}). Loss of the electrostatic contribution at precisely 1 M salt concentration seems somewhat arbitrary but in fact simply reflects taking 1 M salt as the standard condition from which the electrostatic component of the binding energy is counted (Privalov et al., 2011). The salt-independent part of the binding free energy, which is determined by excluding the electrostatic component from the total binding energy, is usually regarded as nonelectrostatic. Labeling the salt-independent part of the binding free energy as entirely nonelectrostatic could be regarded as not entirely appropriate because it might include the Coulombic electrostatic effects that are independent of the salt concentration. However, its main contributions come from dehydration effects, van der Waals interactions, and hydrogen bonding, which are usually regarded as nonelectrostatic. In fact, the van der Waals interactions and hydrogen bonding are also of electrostatic origin: thus it is rather a matter of semantics to call their collective contribution nonelectrostatic, bearing in mind that none of them can be measured separately.

Is the salt-dependent part of the Gibbs energy of binding, that is, its electrostatic component, indeed purely entropic? If this is the case, the logarithmic salt dependence of the binding constants should not depend on temperature. This is just what is observed for the AT-hooks: the salt titration of these peptides was carried out at three temperatures, 10°C, 20°C, and 30°C, and it was observed that for a given peptide (ATH3, ATH2, or CoreATH2) all three titrations overlapped (Fig. 12.17).

> *Thus, the dependence of binding constant on salt is indeed invariant of temperature, which means that the salt-dependent electrostatic component of binding energy is indeed purely entropic.*

A further example studied in detail was the binding of the bZIP fragment from yeast GCN4 to its target ATF/CREB recognition sequence, over the temperature range from 5°C to 40°C. Table 12.3 gives the variation of the total Gibbs energy of binding and its nonelectrostatic and electrostatic components, separated as suggested by Equation (12.1). It is striking that:

> *While the nonelectrostatic component of the Gibbs energy becomes substantially more negative with increasing temperature, the electrostatic component shows no dependence on temperature at all. This is to be expected if the electrostatic component is purely entropic and its enthalpy is close to zero.*

As already mentioned, according to the CC concept, the slope of the plots of Equation (12.1) represents the number of tightly bound counterions (cations) released from the DNA on binding protein and this is assumed to be proportional to the number of ionic contacts made. Therefore, it is particularly interesting to compare the number of contacts made between protein and DNA estimated from

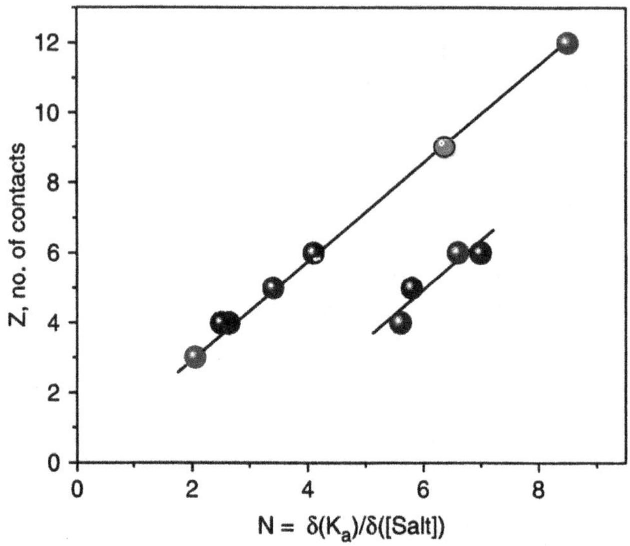

Figure 12.28. The slopes, N, of the salt plots for several protein–DNA complexes (x-axis) graphed against the number of contacts, Z, between DNA phosphates and Lys–Arg residues (y-axis) seen their atomic resolution structures (X-ray or NMR). On the main line that goes through the origin the seven complexes are CoreATH2 (3 contacts), ATH3 (4 contacts), Drosophila D74 (4 contacts), Lef79 (5 contacts), GCN4 (6 contacts), NHP (9 contacts), SRY (12 contacts). The four homeodomains—desNK2 (4 contacts), MATa2 (5 contacts), Engrailed (6 contacts), desAntp (6 contacts)—lie on a parallel line displaced by ~2.5 along the x-axis (Privalov et al., 2011).

thermodynamic studies of their binding in solutions at different concentrations of salt with the data obtained by structural methods, that is, crystallography and NMR (Privalov et al., 2011). These results are summarized in Figure 12.28, which graphs the slopes of the salt plots (N), obtained with increasing concentration of KCl, against the numbers of ionic contacts of DNA phosphate groups, Z, observed in atomic structures of the DNA complexes. Seven cases (four HMG boxes, the bZIP dimer, and two AT-hooks) fall on a single line that goes through the origin, demonstrating that the slopes (N values) are proportional to the number of electrostatic contacts made to the phosphates. In contrast, the data for the four homeodomains fall close to a parallel line shifted horizontally by about 2.5 ions, that is, 2.5 more ions are displaced than expected from the number of phosphate contacts made. This difference comes from release of tightly bound chloride ions from the homeodomains: Figure 12.4d shows that salt titrations of the Engrailed homeodomain using KCl (NaCl acts identically) exhibit a slope $N = 6.6$; however, when NaF is used for dissociation, the slope N is found to be 4.5, which if $\psi = 0.70$ corresponds to six phosphate contacts, the number observed in the Engrailed/DNA complex. Since the difference between KCl/NaCl and NaF is in the anion, it follows that about two tightly bound chloride ions are lost from the Engrailed DBD when it binds to DNA. This additional ion loss can be formally taken into account by writing the slope of Equation (12.1) as $N = (Z \cdot \psi + \beta)$, where β is the number of anions displaced.

Release of anions specifically bound to these homeodomains upon association with DNA is not an exception and has been documented experimentally in the formation several protein–DNA complexes (Lohman et al., 1980; Arosio et al., 2004; Van der Meulen et al., 2008). The specific binding of certain ions to DBDs, which are often rather flexible, can significantly change their structure and thus their ability to interact with DNA. In all such cases it is therefore necessary to test whether the salt affects the state of the free protein in the absence of DNA. Furthermore, in cases where titration of different salts into protein–DNA complexes results in different slopes, the minimal slope must be associated with the lowest level of anion binding to the protein component.

Thus, it appears that thermodynamic analysis of binding might provide useful structural information on the number of contacts made between protein and DNA phosphates in their complex.

12.4.3. Minor versus Major Groove Binding

The energetics of association of various DNA-binding proteins with their target DNAs has been studied calorimetrically by many groups over the years. For the present analysis, selected data are obtained under similar conditions (20°C, near neutral pH, 100 mM NaCl) and analyzed by the same approach, namely, correcting for refolding effects and resolving the electrostatic and nonelectrostatic components of the binding characteristics. Inspection of the data reveals no dramatic differences in the Gibbs free energies of protein binding to the major and minor grooves of DNA (Fig. 12.29a). Therefore, a particularly striking finding is the systematic qualitative difference in binding enthalpies for the two grooves (Fig. 12.29b). Specifically, the enthalpies of binding to the minor groove are always positive, while the enthalpies of binding to the major groove are invariably negative. Since the Gibbs free energies of binding at 20°C do not vary greatly, the enthalpy differences are essentially balanced by entropic factors. Indeed, the entropic contribution (i.e., $T\Delta S$) of minor groove binding is significantly larger in magnitude than that of major groove binding (Fig. 12.29c). Since binding of protein to the minor groove of DNA results in considerable bending of DNA, in contrast to binding to the major groove (Fig. 12.29d), the observation of a different sign in the enthalpies of protein binding to these grooves was initially interpreted as reflecting the work required to bend DNA (Privalov et al., 1999; Jen-Jacobson et al., 2000; Peters et al., 2004). However, the difference in the work on bending should be reflected in the Gibbs energies of binding, but it is not: the total Gibbs energy of binding to the major and minor grooves oscillates around −50 (±10) kJ/mol. This suggests that:

The work required to bend DNA cannot represent the main source of the positive enthalpy associated with protein or ligand binding to the minor groove.

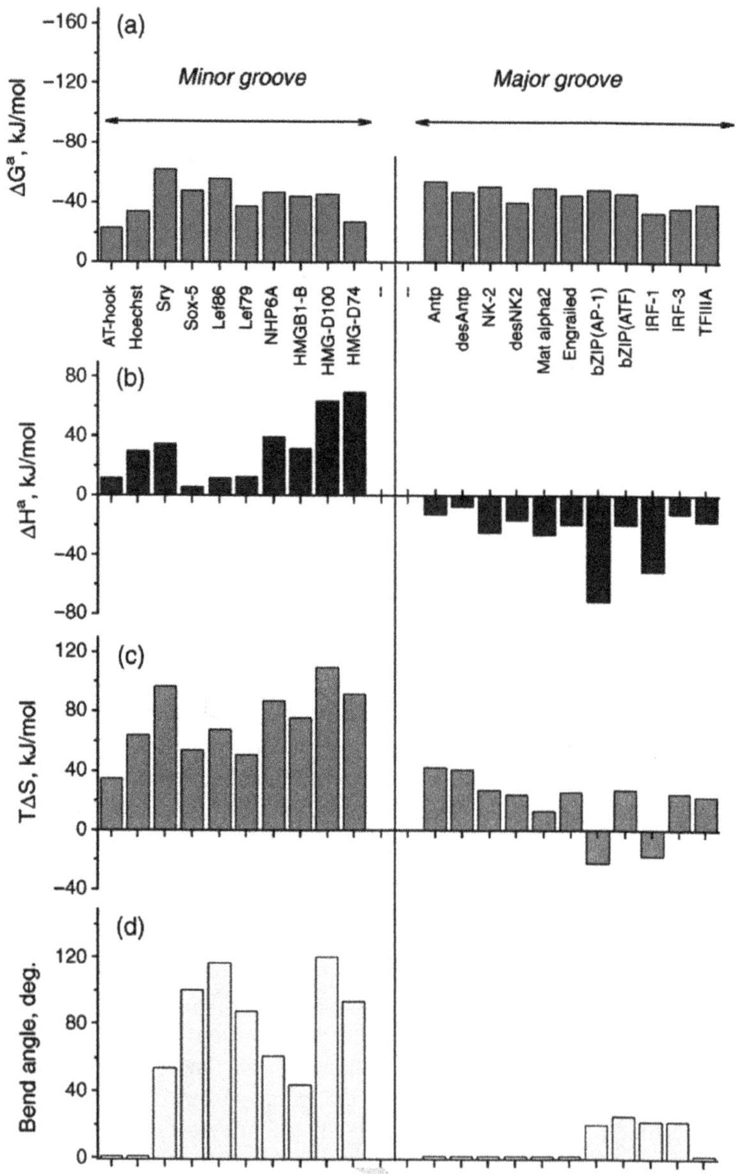

Figure 12.29. Thermodynamic parameters of association for various DBDs to the major and minor grooves of their cognate DNA: (a) Gibbs energies of association; (b) enthalpies of association; (c) entropy factors of association; (d) DNA bending angles (Privalov et al., 2007). (See insert for color representation of the figure.)

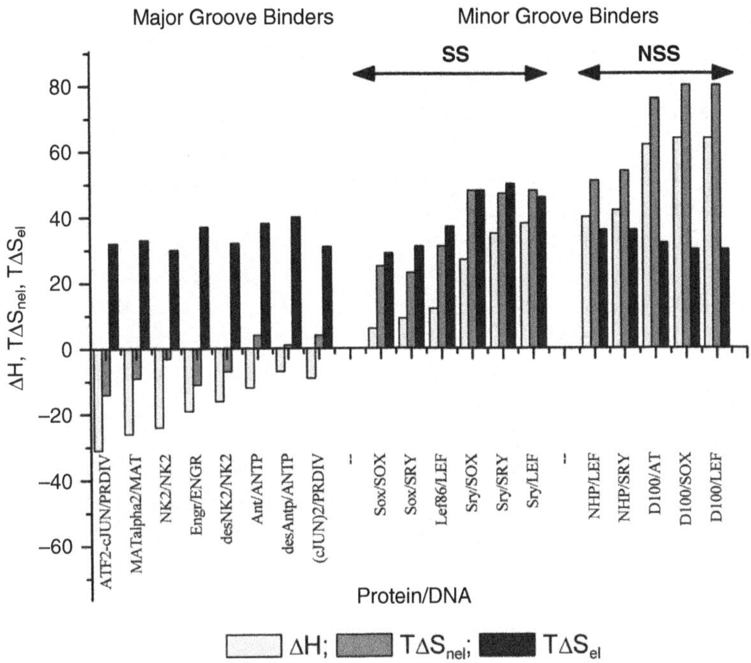

Figure 12.30. Enthalpies and entropy factors (electrostatic and nonelectrostatic) of binding proteins to the minor and major groove of their cognate and semicognate DNAs (Privalov et al., 2011). (See insert for color representation of the figure.)

To understand the nature of the observed qualitative difference in the enthalpies and entropies of protein binding to the major and minor grooves of DNA, analysis of their electrostatic and nonelectrostatic components shown in Figure 12.30 gained particular importance. One can see that there is no dramatic difference in the electrostatic component of the Gibbs energy of binding, $\Delta G^{el} = -T\Delta S^{el}$: for all studied cases its value varies between −30 and −40 kJ/mol, showing that the numbers of released counterions upon protein binding to the minor or major grooves are not very different.

So, although the electrostatic component is the most prevalent in binding, it does not determine the specificity of binding.

Therefore, the systematic qualitative difference in nonelectrostatic components of binding energy, that is, nonelectrostatic entropies and binding enthalpies for these two grooves, is particularly striking. Specifically:

The enthalpies and nonelectrostatic entropies of binding to the major groove are always negative, while for binding to the major groove they are invariably positive.

The negative enthalpy and entropy is just what is usually expected for the binding reaction: it proceeds with heat release, that is, with negative enthalpy, resulting from formation of new interactions, which promotes binding, and negative entropy, resulting from decrease of conformational and translational freedom upon complex formation; the negative entropy factor opposes binding, but its effect is overbalanced by the enthalpy. Thus, **binding to the major grove is enthalpy driven!**

The situation with binding to the minor groove is just the opposite: it proceeds with heat absorption, that is, with positive enthalpy and positive entropy. The positive enthalpy opposes binding, but it is overbalanced by the positive entropy factor. Thus, **binding to the minor groove is entropy driven** (Privalov et al., 2007, 2011). One can also see that the magnitudes of the enthalpy and the nonelectrostatic entropy factor vary considerably, showing that they are sequence dependent and, therefore:

> *These nonelectrostatic components of binding energy determine the specificity of binding. What is remarkable, however, is that the enthalpy and the entropy factor are especially large for the non-sequence-specific bindings (NSS) (see Section 12.4.1).*

The only possible source of such a large positive enthalpy and positive nonelectrostatic entropy at binding to the minor groove might be only the removal of water from this groove at protein binding. However, since the bulk water in both cases is the same, we have to conclude that:

> *The water hydrating the major and minor grooves is different: it should be more ordered in the minor groove.*

In the case of the minor groove, judging from the predominance of AT sequences among its DBD binding sites and bearing in mind the absence of the 2-amino group in adenosine (Fig. 4.16), one could suppose that this groove is more apolar than the major groove; also, since it is known that apolar groups promote water ordering, it was tempting to assume that the large positive nonelectrostatic entropy of protein binding to the minor groove results from the dehydration of apolar groups on forming hydrophobic contacts between protein and DNA. This hypothesis, however, does not explain the large positive enthalpy of binding to the minor groove since the enthalpy of formation of hydrophobic contacts is known to be close to zero at ~20°C (see Section 2.4.3).

The presumption that water ordering in the AT-rich minor groove differs significantly from water hydrating the major groove has been a subject of discussion for quite some time (Drew and Dickerson, 1981; Kopka et al., 1983; Dickerson, 1998). The presence of a spine of well-ordered water molecules in the minor groove of AT-rich DNA sequences has been derived from high-resolution X-ray crystallography and neutron diffraction studies (Shui et al., 1998; Arai et al., 2005; Chiu et al., 1999; Goodsell et al., 1994). The minor groove is especially narrow in AT stretches and a primary shell of water molecules runs across the groove,

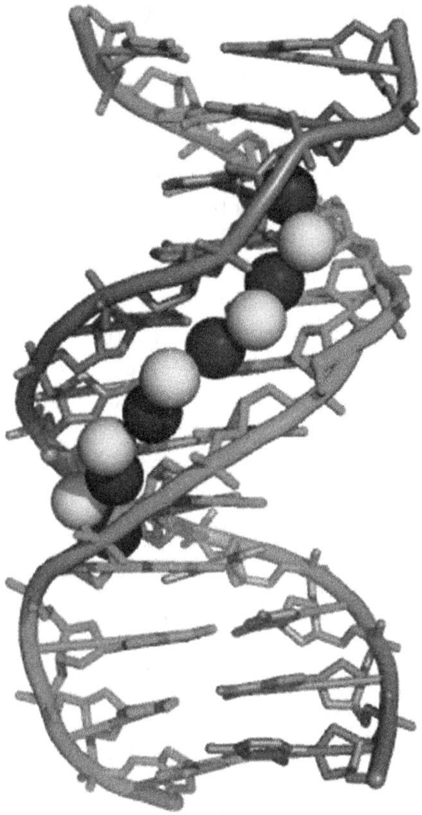

Figure 12.31. Display of primary (blue) and secondary (yellow) layers of the spine of water in the minor groove of the X-link dodecamer CGCAATTCGCG generated from the coordinates of NDB accession number bd0008 (Privalov et al., 2007). (See insert for color representation of the figure.)

bridging acceptor sites on adjacent AT/TA base pairs (Fig. 12.31). A secondary shell of water molecules runs along the groove, donating hydrogen bonds to the primary shell of oxygen atoms that assume the tetrahedral coordination characteristic of ice (Chiu et al., 1999). There is further evidence of an "outer spine" of third- and fourth-shell water molecules in a pattern of fused hexagons (Egli et al., 1998; Tereshko et al., 1999). Exchange of the bound water with bulk solvent has been studied by nuclear magnetic resonance (NMR) and its residence time is longer than that of water residing in the major groove (Jacobson et al., 1996; Liepinsh et al., 1992). Studies of the interaction of bisbenzimide with DNA reveal that the dielectric constant in the minor groove is less than that estimated for the major groove, consistent with stronger hydrogen bonding in the minor groove (Barawkar and Ganesh, 1995; Jin and Breslauer, 1988). In the major groove, the distribution of hydrogen-bond donors and acceptors exhibits greater irregularity than in the minor groove, and no regular water superstructure has been identified (Chiu et al., 1999).

If water hydrating the minor groove of AT-rich DNA is more ordered than water hydrating the major groove, its removal is expected to require a greater enthalpy and thereby provides a larger entropy increase than removal of water hydrating the major groove. If water in the AT-rich minor groove is in an "ordered," icelike state and its removal requires about 6 kJ/mol (the heat of ice melting at 0°C), the loss of 11 water molecules coordinated in the first and second layers of the 7-bp binding site of an HMG DBD (i.e., 6 in the first hydration layer and 5 in the second according to Kopka et al. (1983), would require approximately 70 kJ/mol of enthalpy at 20°C, the temperature at which the enthalpy of formation of hydrophobic contacts is zero. This value is in remarkable agreement with that observed for binding of the non-sequence-specific HMG-D74 DBD to the minor groove, an interaction involving relatively few hydrogen bonds (Churchill et al., 1999; Murphy and Churchill, 2000; Murphy et al., 1999).

Thus, inspection of the thermodynamic data summarized herein on structurally characterized protein–DNA complexes reveals a clear qualitative difference in the energetic signatures of binding to the major and minor grooves that appears to be a consequence of their distinct hydration properties.

It is important to note that protein binding to the minor groove, as well as the binding of small molecules to this groove (Haq et al., 1997), occurs principally at AT-rich DNA sequences, the latter corresponding to those very regions in which water ordering is the most prevalent. The consequence of this is that minor-groove binding is normally driven by the very large entropy of releasing the ordered water, despite an unfavorable enthalpy. However, this does not represent a hydrophobic force since water ordering in the AT-rich minor groove is determined not by the apolar groups of the DNA but by the regular arrangement of its polar groups that stabilize the icelike organization of the water in this groove (Shui et al., 1998; Arai et al., 2005; Chiu et al., 1999; Goodsel et al., 1994).

One can expect that the stretch of highly ordered water in the AT-rich minor groove provides rigidity to the DNA. If so, with removal of this water, this DNA might lose its rigidity. One could expect that this might not require much work since, because of enthalpy–entropy compensation, the Gibbs energy of melting of ice is zero. This might explain why the minor-groove binder proteins choose the AT-rich sequences when it is required to bend DNA considerably.

Thus, according to the results of crystallography and thermodynamics of protein binding to the major and minor grooves of DNA, water in the AT-rich minor groove is in a highly ordered state.

If so the enthalpy of melting the AT-rich DNA should be higher than the GC-rich DNA. This conclusion, however, is in conflict with the widespread opinion that the enthalpy of melting the more stable GC-rich DNA is higher than that of the less stable AT-rich DNA. This confusing contradiction is discussed in the next chapter (see Section 13.3.2).

12.5. CONCLUDING REMARKS

12.5.1. Assembling Multicomponent Protein–DNA Complex

Proteins usually interact specifically with the DNA in certain groups, forming a specific multicomponent complex with no simple function. One of the most studied examples presents the IFN-β enhanceosome, which includes ATF2, cJun, IRF-3, p50, and p65 transcription factors assembled on the enhancer DNA and somehow controlled by the HMGA1 protein (Fig. 12.32). All these transcription factors, as discussed previously, form homo or hetero dimers and interact with their cognate sites, the positive regulatory domains (PDRs): PRDIV, PRDIII, PRDI, and PRDII. Among these sites two, PRDIV and PRDII, are asymmetric and preferentially bind the heterodimers in certain orientations; namely, at PRDIV sites ATF2/cJun (see Sections 12.2.3 and 12.2.5).

When several proteins bind to the DNA one would expect that they might interfere, increasing (positive cooperation) or decreasing (negative cooperation) the binding abilities of the neighbors. In particular, based on crystallographic data, it was expected that IRF-3 dimer and ATF2/cJun bZIP sited on the enhancer DNA interact and that determines orientation of this hetero bZIP (Panne et al., 2004, 2007; Panne, 2008). However, calorimetric analysis of binding of these transcription factors to the IFN-β enhancer DNA showed that they do not interact and the crystallographically found binding of the IRF-3 dimer to the nonconsensus sites is an artifact of crystallization: this dimer binds only to its consensus tandem sites, PRDIII

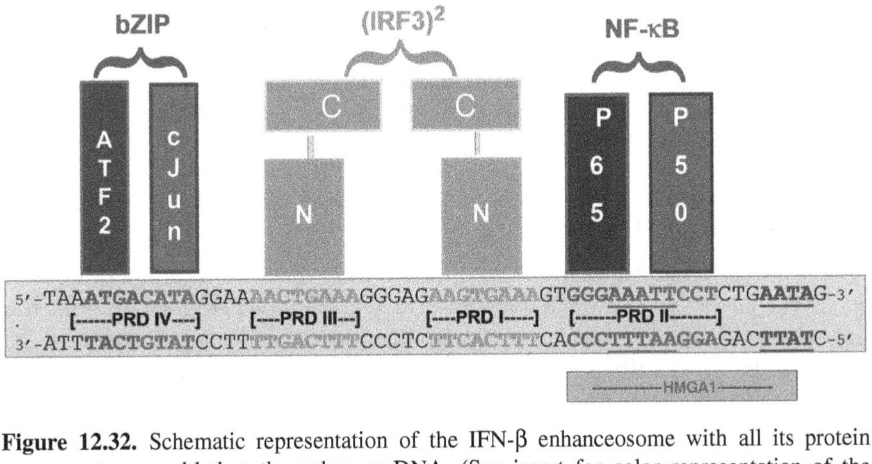

Figure 12.32. Schematic representation of the IFN-β enhanceosome with all its protein components assembled at the enhancer DNA. (See insert for color representation of the figure.)

and PRDI (Dragan et al., 2008). On the other hand, it was found in the fluorescence anisotropy titration experiment that the Gibbs energy of p50/p50 binding to the PRDII site increases from −46.6 5 to −52.5 kJ/mol in the presence of the IRF-3* dimer at the neighboring PRDIII–PRDI sites (Dragan et al., 2008). It appeared thus that NF-κB (p50/p50) forms significant contacts with the IRF-3* dimer upon binding to the IFN-β enhancer. The most intriguing, however, is interference of NF-κB and AT-hook: both these proteins bind to the same site PRDII (shown in blue, Fig. 12.32). However, while the AT-hook enters into the minor groove of this site (Section 12.3.1) the NF-κB binds to the major groove of this site. Thus, these two proteins do not touch each other (Berkowitz et al., 2002). Nevertheless, a calorimetric titration experiment carried out at 25°C showed that association of NF-κB (p50/p50 homodimer) with the major groove of PRDII DNA largely increases in the presence of the AT-Hook 2 of HMG-I at the minor groove of this binding site (Dragan et al., 2008). Specifically, at 25°C in the absence of the AT-hook the NF-kB binds to this site with $K_d = 140$ nM, $\Delta G^a = -39.2$ kJ/mol, $\Delta H^a = -62.7$ kJ/mol, and $T\Delta S = -23.15$ kJ/mol, whereas in the presence of the AT-hook it binds with $K_d = 27$ nM, $\Delta G^a = -43.1$ kJ/mol, $\Delta H^a = -83.0$ kJ/mol, and $T\Delta S = -39.9$ kJ/mol. The large increase in the negative binding enthalpy and the corresponding increase in the negative entropy factor are the most remarkable. They show that:

The AT-hook, which penetrates deep into the minor groove of DNA, significantly changes the NF-κB binding properties of the major groove, perhaps by making this groove more adaptable to NF-κB.

Since formation of this trimeric complex is a highly reversible process, one would expect that binding of the NF-κB to the major groove of PRDII would also change the minor groove, accommodating it for the AT-hook and thus increasing its binding ability.

The AT-hook is just the segment of the HMGA1 protein that enters easily into the minor groove of the AT-rich DNA and this protein, which is completely unfolded, has three such segments. It is supposed that HMGA1 somehow orchestrates assembling of the IFN-β enhanceosome by interacting specifically with its components, certain DNA and proteins. This, however, is a problem deserving thorough investigation.

12.5.2. CC Approach versus PB Theory

Although from the purist's point of view the CC theory presents a certain simplification of the interactions between protein and DNA, as follows from the calorimetric and optical titration experiments considered in previous sections, it appears to be very effective in describing these complexes (Privalov et al., 2011). In particular, direct experiments show that:

(a) The logarithm of the protein–DNA association constant $\log K^a$ is indeed a linear function of log[salt].

(b) The protein–DNA binding energy can be split unequivocally into two qualitatively different components: the salt-dependent and salt-independent parts, which are usually considered as the electrostatic and nonelectrostatic components of the binding free energy. At present, observation of the salt dependence of the binding constant is the only experimentally accessible way of quantitatively evaluating these two components.

(c) The enthalpy of protein–DNA association is indeed independent of the salt concentration, as predicted by the CC theory, and the effect of salt is reflected solely in the entropy of binding. Thus, the electrostatic component of the binding energy is fully entropic.

(d) The specificity of interactions is manifested entirely in the nonelectrostatic component, despite the electrostatic component typically contributing the majority of the affinity.

(e) The CC methodology gives reliable binding characteristics and thus has considerable practical value for elucidating the forces contributing to the formation of protein–DNA complexes and determining the number of contacts formed between protein and DNA.

As for the PB approach to the problem of protein–DNA interactions, it is still not clear yet how efficient it might be for quantitatively understanding such complicated molecular systems. Here the problem is not in the power of the computers, but in the lack of experimental information, particularly on water at the protein–DNA interface. As we saw, unusual states of water in these systems were absolutely unexpectable and its role in the formation and functioning of protein–DNA complexes is far from being understood.

REFERENCES

Arai S., Chatake T., Ohhara T., Kurihara K., Tanaka I., Suzuki N., et al. (2005). *Nucleic Acids Res.*, **33**, 3017–3024.

Aravind L. and Landsman D. (1998). *Nucleic. Acids Res.*, **26**, 4413–4421.

Arosio D., Constantini S., Kong Y., and Vindigni A. (2004). *J. Biol. Chem.*, **279**, 42826–42835.

Barawkar D.A. and Ganesh K.N. (1995). *Nucleic Acids Res.*, **23**, 159–164.

Baumann C.G., Bloomfield V.A, Smith S.B, Bustamante C., Wang M.D, and Block S.M. (2000). *Biophys. J.*, **78**, 1965–1978.

Berger C., Jelesarov I., and Bosshard H.R. (1996). *Biochemistry*, **35**, 14984–14991.

Berkowitz B., Huang D.H., Chen-Park F.E., Sigler P.B., and Ghosh G. (2002). *J. Biol. Chem.*, **277**, 24649–24700.

Bustin M. and Reeves R. (1996). *Prog. Nucleic Acid Res. Mol. Biol.*, **54**, 35–100.

Carra J.H. and Privalov P.L. (1997). *Biochemistry*, **36**, 526–535.

Carrillo R.J., Dragan A.I., and Privalov P.L. (2010). *J. Mol. Biol.*, **396**, 431–440.

Chiu T.K., Kaczor-Grzeskowiak M., and Dickerson R.E. (1999). *J. Mol. Biol.*, **292**, 589–608.

Churchill M.E.A., Changela A., Dow L.K., and Krieg A.J. (1999). *Methods Enzymol.*, **304**, 99–133.

Cloutier T.E. and Widom J. (2004). *Mol. Cell*, **14**, 355–362.

Crane-Robinson C., Dragan A.I., and Privalov P.L. (2006). *Trends Biochem. Sci.*, **31**, 547–552.

Cranz S., Berger C., Baici A., Jelesarov I., and Bosshard R. (2004). *Biochemistry*, **43**, 718–727.

Dickerson R.E. (1998). *Nucleic Acids Res.*, **26**, 1906–1926.

Dragan A.I. and Privalov P.L. (2008). *Methods Enzymol.*, **450**, 185–199.

Dragan A.I., Carillo R., Gerasimova T.I., and Privalov P.L. (2008). *J. Mol. Biol.*, **384**, 335–348.

Dragan A.I., Frank L., Liu Y., Makeyeva E.N., Crane-Robinson C., and Privalov P.L. (2004a). *J. Mol. Biol.*, **343**, 865–878.

Dragan A.I., Hargreaves V.V., Makeyeva E.N., and Privalov P.L. (2007). *Nucleic Acids Res.*, **35**, 3525–3534.

Dragan A.I., Klass J., Read, C., Churchill M.E., Crane-Robinson C., and Privalov P.L. (2003a). *J. Mol. Biol.*, **331**, 795–813.

Dragan A.I., Li Z., Makeyeva E.N., Milgotina E.I., Liu Y., Crane-Robinson C., and Privalov P.L. (2006). *Biochemistry*, **45**, 141–151.

Dragan A.I., Liggins J.R., Crane-Robinson C., and Privalov P.L. (2003b). *J. Mol. Biol.*, **327**, 393–411.

Dragan A.I., Read C.M., Makeyeva E.N., Milgotina E.I., Churchill M.E.A., Crane-Robinson C., and Privalov P.L. (2004b). *J. Mol. Biol.*, **343**, 371–393.

Drew H.R. and Dickerson R.E. (1981). *J. Mol. Biol.*, **151**, 535–556.

Egli M., Tereshko V., Teplova M., Minasov G., Joachimiak A., Sanishvili R., Weeks C.M., Milen R., Maier M.A., An H., Dan C.P., and Manoharan M. (1998). *Biopolymers*, **48**, 234–252.

Ellenberger T.E. (1994). *Curr. Opin. Struct. Biol.*, **4**, 12–21.

Ellenberger T.E., Brandl C.J., Struhl K., and Harrison C. (1992). *Cell*, **71**, 1223–1237.

Escalante C.R., Shen L., Thanos D., and Aggarwal A.K. (2002). *Structure*, **10**, 383–391.

Escalante C.R., Yie J., Thanos D., and Aggarwal A.K. (1998). *Nature*, **391**, 103–106.

Ferrari S., Harley V.R., Pontiggia A., Goodfellow P.N., Lowell-Badge R., and Bianchi M.E. (1992). *EMBO J.*, **11**, 4497–4506.

Fraenkel E. and Pabo C.O. (1998). *Nature Struct. Biol.*, **5**, 692–697.

Fraenkel E., Rould M.A., Chambers K.A., and Pabo C.O. (1998). *J. Mol. Biol.*, **284**, 351–361.

Fujii Y., Shimizu T., Kusumoto M., Kyogoku Y., Tanigushi T., and Hakoshima T. (1999). *EMBO J.*, **18**, 5028–5041.

Giese K., Amsterdam A., and Grosschedl R. (1991). *Genes Dev.*, **5**, 2567–2578.

Giese K., Cox J., and Grosschedl R. (1992). *Cell*, **69**, 185–195.

Glover J.N. and Harrison, S.C. (1995). *Nature*, **373**, 257–261.

Gonzales M., Weiler S., Ferretti J.A., and Ginsburg A. (2001). *Biochemistry*, **40**, 4923–4931.

Goodsell D.S., Kaszor-Grzeskowiak M., and Dickerson R.E. (1994). *J. Mol. Biol.*, **239**, 79–96.

Grosschedl R., Giese K., and Pagel J. (1994). *Trends Genet.*, **10**, 94–100.

Gruschus J.M., Tsao D.H., Wang L.H., Nirenberg M., and Ferretti J.A. (1997). *Biochemistry*, **36**, 5372–5380.

Haq I., Ladbury J.E., Chowdhry B.Z., Jenkins T.C., and Chaires J.B. (1997). *J. Mol. Biol.*, **271**, 244–257.

Hargreaves V.V., Makeyeva E.N., Dragan A.I., and Privalov P.L. (2005). *Biochemistry*, **44**, 14202–14209.

Hecht J.L., Honig B., Shin Y.-K., and Hubbell W.L. (1995). *J. Phys. Chem.*, **99**, 7782–7786.

Huth J.R., Bewley C.A., Nissen M.S., Evans J.N.S., Reeves R., Gronenborn A.M., and Clore G.M (1997). *Nature Struct. Biol.*, **4**, 657–665.

Jacobson A., Laupin W., Liepinsh E., and Otting F. (1996). *Nucleic Acids Res.*, **24**, 2911–2918.

Jen-Jacobson L., Engler L.E., and Jacobson L.A. (2000). *Structure*, **8**, 1015–1023.

Jin R. and Breslauer K.J. (1988). *Proc. Natl. Acad. Sci. USA*, **85**, 8939–8942.

Johnson K.R., Lehn D.A., Elton T.S., Barr P.J., and Reeves R. (1988). *J. Biol. Chem.*, **263**, 18338–18342.

Keller W., König P., and Richmond T.J. (1995). *J. Mol. Biol.*, **254**, 657–667.

Kim J. and Struhl K. (1995). *Nucleic Acids Res.*, **23**, 2531–2537.

Kissinger C.R., Liu B., Martin-Blanco E., Kornberg T.B., and Pabo C.O. (1990). *Cell*, **63**, 579–590.

Klass J., Murphy F.V. IV, Fouts S., Serenil M., Changela A., Siple J., and Churchill M.E. (2003). *Nucleic Acids Res.*, **31**, 2852–2864.

Kohler J.J. and Schepartz A. (2001). *Biochemistry*, **40**, 130–142.

König P. and Richmond T.J. (1993). *J. Mol. Biol.*, **233**, 139–154.

Kopka M.L., Fratini A.V., Drew H.R., and Dickerson R.E. (1983). *J. Mol. Biol.*, **163**, 129–146.

Ladbury J.E., Wright J.G., Sturtevant J.M., and Sigler P.B. (1994). *J. Mol. Biol.*, **238**, 669–681.

Landau L.D. and Lifshitz E.M. (1970). Theory of elasticity. In: *Course of Theoretical Physics* **7**, p. 72. Pergamon, Oxford.

Landschulz W.H., Johnson P.F., and McKnight S.L. (1988). *Science*, **240**, 1759–1764.

Liepinsh E., Otting G., and Wuthrich K. (1992). *Nucleic Acids Res.*, **20**, 6549–6553.

Liggins J.R. and Privalov P.L. (2000). *Biophys. J.*, **81**, 1632–1642.

Lin R., Genin P., Mamane Y., and Hiscott J. (2000). *Mol. Cell Biol.*, **20**, 6342–6353.

Lin R., Heylbroeck C., Potha P.M., and Hiscott J. (1998). *Mol. Cell. Biol.*, **18**, 2986–2996.

Lohman T.M, deHaseth P.L, and Record M.T., Jr. (1980). *Biochemistry*, **19**, 3522–3530.

Lohman T.M., Overman L.B., Ferrari M.E., and Kozlov A.G. (1996). *Biochemistry*, **35**, 5272–5279.

Love J.J., Li X., Case D.A., Giese K., Grosschedl R., and Wright P.E. (1995). *Nature*, **376**, 791–795.

Manning G.S. (1969). *J. Chem. Phys.*, **51**, 924–933.

Manning G.S. (1978). *Quart. Rev. Biophys.*, **11**, 179- 246.

Masse J.E., Wong B., Yen Y.M., Allain F.H.T., Johnson R.C., and Feigon J. (2002). *J. Mol. Biol.*, **323**, 263–284.

Merabet E. and Ackers G.K. (1995). *Biochemistry*, **34**, 8554–8563.

Miller M., Shuman J.D., Sebastian T., Dauter Z., and Johnson P.F. (2003). *J. Biol. Chem.*, **278**, 15178–15184.

Misra V.K. and Honig B. (1995). *Proc. Natl. Acad. Sci. USA*, **92**, 4691–4695.

Misra V.K., Hecht J.L., Sharp K.A., Friedman R.A., and Honig B. (1994a). *J. Mol. Biol.*, **238**, 264–280.

Misra V.K., Sharp K.A., Friedman R.A., and Honig B. (1994b). *J. Mol. Biol.*, **238**, 245–263.

Murphy F.V. IV and Churchill M.E. (2000). *Structure Fold Des.*, **8**, R83-R89.

Murphy F.V. IV, Sweet R.M., and Churchill M.E.A. (1999). *EMBO J.*, **18**, 6610–6618.

Panne D. (2008). *Curr. Opin. Struct. Biol.*, **18**, 236–242.

Panne D., Maniatis T., and Harrison S.C. (2004). *EMBO J.*, **23**, 4384–4393.

Panne D., Maniatis T., and Harrison S.C. (2007). *Cell*, **129**, 1111–1123.

Patel L.R., Curran T., and Kerppola T.K. (1994). *Proc. Natl. Acad. Sci. USA*, **91**, 7360–7364.

Peters W.B., Edmondson S.P., and Shriver J.W. (2004). *J. Mol. Biol.*, **343**, 339–360.

Privalov P.L. and Makhatadze G.I. (1992). *J. Mol. Biol.*, **244**, 715–723.

Privalov P.L., Dragan A.I., and Crane-Robinson C. (2009). *Trends Biochem. Sci.*, **34**, 464–470.

Privalov P.L., Dragan A.I., and Crane-Robinson C. (2011). *Nucleic Acids Res.*, **39**, 2483–2491.

Privalov P.L., Dragan A. I. Crane-Robinson C., Breslauer K.J., Remeta D.P., and Minetti C.A.S. (2007). *J. Mol. Biol.*, **365**, 1–9.

Privalov P.L., Jelesarov I., Read C.M., Dragan A.I., and Crane-Robinson C. (1999). *J. Mol. Biol.*, **294**, 997–1013.

Qin B.Y., Liu C., Srinath H., Lam S.S., Correia J.J., Dernick R., and Lin K. (2005). *Structure*, **13**, 1269–1277.

Record M.T., Jr., Anderson C.F., and Lohman T.M. (1978). *Quart. Rev. Biophys.*, **11**, 103–178.

Record M.T., Jr., Zhang W., and Anderson C.F. (1998). *Adv. Protein Chem.*, **51**, 281–353.

Reeves R. and Nissen M.S. (1990). *J. Biol. Chem.*, **265**, 8573–8582.

Schumacher M.A., Goodman R.H., and Brennan R.G. (2000). *J. Biol. Chem.*, **275**, 35242–35247.

Servant M.J., Grandvaux N., Tenoever B.R., Duguay D., Lin R., and Hiscott J. (2003). *J. Biol. Chem.*, **278**, 9441–9447.

Sharp K.A. and Honig B. (1990). *J. Phys. Chem.*, **94**, 7684–7692.

Shui X., Sines C.C., McFail-Isom L., VanDerveer D., and Williams L.D. (1998). *Biochemistry*, **37**, 16877–16887.

Takahashi K., Suzuki N.N., Horiuchi M., et al. (2003). *Nature Struct. Biol.*, **10**, 922–927.

Takeda Y., Ross P.D., and Mudd C.P. (1992). *Proc. Natl. Acad. Sci. USA*, **89**, 8180–8184.

Tereshko V., Minasov G., and Egli M. (1999). *J. Am. Chem. Soc.*, **121**, 470–471.

Tsao D.H., Gruschus J.M., Wang L.H., Nirenberg M., and Ferretti J.A. (1994). *Biochemistry*, **33**, 15053–15060.

Van der Meulen K.A., Saecker R.M., and Record M.T., Jr. (2008). *J. Mol. Biol.*, **377**, 9–27.

Vinson C.R., Sigler P.B., and McKnight S.L. (1989). *Science*, **246**, 911–916.

Weiss M.A. (1990). *Biochemistry*, **29**, 8020–8024.

Weiss M.A., Ellenberger T., Wobbe C.R., Lee J.P., Harrison S.C., and Struhl K. (1990). *Nature*, **347**, 575–578.

Werner M.H., Huth J.R., Gronenborn A.M., and Clor G.M. (1995). *Cell*, **81**, 705–714.

Wiggins P.A., Phillips R., and Nelson P.C. (2005). *Phys. Rev. E*, **71**, 021909.

Wolberger C., Vershon A.K., Liu B., Johnson A.D., and Pabo C.O. (1991). *Cell*, **67**, 517–528.

13

NUCLEIC ACIDS

13.1. DNA

13.1.1. Problems

Discovery that two complementary strands of DNA are wound into a double helix and their separation is a key process in multiplication of the genetic information coded in the sequences of nucleotides forming these strands (Watson and Crick, 1953) had raised immediate interest in the energetic bases of this molecular construction, that is, the forces between the complementary strands of DNA, and the work required for their separation. However, the very first experiments showed that thermodynamic investigation of the unfolding–refolding of the natural DNA, which is long and heterogeneous in sequences, was far from easy because the process of dissociation of its strands is not simple and, moreover, is irreversible since the dissociated strands can no longer find their proper adjustment. Therefore, all attention was focused on the synthetic homo-polynucleotides, which appeared to have a rather simple and reversible process of association and dissociation. Long homo-polynucleotides can be easily synthesized, but unfortunately they were not identical in length and, correspondingly, the resulting duplexes were not 100%

Microcalorimetry of Macromolecules: The Physical Basis of Biological Structures, First Edition.
Peter L. Privalov.

helical. Synthesis of hetero-polynucleotides with a definite sequence is much more expensive; therefore they cannot be too long. Also, although they might be exactly complementary and associate–dissociate reversibly with 100% pairing, for short duplexes the contribution of fraying ends is no longer negligible. The other problem, concerning both these cases, was that separated strands are usually not completely unfolded but have some residual structure, which has to be taken into account in studying the thermodynamics of formation of the double helix from the two completely unfolded complementary strands.

A certain obstacle to developing the thermodynamics of double-helix formation by the poly and oligo nucleotides was that this process can hardly be approximated by a two-state transition that would permit application of equilibrium analysis to the observed process. Several theories of unfolding–dissociation of regular double helices under changing environmental conditions appeared for obtaining thermodynamic characteristics by equilibrium analysis of this process. However, verification of these theories required the results obtained to be compared with the directly measured basic thermodynamic characteristics, that is, the enthalpy and entropy of association–dissociation of the strands. The main practical problem in the realization of such experiments was the precision of the calorimetric measurements of the highly viscous solutions, moreover the solutions which drastically change their viscosity upon dissociation of the strands. The first differential scanning microcalorimeter (DSC) was actually designed especially to measure the heat effect of melting the DNA double helix in dilute solutions (see Section 3.2.2). The need to minimize the artifacts associated with changing viscosity of the DNA solutions upon heating led later to the appearance of the DSC with capillary cells.

13.1.2. Factors Affecting DNA Melting

The very first DSC experiments with natural DNA revealed remarkable features of the DNA (Fig. 13.1): the heat capacity of the double-stranded state extrapolated to higher temperatures comes close to the heat capacity of dissociated strands (Privalov

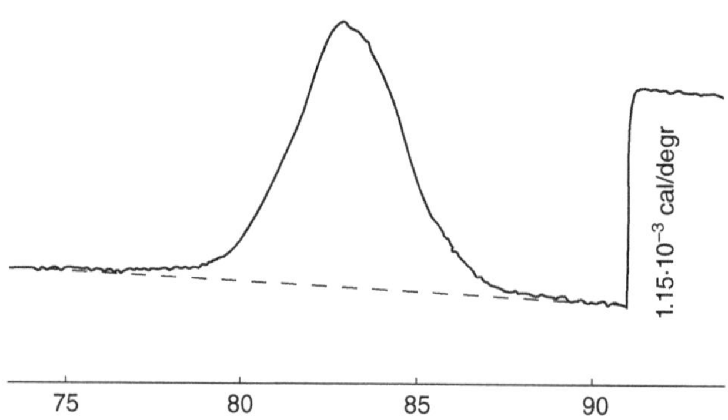

Figure 13.1. One of the first DSC recordings of the heat effect of phage T_2 DNA melting in 200 mM NaCl, pH 8.5 solution. Concentration of DNA is 0.75 mg/mL (Privalov et al., 1969).

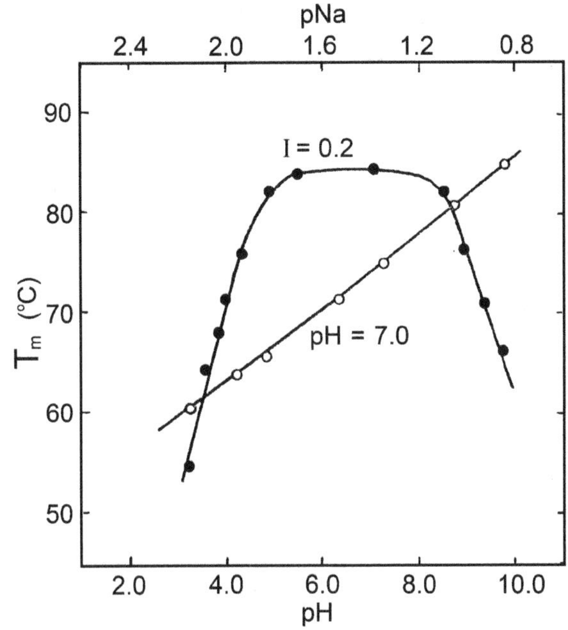

Figure 13.2. Dependence of the melting temperature of phage T_2 DNA on the pH and p(NaCl) of solution (Privalov et al., 1969).

et al., 1969). It appeared as though unfolding–dissociation of the DNA strands does not result in significant heat capacity increment, in contrast to what was found with proteins (see Section 7.2.1). Thus, it was concluded that the enthalpy of unfolding–dissociation of the DNA strands does not depend on temperature. On the other hand, studies of DNA melting at different solvent conditions showed that the melting temperature and enthalpy depend on the pH and salt concentration (Figs. 13.2 and 13.3).

The change of enthalpy upon pH variation was expectable, since upon shifting from neutral pH the groups of DNA and buffer are protonated–deprotonated with nonzero enthalpy, but the slight decrease of enthalpy with the increase of salt concentration seemed confusing since that did not match the entropic hypothesis of salt effect on DNA stability.

According to the theoretical model, due to the negatively charged phosphates, DNA represents a highly charged polyelectrolyte that is coated by the positively charged counterions of the salt present in solution (Mock and Marshall, 1955; Oosawa, 1968; Manning, 1972; Record et al., 1978). Upon dissociation of strands the linear density of negative charges drops. This results in decrease of the counterions coating the DNA; that is, they move from the vicinity of DNA and mix with the ions of the bulk solution. The entropy of mixing is proportional to the logarithm of the activity of this ion in solution, that is, practically to the concentration of salt. Therefore, the entropy of mixing decreases in magnitude with the increase of concentration of ions in the solution and that results in an increase of the Gibbs energy of DNA unfolding–dissociation (as $\Delta G = \Delta H - T\Delta S$), that is, the increase of DNA

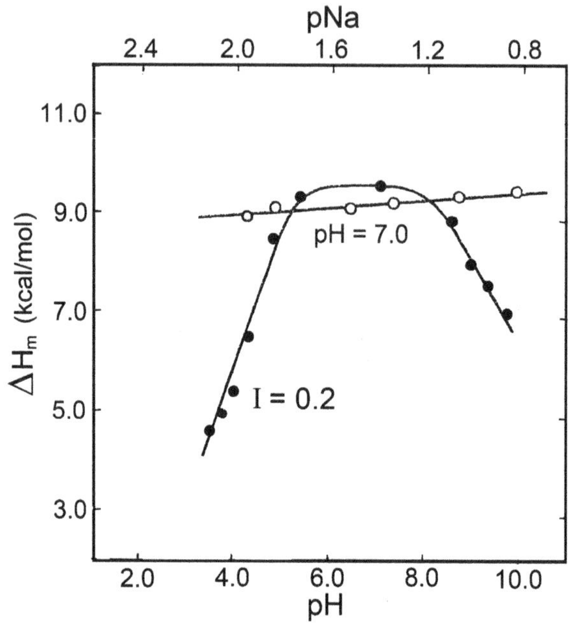

Figure 13.3. Dependence of the enthalpy of phage T_2 DNA melting on the pH and pNa of solution (Privalov et al., 1969).

stability. Thus, the melting temperature of the DNA appears to be inversely proportional to the negative logarithm of the salt activity, $-\log(a) = p(a)$, or just the salt concentration (p[salt]), while the melting enthalpy should not depend on the salt concentration.

If ν is the number of counterions associated with DNA, its change upon variation of counterion activity **a** is determined by

$$\Delta\nu = -\partial\Delta G(\mathbf{a})/\partial\ln(\mathbf{a}) = \Delta H_t[\partial(1/RT_t)]/\partial\ln(\mathbf{a}) \tag{13.1}$$

or, bearing in mind that $\ln(\mathbf{a}) = 2.3 \cdot \log(\mathbf{a}) = -2.3 \cdot \text{p[NaCl]}$,

$$\Delta\nu = -(\Delta H_t/2.3)\cdot\partial(1/RT_t)/\partial\text{p[NaCl]} \tag{13.2}$$

Analysis of the dependence of $1/RT_t$ on p(a) for the phage T_2 DNA gave a value for $\Delta\nu$ of about 0.34 per base pair (Privalov et al., 1969; Shiao and Sturtevant, 1973).

Another remarkable feature of DNA was that its melting temperature, and thus stability, appeared to be proportional to the averaged content of the GC base pairs. In solutions with 100 mM NaCl and neutral pH, this dependence of the half-melting temperature T_m appeared to be expressed by the following equation (Blackburn and Gait, 1996):

$$T_m = X + 0.41\cdot[\%(\text{C}\cdot\text{G})] \quad ^\circ\text{C} \tag{13.3}$$

where X is dependent on salt concentration and pH and has a value of 69.3°C for 0.3 M sodium ions at pH 7.0.

A more detailed equation, which takes into account the activity **a** of Na^+ ion in 1 mM citric acid buffer, was suggested (Tong and Battersby, 1979):

$$T_m(\mathbf{a}_{Na}, G\cdot C) = -\{5.2\log[\mathbf{a}_{Na}\cdot(G\cdot C)] + 18.4\log[\mathbf{a}_{Na}] + 41\cdot(G\cdot C) + 342.5\} \quad K \tag{13.4}$$

However, the melting profiles of the natural DNAs are rather broad and complicated in shape, due to the uneven clustering of the G·C base pairs along the DNA molecule. The phage T_2 DNA, (the melting profile of which is shown in Fig. 13.1) represents a certain exception: the melting of the DNA of bacterial—and, more so, animal—species spreads over a much wider temperature range and with rather complicated profiles. Therefore characterization of the natural DNA thermostability in transition temperatures is of rather limited value.

13.2. POLYNUCLEOTIDES

13.2.1. Melting of Polynucleotides

The advantages of using long synthetic polynucleotides in studying thermodynamics of formation of a double helix are that association–dissociation of the complementary strands is a reversible process and it proceeds in a short temperature range with sharp heat absorption. Therefore, calorimetric studies of polynucleotide melting do not require long extrapolations of the initial and final slopes of the heat capacity function to the transition zone as needed for the unfolding enthalpy estimation; this was important to reveal the heat capacity increment upon dissociation of the polynucleotides.

However, thermodynamic studies of the polynucleotides were also not without complications: synthetic homo- and hetero-polynucleotides can form complexes of the higher order structures, for example, triple helices, quadruplexes, and junctions (for reviews see Pilch et al., 1995; Plum et al., 1995). At neutral pH poly(A) and poly(U) can form either the double helix, poly(A)·poly(U), or the triple helix, poly(A)·2poly(U), depending on the ionic strength, temperature, and polymer concentrations (Poland et al., 1966; Michelson et al., 1967). When concentrations of both bases are equal and the salt concentration is below 0.2 M, the transition

$$\text{poly(A)}\cdot\text{poly(U)} \leftrightarrow \text{poly(A)} + \text{poly(U)} \tag{13.5}$$

can be induced by temperature increase. At higher salt concentrations the triple-helical complex is formed from the double-stranded one upon heating:

$$\text{poly(A)}\cdot\text{poly(U)} \leftrightarrow 0.5\cdot\text{poly(A)}\cdot\text{poly(U)} + 0.5\text{poly(A)} \tag{13.6}$$

On further heating the complex 0.5·poly(A)·poly(U) dissociates completely:

$$0.5\cdot\text{poly(A)}\cdot\text{poly(U)} + 0.5\text{poly(A)} \leftrightarrow \text{poly(A)} + \text{poly(U)} \tag{13.7}$$

All three transitions are highly cooperative, but the products of these reactions, and particularly poly(A), have extended residual structure. Therefore the temperature-induced changes of the poly(A)·poly(U) complex represent a rather complicated reaction:

$$\begin{array}{c} \text{poly(A)} \cdot \text{poly(U)}^{\text{helix}} \leftrightarrow \text{poly(A)}^{\text{helix}} + \text{poly(U)}^{\text{coil}} \\ \updownarrow \\ \text{poly(A)}^{\text{coil}} \end{array} \tag{13.8}$$

Analysis of this complicated reaction is not simple. It certainly requires thermodynamic characteristics of the poly(A)$^{\text{helix}} \leftrightarrow$ poly(A)^{coil} process.

According to the optical evidence, the poly(A)$^{\text{helix}} \leftrightarrow$ poly(A)^{coil} process is spread over a very broad temperature range, from 263 to 363 K (Brahms et al., 1966; Leng and Felsenfeld, 1966; Applequist and Damie, 1966; Rawitcher et al., 1963). The estimates of the enthalpy of this process, based on the analysis of optically observed melting profiles, were absolutely unrealistic since it certainly does not represent a two-state process, while all other approximations are highly uncertain and led to very broad spread of values, from 55 to 10 kJ/mol. Therefore, the heat effect of poly(A)^{helix} melting could be determined only by calorimetry.

13.2.2. Calorimetry of Poly(A)·Poly(U)

Figure 13.4 shows the DSC recording of the heat effect at heating the long poly(A)·poly(U) at two largely different concentrations: 0.3 mg/mol (left) and 20 times larger concentration, 5.0 mg/mL (right). The sharpness of transition and the contribution—negligible for the long polynucleotide—of fraying of the terminal groups permitted determination of the heat capacity of folded and unfolded complex. Assuming the partial specific volume of polynucleotides at 25°C being 0.55 mL/g, the partial specific heat capacity of polynucleotides at this temperature appeared to be $C_p^{(AU)} = (1.08 \pm 0.05)$ J/K·g and its temperature dependence to be $C_p/dT = 3.6 \times 10^{-3}$ J/K^2·g; while the melting of poly(A)·poly(U) results in heat capacity increment $\Delta C_p = (125 \pm 15)$ J/K·mol.

The apparent melting enthalpy values estimated from the peak area above the heat capacity lines extrapolated to the temperature of the peak maximum, T_m, in solutions with various concentrations of salt are shown by filled squares in Figure 13.5. The same figure shows the apparent enthalpies obtained with similar polynucleotides but at one order magnitude larger concentrations by Krakauer and Sturtevant (1968) and Neuman and Ackerman (1969). The close agreement of these results obtained at very different concentrations of polynucleotides is evidence that the concentration effects are negligible. The problem is, however, that poly(A) is not completely unfolded but retains the residual structure, which has to be taken into account in estimating the net enthalpy of complete unfolding–dissociation of poly(A)·poly(U).

Figure 13.6 shows the temperature dependence of the partial heat capacity for poly(A). This function was obtained calorimetrically in solutions with different salt

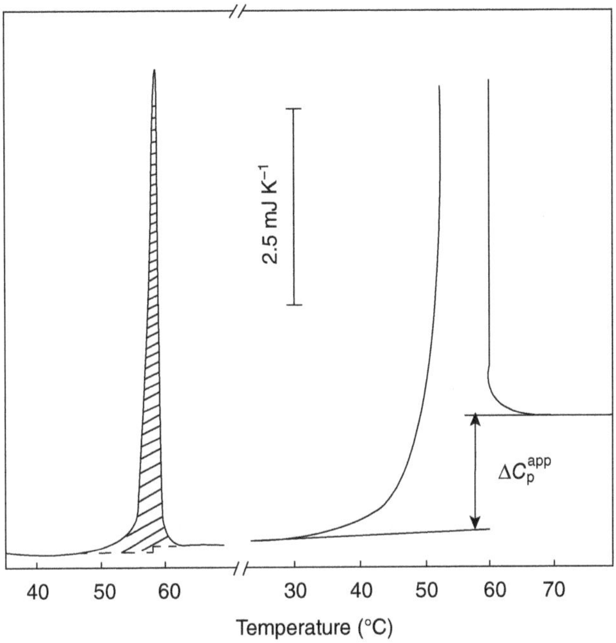

Figure 13.4. DSC recording of poly(A)poly(U) melting at 0.3 mM concentration (left) and a fragment of a recording at 5.0 mM concentration (right) in the presence of 100 mM NaCl. The hatched area corresponds to the apparent melting enthalpy; arrows indicate the observed heat capacity change; NaCl concentration is 0.1 M (Filimonov and Privalov, 1978).

concentration (from 10 to 100 mM NaCl) and was found to be independent of the salt content. This heat capacity was measured from 2°C (i.e., 275 K). Therefore it should be extrapolated down to the beginning of the poly(A) coil melting, which appeared to be close to −20°C ≈ 263 K. This extrapolation procedure, however, was not arbitrary since the starting temperature and the specific heat capacity of folded poly(A) calculated per base or per gram should be close to that of poly(AU), which was known. In any case, it is clear that such an extrapolation cannot lead to significant errors in determining the total enthalpy of poly(A) melting and, particularly, its high-temperature part. Assuming that the heat capacity change upon poly(A) melting is not large, and approximating the dependence of heat capacity on temperature by a straight line, one can obtain the excess heat absorption function shown in Figure 13.6 by the dashed-and-dotted line. Its rather symmetrical shape provides evidence that the poly(A) melting does not depend much on temperature, that is, confirms the above assumption that the heat capacity change at melting is small. It appears that for poly(A) melting $T_m = 313$ K, $\Delta H_m = (12.6 \pm 2.1)$ kJ/mol.

Considering correction for the apparent heat effects of poly(A)·poly(U) melting associated with the residual structure in the poly(A), it is evident that it corresponds to the area bounded by the heat absorption function of poly(A) melting from T_m to the end, as shown by hatching in Figure 13.6. The enthalpy of base interaction

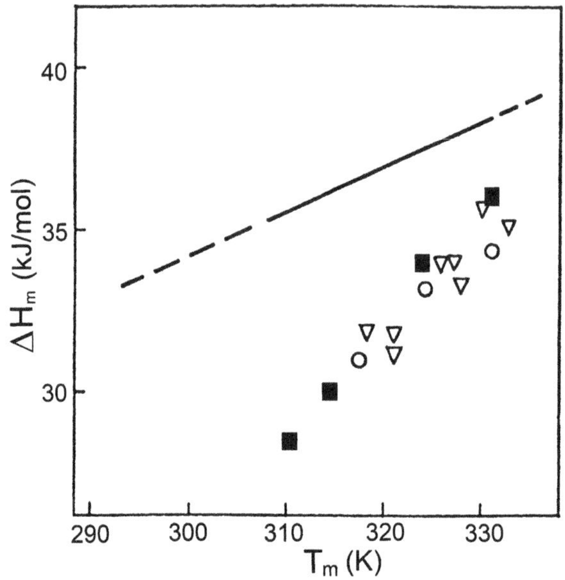

Figure 13.5. Apparent molar enthalpy of poly(A)·poly(U) melting in solutions with different concentration of salt (from 10 to 100 mM NaCl): obtained by (■) Filimonov and Privalov (1978), (○) Krakauer and Sturtevant (1968), and (Δ) Neumen and Ackerman (1969). Solid line shows the enthalpy corrected for the residual structure in poly(A) strand (Filimonov and Privalov, 1978).

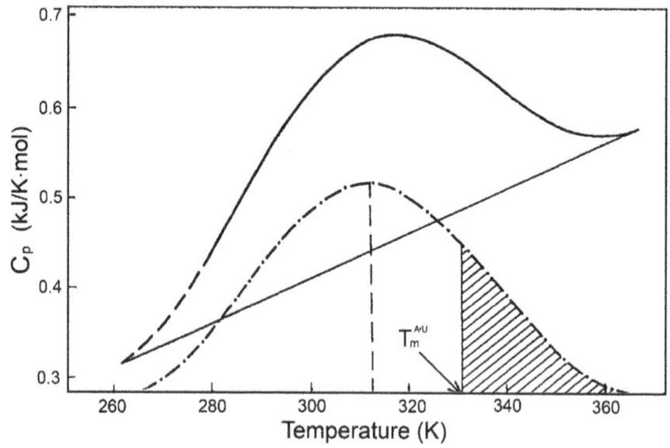

Figure 13.6. Molar partial heat capacity function for poly(A) (solid line) and the excess heat capacity function (dot-and-dashed line). The hatched area corresponds to the correction for apparent poly(A)poly(U) melting enthalpy at the transition temperature T_m (Filimonov and Privalov, 1978).

TABLE 13.1. Thermodynamic Characteristics of AU base Pairing[a]

[NaCl] mM	T_m (K)	ΔH^{AU} (kJ/mol)	ΔS^{AU} (J/K·mol)	ΔG^{AU} (kJ/mol)
10	310.5	34.0	110	1.3
50	324.0	34.0	105	2.8
100	331.0	34.0	102	3.5

[a]At 25°C and three NaCl concentrations; according to Filimonov and Privalov (1978).

obtained, after correction, from the poly(A)·poly(U) complex is presented in Figure 13.5 by a solid line. It can be seen that this corrected enthalpy is significantly larger than the apparent melting enthalpy at the corresponding temperature and depends much less on temperature. The dependence of the net melting enthalpy of poly(A·U) on temperature appears now to be $\partial\Delta H_m(\text{A}\cdot\text{U})/\partial T_m = (134 \pm 10)$ J/K·mol. It is remarkable that the dependence found for the melting enthalpy of poly(A)·poly(U) is close to the directly measured heat capacity increment $\Delta C_p^{(\text{A}\cdot\text{U})} = (125 \pm 15)$ J/K × mol_{bp} (Fig. 13.4). The correspondence of the $d\Delta H_m(\text{A}\cdot\text{U})/dT_m$ and $\Delta C_p^{(\text{A}\cdot\text{U})}$ values is remarkable. It means that:

> The base-pairing enthalpy is a function of temperature only and not of ionic strength (Table 13.1).

13.3. SHORT DNA DUPLEXES

13.3.1. Calorimetry of Short DNA Duplexes

The advantage of short DNA duplexes (i.e., duplexes with fewer than 20 bp), in comparison with long ones, is that their complementary strands can be easily synthesized in any desired sequence and they dissociate–associate reversibly upon heating and consequent cooling of their solution (Fig. 13.7). It is remarkable, however, that the linear extrapolation of the initial apparent heat capacity of the DNA duplex to higher temperatures does not show a visible heat capacity increment associated with the duplex unfolding–dissociation. Therefore, in earlier DSC studies of the DNA duplexes it was just assumed that they melt without heat capacity increment, that is, that the enthalpy of their melting does not depend on temperature (Marky and Breslauer, 1982; Breslauer et al., 1986; Vesnaver and Breslauer, 1991; Chalikian et al., 1999).

However, the enthalpies of DNA melting determined on the assumption that it proceeds without heat capacity increment appeared to be in conflict with the enthalpies of association of complementary strands measured by isothermal titration calorimetry (ITC) at various fixed temperatures: the ITC-measured enthalpies of association of complementary strands were much smaller in magnitude than the DSC-measured enthalpies of melting of the formed duplexes and strongly depended

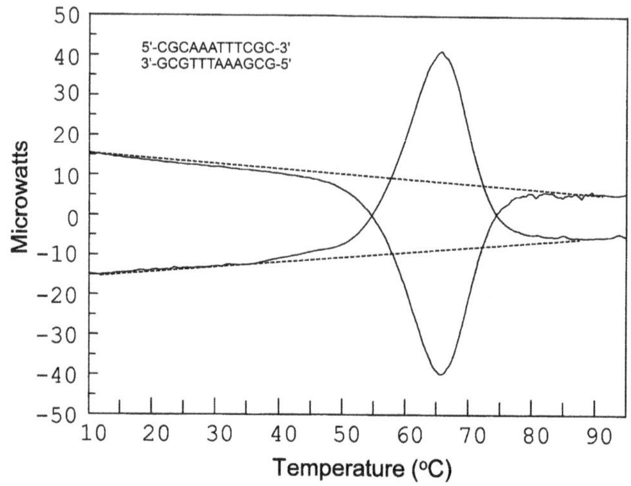

Figure 13.7. Original DSC recordings on heating and consequent cooling of the 12-bp DNA duplex at a rate 1 K/min, demonstrating reversibility of the duplex unfolding and the steep increase of its heat capacity with increasing temperature. Concentration of the duplex is 2.23 mg/mL; solvent 150 mM NaCl, 5 mM sodium phosphate, pH 7.0.

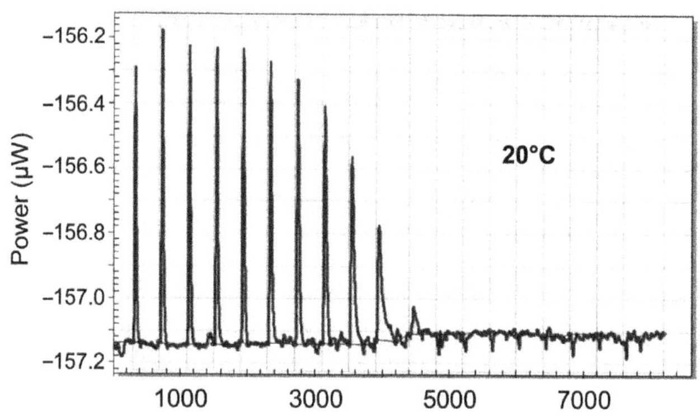

Figure 13.8. (a) ITC titration of the 5′-CGC**AAATTT**CGC-3′ strand by the 3′-GCG**TTTAAA**GCG-5′, which is injected in 5-μL portions into 1-mL cell at 20°C. Concentration of the oligonucleotide in the cell is 1.5 μM = 0.005 mg/mL; of the titrant, 25 μM = 0.092 mg/mL; solvent 150 mM NaCl, 5 mM sodium phosphate, pH 7.4.

on temperature (Zieba et al., 1991; Ladbury et al., 1994; Holbrook et al., 1999). For example, while the DSC-measured enthalpy of temperature-induced cooperative dissociation of the 12-bp DNA duplex shown in Figure 13.7, amounts to 450 kJ/mol, the ITC-measured enthalpy of association of its strands at 20°C is much smaller, only −130 kJ/mol (Fig. 13.8).

The discrepancy between the DSC- and ITC-obtained enthalpies of formation of the DNA duplex might be caused by the residual structure in the strands. The DSC

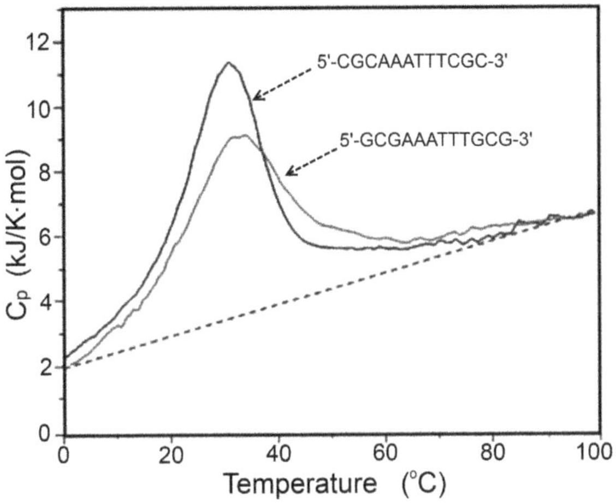

Figure 13.9. DSC-determined melting profiles of the oligonucleotides in solutions containing 150 mM NaCl, 5 mM sodium phosphate, pH7.4.

studies of the separated oligonucleotides showed indeed that their heating proceeds with certain excess heat absorption, indicating that the separated strands possess some structure that breaks down upon heating (Fig. 13.9). This excess enthalpy of the separated strands' unfolding must certainly be taken into account at determination of the enthalpy of duplex formation based on the ITC-measured heat effect of their association.

Also somewhat surprising was the too-steep increase of the initial apparent heat capacity of the DNA duplex with heating: for the 12-bp DNA duplex presented in Figure 13.7 the $\partial C_p(T)/\partial T$ amounts to 0.13 kJ/K^2·mol, or calculated per base weight to be about 20×10^{-3} J/K^2·g. This value is more than 3 times larger than what is observed for the temperature dependence of the specific heat capacity of a stable globular protein, which is about 6×10^{-3} J/K^2·g (Section 7.2.4). This shows that the short DNA duplex is highly flexible; it fluctuates and the fluctuations, particularly of its ends, intensify with increasing temperature, accumulating heat. Therefore, the heat capacity function of the duplex below the extensive heat absorption peak, associated with cooperative separation of strands, cannot be regarded as an intrinsic heat capacity of the fully folded duplex. In fact, calorimetric studies of the melting polynucleotides, which proceed in a very narrow temperature range and are easy to treat, showed that melting of the poly(A)·poly(U) results in a heat capacity increment on the order of 130 J/K·mol$_{bp}$ (see Section 13.2.2). According to Jelesarov et al. (1999), the heat capacity increment upon dissociation of the 12-bp DNA duplex amounts to 2.9 kJ/K·mol, or recalculated per base pair to 240 J/K·mol$_{bp}$. Thus, one would expect that the heat capacity increment upon melting of the 12-bp DNA duplex considered above should be of this order. Plotting the 2.9 kJ/K·mol heat capacity increment at the heat capacity profile of this duplex, one finds the heat capacity function expected for this duplex in the fully folded state (Fig. 13.10). The hatched area indicates the excess heat effect associated with unfolding of this duplex.

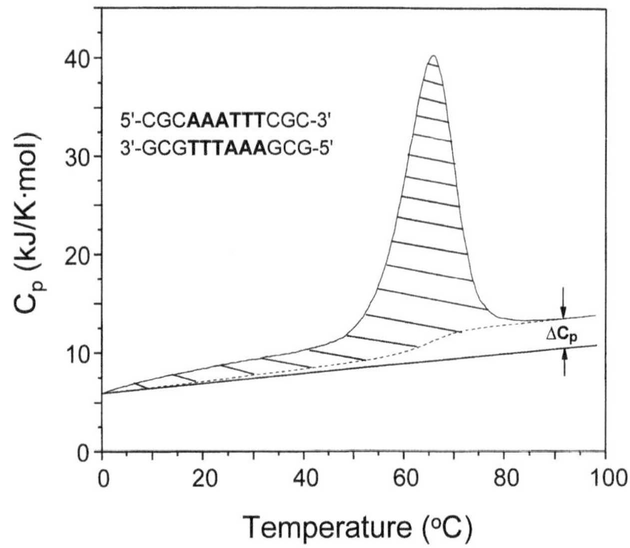

Figure 13.10. Partial molar heat capacity functions of the 12-bp DNA duplex considered in Figure 13.7 and the expected heat capacity function of the fully folded duplex plotted under the assumption that the heat capacity increment, ΔC_p, amounts to 2.9 kJ/K·mol.

Thus, it appears that the total heat effect of duplex unfolding is significantly larger than what the straightforward extrapolation of its initial heat capacity gives and, therefore, is larger than the gap between the ITC and DSC estimates of the enthalpy of DNA duplex formation. The question now is how to bring the DSC and ITC results into correspondence.

This problem was solved by combining the ITC and DSC experiments, that is, by measuring with ITC the enthalpy of association of the complementary strands and correcting it for the contribution of residual structure in the separated strands and for the duplex premelting determined in the DSC experiment (Jelesarov et al., 1999). The idea of such a combined experiment, permitting determination of the intrinsic thermodynamic parameters of formation of the DNA duplexes, is illustrated in Figure 13.11. Panel (a) in this figure shows the ITC measured enthalpy of association of two complementary strands at 40°C. Panel (b) shows the DSC-determined partial molar heat capacity function of the duplex and the hatched area shows the heat of premelting of the duplex upon heating to 40°C at which the ITC experiment is carried. Panels (c) and (d) show the heat capacity profiles of the separated strands and the hatched areas shows the excess enthalpies of the residual structure melting in these strands upon heating to the temperature of the ITC experiment, that is, to 40°C.

The results of such combined calorimetric experiments carried out by Paulius Vaitiecunas at the author's laboratory are presented in Figure 13.12. Here the empty circles show the ITC-measured enthalpies of association of the complementary strands at different fixed temperatures; the squares show the association enthalpies corrected for the enthalpy of residual structure in the single strands, obtained from

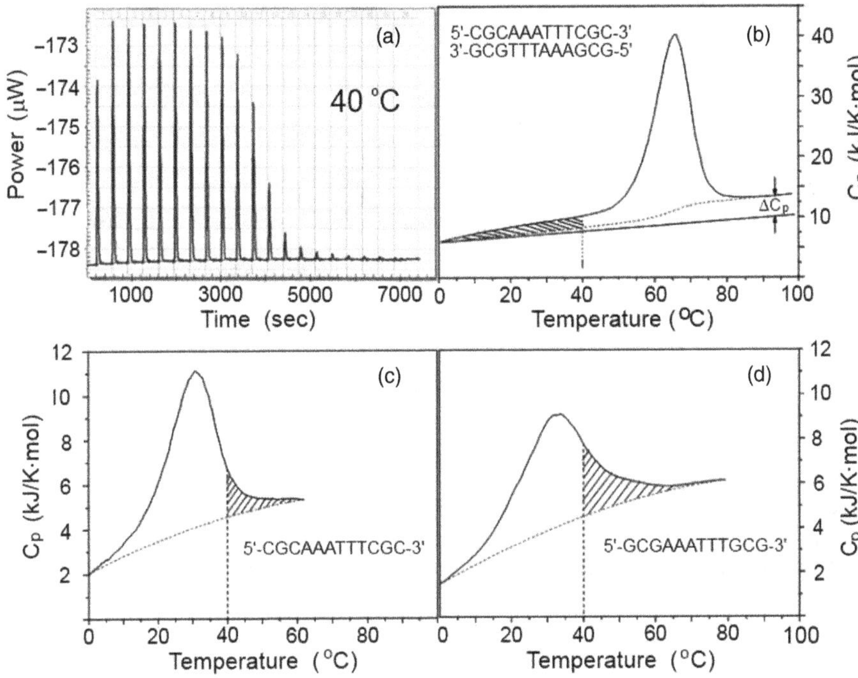

Figure 13.11. (a) ITC titration of the 5′-CGC**AAATTT**CGC-3′ strand by the 3′-GCG**TTTAAA**GCG-5′ strand at 40°C. (b) DSC-measured partial molar heat capacity of Duplex (I); hatched area shows the enthalpy of the duplex premelting upon heating to 40°C. (c, d) Partial heat capacities of the isolated oligonucleotides; the hatched areas show the enthalpy of the residual structures in the duplexes at 40°C. All in 5 mM sodium phosphate, pH 7.0, 5 mM NaCl.

DSC experiments (Fig. 13.11c,d); the triangles show the association enthalpies further corrected for the enthalpy of duplex premelting at the temperature of the ITC experiment (Fig. 13.11b). One can see that the most essential correction comes from the residual structure in the separated strands: it significantly raises the enthalpy of strand association measured at different fixed temperatures by ITC, and decreases its temperature dependence. Correction for duplex premelting is smaller but as a result of this correction the enthalpy function of strand association projects to the DSC-determined full enthalpy of duplex formation (shown by crosses), assigned to the temperature of cooperative dissociation of strands, ΔH^*_m.

It should be noted that, since enthalpy of dissociation–association of the DNA strands depends on temperature, one cannot simply assign all the observed excess heat effect of DNA melting to the temperature of cooperative dissociation of strands, that is, temperature of the heat absorption peak, T_m. We have to extrapolate this widely spread in temperature heat effect of melting up to T_m using the heat capacity increment specific for the considered DNA duplex, particularly 2.9 kJ/kmol as was taken for the first approximation. The extrapolation can be done by dissecting the

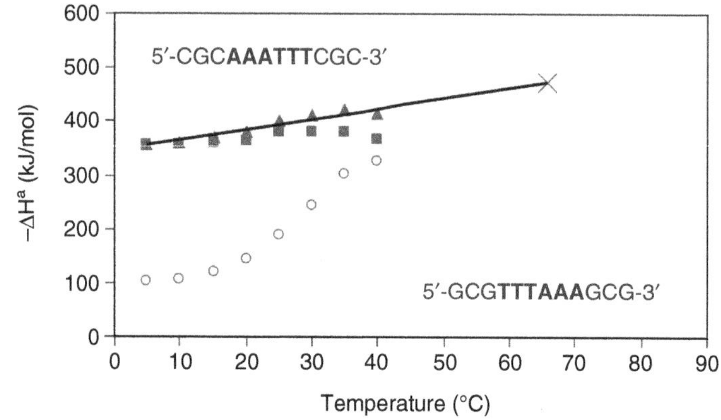

Figure 13.12. The enthalpies of formation of the considered DNA duplex in the presence of 150 mM NaCl in 5 mM sodium phosphate pH 7.4 solution. The open circles give the ITC-measured enthalpies of association of the complementary strands at different fixed temperatures. The squares give these association enthalpies corrected for the enthalpies of residual structure in the separated strands. The triangles give the association enthalpies also corrected for the premelting of the duplexes. The crosses show the enthalpies of formation of these duplexes $\Delta H^*(T_m)$ determined from the DSC-measured excess heat effects attributed to the temperatures of cooperative dissociation of strands, T_m, by Equation (13.9).

whole temperature range of the excess heat absorption into several segments and extrapolating the enthalpy of each segment (ΔH_i) from its mid-temperature T_i to the T_m, assuming that the heat capacity increment of this segment is proportional to its enthalpy, $\Delta C_{p,i} = \Delta C_p \cdot (\Delta H_i/\Delta H^{\text{exc}})$. Thus, the total enthalpy of DNA duplex unfolding at T_m would be

$$\begin{aligned} \Delta H^*(T_m) &= \sum [\Delta H_i + (\Delta H_i/\Delta H^{\text{exc}})\cdot \Delta C_p \cdot (T_m - T_i)] \\ &= \Delta H^{\text{exc}} + \sum (\Delta C_p \times \Delta H_i/\Delta H^{\text{exc}})[(T_m - T_i)] \end{aligned} \tag{13.9}$$

This extrapolation of the excess heat effect of duplex unfolding to T_m raises the melting enthalpy value by about 20 kJ/mol and places the corresponding total enthalpy of duplex formation on the line projecting the corrected association enthalpies obtained at different temperatures. The correspondence of these two approaches, that is, ITC and DSC, is a strong argument that we are on the right track in determining the intrinsic enthalpy of DNA duplex formation.

13.3.2. Specificity of the AT-rich DNA Duplexes

In aqueous media the highly charged DNA molecule is heavily hydrated and one can expect that many properties of DNA are stipulated by the water surrounding it. Particularly, as discussed in Section 12.4.3, water in the minor groove of the AT-rich DNA sequences appears to be in a highly ordered state and this seems to be the reason for positive enthalpy and entropy of protein binding to the minor groove at the AT-rich sequences (see Figs. 12.29 and 12.30). If so, the enthalpy of melting the AT-rich DNA duplex should be large, probably even larger than that of the GC-rich duplex. This sounds surprising in that the GC base pair is maintained by three hydrogen bonds, while the AT base pair is maintained by two hydrogen bonds (see Fig. 4.16). In fact, because of the experimental difficulties considered above, the intrinsic enthalpies of DNA duplex formation have never been accurately determined.

To specify thermodynamically the AT- and GC-rich DNA duplexes, formations of two 12-bp duplexes were studied calorimetrically, with the duplexes differing in their central parts flanked by the identical CGC/GCG sequences to minimize contribution of the fraying ends and to make it similar for both duplexes (Vaitiekunas and Privalov, in preparation):

Duplex (I)	5′–CGC**AAATTT**CGC–3′	and Duplex (II)	5′–CGC**CGCCGC**CGC–3′
	3′–GCG**TTTAAA**GCG–5′		3′–GCG**GCGGCG**GCG–5′

The intrinsic enthalpies of formation of these two duplexes (i.e., enthalpies corrected for the residual structure in the separated strands and for premelting of the duplex) in the presence of 5 and 150 mM NaCl are listed in Table 13.2. It is notable that they depend on temperature but do not depend on the salt concentration if compared at the same temperatures (Fig. 13.13). This is just what is expected if the salt-induced stabilization of the DNA double helix results from the entropy of mixing of the counterions released by DNA upon association of the highly charged DNA strands: this mixing entropy decreases as the salt concentration in the bulk solution rises (see Section 13.1.2).

The most striking observation is that at low temperatures the melting enthalpy of Duplex (I), which contains the continuous stretch of six AT base pairs, is significantly larger than the melting enthalpy of Duplex (II), consisting totally of GC base pairs (Fig. 13.14). At room temperature this difference amounts to 50 kJ/mol, that is, the value expected from studies of the protein–DNA interactions (Section 12.4.3).

It is notable that the difference between the enthalpies of formation of Duplex (I) and Duplex (II) decrease with increasing temperature. It appears as though the water hydrating the minor groove of the AT-rich DNA duplex "melts" gradually with heating. Thus, this gradual process of "melting" the ordered water actually determines the temperature dependence of the enthalpy of DNA duplex formation. The difference in the slopes of the enthalpy functions assumes that the intrinsic heat capacity increments of these two duplexes are different: for the GC-rich Duplex (II), $\partial\Delta H/\partial T = \Delta C_p$ amounts to (3.5 ± 0.2) kJ/K·mol, or recalculated per base pair (290 ± 17) kJ/K·mol-bp, while for the AT-rich Duplex (I) it is (2.0 ± 0.2) kJ/K·mol = (167 ± 17) J/K·mol-bp. It is remarkable that this value of ΔC_p for the AT-rich sequence is close to that found for poly(A)·poly(U) melting, (134 ± 15) J/K·mol (see Section 13.2.2). The similarity of the heat capacity increment observed

TABLE 13.2. Enthalpies of Formation of Duplex (I) and Duplex (II)[a]

	5 mM NaCl		150 mM NaCl	
Temperature (°C)	Duplex I $-\Delta H$ kJ/mol	Duplex II $-\Delta H$ kJ/mol	Duplex I $-\Delta H$ kJ/mol	Duplex II $-\Delta H$ kJ/mol
5	350 ± 14		357 ± 14	
10	375 ± 14	304 ± 22	361 ± 14	301 ± 12
15	390 ± 16	319 ± 22	369 ± 14	328 ± 12
20	401 ± 16	319 ± 12	381 ± 16	344 ± 14
25	**400 ± 16**	**340 ± 14**	**402 ± 16**	**356 ± 14**
30	410 ± 16	364 ± 14	410 ± 16	364 ± 14
35	426 ± 17	392 ± 14	420 ± 16	378 ± 14
40		413 ± 16	415 ± 16	402 ± 16
45		438 ± 16		421 ± 16
50		461 ± 18		458 ± 18
51.1*	443* ± 17			
55				472 ± 19
65.7*			472* ± 19	
70.4*		510* ± 20		
84.9*				562* ± 20
ΔC_p kJ/Kmol	1.9 ± 0.2	3.8 ± 0.2	2.0 ± 0.2	3.5 ± 0.2

*—asterisks indicate the T_m and $\Delta H(T_m)$ values.

[a]In the presence of 5 and 150 mM NaCl in 5 mM Sodium Phosphate, pH 7.0 solution.

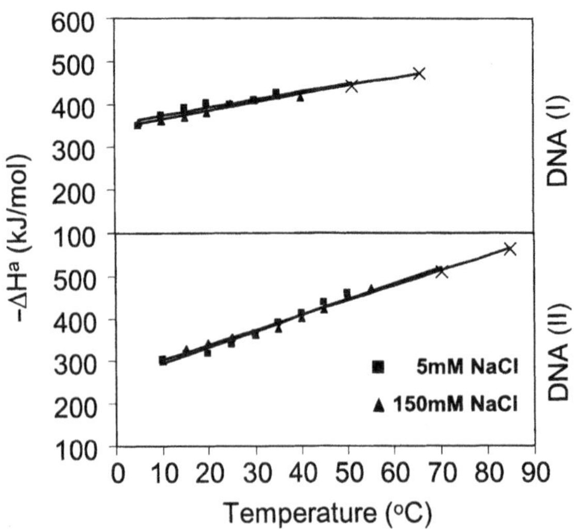

Figure 13.13. Comparison of the enthalpies of formation of Duplex (I) and Duplex (II) in the presence of 5 mM NaCl (squares) and 150 mM NaCl (triangles) in the 5 mM sodium phosphate, pH 7.0 solutions.

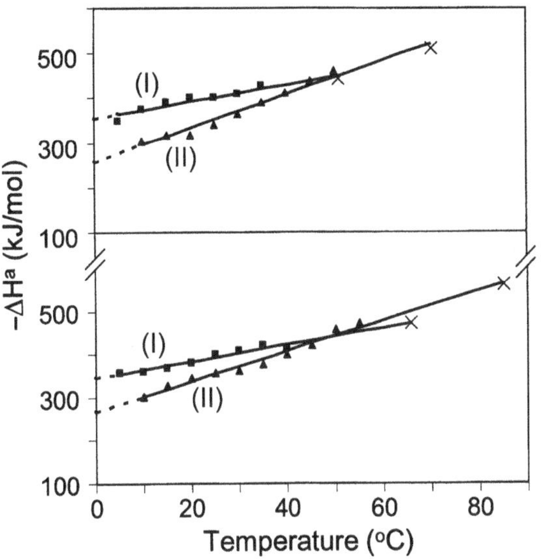

Figure 13.14. Enthalpies of formation of Duplex (I) and Duplex (II) determined from the separated complementary strands as measured by ITC at various fixed temperatures and corrected for the residual structure in the strands and premelting of the duplexes in the presence of 5 and 150 mM NaCl in 5 mM sodium phosphate, pH 7.4. The crosses represent the DSC determined $\Delta H^*(T_m)$ values at the temperature of cooperative dissociation of strands, T_m.

for the AT-rich DNA duplex and poly(A)·poly(U) suggests that highly ordered water may also be present in the minor groove of this polynucleotide.

Using the new heat capacity increments one can determine the heat capacity function of the fully folded duplex and the excess enthalpy of its melting with higher accuracy. The results obtained in this second approximation are listed in Table 13.3. One can see that the second approximation has changed the heat capacity increments and ΔH_m^* values: for Duplex (I) this intrinsic melting enthalpy decreased and for Duplex (II) it increased in the presence of both 150 and 5 mM NaCl. However, all these changes are rather small, suggesting that higher approximation is hardly required.

In considering the entropy of duplex unfolding one also should take into account not only the process of cooperative unfolding–dissociation of strands at T_m but also the gradual disordering of the duplex upon heating to this temperature. The entropy of cooperative dissociation of the heterodimer at T_m is expressed by

$$\Delta S(T_m) = \frac{\Delta H_m}{T_m} + R\ln\left(\frac{[N]}{2}\right) \tag{13.10}$$

where $[N] = [N_0]/[N_{st}]$ is the dimensionless concentration of the duplex, $[N_0]$ is the initial molar concentration of the duplex and $[N_{st}]$ is the standard 1 M concentration (see Section 3.2.7). The entropic effect of the noncooperative changes of DNA

TABLE 13.3. Thermodynamic Parameters Specifying Formation of Duplex (I) and Duplex (II) from the Completely Unfolded Oligonucleotides[a]

	NaCl (mM)	DNA Duplex	$-\Delta C_p$ (kJ/K·mol)	$-\Delta H^*$ (kJ/mol)	$-\Delta C_p^*$ (kJ/K·mol)	$-\Delta H_m^*$ (kJ/mol)	$-T_m\Delta S_m^*$ (kJ/mol)
1st. Approximation	5	(I)	2.9	429	1.9	443	427
		(II)	2.9	487	3.8	510	500
	150	(I)	2.9	453	2	472	460
		(II)	2.9	532	3.5	562	556
2nd. Approximation	5	(I)	1.9	424	1.7	434	416
		(II)	3.8	497	4	530	522
	150	(I)	2	443	1.7	455	440
		(II)	3.5	545	3.7	582	579

[a]In the presence of 5 and 150 mM NaCl in 5 mM sodium phosphate, pH 7.4 solution.

duplex prior to its dissociation again can be determined by dissecting the low-temperature tail of the excess heat absorption into several segments, determining the entropic effect for each segment as $\Delta S_i = \Delta H_i/T_i$, and extrapolating it to T_m, assuming that the heat capacity increment of the segment is proportional to its enthalpy. Thus, the total entropy of DNA duplex unfolding at T_m would be

$$\Delta S^*(T_m) = \Delta H_m/T_m + R\ln([\mathrm{N}_0]/2) + \sum[\Delta H_i/T_i + (\Delta C_{p,i}\ \Delta H_i/\Delta H^{\mathrm{exc}})\ln(T_m/T_i)] \quad (13.11)$$

Assuming that the determined heat capacity increment at DNA duplex unfolding does not depend on temperature, one can estimate the enthalpy, entropy, and Gibbs energy functions of the DNA duplex formation in a broad temperature range as

$$\Delta H(T) = \Delta H_m^* - \Delta C_p(T_m - T) \quad (13.12)$$

$$\Delta S(T) = \Delta S_m^* - \Delta C_p \ln(T_m/T) \quad (13.13)$$

$$\Delta G(T) = \Delta H(T) - T\ \Delta S(T) \quad (13.14)$$

The enthalpy, entropy, and Gibbs energy functions of formation of the AT-rich Duplex (I) and GC-rich Duplex (II) in the presence of 5 and 150 mM NaCl are compared in Figure 13.15. The figure shows that the enthalpy of formation of Duplex (I) is larger than that of Duplex (II), but the entropy factor is also larger for Duplex

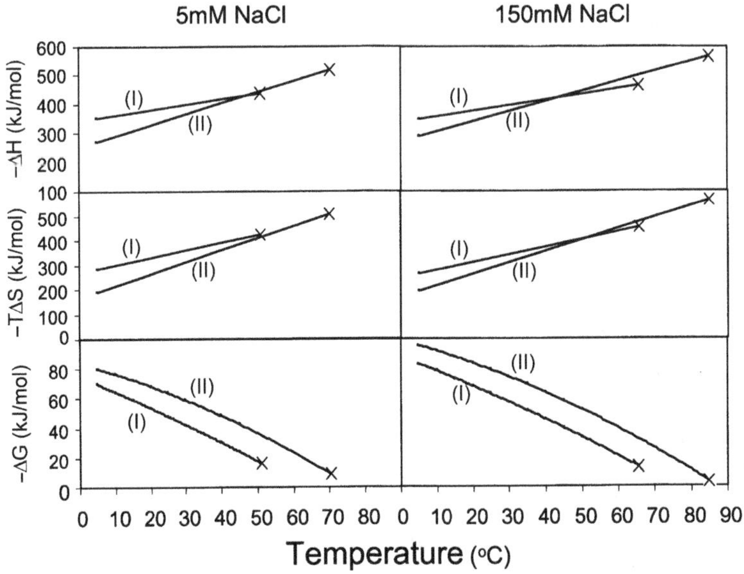

Figure 13.15. The enthalpy, the entropy factor, and the Gibbs energy functions of Duplex (I) and Duplex (II) formation from the completely unfolded complementary strands in the presence of 5 and 150 mM NaCl in 5 mM sodium phosphate, pH7.4 solutions.

(I) and overbalances the enthalpy. As a result the Gibbs energy of dissociation of Duplex (I) is smaller than that of Duplex (II), that is, the AT-rich duplex is less stable than the GC-rich duplex. The observed enthalpy–entropy compensation makes it clear that the large enthalpy and entropy of formation of Duplex (I), significantly exceeding the enthalpy and entropy of formation of Duplex (II), result from water forming a highly ordered icelike stretch in the minor groove of the AT-rich DNA: disordering of this water upon dissociation of the DNA strands resembles the melting of ice, a process that occurs with zero Gibbs energy (Privalov et al., 2007, 2010).

An efficient compensation of the enthalpy and entropy effects associated with the DNA hydration raises the question: does water ordered in the minor groove of DNA have some significance in its functioning? The possible role of this water in the minor groove of the AT-rich DNA is discussed in the subsequent Sections 13.3.3 and 13.3.4.

It is tempting to determine from the calorimetric data obtained on melting of two duplexes the contribution of the individual AT and GC base pairs to the energetic basis of the DNA double helix. Dividing the melting enthalpy of Duplex (II) at 25°C (which amounts to 356 kJ/mol) on 12 one gets the enthalpic contribution of the GC base pair, which appears to be 29.7 kJ/(mol_{bp}). Excluding the contribution of the six-GC base pair (i.e., 6×29.7 kJ/mol = 178 kJ/mol) from the melting enthalpy of Duplex (I), which is 402 kJ/mol at 25°C, we get 224 kJ/mol for the enthalpy of melting the six-AT base-pair stretch. Dividing this value by 6 gives for the enthalpy contribution of each AT base pair the value 37.3 kJ/mol, that is, the value almost 30% higher than the enthalpic contribution of the GC base pair. However, this value can hardly be considered as an individual contribution of the AT base pair to the DNA double helix because water ordering in the AT-rich minor groove of DNA, which is responsible for the superiority of this enthalpy value, is likely to be a cooperative effect of the continuous regular arrangement of certain polar groups in the minor groove of DNA. Understanding of the water ordering mechanism in the minor groove of DNA requires detailed calorimetric investigation of the DNA duplexes differing not only in total content of AT and GC base pairs but also in their composition. This is a big but important piece of work that has to be done to understand the thermodynamic basis of the DNA double helix.

13.3.3. DNA Hydration Studied by Pressure Perturbation Calorimetry

The preceding results demonstrating a direct connection between the thermodynamic properties of DNA and water hydrating its grooves raise the question: how closely does the water ordered in the minor groove of the AT-rich DNA resemble an ice, particularly in its density? One could get certain information on this matter by pressure perturbation (PP) calorimetry (see Section 3.3).

Studies of the previously considered 12-bp DNA duplexes by the PP Nano-DSC showed that at low temperatures their thermal expansion coefficients are rather similar and decrease significantly with increasing temperature, up to the point at which the duplexes start to melt, a process associated with a peak in the thermal expansion coefficient (Fig. 13.16). Melting of the two duplexes, which have differing

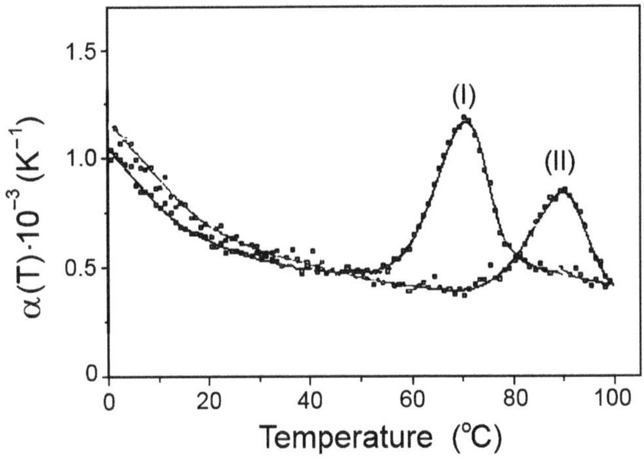

Figure 13.16. The thermal expansion coefficient of DNA Duplexes (I) and (II) in the temperature range from 0°C to 100°C estimated in a pressure perturbation calorimetry (PPC) experiment in 100 mM NaCl, 20 mM sodium phosphate, pH 7.4 solution. Concentration of the DNA duplexes is 13.4 mg/mL. The PPC experiment was carried out at a scanning rate 0.1 K/min with 300-sec intervals between pressure jumps from 1 to 5 atm (Dragan et al., 2008).

stabilities, takes place in different temperature ranges and it is notable that the area of the peaks of the thermal expansion coefficients is different for these two duplexes, being significantly larger for Duplex (I), which contains the stretch of –AAATTT– bases.

By integrating the temperature function of the thermal expansion coefficients one can determine the relative change of the partial molar volume of the DNA over the temperature range considered. Since the change in the thermal expansion coefficients upon melting of the two duplexes is very different, the increase in their partial molar volumes must be also different. It appears that melting of Duplex (I) results in a particularly large increase of partial volume. To simplify a comparison of the volume effects associated with unfolding these duplexes, we can assume that in the unfolded state their partial specific volumes are identical. This is a justified assumption if the differences in the partial specific volumes of these duplexes result mainly from their hydration and this difference disappears at 100°C, where these duplexes are unfolded. Furthermore, for graphical representation of these volume effects we need to choose a standard. This can be the partial molar volume of Duplex (II) at 25°C. The advantage of this 100% GC duplex is that the partial molar volume of poly[d(G-C)] has been densimetrically measured at 25°C and was found to be (152 ± 1.5) mL per mole of averaged nucleotide (Chalikian et al., 1994). Although this polynucleotide differs from Duplex (II) in size and base sequence, this is unessential for us since their partial volumes hardly differ significantly, while we are interested only in relative changes of partial molar volumes upon unfolding and not in their absolute values. The densimetrically measured partial molar volume of GC deoxypolynucleotides is therefore

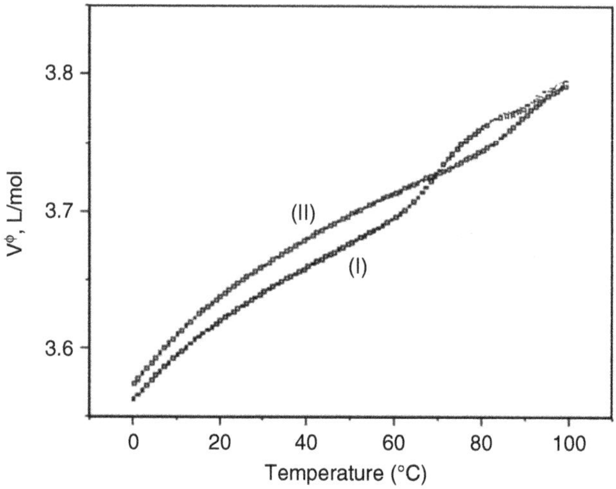

Figure 13.17. Temperature dependence of the partial molar volumes of DNA Duplexes (I) and (II) determined by integrating the thermal expansion coefficients with respect to temperature and scaling them assuming that at 25°C the partial volume of the 100% GC duplex is 152 mL/mol and that at 100°C the partial specific volumes of these two duplexes are equal (Dragan et al., 2008).

used only for presenting the observed relative changes in the volumes of the three studied duplexes on an appropriate scale (Fig. 13.17).

It appears that temperature decrease from 100°C results in a reduction of the partial volumes of the both duplexes, especially in the temperature range of their refolding, and this volume decrease is particularly large for Duplex (I), which includes the contiguous –AAATTT– stretch. The decrease of the partial volume on formation of Duplex (I) is about 20 mL/mol greater than that of the Duplex (II).

Since the volumes of the anhydrous base pairs of DNA do not differ greatly, the observed difference in partial volumes must be attributed to differences in the volumes of the hydrating water. It follows from the above results that the volume of water hydrating the duplex containing the –AAATTT– stretch is considerably smaller than the volume of water hydrating the two other duplexes. Earlier it was shown by precise direct densimetric measurements that the density of water hydrating DNA duplexes is significantly higher than that of the bulk water (Chalikian et al., 1994). The decrease of the partial volume of the DNA duplexes observed by PP-DSC upon lowering the temperature shows that the folded duplexes affect the state of the surrounding water more extensively than do the unfolded oligonucleotides and this is particularly true for the duplex containing the –AAATTT– sequence. This is surprising since the water accessible surface area of the unfolded oligonucleotides is larger than that of the folded duplex. One can conclude then that the influence of DNA on the state of the water comes mostly from its helical conformation. This is just what has been shown by structural studies of the B-DNA dodecamer CGCGAATTCGCG (Drew and Dickerson, 1981; Shui et al., 1998; Arai et al., 2005).

According to these studies the minor groove at the AATT sequence is particularly narrow and the water in this groove is in a highly ordered state.

The PP-DSC studies show that at room temperature the duplex containing the –AAATTT– stretch occupies less volume than its dissociated oligonucleotides, which do not order water. It follows that disordering the regularly ordered water in the minor groove of AT-rich DNA proceeds with an increase of its volume. However, the melting of normal ice at normal pressure proceeds with a decrease of volume; that is, the density of ice at normal pressure is lower than that of liquid water—it is higher only for ice formed at high pressures. It appears, therefore, that:

The stretch of highly ordered water in the AT-rich minor groove of DNA thermodynamically resembles ice at high pressure.

13.3.4. The Cost of DNA Bending

Classical studies characterized DNA as an elastic rod with a persistence length L_p of the order of 50–60 nm (~150 bp) (Wiggins et al., 2005; Bustamante et al., 1994; Baumann et al., 2000). The wormlike chain model predicts a free energy expenditure of ~70 kJ/mol in bending a 10-bp duplex through 50°, that is, ~1.5 kJ per degree of bend (Landau and Lifshitz, 1970). This provided an explanation for the low rates of ligation-mediated cyclization of linear 200- to 350-bp DNA with short cohesive ends (Vologodskaia and Vologodskii, 2002; Cloutier and Widom, 2005).

The bending of naked DNA specifies the flexibility and rigidity of the double helix but of particular interest is the substantial bending induced by special DNA-binding proteins, in particular transcription factors, an action of functional significance in processing genetic information. At present, two major mechanisms of protein-induced DNA bending are considered: the asymmetric neutralization of DNA phosphates and the insertion of protein side chains between DNA bases to generate kinks.

The idea that asymmetric shielding of DNA phosphates by positive charges of bound protein might be a driving force in bending DNA was originally advanced by Mirzabekov and Rich (1979). All-atom simulations of 12-mer duplexes predicted a bend of ~11° when the neutralized phosphates lie across the major groove (Manning et al., 1989; Kosikov et al., 2002). The most obvious example of bending by asymmetric neutralization of phosphates is the DNA in the nucleosome that forms a continuous superhelix of 80 base pairs per turn (Luger et al., 1997). A further example of bending induced by neutralization of DNA phosphates is provided by the unfolded N- or C-terminal tails of the DNA-binding domains of certain HMG proteins that carry multiple positive charges and lie in the major groove on the inside of the bent DNA in the complex (Dragan et al., 2003, 2004).

The possible involvement of kinks in generating significantly bent DNA was first pointed out by Crick and Klug in 1975 (Crick and Klug, 1975) and many subsequent studies have shown that kinking is the typical mechanism by which proteins induce a sharp bend into DNA without loss of base pairing (Shultz et al., 1991; Parkinson et al., 1996; Dickerson, 1998). It has also been found that to bend DNA through substantial angles over short lengths, proteins most typically bind in the minor groove at AT-rich sites forcing the DNA to bend away from the protein toward the major groove (Giese et al., 1992; Ferrari et al., 1992; Paull et al., 1993; Love et al., 1995; Werner et al., 1995; Murphy et al., 1999). The question then arises: what is the reason for using the AT-rich minor groove for sharp bending of DNA and how much work does that require?

As was discussed in Section 12.4.3, while binding of proteins to the major groove proceeds with negative enthalpy (i.e., with heat release) as one would expect for a normal enthalpy-driven reaction, binding to the minor groove proceeds with a positive enthalpy (i.e., with heat absorption). The positive enthalpy does not favor binding but is overcompensated by the positive entropy factor ($T\Delta S$). Binding to the minor groove is thus fully entropy driven, in contrast to the largely enthalpy-driven binding to the major groove (Fig. 12.29). This difference in thermodynamics of protein binding to the minor and major grooves was explained by the state of water that hydrates the DNA grooves and is in a particularly ordered state in the minor groove of the AT-rich sequences, as discussed in previous sections. According to the previously described PP-DSC experiments, density of water in the minor groove of AT-rich DNA is higher than of the bulk water, in contrast to ice at normal pressures, the density of which is lower than that of bulk water. Density of ice is higher than that of water only for ice formed at high pressure. It appears, therefore, that the stretch of highly ordered water in the AT-rich minor groove of DNA thermodynamically resembles ice at high pressure. One might expect this dense ice to be more rigid than normal ice and much more so than bulk water. This suggests that the stretch of ordered water in the minor groove at AT-rich sequences provides rigidity to its helical conformation (Privalov et al., 2009). With removal of this stretch of ordered water the DNA double helix might lose its rigidity, and this should not cost much in Gibbs energy if the "melting" of the ordered water resembles a phase transition. This appears as one of genius inventions of living nature:

> If a functional reason requires sharp bending of the DNA, Nature uses AT-rich sequences and proteins that bind specifically to the minor groove of these sequences.

13.4. RNA

13.4.1. Calorimetry of RNA

The main specificity of natural RNA, which distinguishes it from DNA, is that it works in living cells as a single strand (see Section 4.4). Thus, it does not have a complementary strand with which it could form a double helix, but it finds short complementary sequences within itself, not necessarily in the close vicinity along the chain; pairing of these sequences results in formation of rather complicated spatial constructions. Correspondingly, RNA unfolds and folds in discrete cooperative steps.

RNA can be unfolded and refolded by changing the environmental conditions, particularly temperature. In most cases unfolding–refolding of RNA is a highly reversible process because the RNA does not need to search for the complementary partner: its folding is an intramolecular process. Therefore, DSC has gained especial importance in the analysis of temperature-induced unfolding–refolding of RNA. This is because of a fundamental connection between temperature and enthalpy; that is, the functional dependence of the conjugate intensive and extensive thermodynamic parameters includes all information on the population of macroscopic states realized in the temperature range being considered (see Chapter 3).

The long RNA molecule usually has many complementary regions (see Fig. 4.18). Therefore it unfolds and refolds in many discrete steps. On the other hand, since complementary regions are usually rather short, their enthalpies of unfolding are small. Correspondingly they fold and unfold in a rather broad temperature range. This results in overlap of the discrete steps in the temperature scale. Therefore, as these discrete steps increase in number, so do the difficulties of revealing them (Cammack, 1970; Matveyev and Privalov, 1979; Matveyev et al., 1982). Because of that, most calorimetric studies were being done with short transfer RNAs. However, with the discovery of micro-RNA and rapid increase of interest in these molecules, they might become the most attractive objects for calorimetric studies.

Figure 13.18 presents the secondary and tertiary structures of transfer RNA (tRNA), a small molecule specialized to transfer amino acids to the protein-synthesizing ribosome. Since each amino acid has its own transfer RNA, there are

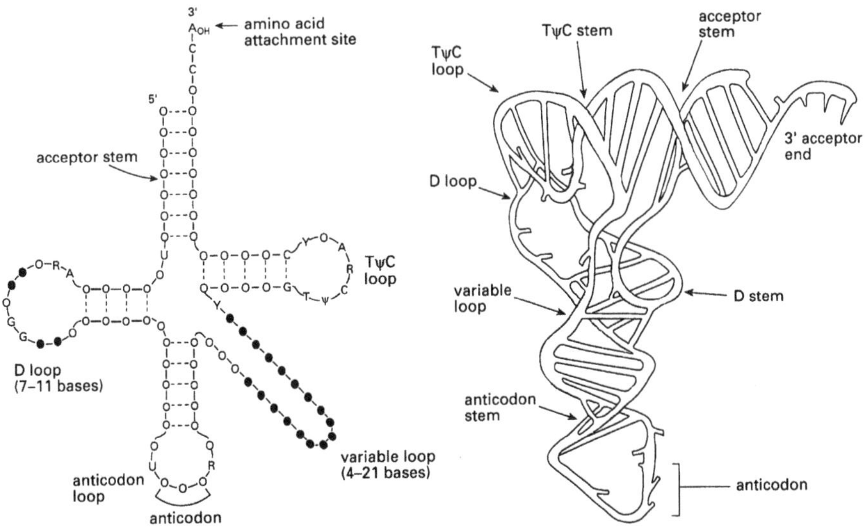

Figure 13.18. Cloverleaf secondary structure and three-dimensional structure of tRNA according to Kim et al. (1974). The size of the different tRNAs varies from 74 to 95, but the tRNAs can all be folded into a common secondary structure, a cloverleaf.

as many tRNA molecules as there are amino acids used for protein synthesis. Each of these tRNAs is designed to transfer a certain amino acid. However, they all have common structural features: they can all be folded into a common secondary structure, the cloverleaf, in which four- or five-bp regions are separated by single-stranded loops. Correspondingly, the tRNAs unfold and refold in discrete cooperative steps. Analysis of these steps could help in understanding the folded RNA construction and the technology of its formation.

13.4.2. Calorimetric Studies of Transfer RNAs

Calorimetric studies of tRNA unfolding have been reported by many groups (e.g., Brandts et al., 1974; Privalov et al., 1975; Filimonov and Privalov, 1976; Hinz et al., 1977), but comparison of results obtained on different tRNAs at different conditions using DSC instruments differing in sensitivity is not easy. Therefore, here we consider results obtained in one laboratory at strictly identical conditions (Privalov and Filimonov, 1978).

Figure 13.19 presents two DSC recordings of consecutive heating of $tRNA^{Val}$ in salt-free solution. It shows that melting of this transfer RNA is a highly reversible complex process that extends over a wide temperature range. These features are general for all types of tRNA. Melting profiles of some of them at various salt conditions are presented in Figure 13.20. These results show that the excess heat absorption on heating all the RNA solutions in the absence of divalent ions starts at about 10°C and ends at 100°C. All observed heat capacity profiles are complicated, but for any given tRNA species they are very specific.

In all cases, increase in sodium chloride concentration shifts the tRNA melting profile to a higher temperature, without much change in its overall shape. However, the addition of magnesium chloride significantly stabilizes tRNA and sharpens its melting profile. It appears that the Mg^{2+} ion leads to cooperation of different stages

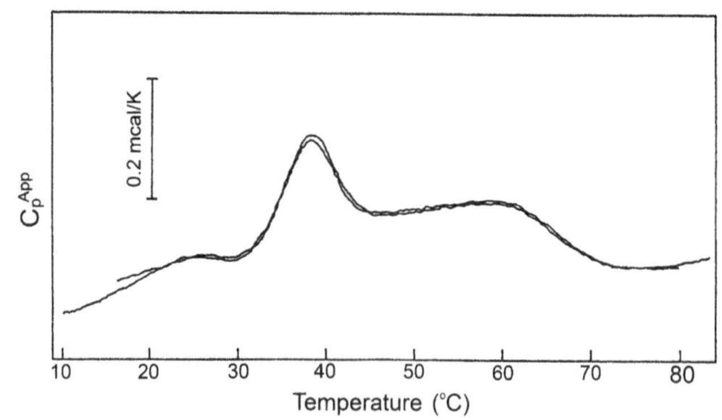

Figure 13.19. Microcalorimetric recordings of the apparent heat capacity effect upon heating and consecutive cooling of $tRNA^{Val}$ in 10 mM sodium phosphate, pH 7.0 solution. The $tRNA^{Val}$ (yeast) concentration is 30 μM (Privalov et al., 1975).

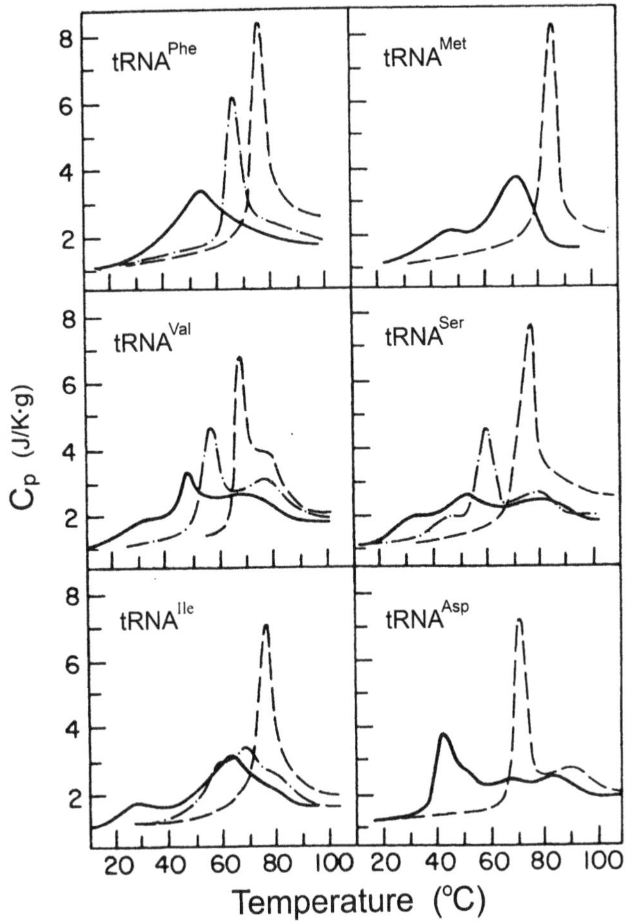

Figure 13.20. Partial specific heat capacities of various specific tRNAs under different salt conditions: continuous line, 150 mM NaCl; dot-and-dashed line, 150 mM NaCl and 1 mM $MgCl_2$; dashed line, 1 mM $MgCl_2$ (Privalov and Filimonov, 1978).

of tRNA unfolding but, even at 1 mM $MgCl_2$, the process cannot be considered as a two-state transition.

In the absence of magnesium ions, temperature-induced unfolding of all tRNAs studied is a highly reversible process. However, in the presence of magnesium it becomes irreversible, and the degree of irreversibility increases with increasing magnesium concentration and incubation time at high temperature (see also Cammack et al., 1970). Electrophoresis analysis showed a marked degradation of polynucleotides heated in the presence of magnesium. Thus, it appears that:

The irreversibility observed upon heat treatment of tRNA in the presence of magnesium ions results from degradation of the tRNA.

The mechanism of the degradation caused by magnesium is not clear, but it does not result from magnesium-activated ribonuclease impurities, since that occurs only at the elevated temperatures where the tRNA is unfolded and the chance of the ribonuclease not being denatured is not great.

For all the tRNAs studied, the partial specific heat capacity increases upon heating from (1.10 ± 0.12) J/K·g at 20°C to (1.67 ± 0.12) J/K·g at 100°C. This increase in heat capacity results both from the trivial increase of the heat capacity with temperature and from the unfolding of the compact native state (Filimonov and Privalov, 1976; Hinz et al., 1977). Because of the very broad temperature range of tRNA unfolding, the heat capacity increment of unfolding cannot be reliably determined, however, judging by the results obtained on the DNA duplexes, the heat capacity increment upon tRNA unfolding should not be large and thus the enthalpy of unfolding should not depend much on temperature. Indeed, the overall excess heat effect of tRNA unfolding in the presence of different concentrations of monovalent salt, which changes its thermal stability, does not noticeably change. For all the tRNAs studied, the overall melting enthalpy is on the order of 1300 (±50) kJ/mol (Privalov and Filimonov, 1978). This fact, that the specific melting enthalpies of all studied tRNAs are similar, means that all of these tRNAs are equally saturated with secondary bonds maintaining their structure.

The complicated melting profiles of the specific tRNAs suggest that unfolding of their structure proceeds in several discrete steps. Since the overall process of unfolding in the absence of magnesium is reversible, one can analyze it thermodynamically (see Chapter 3). To simplify the deconvolution analysis of the partial molar heat capacity functions of RNA, it was assumed that (a) all transitions are independent, (b) all transitions represent a two-state unfolding, and (c) the transition enthalpy does not depend on temperature.

Figure 13.21 summarizes the results obtained for all the tRNA species studied so far, examined under the same salt conditions: 150 mM NaCl. One can see that, although the overall melting profiles are very different, there are many common features in the unfolding of all these molecules, namely: (a) they all have the same number of components, except $tRNA^{Ser}$, which has one extra component, and $tRNA^{Asp}$, which has one component less; (b) each component of one tRNA has an analog in the other tRNAs; (c) in all cases, except $tRNA^{Asp}$, the sequence of the first two components is the same.

One can try to assign the component transitions to the melting of certain parts of the tRNA structure assuming that the cloverleaf model is applicable to all of them (see insets in Fig. 13.21). Very helpful information for assigning the transitions is that in $tRNA^{Asp}$ the rTΨ branch is the most stable, while the anticodon branch is less stable (Coutts et al., 1974, 1975). It is remarkable that, after disruption of the tertiary structure of tRNA, the dhU branch breaks next in all the tRNAs studied, and the tight L-shaped tRNA molecule unfolds into a topologically flat structure with separate helical loops that melt independently. Thus, after the first transition, the compact tRNA molecule breaks down into independent cooperative subsystems. This might be important for the technology of self-assembly of the tRNA three-dimensional structure. Indeed, without this, it is impossible to understand how

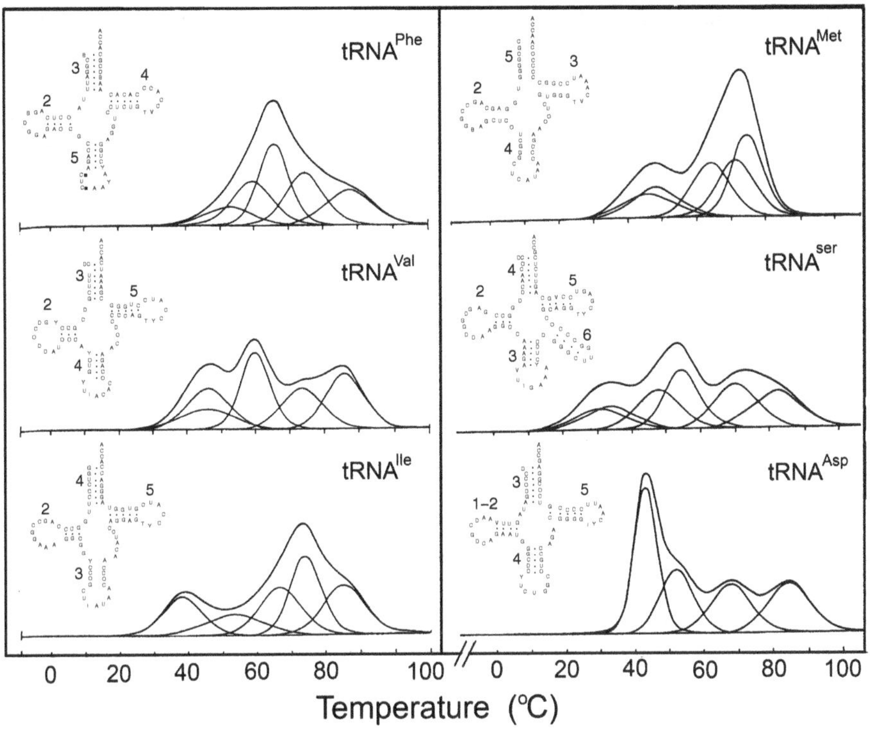

Figure 13.21. Deconvolution of the heat capacity profiles of various specific tRNAs in 150 mM NaCl solution on the component two-state transitions; their nucleotide sequences are presented as cloverleafs (Privalov and Filimonov, 1978).

different polynucleotide chains of various tRNAs can assemble into a unified structure:

> *The folding of the tertiary structure is the final step of self-assembly of the tRNA native structure, and this is not trivial: it functions as a "lock" for the other parts of this macromolecule, and the magnesium ion seems to be a key to this lock.*

The stability conferred by magnesium is especially strong only for the tertiary structure of tRNAs and this locks all the other domains: they cannot unfold even if their stability is less than the stability of the tertiary structure. It follows that disruption of the tRNA structure appears to be a two-state process when the stability of its tertiary structure exceeds the stability of any of its branches. This is possible only at a substantial concentration of magnesium in solution.

Since after unfolding of the tertiary structure the tRNA branches unfold independently, using the thermodynamic parameters of each component transition estimated by deconvolution analysis of the melting profile and the structural information, one

TABLE 13.4. Observed and Calculated Enthalpies for Different Structural Elements of tRNAs[a]

Structure	tRNA	Peak Number	T_m (K)	ΔH_m (kJ/mol)	ΔH_{calc} (kJ/mol)
dhU branch	Phe	2	323	251	220
	Val	2	310	167	150
	Ile	2	317	176	180
	fMet	2	321	206	220
	Ser	2	307	178	180
	Asp	1–2			150
Acceptor stem	Phe	3	330	343	330
	Val	3	324	335	330
	Ile	4	338	356	380
	fMet	5	348	367	360
	Ser	4	328	294	330
	Asp	3	325	305	350
TΨC branch	Phe	4	338	285	260
	Val	5	350	305	280
	Ile	5	349	289	260
	fMet	3	337	287	270
	Ser	5	344	270	280
	Asp	5	358	291	300
Extra branch	Ser	6	356	245	240
Anticodon branch	Phe	5	351	243	260
	Val	4	337	264	260
	Ile	3	330	264	250
	Met	4	344	299	280
	Ser	3	321	230	220
	Asp	4	341	273	270

[a]At 25°C in 150 mM NaCl solution (Privalov and Filimonov, 1978).

can estimate the contribution of the individual base pairs in this process. Table 13.4 illustrates the fitness of the calorimetrically determined and calculated enthalpies of unfolding of each structural element in the considered tRNAs. Table 13.5 presents the optimal thermodynamic parameters of the base–base interactions in these RNAs, obtained from the best-fit analysis. It shows that the largest helix-stabilizing effect is specific for the G-C base pair and it is enthalpic in nature: the enthalpy of formation of this base pair significantly exceeds the enthalpies of formation of other base pairs.

The question is: how reliable are these parameters obtained by the best-fit analysis of the melting profiles of the various tRNAs? This can be judged only by comparing the enthalpy values obtained with the values directly measured by calorimeter for the enthalpies of melting of regular RNA duplexes. However, these are known only for the $(A{\cdot}U)_n$ duplex. As considered previously (Fig. 13.5), $\Delta H^{A\cdot U}$ corrected for

TABLE 13.5. Optimal Parameters Providing Thermodynamic Characteristics of the Structural Elements of tRNAs[a]

Base Pair	ΔH (kJ/mol)	ΔS (J/K·mol)	ΔG (kJ/mol)
G-C	60	150	15
A-U	40	120	4
A-Ψ	40	120	4
G-U	30	90	3
G-A	30	90	3

[a]Privalov and Filimonov (1978).

poly(A) helicity is 38 kJ/mol at the mid-temperature of tRNA unfolding (60°C). This value of $\Delta H^{A\cdot U}$ is in perfect correspondence with that presented in Table 13.5. Therefore, we can expect that the contributions of other base pairs to the stabilization of the RNA double helixes that have been found by the analysis of their melting profiles are also reliable. Nevertheless, the direct calorimetric measurements of thermodynamic characteristics of formation of the pairings and loops in all possible combinations in the RNA molecule is certainly needed for clear understanding of the mechanism of formation of their not simple spatial structure.

REFERENCES

Applequist J. and Damie V. (1966). *J. Am. Chem. Soc.*, **88**, 3895–3900.

Arai S., Chatake T., Ohhara T., Kurihara K., Tanaka I., Suzuki N., et al. (2005). *Nucleic Acids Res.*, **33**, 3017–3024.

Baumann C.G., Bloomfield V.A., Smith B.B., Bustamante C., Wang M.D., and Block S.M. (2000). *Biophys J.*, **78**, 1965–1978.

Blackburn G.M. and Gait M.J., Eds. (1996). *Nucleic Acids in Chemistry and Biology*. Oxford University Press, Oxford.

Brahms J., Michelson A.M., and van Holde K.E. (1966). *J. Mol. Biol.*, **15**, 467–488.

Brandts J.F., Jackson W.M., and Ting T.Y. (1974). *Biochemistry*, **13**, 3595–3600.

Breslauer K.J., Frank R., Blocker H., and Marky L.A. (1986). *Proc. Natl. Acad. Sci. USA*, **83**, 3746–3750.

Bustamante C., Marko J.F., Siggia E.D., and Smith S. (1994). *Science*, **265**, 1599–1600.

Cammack K.A., Miller D.S., and Grinstead K.H. (1970). *Biochem. J.*, **117**, 745–755.

Chalikian T.V., Sarvazian A.P., Plum G.E., and Breslauer K.J. (1994). *Biochemistry*, **33**, 2394–2401.

Chalikian T.V., Volker J., Plum G.E., and Breslauer K.J. (1999). *Proc. Natl. Acad. Sci. USA*, **96**, 7853–7858.

Cloutier T.E. and Widom J. (2005). *Proc. Natl. Acad. Sci. USA*, **102**, 3645–3650.

Coutts S.M., Gangloff J., and Dirheimer G. (1974). *Biochemistry*, **13**, 3938–3948.

Coutts S.M., Riesner D., Romer R., Rabl C.R., and Maass G. (1975). *Biophys Chem.*, **3**, 275–289.

Crick F.H.C. and Klug A. (1975). *Nature*, **255**, 530–533.

Dickerson R.E. (1998). *Nucleic Acids Res.*, **26**, 1906–1926.

Dragan A.I., Klass J., Read C., Churchill M.E.A., Crane-Robinson C., and Privalov P.L. (2003). *J. Mol. Biol.*, **331**, 795–813.

Dragan A.I., Read C.M., Makeyeva E.N., Milgotina E.I., Churchill M.E.A., Crane-Robinson C., and Privalov P.L. (2004). *J. Mol. Biol.*, **343**, 371–393.

Dragan A.I., Russell D.J., and Privalov P.L. (2008). *Biopolymers*, **91**, 95–101.

Drew H.R. and Dickerson R.E. (1981). *J. Mol. Biol.*, **151**, 535–556.

Ferrari S., Harley V.R., Pontiggia A., Goodfellow P.N., Lovell-Badge R., and Bianchi M.E. (1992). *EMBO J.*, **11**, 4497–4506.

Filimonov V.V. and Privalov P.L. (1976). *Eur. J. Biochem.*, **70**, 25–31.

Filimonov V.V. and Privalov P.L. (1978). *J. Mol. Biol.*, **122**, 465–470.

Giese K., Cox J., and Grosschedl R. (1992). *Cell*, **69**, 185–195.

Hinz H.-J., Filimonov V.V., and Privalov P.L. (1977). *Eur. J. Biochem.*, **72**, 79–86.

Holbrook J.A., Capp M.W., Saecker R.M., and Record M.T., Jr. (1999). *Biochemistry*, **38**, 8409- 8422.

Jelesarov I., Crane-Robinson C., and Privalov P.L. (1999). *J. Mol. Biol.*, **294**, 981–995.

Kim S.H., Sussman J.L., Suddath F.L., Quigley G.J., McPherson A., Wang A.H., Seeman N.C., and Rich A. (1974). *Proc. Natl. Acad. Sci. USA*, **71**, 4970–4974.

Kosikov K.M., Gorin A.A., Lu X.J., Olson W.K., and Manning G.S. (2002). *J. Am. Chem. Soc.*, **124**, 4838–4847.

Krakauer H. and Sturtevant J.M. (1968). *Biopolymers*, **6**, 491–512.

Ladbury J.E., Sturtevant J.M., and Leontis N.B. (1994). *Biochemistry*, **33**, 6828–6833.

Landau L.D. and Lifshitz E.M. (1970). Theory of elasticity. In: *Course of Theoretical Physics*, 2nd ed., p. 82. Pergamon, Oxford.

Leng M. and Felsenfeld G. (1966). *J. Mol Biol.*, **15**, 455–466.

Love J.J., Li X., Case D.A., Giese K., Grosschedl R., and Wright P.E. (1995). *Nature*, **376**, 791–795.

Luger K., Mader A.W., Richmond R.K., Sargent D.F., and Richmond T.J. (1997). *Nature*, **389**, 251–260.

Manning G.S. (1972). *Biopolymers*, **11**, 937–949.

Manning G.S., Ebralidze K.K., Mirzabekov A.D., and Rich A. (1989). *J. Biomol. Struct. Dynam.*, **6**, 877–889.

Marky L.A. and Breslauer K.J. (1982). *Biopolymers*, **21**, 2185–2194.

Matveyev S.V. and Privalov P.L. (1979). *Proc. Acad. Sci. USSR*, **247**, 985–989.

Matveyev S.V., Filimonov V.V., and Privalov P.L. (1982). *Molek. Biol. (USSR)*, **16**, 1234–1244.

Michelson A.M., Massoulie J., and Guschlbauer W. (1967). *Prog. Nucleic Acid Res. Mol. Biol.*, **6**, 83–141.

Mirzabekov A.D. and Rich A. (1979). *Proc. Natl. Acad. Sci. USA*, **76**, 1118–1121.

Mock R.A. and Marshall C.A. (1955). *J. Polymer Sci.*, **13**, 203–210.

Murphy F.V. IV, Sweet R.M., and Churchill M.E. (1999). *EMBO J.*, **18**, 6610–6618.

Neuman E. and Ackerman T. (1969). *J. Phys. Chem.*, **73**, 2170–2178.

Oosawa F. (1968). *Biopolymers*, **6**, 145–150.

Parkinson G., Wilson C., Gunasekera A., Ebright Y.W., Ebright R.E., and Berman H.M. (1996). *J. Mol. Biol.*, **260**, 395–408.

Paull T.T., Haykinson M.J., and Johnson R.C. (1993). *Genes Dev.*, **7**, 1521–1534.

Pilch D.S., Plum G.E., and Breslauer K.J. (1995). *Curr. Opin. Struct. Biol.*, **5**, 334–342.

Plum G.E., Pilch D.S., Singleton S.F., and Breslauer K.J. (1995). *Annu. Rev. Biophys. Biomol. Struct.*, **24**, 319–350.

Poland D., Vournakis J.N., and Scheraga H.A. (1966). *Biopolymers*, **4**, 223–235.

Privalov P.L. and Filimonov V.V. (1978). *J. Mol. Biol.*, **122**, 447–464.

Privalov P.L., Dragan A.I., and Crane-Robinson C. (2009). *Trends Biochem. Sci.*, **34**, 464–470.

Privalov P.L., Dragan A.I., and Crane-Robinson C. (2010). *Nucleic Acids Res.*, **39**, 2483–2491.

Privalov P.L., Dragan A.I., Crane-Robinson C., Breslauer K.J., Remeta D.P., and Minetti C.A.S. (2007). *J. Mol. Biol.*, **365**, 1–9.

Privalov P.L., Filimonov V.V., Venkstern T.V., and Bayev A.A. (1975). *J. Mol. Biol.*, **97**, 279–288.

Privalov P.L., Pritsyn O.B., and Birshtein T.M. (1969). *Biopolymers*, **8**, 559–571.

Rawitcher M.A., Ross P.D., and Sturtevant J.M. (1963). *J. Am. Chem. Soc.*, **85**, 1915–1918.

Record M.T., Jr., Anderson C.F., and Lohman T.M. (1978). *Quart. Rev. Biophys.*, **11**, 103–178.

Schultz S.C., Shields G.C., and Steitz T.A. (1991). *Science*, **253**, 1001–1007.

Shiao D.D. and Sturtevant J.M. (1973). *Biopolymers*, **12**, 1829–1836.

Shui X., McFail-Isom L., Hu G.G., and Williams L.D. (1998). *Biochemistry*, **37**, 8341–8355.

Tong B.Y. and Battersby S.J. (1979). *Biopolymers*, **18**, 1917–1936.

Vaitiekunas P. and Privalov P.L. (in preparation).

Vesnaver G. and Breslauer K.J. (1991). *Proc. Natl. Acad. Sci. USA*, **88**, 3569–3573.

Vologodskaia M. and Vologodskii A. (2002). *J. Mol. Biol.*, **317**, 205–213.

Watson J.D. and Crick F.H.C. (1953). *Nature*, **171**, 964–967.

Werner M.H., Huth J.R., Gronenborn A.M., and Clore G.M. (1995). *Cell*, **81**, 705–714.

Wiggins P.A., Phillips R., and Nelson P.C. (2005). *Phys. Rev. E*, **71**, 021909.

Zieba K., Chu T.M., Kupke D.W., and Marky L.A. (1991). *Biochemistry*, **30**, 8018–8026.

INDEX

Microcalorimetry of Macromolecules: The Physical Basis of Biological Structures, First Edition.
Peter L. Privalov.

www.ingramcontent.com/pod-product-compliance
Lightning Source LLC
LaVergne TN
LVHW011832200926
848484LV00002B/346